Toyota Tundra & Sequoia Automotive Repair Manual

by Jeff Killingsworth and John H Haynes

Member of the Guild of Motoring Writers

Models covered:

Toyota Tundra - 2007 through 2019
Toyota Sequoia - 2008 through 2019
All 2WD and 4WD models

(92179-3X5)

Haynes Group Limited
Haynes North America, Inc.
www.haynes.com

Acknowledgements

Wiring diagrams provided exclusively for Haynes North America, Inc. by Valley Forge Technical Information Services.

© **Haynes North America, Inc. 2013, 2015, 2019**

With permission from Haynes Group Limited

A book in the Haynes Automotive Repair Manual Series

ISBN-10: 1-62092-367-X
ISBN-13: 978-1-62092-367-2

Library of Congress Control Number: 2019937615

While every attempt is made to ensure that the information in this manual is correct, no liability can be accepted by the authors or publishers for loss, damage or injury caused by any errors in, or omissions from, the information given.

Contents

Haynes mechanic and photographer with a 2010 Toyota Tundra

About this manual

Its purpose

The purpose of this manual is to help you get the best value from your vehicle. It can do so in several ways. It can help you decide what work must be done, even if you choose to have it done by a dealer service department or a repair shop; it provides information and procedures for routine maintenance and servicing; and it offers diagnostic and repair procedures to follow when trouble occurs.

We hope you use the manual to tackle the work yourself. For many simpler jobs, doing it yourself may be quicker than arranging an appointment to get the vehicle into a shop and making the trips to leave it and pick it up. More importantly, a lot of money can be saved by avoiding the expense the shop must pass on to you to cover its labor and overhead costs. An added benefit is the sense of satisfaction and accomplishment that you feel after doing the job yourself.

Using the manual

The manual is divided into Chapters. Each Chapter is divided into numbered Sections, which are headed in bold type between horizontal lines. Each Section consists of consecutively numbered paragraphs.

At the beginning of each numbered Section you will be referred to any illustrations which apply to the procedures in that Section. The reference numbers used in illustration captions pinpoint the pertinent Section and the Step within that Section. That is, illustration 3.2 means the illustration refers to Section 3 and Step (or paragraph) 2 within that Section.

Procedures, once described in the text, are not normally repeated. When it's necessary to refer to another Chapter, the reference will be given as Chapter and Section number. Cross references given without use of the word "Chapter" apply to Sections and/or paragraphs in the same Chapter. For example, "see Section 8" means in the same Chapter.

References to the left or right side of the vehicle assume you are sitting in the driver's seat, facing forward.

Even though we have prepared this manual with extreme care, neither the publisher nor the author can accept responsibility for any errors in, or omissions from, the information given.

NOTE

A **Note** provides information necessary to properly complete a procedure or information which will make the procedure easier to understand.

CAUTION

A **Caution** provides a special procedure or special steps which must be taken while completing the procedure where the Caution is found. Not heeding a Caution can result in damage to the assembly being worked on.

WARNING

A **Warning** provides a special procedure or special steps which must be taken while completing the procedure where the Warning is found. Not heeding a Warning can result in personal injury.

Introduction to the Toyota Tundra and Sequoia

The second generation Tundra was manufactured beginning in 2007 and is equipped with either a 4.0L V6 engine, a 4.7L V8 engine, a 4.6L V8 engine or a 5.7 liter V8 engine. The Sequoia is only available with the 4.7L V8 engine, a 4.6L V8 engine or a 5.7 liter V8 engine.

All engines are equipped with the Electronic Fuel Injection (EFI) system. Some 2009 and later V8 engines are Flex Fuel engines.

The engine drives the rear wheels through either a 5-speed or 6-speed automatic transmission via a driveshaft and solid rear axle. A transfer case and driveshaft are used to drive the front axle on 4WD models. All models are available with either 2WD or 4WD.

The front suspension is fully independent: It consists of upper and lower control arms, a stabilizer bar, and integral coil spring/shock absorber assemblies. A solid axle at the rear is suspended by leaf springs and shock absorbers (Tundra models) or coil springs and shock absorbers (Sequoia models). Sequoia models have four rear suspension arms (two upper and two lower arms) and a lateral control rod.

The steering gear is a rack-and-pinion type and is connected to the steering knuckles by tie-rods.

The front and rear brakes are disc type on all models. Power assist is standard on all models. Most models are equipped with anti-lock brakes.

Vehicle identification numbers

Modifications are a continuing and unpublicized process in vehicle manufacturing. Since spare parts manuals and lists are compiled on a numerical basis, the individual vehicle numbers are essential to correctly identify the component required.

Vehicle Identification Number (VIN)

This very important identification number is stamped on a plate attached to the left side of the dashboard just inside the windshield on the driver's side of the vehicle (see illustration). The VIN also appears on the Vehicle Certificate of Title and Registration. It contains information such as where and when the vehicle was manufactured, the model year and the body style.

Counting from the left, the engine code is the eighth digit and the model year code is the tenth digit.

Model year codes:

7	2007
8	2008
9	2009
A	2010
B	2011
C	2012
D	2013
E	2014
F	2015
G	2016
H	2017
J	2018
K	2019

Engine serial number

The location of the engine identification number is as follows:

V6 engine: Stamped on a pad at the right rear portion of the engine block.

4.6L V8 engine: Stamped on a pad at the left rear portion of the engine block.

4.7L V8 engine: Stamped on a pad at the top front of the engine block, between the cylinder heads.

5.7L V8 engine: Stamped on a pad at the right front of the engine block (see illustration).

Manufacturer Certification label

The Manufacturer Certification label is affixed to the front door pillar. The plate contains the name of the manufacturer, the month and year of production, the Gross Vehicle Weight Rating (GVWR) and the certification statement (see illustration).

Vehicle Emissions Control Information (VECI) label

The emissions control information label is found under the hood. This label contains information on the emissions control equipment installed on the vehicle, as well as tune-up specifications (see illustration).

Transfer case and transmission identification number

The transfer case identification number is stamped into the case of the component. Automatic transmission numbers are stamped onto ID plates (see illustration).

The Vehicle Identification Number (VIN) is visible through the driver's side of the windshield

The engine serial number on the 5.7L V8 engine is stamped onto a pad on the right side of the engine block, just behind the oil filter cartridge

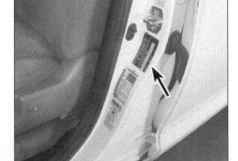

The manufacturer's certification label is affixed to the driver's side door jamb

The Vehicle Emissions Control Label (VECI) is located on the underside of the hood

The identification number on automatic transmissions is stamped into a plate located on the side of the transmission

Buying parts

Replacement parts are available from many sources, which generally fall into one of two categories - authorized dealer parts departments and independent retail auto parts stores. Our advice concerning these parts is as follows:

Retail auto parts stores: Good auto parts stores will stock frequently needed components which wear out relatively fast, such as clutch components, exhaust systems, brake parts, tune-up parts, etc. These stores often supply new or reconditioned parts on an exchange basis, which can save a considerable amount of money. Discount auto parts stores are often very good places to buy materials and parts needed for general vehicle maintenance such as oil, grease, filters, spark plugs, belts, touch-up paint, bulbs, etc. They also usually sell tools and general accessories, have convenient hours, charge lower prices and can often be found not far from home.

Authorized dealer parts department: This is the best source for parts which are unique to the vehicle and not generally available elsewhere (such as major engine parts, transmission parts, trim pieces, etc.).

Warranty information: If the vehicle is still covered under warranty, be sure that any replacement parts purchased - regardless of the source - do not invalidate the warranty!

To be sure of obtaining the correct parts, have engine and chassis numbers available and, if possible, take the old parts along for positive identification.

Recall information

Vehicle recalls are carried out by the manufacturer in the rare event of a possible safety-related defect. The vehicle's registered owner is contacted at the address on file at the Department of Motor Vehicles and given the details of the recall. Remedial work is carried out free of charge at a dealer service department.

If you are the new owner of a used vehicle which was subject to a recall and you want to be sure that the work has been carried out, it's best to contact a dealer service department and ask about your individual vehicle - you'll need to furnish them your Vehicle Identification Number (VIN).

The table below is based on information provided by the National Highway Traffic Safety Administration (NHTSA), the body which oversees vehicle recalls in the United States. The recall database is updated constantly. For the latest information on vehicle recalls, check the NHTSA website at www.nhtsa.gov, www.safercar.gov, or call the NHTSA hotline at 1-888-327-4236.

Recall date	Recall campaign number	Model(s) affected	Concern
January 18, 2007	07V013000	2007 Sequoia	On some models, the lower balljoints may experience an incidental deterioration of the internal lubrication. This may cause the balljoint to wear and loosen prematurely, which could result in increased steering effort, reduced vehicle self-centering, and noise in the front suspension. In extreme cases, the lower balljoint may separate from the knuckle and could cause a loss of vehicle control.
December 14, 2007	07V579000	2007 Tundra	On some 4-wheel drive models, the hardness of the driveshaft slip yoke may be insufficient. In this condition, there is a possibility that the yoke may break, causing the propeller shaft to separate at the joint and come into contact with the road surface or the fuel tank which could result in a loss of vehicle control or a fuel leak.
February 22, 2008	08E029000	2007 Tundra	Southeast Toyota is recalling alloy wheel lug nuts used for the 18" diamond finish 5-spoke alloy wheels and 20" painted alloy wheels with 6-dual spokes sold as an upgrade package for some models. The wheel nuts can crack and break if the wheels are not installed carefully or not carefully tightened. This wheel nut condition could occur at the time of wheel installation, tire rotation, or service. Cracked or broken wheel nuts could result in a vehicle crash.

Recall date	Recall campaign number	Model(s) affected	Concern
February 25, 2008	08V0800000	2007 Tundra	Southeast Toyota is recalling some models due to a wheel nut issue, equipped with 18" installed diamond finish 5-spoke alloy wheels (option code WN1 or WN4) or 20" installed painted alloy wheels with 6 dual spokes (option code WV1 or WV4). The upgraded alloy wheels were fastened with new wheel nuts. The wheel nuts can crack and break if they are not installed carefully or not carefully tightened. This wheel nut condition could occur at the time of wheel installation, tire rotation, or service. If a vehicle's wheel has a cracked or broken wheel nut condition and the wheel becomes loose and falls off the vehicle without warning, a crash could occur.
June 10, 2009	09V223000	2007, 2008, 2009, 2010 Tundra	Some models were not equipped with load carrying capacity modification labels which fails to conform with the requirements of Federal Motor Vehicle Safety Standard No. 110, "tire selection and rims." Missing or incorrect load carrying capacity modification labels could result in the vehicle being overloaded, increasing the risk of a crash.
October 05, 2009	09V388000	2007, 2008, 2009, 2010 Tundra	On some models, the accelerator pedal can get stuck in the wide open position due to its being trapped by an unsecured or incompatible driver's floor mat. A stuck open accelerator pedal may result in very high vehicle speeds and make it difficult to stop the vehicle, which could cause a crash, serious injury or death.
January 21, 2010	10V017000	2007, 2008, 2009, 2010 Tundra and Sequoia	On some models, due to the manner in which the friction lever interacts with the sliding surface of the accelerator pedal inside the pedal sensor assembly, the sliding surface of the lever may become smooth during vehicle operation. In this condition, if condensation occurs on the surface, as may occur from heater operation (without a/c) when the pedal assembly is cold, the friction when the accelerator pedal is operated may increase, which may result in the accelerator pedal becoming harder to depress, slower to return, or, in the worst case, mechanically stuck in a partially depressed position, increasing the risk of a crash.
January 26, 2010	10V035000	2007, 2008, 2009, 2010 Tundra and Sequoia	Gulf States Toyota is recalling certain models for failing to comply with the requirements of Federal Motor Vehicle Safety Standard No. 110, "tire selection and rims." These vehicles were sold without the requisite load carrying capacity modification labels. Dealers will mail to consumers the corrected label or the customer will have the option for dealers to install the label free of charge. Dealers will also correct the owner's manual.
March 09, 2010	10V091000	2010 Tundra and Sequoia	On some vehicles with four wheel drive, there is a possibility that an improper weld exists at the union of the propeller shaft and yoke. Due to this improper weld, this joint may separate, and the separated shaft may come into contact with the road surface. This may result in a loss of vehicle control increasing the risk of a crash.
October 01, 2010	10V036000	2007, 2008, 2009, 2010 Tundra and Sequoia	Southeast Toyota is recalling certain models for failing to comply with the requirements of Federal Motor Vehicle Safety Standard No. 110, "tire selection and rims." These vehicles were sold without the requisite load carrying capacity modification labels. A driver may overload a vehicle which may increase the risk of a crash.
March 04, 2011	11V148000	2008, 2009, 2010, 2011 Tundra and Sequoia	Some models fail to comply with the requirements of Federal Motor Vehicle Safety Standard No. 138, "tire pressure monitoring system (TPMS)." When factory-installed wheels and tires were replaced with Toyota authorized accessory wheels and tires prior to first sale, the tire pressure monitoring systems were not re-calibrated correctly and therefore do not start illuminating the low tire pressure warning lamp at the required minimum activation pressure. Failure to warn of tire deflations is a non-compliance with FMVSS 138 and could lead to tire failure increasing the risk of a crash.

Recall date	Recall campaign number	Model(s) affected	Concern
March 16, 2011	11V185000	2009, 2010, 2011 Tundra 2009 and 2010 Sequoia	Gulf States Toyota is recalling some models for failing to comply with the requirements of Federal Motor Vehicle Safety Standard No. 138, "tire pressure monitoring systems." The TPMS on some vehicles may not have been properly calibrated and as a result the low tire pressure warning lamp may not illuminate should the inflation pressure in one or more of the vehicle's tires fall below the threshold for when the low tire pressure warning lamp should illuminate. Drivers will not receive a warning from the tire pressuring monitor that one or more tires are underinflated, increasing the risk that a vehicle will be driven with one or more underinflated tires and increasing the risk of a tire failure that may lead to a crash.
April 26, 2011	11V254000	2011 Tundra	On some models equipped with a 3-joint type propeller shaft, due to improper casting of the slip yokes, there is a possibility that the slip yoke may break, causing the propeller shaft to separate at the joint and come into contact with the road surface. The driver could experience loss of motive power and vehicle control, increasing the risk of a crash.
May 12, 2011	11V278000	2007, 2008, 2009, 2010, 2011 Tundra	Southeast Toyota is recalling some models for failing to comply with the requirements of Federal Motor Vehicle Safety Standard No. 138, "tire pressure monitoring systems." The TPMS on some vehicles may not have been properly calibrated, and as a result, the low tire pressure warning lamp may not illuminate should the inflation pressure in one or more of the vehicle's tires fall below the threshold for when the low tire pressure warning lamp should illuminate. Drivers will not receive a warning from the tire pressure monitor that one or more tires are underinflated, increasing the risk that the vehicle will be driven with one or more underinflated tires, increasing the risk of a tire failure that may lead to a crash.
December 14, 2011	11V588000	2011 Tundra	Gulf States Toyota is recalling some models with an inaccurate load carrying capacity label. These vehicles fail to comply with the requirements of Federal Motor Vehicle Safety Standard No. 110, "tire selection and rims." An inaccurate label could lead to vehicle overloading which could result in tire failure, increasing the risk of a crash.
October 10, 2012	12V491000	2007, 2008, 2009 Tundra and Sequoia	Toyota is recalling certain models with power window master switch assemblies that were built using a less precise process for lubricating the internal components of the switch assemblies. Irregularities in this lubrication process may cause the power window master switch assemblies to malfunction and overheat. If the switch overheats, it may melt, possibly resulting in a fire.
January 16, 2013	13V014000	2009, 2010, 2011 2012 Tundra and 2010, 2011, 2012 Sequoia	Some models with accessories such as leather seat covers, seat heaters or headrest DVD systems may not have had the passenger seat occupant sensing system calibration tested. Without passing the calibration test, the occupant sensing system may not operate as designed. If the front passenger seat occupant sensing system is out of calibration, the front passenger airbags may not deploy or they may deploy inappropriately for the passenger's size and position. This could increase the risk of personal injury during the event of a vehicle crash necessitating airbag deployment.
April 09, 2013	13V123000	2008, 2010, 2011, 2012, 2013 Tundra 2012 Sequoia	Southeast Toyota is recalling some models that were sold with labels that were outside the allowable one percent of accuracy of actual weight added. Thus, these vehicles fail to comply with the requirements of FMVSS Number 110, "Tire Selection and Rims." An inaccurate label could lead to owners overloading their vehicles and tires. An overloaded vehicle can result in a tire failure which may result in a vehicle crash, personal injury, or property damage.

Recall date	Recall campaign number	Model(s) affected	Concern
July 16, 2014	**14V429000**	2013, 2014 Tundra	Gulf States Toyota is recalling some models that may be equipped with a combination of non-Toyota-brand 20-inch alloy wheels and chrome plated lug nuts. The coating on the lug nuts may give, causing the lug nuts to loosen or the wheel studs to fracture. If the lug nuts loosen, or the wheel studs fracture, the wheel may separate from the vehicle, increasing the risk of a crash.
September 11, 2014	**14V556000**	2014 Tundra	Toyota is recalling some models that had the improper installation of the upper tab of the B-pillar interior trim, the Curtain-Shield Air Bags may not deploy properly in the event of a crash. If the Curtain-Shield Air Bags do not deploy properly there is an increased risk of occupant injury during a crash.
November 19, 2014	**14V743000**	2006, 2007, 2008, 2009, 2010, 2011 Tundra 2008, 2009, 2010, 2011 Sequoia	Southeast Toyota is recalling some models that may experience compression of the seat cushion which may damage the seat heater wiring. Damage to the seat heater wiring could cause the wires to short, increasing the risk of the seat burning and causing personal injury to the occupant.
May 13, 2015	**15V285000**	2007 Toyota Tundra, Sequoia	This is to address a safety defect in the passenger side frontal airbag inflator which may produce excessive internal pressure causing the inflator to rupture upon deployment of the airbag. This recall addresses both the passenger side frontal airbags that were originally installed in the vehicles, as well as replacement airbags that may have been installed as replacement service parts. A replacement ai bag may have been installed, as one example, if a vehicle had been in a crash necessitating the replacement of the passenger side frontal airbag. In the event of a crash necessitating deployment of the passenger's frontal airbag, the inflator could rupture with metal fragments striking the vehicle occupants potentially resulting in serious injury or death.
May 27, 2015	**15V315000**	2015 Toyota Tundra	On some models, equipped with Nitto Terra Grappler G2 275/60R20 116S XL tires. These vehicles have a tire placard that incorrectly states the recommended cold tire inflation pressure. If the operator inflates the tires according to the information on the tire placard, the tires may be underinflated, increasing the risk of tire failure which could result in a crash.
October 22, 2015	**15V689000**	2009, 2010, 2011 Toyota Tundra and Sequoia	On some models, during the manufacturing of the Power Window Master Switch (PWMS), grease may have been inconsistently applied to the sliding electrical contacts. If the sliding electrical contacts are not protected by lubricant, debris and moisture that get into the switch may cause a short circuit and the switch assembly may overheat and melt, increasing the risk of a fire.
June 2, 2016	**16V396000**	2007, 2008 Toyota Tundra and Sequoia	On some models, equipped with aftermarket accessory seat heaters with a copper strand heating element, the wiring in the seat heaters may be damaged when the seat cushion is compressed. If damaged, the copper strand heating element may short circuit, increasing the risk of a fire.
June 9, 2016	**16V420000**	2015, 2016 Tundra	On the affected vehicles, the Load Carrying Capacity Modification Label may not reflect the correct added weight of the installed accessories. An incorrect label may lead an owner to overload the vehicle, increasing the risk of a crash.
January 24, 2017	**17V051000**	2016, 2017 Tundra, Sequoia	On the affected vehicles equipped with a resin rear step bumper and resin reinforcement brackets (vehicles with chrome step bumpers are not affected), in the event of an impact to the corner of the bumper, the resin bracket may be damaged but not be noticed. If a person steps on the corner of the bumper that is damaged, a portion of the bumper may break away, increasing the risk of injury.

Recall date	Recall campaign number	Model(s) affected	Concern
May 11, 2017	17V311000	2013, 2014, 2015, 2016, 2017 Tundra and Sequoia	On some models equipped with Southeast Toyota accessory 20-inch Rockstar wheels, the Rockstar wheels were installed with lugnuts that may crack and detach. Lugnuts that crack and detach may cause the wheels to separate from the vehicle, increasing the risk of a crash.
June 29, 2017	17V416000	2016 Tundra	The affected vehicles have a passenger knee airbag module that was attached to the instrument panel mounting brackets with incorrect bolts. The incorrect bolts may cause the airbag module to loosen over time, affecting the performance of the knee airbag, increasing the risk of injury during a crash.
December 20, 2017	17V831000	2017, 2018 Tundra	The affected vehicles may have incorrect load carrying capacity modification labels. An incorrect load information label can result in the operator overloading the vehicle and increasing the risk of a crash.
February 20, 2018	18V122000	2018 Tundra and Sequoia	On the affected vehicles, the electrical interference within the power supply circuit may cause the vehicle's electronic stability control system to be deactivated. Deactivation of the vehicle stability control system can increase the risk of a crash in certain conditions.
February 20, 2018	18V123000	2017 Tundra	On the affected vehicles, one of the rear split bench seat leg brackets may not have been properly tightened to the vehicle's floor pan, possibly allowing the seat to move in a crash. If the rear split bench seat moves in a crash, the seat occupant has an increased risk of injury.
October 4, 2018	18V685000	2018, 2019 Tundra and Sequoia	These vehicles are being recalled because the airbag electronic control unit (ECU) may erroneously detect a fault during the vehicle start-up self-check. If this occurs, the ECU may not deploy the airbags as intended in the event of a crash. If the airbags do not deploy as intended, it can increase the risk of injury in a crash.
October 11, 2018	18V711000	2018, 2019 Tundra	These vehicles are being recalled to because the accessory all-weather floor mats may have been counted twice when creating the load carrying capacity modification label, resulting in the capacity modification label incorrectly overstating the additional weight by 10 pounds. Since the incorrect label is not within 1 percent of the additional weight, the vehicles are not compliant with the FMVSS.
December 19, 2018	18V900000	2019 Tundra	On these models, the load carrying capacity modification label may have been incorrectly calculated. If the label is followed, the vehicle may be overloaded. An overloaded vehicle can increase the risk of a crash.

Maintenance techniques, tools and working facilities

Maintenance techniques

There are a number of techniques involved in maintenance and repair that will be referred to throughout this manual. Application of these techniques will enable the home mechanic to be more efficient, better organized and capable of performing the various tasks properly, which will ensure that the repair job is thorough and complete.

Fasteners

Fasteners are nuts, bolts, studs and screws used to hold two or more parts together. There are a few things to keep in mind when working with fasteners. Almost all of them use a locking device of some type, either a lockwasher, locknut, locking tab or thread adhesive. All threaded fasteners should be clean and straight, with undamaged threads and undamaged corners on the hex head where the wrench fits. Develop the habit of replacing all damaged nuts and bolts with new ones. Special locknuts with nylon or fiber inserts can only be used once. If they are removed, they lose their locking ability and must be replaced with new ones.

Rusted nuts and bolts should be treated with a penetrating fluid to ease removal and prevent breakage. Some mechanics use turpentine in a spout-type oil can, which works quite well. After applying the rust penetrant, let it work for a few minutes before trying to loosen the nut or bolt. Badly rusted fasteners may have to be chiseled or sawed off or removed with a special nut breaker, available at tool stores.

If a bolt or stud breaks off in an assembly, it can be drilled and removed with a special tool commonly available for this purpose. Most automotive machine shops can perform this task, as well as other repair procedures, such as the repair of threaded holes that have been stripped out.

Flat washers and lockwashers, when removed from an assembly, should always be replaced exactly as removed. Replace any damaged washers with new ones. Never use a lockwasher on any soft metal surface (such as aluminum), thin sheet metal or plastic.

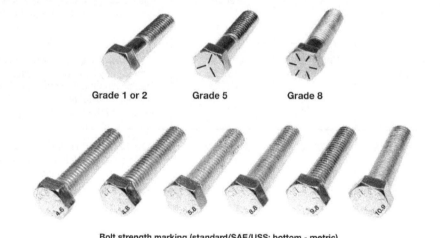

Grade 1 or 2   Grade 5   Grade 8

Bolt strength marking (standard/SAE/USS; bottom - metric)

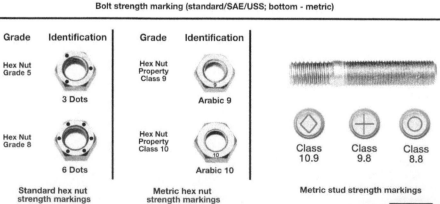

Grade	Identification	Grade	Identification
Hex Nut Grade 5	3 Dots	Hex Nut Property Class 9	Arabic 9
Hex Nut Grade 8	6 Dots	Hex Nut Property Class 10	Arabic 10

Standard hex nut strength markings

Metric hex nut strength markings

Class 10.9   Class 9.8   Class 8.8

Metric stud strength markings

00-1 HAYNES

Fastener sizes

For a number of reasons, automobile manufacturers are making wider and wider use of metric fasteners. Therefore, it is important to be able to tell the difference between standard (sometimes called U.S. or SAE) and metric hardware, since they cannot be interchanged.

All bolts, whether standard or metric, are sized according to diameter, thread pitch and length. For example, a standard 1/2 - 13 x 1 bolt is 1/2 inch in diameter, has 13 threads per inch and is 1 inch long. An M12 - 1.75 x 25 metric bolt is 12 mm in diameter, has a thread pitch of 1.75 mm (the distance between threads) and is 25 mm long. The two bolts are nearly identical, and easily confused, but they are not interchangeable.

In addition to the differences in diameter, thread pitch and length, metric and standard bolts can also be distinguished by examining the bolt heads. To begin with, the distance across the flats on a standard bolt head is measured in inches, while the same dimension on a metric bolt is sized in millimeters (the same is true for nuts). As a result, a standard wrench should not be used on a metric bolt and a metric wrench should not be used on a standard bolt. Also, most standard bolts have slashes radiating out from the center of the head to denote the grade or strength of the bolt, which is an indication of the amount of torque that can be applied to it. The greater the number of slashes, the greater the strength of the bolt. Grades 0 through 5 are commonly used on automobiles. Metric bolts have a property class (grade) number, rather than a slash, molded into their heads to indicate bolt strength. In this case, the higher the number, the stronger the bolt. Property class numbers 8.8, 9.8 and 10.9 are commonly used on automobiles.

Strength markings can also be used to distinguish standard hex nuts from metric hex nuts. Many standard nuts have dots stamped into one side, while metric nuts are marked with a number. The greater the number of dots, or the higher the number, the greater the strength of the nut.

Metric studs are also marked on their ends according to property class (grade). Larger studs are numbered (the same as metric bolts), while smaller studs carry a geometric code to denote grade.

It should be noted that many fasteners, especially Grades 0 through 2, have no distinguishing marks on them. When such is the case, the only way to determine whether it is standard or metric is to measure the thread pitch or compare it to a known fastener of the same size.

Standard fasteners are often referred to as SAE, as opposed to metric. However, it should be noted that SAE technically refers to a non-metric fine thread fastener only. Coarse thread non-metric fasteners are referred to as USS sizes.

Since fasteners of the same size (both standard and metric) may have different strength ratings, be sure to reinstall any bolts,

Metric thread sizes	Ft-lbs	Nm
M-6	6 to 9	9 to 12
M-8	14 to 21	19 to 28
M-10	28 to 40	38 to 54
M-12	50 to 71	68 to 96
M-14	80 to 140	109 to 154

Pipe thread sizes		
1/8	5 to 8	7 to 10
1/4	12 to 18	17 to 24
3/8	22 to 33	30 to 44
1/2	25 to 35	34 to 47

U.S. thread sizes		
1/4 - 20	6 to 9	9 to 12
5/16 - 18	12 to 18	17 to 24
5/16 - 24	14 to 20	19 to 27
3/8 - 16	22 to 32	30 to 43
3/8 - 24	27 to 38	37 to 51
7/16 - 14	40 to 55	55 to 74
7/16 - 20	40 to 60	55 to 81
1/2 - 13	55 to 80	75 to 108

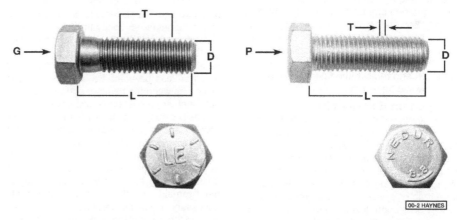

Standard (SAE and USS) bolt dimensions/ grade marks

G   Grade marks (bolt strength)
L   Length (in inches)
T   Thread pitch (number of threads per inch)
D   Nominal diameter (in inches)

Metric bolt dimensions/grade marks

P   Property class (bolt strength)
L   Length (in millimeters)
T   Thread pitch (distance between threads in millimeters)
D   Diameter

studs or nuts removed from your vehicle in their original locations. Also, when replacing a fastener with a new one, make sure that the new one has a strength rating equal to or greater than the original.

Tightening sequences and procedures

Most threaded fasteners should be tightened to a specific torque value (torque is the twisting force applied to a threaded component such as a nut or bolt). Overtightening the fastener can weaken it and cause it to break, while undertightening can cause it to eventually come loose. Bolts, screws and studs, depending on the material they are made of and their thread diameters, have specific

torque values, many of which are noted in the Specifications at the beginning of each Chapter. Be sure to follow the torque recommendations closely. For fasteners not assigned a specific torque, a general torque value chart is presented here as a guide. These torque values are for dry (unlubricated) fasteners threaded into steel or cast iron (not aluminum). As was previously mentioned, the size and grade of a fastener determine the amount of torque that can safely be applied to it. The figures listed here are approximate for Grade 2 and Grade 3 fasteners. Higher grades can tolerate higher torque values.

Fasteners laid out in a pattern, such as cylinder head bolts, oil pan bolts, differential cover bolts, etc., must be loosened or tightened in sequence to avoid warping the com-

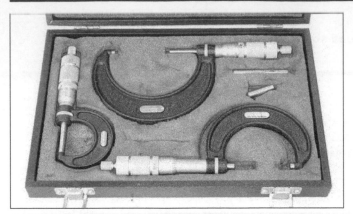

Micrometer set

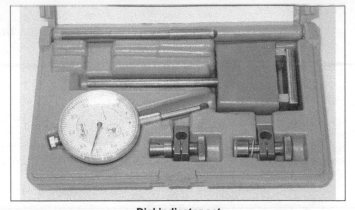

Dial indicator set

ponent. This sequence will normally be shown in the appropriate Chapter. If a specific pattern is not given, the following procedures can be used to prevent warping.

Initially, the bolts or nuts should be assembled finger-tight only. Next, they should be tightened one full turn each, in a crisscross or diagonal pattern. After each one has been tightened one full turn, return to the first one and tighten them all one-half turn, following the same pattern. Finally, tighten each of them one-quarter turn at a time until each fastener has been tightened to the proper torque. To loosen and remove the fasteners, the procedure would be reversed.

Component disassembly

Component disassembly should be done with care and purpose to help ensure that the parts go back together properly. Always keep track of the sequence in which parts are removed. Make note of special characteristics or marks on parts that can be installed more than one way, such as a grooved thrust washer on a shaft. It is a good idea to lay the disassembled parts out on a clean surface in the order that they were removed. It may also be helpful to make sketches or take instant photos of components before removal.

When removing fasteners from a component, keep track of their locations. Sometimes threading a bolt back in a part, or putting the washers and nut back on a stud, can prevent mix-ups later. If nuts and bolts cannot be returned to their original locations, they should be kept in a compartmented box or a series of small boxes. A cupcake or muffin tin is ideal for this purpose, since each cavity can hold the bolts and nuts from a particular area (i.e. oil pan bolts, valve cover bolts, engine mount bolts, etc.). A pan of this type is especially helpful when working on assemblies with very small parts, such as the carburetor, alternator, valve train or interior dash and trim pieces. The cavities can be marked with paint or tape to identify the contents.

Whenever wiring looms, harnesses or connectors are separated, it is a good idea to identify the two halves with numbered pieces of masking tape so they can be easily reconnected.

Gasket sealing surfaces

Throughout any vehicle, gaskets are used to seal the mating surfaces between two parts and keep lubricants, fluids, vacuum or pressure contained in an assembly.

Many times these gaskets are coated with a liquid or paste-type gasket sealing compound before assembly. Age, heat and pressure can sometimes cause the two parts to stick together so tightly that they are very difficult to separate. Often, the assembly can be loosened by striking it with a soft-face hammer near the mating surfaces. A regular hammer can be used if a block of wood is placed between the hammer and the part. Do not hammer on cast parts or parts that could be easily damaged. With any particularly stubborn part, always recheck to make sure that every fastener has been removed.

Avoid using a screwdriver or bar to pry apart an assembly, as they can easily mar the gasket sealing surfaces of the parts, which must remain smooth. If prying is absolutely necessary, use an old broom handle, but keep in mind that extra clean up will be necessary if the wood splinters.

After the parts are separated, the old gasket must be carefully removed and the gasket surfaces cleaned. If you're working on cast iron or aluminum parts, stubborn gasket material can be soaked with rust penetrant or treated with a special chemical to soften it so it can be easily scraped off. **Caution:** *Never use gasket removal solutions or caustic chemicals on plastic or other composite components.* A scraper can be fashioned from a piece of copper tubing by flattening and sharpening one end. Copper is recommended because it is usually softer than the surfaces to be scraped, which reduces the chance of gouging the part. Some gaskets can be removed with a wire brush, but regardless of the method used, the mating surfaces must be left clean and smooth. If for some reason the gasket surface is gouged, then a gasket sealer thick enough to fill scratches will have to be used during reassembly of the components. For most applications, a non-drying (or semi-drying) gasket sealer should be used.

Hose removal tips

Warning: *If the vehicle is equipped with air conditioning, do not disconnect any of the A/C hoses without first having the system depressurized by a dealer service department or a service station.*

Hose removal precautions closely parallel gasket removal precautions. Avoid scratching or gouging the surface that the hose mates against or the connection may leak. This is especially true for radiator hoses. Because of various chemical reactions, the rubber in hoses can bond itself to the metal spigot that the hose fits over. To remove a hose, first loosen the hose clamps that secure it to the spigot. Then, with slip-joint pliers, grab the hose at the clamp and rotate it around the spigot. Work it back and forth until it is completely free, then pull it off. Silicone or other lubricants will ease removal if they can be applied between the hose and the outside of the spigot. Apply the same lubricant to the inside of the hose and the outside of the spigot to simplify installation.

As a last resort (and if the hose is to be replaced with a new one anyway), the rubber can be slit with a knife and the hose peeled from the spigot. If this must be done, be careful that the metal connection is not damaged.

If a hose clamp is broken or damaged, do not reuse it. Wire-type clamps usually weaken with age, so it is a good idea to replace them with screw-type clamps whenever a hose is removed.

Tools

A selection of good tools is a basic requirement for anyone who plans to maintain and repair his or her own vehicle. For the owner who has few tools, the initial investment might seem high, but when compared to the spiraling costs of professional auto maintenance and repair, it is a wise one.

To help the owner decide which tools are needed to perform the tasks detailed in this manual, the following tool lists are offered: *Maintenance and minor repair, Repair/overhaul* and *Special.*

The newcomer to practical mechanics should start off with the *maintenance and minor repair* tool kit, which is adequate for the

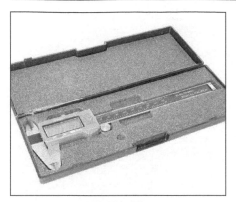

Dial caliper

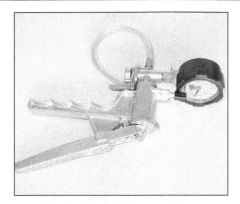

Hand-operated vacuum pump

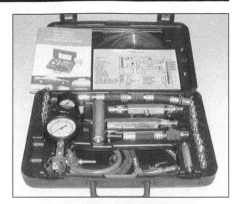

Fuel pressure gauge set

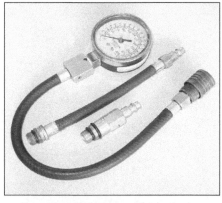

Compression gauge with spark plug
hole adapter

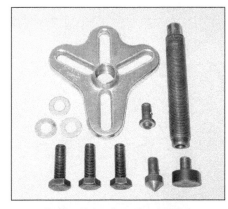

Damper/steering wheel puller

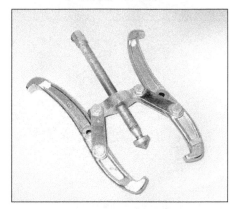

General purpose puller

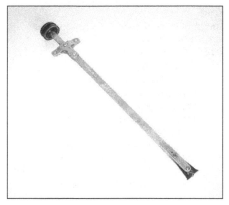

Hydraulic lifter removal tool

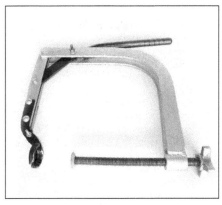

Valve spring compressor

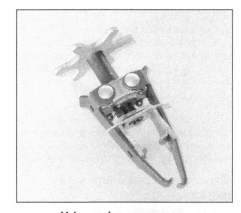

Valve spring compressor

Ridge reamer

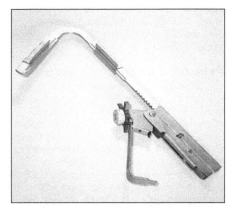

Piston ring groove cleaning tool

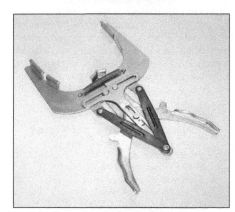

Ring removal/installation tool

Ring compressor

Cylinder hone

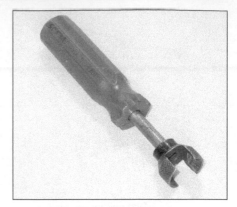

Brake hold-down spring tool

Torque angle gauge

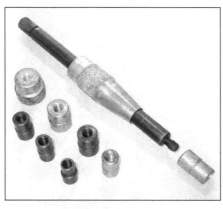

Clutch plate alignment tool

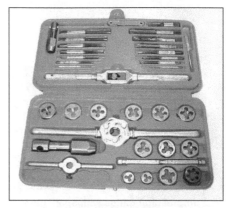

Tap and die set

simpler jobs performed on a vehicle. Then, as confidence and experience grow, the owner can tackle more difficult tasks, buying additional tools as they are needed. Eventually the basic kit will be expanded into the *repair and overhaul* tool set. Over a period of time, the experienced do-it-yourselfer will assemble a tool set complete enough for most repair and overhaul procedures and will add tools from the special category when it is felt that the expense is justified by the frequency of use.

Maintenance and minor repair tool kit

The tools in this list should be considered the minimum required for performance of routine maintenance, servicing and minor repair work. We recommend the purchase of combination wrenches (box-end and open-end combined in one wrench). While more expensive than open end wrenches, they offer the advantages of both types of wrench.

> *Combination wrench set (1/4-inch to*
>   *1 inch or 6 mm to 19 mm)*
> *Adjustable wrench, 8 inch*
> *Spark plug wrench with rubber insert*
> *Spark plug gap adjusting tool*
> *Feeler gauge set*
> *Brake bleeder wrench*
> *Standard screwdriver*
>   *(5/16-inch x 6 inch)*
> *Phillips screwdriver (No. 2 x 6 inch)*
> *Combination pliers - 6 inch*

> *Hacksaw and assortment of blades*
> *Tire pressure gauge*
> *Grease gun*
> *Oil can*
> *Fine emery cloth*
> *Wire brush*
> *Battery post and cable cleaning tool*
> *Oil filter wrench*
> *Funnel (medium size)*
> *Safety goggles*
> *Jackstands (2)*
> *Drain pan*

Note: *If basic tune-ups are going to be part of routine maintenance, it will be necessary to purchase a good quality stroboscopic timing light and combination tachometer/dwell meter. Although they are included in the list of special tools, it is mentioned here because they are absolutely necessary for tuning most vehicles properly.*

Repair and overhaul tool set

These tools are essential for anyone who plans to perform major repairs and are in addition to those in the maintenance and minor repair tool kit. Included is a comprehensive set of sockets which, though expensive, are invaluable because of their versatility, especially when various extensions and drives are available. We recommend the 1/2-inch drive over the 3/8-inch drive. Although the larger drive is bulky and more expensive, it has the capacity of accepting a very wide

range of large sockets. Ideally, however, the mechanic should have a 3/8-inch drive set and a 1/2-inch drive set.

> *Socket set(s)*
> *Reversible ratchet*
> *Extension - 10 inch*
> *Universal joint*
> *Torque wrench (same size drive as*
>   *sockets)*
> *Ball peen hammer - 8 ounce*
> *Soft-face hammer (plastic/rubber)*
> *Standard screwdriver (1/4-inch x 6 inch)*
> *Standard screwdriver (stubby -*
>   *5/16-inch)*
> *Phillips screwdriver (No. 3 x 8 inch)*
> *Phillips screwdriver (stubby - No. 2)*
> *Pliers - vise grip*
> *Pliers - lineman's*
> *Pliers - needle nose*
> *Pliers - snap-ring (internal and external)*
> *Cold chisel - 1/2-inch*
> *Scribe*
> *Scraper (made from flattened copper*
>   *tubing)*
> *Centerpunch*
> *Pin punches (1/16, 1/8, 3/16-inch)*
> *Steel rule/straightedge - 12 inch*
> *Allen wrench set (1/8 to 3/8-inch or*
>   *4 mm to 10 mm)*
> *A selection of files*
> *Wire brush (large)*
> *Jackstands (second set)*
> *Jack (scissor or hydraulic type)*

Note: *Another tool which is often useful is an electric drill with a chuck capacity of 3/8-inch and a set of good quality drill bits.*

Special tools

The tools in this list include those which are not used regularly, are expensive to buy, or which need to be used in accordance with their manufacturer's instructions. Unless these tools will be used frequently, it is not very economical to purchase many of them. A consideration would be to split the cost and use between yourself and a friend or friends. In addition, most of these tools can be obtained from a tool rental shop on a temporary basis.

This list primarily contains only those tools and instruments widely available to the public, and not those special tools produced by the vehicle manufacturer for distribution to dealer service departments. Occasionally, references to the manufacturer's special tools are included in the text of this manual. Generally, an alternative method of doing the job without the special tool is offered. However, sometimes there is no alternative to their use. Where this is the case, and the tool cannot be purchased or borrowed, the work should be turned over to the dealer service department or an automotive repair shop.

Valve spring compressor
Piston ring groove cleaning tool
Piston ring compressor
Piston ring installation tool
Cylinder compression gauge
Cylinder ridge reamer
Cylinder surfacing hone
Cylinder bore gauge
Micrometers and/or dial calipers
Hydraulic lifter removal tool
Balljoint separator
Universal-type puller
Impact screwdriver
Dial indicator set
Stroboscopic timing light (inductive pick-up)
Hand operated vacuum/pressure pump
Tachometer/dwell meter
Universal electrical multimeter
Cable hoist
Brake spring removal and installation tools
Floor jack

Buying tools

For the do-it-yourselfer who is just starting to get involved in vehicle maintenance and repair, there are a number of options available when purchasing tools. If maintenance and minor repair is the extent of the work to be done, the purchase of individual tools is satisfactory. If, on the other hand, extensive work is planned, it would be a good idea to purchase a modest tool set from one of the large retail chain stores. A set can usually be bought at a substantial savings over the individual tool prices, and they often come with a tool box. As additional tools are

needed, add-on sets, individual tools and a larger tool box can be purchased to expand the tool selection. Building a tool set gradually allows the cost of the tools to be spread over a longer period of time and gives the mechanic the freedom to choose only those tools that will actually be used.

Tool stores will often be the only source of some of the special tools that are needed, but regardless of where tools are bought, try to avoid cheap ones, especially when buying screwdrivers and sockets, because they won't last very long. The expense involved in replacing cheap tools will eventually be greater than the initial cost of quality tools.

Care and maintenance of tools

Good tools are expensive, so it makes sense to treat them with respect. Keep them clean and in usable condition and store them properly when not in use. Always wipe off any dirt, grease or metal chips before putting them away. Never leave tools lying around in the work area. Upon completion of a job, always check closely under the hood for tools that may have been left there so they won't get lost during a test drive.

Some tools, such as screwdrivers, pliers, wrenches and sockets, can be hung on a panel mounted on the garage or workshop wall, while others should be kept in a tool box or tray. Measuring instruments, gauges, meters, etc. must be carefully stored where they cannot be damaged by weather or impact from other tools.

When tools are used with care and stored properly, they will last a very long time. Even with the best of care, though, tools will wear out if used frequently. When a tool is damaged or worn out, replace it. Subsequent jobs will be safer and more enjoyable if you do.

How to repair damaged threads

Sometimes, the internal threads of a nut or bolt hole can become stripped, usually from overtightening. Stripping threads is an all-too-common occurrence, especially when working with aluminum parts, because aluminum is so soft that it easily strips out.

Usually, external or internal threads are only partially stripped. After they've been cleaned up with a tap or die, they'll still work. Sometimes, however, threads are badly damaged. When this happens, you've got three choices:

1) *Drill and tap the hole to the next suitable oversize and install a larger diameter bolt, screw or stud.*
2) *Drill and tap the hole to accept a threaded plug, then drill and tap the plug to the original screw size. You can also buy a plug already threaded to the original size. Then you simply drill a hole to the specified size, then run the threaded plug into the hole with a bolt and jam nut.*

Once the plug is fully seated, remove the jam nut and bolt.
3) *The third method uses a patented thread repair kit like Heli-Coil or Slimsert. These easy-to-use kits are designed to repair damaged threads in straight-through holes and blind holes. Both are available as kits which can handle a variety of sizes and thread patterns. Drill the hole, then tap it with the special included tap. Install the Heli-Coil and the hole is back to its original diameter and thread pitch.*

Regardless of which method you use, be sure to proceed calmly and carefully. A little impatience or carelessness during one of these relatively simple procedures can ruin your whole day's work and cost you a bundle if you wreck an expensive part.

Working facilities

Not to be overlooked when discussing tools is the workshop. If anything more than routine maintenance is to be carried out, some sort of suitable work area is essential.

It is understood, and appreciated, that many home mechanics do not have a good workshop or garage available, and end up removing an engine or doing major repairs outside. It is recommended, however, that the overhaul or repair be completed under the cover of a roof.

A clean, flat workbench or table of comfortable working height is an absolute necessity. The workbench should be equipped with a vise that has a jaw opening of at least four inches.

As mentioned previously, some clean, dry storage space is also required for tools, as well as the lubricants, fluids, cleaning solvents, etc. which soon become necessary.

Sometimes waste oil and fluids, drained from the engine or cooling system during normal maintenance or repairs, present a disposal problem. To avoid pouring them on the ground or into a sewage system, pour the used fluids into large containers, seal them with caps and take them to an authorized disposal site or recycling center. Plastic jugs, such as old antifreeze containers, are ideal for this purpose.

Always keep a supply of old newspapers and clean rags available. Old towels are excellent for mopping up spills. Many mechanics use rolls of paper towels for most work because they are readily available and disposable. To help keep the area under the vehicle clean, a large cardboard box can be cut open and flattened to protect the garage or shop floor.

Whenever working over a painted surface, such as when leaning over a fender to service something under the hood, always cover it with an old blanket or bedspread to protect the finish. Vinyl covered pads, made especially for this purpose, are available at auto parts stores.

Jacking and towing

Jacking

The jack supplied with the vehicle should only be used for raising the vehicle when changing a tire or placing jackstands under the frame.

Warning: *Never work under the vehicle or start the engine while this jack is being used as the only means of support.*

The vehicle should be on level ground with the hazard flashers on, the wheels blocked, the parking brake applied and the transmission in Park (automatic). If a tire is being changed, loosen the lug nuts one-half turn and leave them in place until the wheel is raised off the ground.

Place the jack under the vehicle **(see illustrations)**:

Front: Under the frame rail, where the crossmember is attached.

Rear: Under the rear axle housing.

Operate the jack with a slow, smooth motion until the wheel is raised off the ground. Remove the lug nuts, pull off the wheel, install the spare and thread the lug nuts back on with the beveled sides facing in. Tighten them snugly, but wait until the vehicle is lowered to tighten them completely.

Lower the vehicle, remove the jack and tighten the nuts (if loosened or removed) in a criss-cross pattern.

Towing

As a general rule, the vehicle should be towed with professional towing equipment of the flatbed or wheel-lift type. If towed from the front, the rear wheels should be placed on a towing dolly. If a 4WD model is towed from the rear, the front wheels should be placed on a towing dolly; if a dolly is not available, turn the ignition key to the ACC position, place the transfer case in the 2H position and the transmission in Neutral.

When a vehicle is towed with the rear wheels raised, the steering wheel must be clamped in the straight ahead position with a special device designed for use during towing. The ignition key must not be in the LOCK position, since the steering lock mechanism isn't strong enough to hold the front wheels straight while towing.

Equipment specifically designed for towing should be used. It should be attached to the main structural members of the vehicle, not the bumpers or brackets. Safety is a major consideration when towing and all applicable state and local laws must be obeyed. A safety chain system must be used at all times. Remember that power steering and power brakes will not work with the engine off.

Front jacking point (under the frame rail)

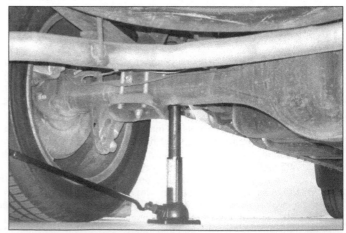

Rear jacking point - Tundra models (under the axle tube).
On Sequoia models, place the jack under the frame rail
nearest the wheel to be changed

Booster battery (jump) starting

Observe these precautions when using a booster battery to start a vehicle:

a) *Before connecting the booster battery, make sure the ignition switch is in the Off position.*

b) *Turn off the lights, heater and other electrical loads.*

c) *Your eyes should be shielded. Safety goggles are a good idea.*

d) *Make sure the booster battery is the same voltage as the dead one in the vehicle.*

e) *The two vehicles MUST NOT TOUCH each other!*

f) *Make sure the transaxle is in Neutral (manual) or Park (automatic).*

g) *If the booster battery is not a maintenance-free type, remove the vent caps and lay a cloth over the vent holes.*

Connect the red jumper cable to the positive (+) terminals of each battery **(see illustration)**.

Connect one end of the black jumper cable to the negative (-) terminal of the booster battery. The other end of this cable should be connected to a good ground on the vehicle to be started, such as a bolt or bracket on the body.

Start the engine using the booster battery, then, with the engine running at idle speed, disconnect the jumper cables in the reverse order of connection.

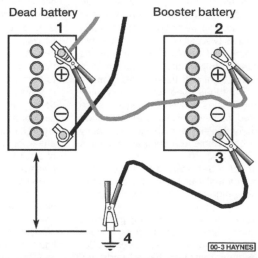

Dead battery   Booster battery

Make the booster battery cable connections in the numerical order shown (note that the negative cable of the booster battery is NOT attached to the negative terminal of the dead battery)

Automotive chemicals and lubricants

A number of automotive chemicals and lubricants are available for use during vehicle maintenance and repair. They include a wide variety of products ranging from cleaning solvents and degreasers to lubricants and protective sprays for rubber, plastic and vinyl.

Cleaners

Carburetor cleaner and choke cleaner is a strong solvent for gum, varnish and carbon. Most carburetor cleaners leave a dry-type lubricant film which will not harden or gum up. Because of this film it is not recommended for use on electrical components.

Brake system cleaner is used to remove brake dust, grease and brake fluid from the brake system, where clean surfaces are absolutely necessary. It leaves no residue and often eliminates brake squeal caused by contaminants.

Electrical cleaner removes oxidation, corrosion and carbon deposits from electrical contacts, restoring full current flow. It can also be used to clean spark plugs, carburetor jets, voltage regulators and other parts where an oil-free surface is desired.

Demoisturants remove water and moisture from electrical components such as alternators, voltage regulators, electrical connectors and fuse blocks. They are non-conductive and non-corrosive.

Degreasers are heavy-duty solvents used to remove grease from the outside of the engine and from chassis components. They can be sprayed or brushed on and, depending on the type, are rinsed off either with water or solvent.

Lubricants

Motor oil is the lubricant formulated for use in engines. It normally contains a wide variety of additives to prevent corrosion and reduce foaming and wear. Motor oil comes in various weights (viscosity ratings) from 0 to 50. The recommended weight of the oil depends on the season, temperature and the demands on the engine. Light oil is used in cold climates and under light load conditions. Heavy oil is used in hot climates and where high loads are encountered. Multi-viscosity oils are designed to have characteristics of both light and heavy oils and are available in a number of weights from 0W-20 to 20W-50.

Gear oil is designed to be used in differentials, manual transmissions and other areas where high-temperature lubrication is required.

Chassis and wheel bearing grease is a heavy grease used where increased loads and friction are encountered, such as for wheel bearings, balljoints, tie-rod ends and universal joints.

High-temperature wheel bearing grease is designed to withstand the extreme temperatures encountered by wheel bearings in disc brake equipped vehicles. It usually contains molybdenum disulfide (moly), which is a dry-type lubricant.

White grease is a heavy grease for metal-to-metal applications where water is a problem. White grease stays soft under both low and high temperatures (usually from -100 to +190-degrees F), and will not wash off or dilute in the presence of water.

Assembly lube is a special extreme pressure lubricant, usually containing moly, used to lubricate high-load parts (such as main and rod bearings and cam lobes) for initial start-up of a new engine. The assembly lube lubricates the parts without being squeezed out or washed away until the engine oiling system begins to function.

Silicone lubricants are used to protect rubber, plastic, vinyl and nylon parts.

Graphite lubricants are used where oils cannot be used due to contamination problems, such as in locks. The dry graphite will lubricate metal parts while remaining uncontaminated by dirt, water, oil or acids. It is electrically conductive and will not foul electrical contacts in locks such as the ignition switch.

Moly penetrants loosen and lubricate frozen, rusted and corroded fasteners and prevent future rusting or freezing.

Heat-sink grease is a special electrically non-conductive grease that is used for mounting electronic ignition modules where it is essential that heat is transferred away from the module.

Sealants

RTV sealant is one of the most widely used gasket compounds. Made from silicone, RTV is air curing, it seals, bonds, waterproofs, fills surface irregularities, remains flexible, doesn't shrink, is relatively easy to remove, and is used as a supplementary sealer with almost all low and medium temperature gaskets.

Anaerobic sealant is much like RTV in that it can be used either to seal gaskets or to form gaskets by itself. It remains flexible, is solvent resistant and fills surface imperfections. The difference between an anaerobic sealant and an RTV-type sealant is in the curing. RTV cures when exposed to air, while an anaerobic sealant cures only in the absence of air. This means that an anaerobic sealant cures only after the assembly of parts, sealing them together.

Thread and pipe sealant is used for sealing hydraulic and pneumatic fittings and vacuum lines. It is usually made from a Teflon compound, and comes in a spray, a paint-on liquid and as a wrap-around tape.

Chemicals

Anti-seize compound prevents seizing, galling, cold welding, rust and corrosion in fasteners. High-temperature anti-seize, usually made with copper and graphite lubricants, is used for exhaust system and exhaust manifold bolts.

Anaerobic locking compounds are used to keep fasteners from vibrating or working loose and cure only after installation, in the absence of air. Medium strength locking compound is used for small nuts, bolts and screws that may be removed later. High-strength locking compound is for large nuts, bolts and studs which aren't removed on a regular basis.

Oil additives range from viscosity index improvers to chemical treatments that claim to reduce internal engine friction. It should be noted that most oil manufacturers caution against using additives with their oils.

Gas additives perform several functions, depending on their chemical makeup. They usually contain solvents that help dissolve gum and varnish that build up on carburetor, fuel injection and intake parts. They also serve to break down carbon deposits that form on the inside surfaces of the combustion chambers. Some additives contain upper cylinder lubricants for valves and piston rings, and others contain chemicals to remove condensation from the gas tank.

Miscellaneous

Brake fluid is specially formulated hydraulic fluid that can withstand the heat and pressure encountered in brake systems. Care must be taken so this fluid does not come in contact with painted surfaces or plastics. An opened container should always be resealed to prevent contamination by water or dirt.

Weatherstrip adhesive is used to bond weatherstripping around doors, windows and trunk lids. It is sometimes used to attach trim pieces.

Undercoating is a petroleum-based, tar-like substance that is designed to protect metal surfaces on the underside of the vehicle from corrosion. It also acts as a sound-deadening agent by insulating the bottom of the vehicle.

Waxes and polishes are used to help protect painted and plated surfaces from the weather. Different types of paint may require the use of different types of wax and polish. Some polishes utilize a chemical or abrasive cleaner to help remove the top layer of oxidized (dull) paint on older vehicles. In recent years many non-wax polishes that contain a wide variety of chemicals such as polymers and silicones have been introduced. These non-wax polishes are usually easier to apply and last longer than conventional waxes and polishes.

Conversion factors

Length (distance)

	X				X		
Inches (in)	X	25.4	= Millimeters (mm)		X	0.0394	= Inches (in)
Feet (ft)	X	0.305	= Meters (m)		X	3.281	= Feet (ft)
Miles	X	1.609	= Kilometers (km)		X	0.621	= Miles

Volume (capacity)

	X				X		
Cubic inches (cu in; in³)	X	16.387	= Cubic centimeters (cc; cm³)		X	0.061	= Cubic inches (cu in; in³)
Imperial pints (Imp pt)	X	0.568	= Liters (l)		X	1.76	= Imperial pints (Imp pt)
Imperial quarts (Imp qt)	X	1.137	= Liters (l)		X	0.88	= Imperial quarts (Imp qt)
Imperial quarts (Imp qt)	X	1.201	= US quarts (US qt)		X	0.833	= Imperial quarts (Imp qt)
US quarts (US qt)	X	0.946	= Liters (l)		X	1.057	= US quarts (US qt)
Imperial gallons (Imp gal)	X	4.546	= Liters (l)		X	0.22	= Imperial gallons (Imp gal)
Imperial gallons (Imp gal)	X	1.201	= US gallons (US gal)		X	0.833	= Imperial gallons (Imp gal)
US gallons (US gal)	X	3.785	= Liters (l)		X	0.264	= US gallons (US gal)

Mass (weight)

	X				X		
Ounces (oz)	X	28.35	= Grams (g)		X	0.035	= Ounces (oz)
Pounds (lb)	X	0.454	= Kilograms (kg)		X	2.205	= Pounds (lb)

Force

	X				X		
Ounces-force (ozf; oz)	X	0.278	= Newtons (N)		X	3.6	= Ounces-force (ozf; oz)
Pounds-force (lbf; lb)	X	4.448	= Newtons (N)		X	0.225	= Pounds-force (lbf; lb)
Newtons (N)	X	0.1	= Kilograms-force (kgf; kg)		X	9.81	= Newtons (N)

Pressure

	X				X		
Pounds-force per square inch (psi; lbf/in²; lb/in²)	X	0.070	= Kilograms-force per square centimeter (kgf/cm²; kg/cm²)		X	14.223	= Pounds-force per square inch (psi; lbf/in²; lb/in²)
Pounds-force per square inch (psi; lbf/in²; lb/in²)	X	0.068	= Atmospheres (atm)		X	14.696	= Pounds-force per square inch (psi; lbf/in²; lb/in²)
Pounds-force per square inch (psi; lbf/in²; lb/in²)	X	0.069	= Bars		X	14.5	= Pounds-force per square inch (psi; lbf/in²; lb/in²)
Pounds-force per square inch (psi; lbf/in²; lb/in²)	X	6.895	= Kilopascals (kPa)		X	0.145	= Pounds-force per square inch (psi; lbf/in²; lb/in²)
Kilopascals (kPa)	X	0.01	= Kilograms-force per square centimeter (kgf/cm²; kg/cm²)		X	98.1	= Kilopascals (kPa)

Torque (moment of force)

	X				X		
Pounds-force inches (lbf in; lb in)	X	1.152	= Kilograms-force centimeter (kgf cm; kg cm)		X	0.868	= Pounds-force inches (lbf in; lb in)
Pounds-force inches (lbf in; lb in)	X	0.113	= Newton meters (Nm)		X	8.85	= Pounds-force inches (lbf in; lb in)
Pounds-force inches (lbf in; lb in)	X	0.083	= Pounds-force feet (lbf ft; lb ft)		X	12	= Pounds-force inches (lbf in; lb in)
Pounds-force feet (lbf ft; lb ft)	X	0.138	= Kilograms-force meters (kgf m; kg m)		X	7.233	= Pounds-force feet (lbf ft; lb ft)
Pounds-force feet (lbf ft; lb ft)	X	1.356	= Newton meters (Nm)		X	0.738	= Pounds-force feet (lbf ft; lb ft)
Newton meters (Nm)	X	0.102	= Kilograms-force meters (kgf m; kg m)		X	9.804	= Newton meters (Nm)

Vacuum

	X				X		
Inches mercury (in. Hg)	X	3.377	= Kilopascals (kPa)		X	0.2961	= Inches mercury
Inches mercury (in. Hg)	X	25.4	= Millimeters mercury (mm Hg)		X	0.0394	= Inches mercury

Power

	X				X		
Horsepower (hp)	X	745.7	= Watts (W)		X	0.0013	= Horsepower (hp)

Velocity (speed)

	X				X		
Miles per hour (miles/hr; mph)	X	1.609	= Kilometers per hour (km/hr; kph)		X	0.621	= Miles per hour (miles/hr; mph)

Fuel consumption*

	X				X		
Miles per gallon, Imperial (mpg)	X	0.354	= Kilometers per liter (km/l)		X	2.825	= Miles per gallon, Imperial (mpg)
Miles per gallon, US (mpg)	X	0.425	= Kilometers per liter (km/l)		X	2.352	= Miles per gallon, US (mpg)

Temperature

Degrees Fahrenheit = (°C x 1.8) + 32   Degrees Celsius (Degrees Centigrade; °C) = (°F - 32) x 0.56

*It is common practice to convert from miles per gallon (mpg) to liters/100 kilometers (l/100km), where mpg (Imperial) x l/100 km = 282 and mpg (US) x l/100 km = 235

Safety first!

Regardless of how enthusiastic you may be about getting on with the job at hand, take the time to ensure that your safety is not jeopardized. A moment's lack of attention can result in an accident, as can failure to observe certain simple safety precautions. The possibility of an accident will always exist, and the following points should not be considered a comprehensive list of all dangers. Rather, they are intended to make you aware of the risks and to encourage a safety conscious approach to all work you carry out on your vehicle.

Essential DOs and DON'Ts

DON'T rely on a jack when working under the vehicle. Always use approved jackstands to support the weight of the vehicle and place them under the recommended lift or support points.

DON'T attempt to loosen extremely tight fasteners (i.e. wheel lug nuts) while the vehicle is on a jack - it may fall.

DON'T start the engine without first making sure that the transmission is in Neutral (or Park where applicable) and the parking brake is set.

DON'T remove the radiator cap from a hot cooling system - let it cool or cover it with a cloth and release the pressure gradually.

DON'T attempt to drain the engine oil until you are sure it has cooled to the point that it will not burn you.

DON'T touch any part of the engine or exhaust system until it has cooled sufficiently to avoid burns.

DON'T siphon toxic liquids such as gasoline, antifreeze and brake fluid by mouth, or allow them to remain on your skin.

DON'T inhale brake lining dust - it is potentially hazardous (see *Asbestos* below).

DON'T allow spilled oil or grease to remain on the floor - wipe it up before someone slips on it.

DON'T use loose fitting wrenches or other tools which may slip and cause injury.

DON'T push on wrenches when loosening or tightening nuts or bolts. Always try to pull the wrench toward you. If the situation calls for pushing the wrench away, push with an open hand to avoid scraped knuckles if the wrench should slip.

DON'T attempt to lift a heavy component alone - get someone to help you.

DON'T *rush or take unsafe shortcuts to finish a job.*

DON'T allow children or animals in or around the vehicle while you are working on it.

DO wear eye protection when using power tools such as a drill, sander, bench grinder, etc. and when working under a vehicle.

DO keep loose clothing and long hair well out of the way of moving parts.

DO make sure that any hoist used has a safe working load rating adequate for the job.

DO get someone to check on you periodically when working alone on a vehicle.

DO carry out work in a logical sequence and make sure that everything is correctly assembled and tightened.

DO keep chemicals and fluids tightly capped and out of the reach of children and pets.

DO remember that your vehicle's safety affects that of yourself and others. If in doubt on any point, get professional advice.

Steering, suspension and brakes

These systems are essential to driving safety, so make sure you have a qualified shop or individual check your work. Also, compressed suspension springs can cause injury if released suddenly - be sure to use a spring compressor.

Airbags

Airbags are explosive devices that can **CAUSE** injury if they deploy while you're working on the vehicle. Follow the manufacturer's instructions to disable the airbag whenever you're working in the vicinity of airbag components.

Asbestos

Certain friction, insulating, sealing, and other products - such as brake linings, brake bands, clutch linings, torque converters, gaskets, etc. - may contain asbestos or other hazardous friction material. Extreme care must be taken to avoid inhalation of dust from such products, since it is hazardous to health. If in doubt, assume that they do contain asbestos.

Fire

Remember at all times that gasoline is highly flammable. Never smoke or have any kind of open flame around when working on a vehicle. But the risk does not end there. A spark caused by an electrical short circuit, by two metal surfaces contacting each other, or even by static electricity built up in your body under certain conditions, can ignite gasoline vapors, which in a confined space are highly explosive. Do not, under any circumstances, use gasoline for cleaning parts. Use an approved safety solvent.

Always disconnect the battery ground (-) cable at the battery before working on any part of the fuel system or electrical system. Never risk spilling fuel on a hot engine or exhaust component. It is strongly recommended that a fire extinguisher suitable for use on fuel and electrical fires be kept handy in the garage or workshop at all times. Never try to extinguish a fuel or electrical fire with water.

Fumes

Certain fumes are highly toxic and can quickly cause unconsciousness and even death if inhaled to any extent. Gasoline vapor falls into this category, as do the vapors from some cleaning solvents. Any draining or pouring of such volatile fluids should be done in a well ventilated area.

When using cleaning fluids and solvents, read the instructions on the container carefully. Never use materials from unmarked containers.

Never run the engine in an enclosed space, such as a garage. Exhaust fumes contain carbon monoxide, which is extremely poisonous. If you need to run the engine, always do so in the open air, or at least have the rear of the vehicle outside the work area.

The battery

Never create a spark or allow a bare light bulb near a battery. They normally give off a certain amount of hydrogen gas, which is highly explosive.

Always disconnect the battery ground (-) cable at the battery before working on the fuel or electrical systems.

If possible, loosen the filler caps or cover when charging the battery from an external source (this does not apply to sealed or maintenance-free batteries). Do not charge at an excessive rate or the battery may burst.

Take care when adding water to a non maintenance-free battery and when carrying a battery. The electrolyte, even when diluted, is very corrosive and should not be allowed to contact clothing or skin.

Always wear eye protection when cleaning the battery to prevent the caustic deposits from entering your eyes.

Household current

When using an electric power tool, inspection light, etc., which operates on household current, always make sure that the tool is correctly connected to its plug and that, where necessary, it is properly grounded. Do not use such items in damp conditions and, again, do not create a spark or apply excessive heat in the vicinity of fuel or fuel vapor.

Secondary ignition system voltage

A severe electric shock can result from touching certain parts of the ignition system (such as the spark plug wires) when the engine is running or being cranked, particularly if components are damp or the insulation is defective. In the case of an electronic ignition system, the secondary system voltage is much higher and could prove fatal.

Hydrofluoric acid

This extremely corrosive acid is formed when certain types of synthetic rubber, found in some O-rings, oil seals, fuel hoses, etc. are exposed to temperatures above 750-degrees F (400-degrees C). The rubber changes into a charred or sticky substance containing the acid. *Once formed, the acid remains dangerous for years. If it gets onto the skin, it may be necessary to amputate the limb concerned.*

When dealing with a vehicle which has suffered a fire, or with components salvaged from such a vehicle, wear protective gloves and discard them after use.

DECIMALS to MILLIMETERS

Decimal	mm	Decimal	mm
0.001	0.0254	0.500	12.7000
0.002	0.0508	0.510	12.9540
0.003	0.0762	0.520	13.2080
0.004	0.1016	0.530	13.4620
0.005	0.1270	0.540	13.7160
0.006	0.1524	0.550	13.9700
0.007	0.1778	0.560	14.2240
0.008	0.2032	0.570	14.4780
0.009	0.2286	0.580	14.7320
0.010	0.2540	0.590	14.9860
0.020	0.5080		
0.030	0.7620		
0.040	1.0160	0.600	15.2400
0.050	1.2700	0.610	15.4940
0.060	1.5240	0.620	15.7480
0.070	1.7780	0.630	16.0020
0.080	2.0320	0.640	16.2560
0.090	2.2860	0.650	16.5100
		0.660	16.7640
0.100	2.5400	0.670	17.0180
0.110	2.7940	0.680	17.2720
0.120	3.0480	0.690	17.5260
0.130	3.3020		
0.140	3.5560		
0.150	3.8100		
0.160	4.0640	0.700	17.7800
0.170	4.3180	0.710	18.0340
0.180	4.5720	0.720	18.2880
0.190	4.8260	0.730	18.5420
		0.740	18.7960
0.200	5.0800	0.750	19.0500
0.210	5.3340	0.760	19.3040
0.220	5.5880	0.770	19.5580
0.230	5.8420	0.780	19.8120
0.240	6.0960	0.790	20.0660
0.250	6.3500		
0.260	6.6040		
0.270	6.8580	0.800	20.3200
0.280	7.1120	0.810	20.5740
0.290	7.3660	0.820	21.8280
		0.830	21.0820
0.300	7.6200	0.840	21.3360
0.310	7.8740	0.850	21.5900
0.320	8.1280	0.860	21.8440
0.330	8.3820	0.870	22.0980
0.340	8.6360	0.880	22.3520
0.350	8.8900	0.890	22.6060
0.360	9.1440		
0.370	9.3980		
0.380	9.6520		
0.390	9.9060		
		0.900	22.8600
0.400	10.1600	0.910	23.1140
0.410	10.4140	0.920	23.3680
0.420	10.6680	0.930	23.6220
0.430	10.9220	0.940	23.8760
0.440	11.1760	0.950	24.1300
0.450	11.4300	0.960	24.3840
0.460	11.6840	0.970	24.6380
0.470	11.9380	0.980	24.8920
0.480	12.1920	0.990	25.1460
0.490	12.4460	1.000	25.4000

FRACTIONS to DECIMALS to MILLIMETERS

Fraction	Decimal	mm	Fraction	Decimal	mm
1/64	0.0156	0.3969	33/64	0.5156	13.0969
1/32	0.0312	0.7938	17/32	0.5312	13.4938
3/64	0.0469	1.1906	35/64	0.5469	13.8906
1/16	0.0625	1.5875	9/16	0.5625	14.2875
5/64	0.0781	1.9844	37/64	0.5781	14.6844
3/32	0.0938	2.3812	19/32	0.5938	15.0812
7/64	0.1094	2.7781	39/64	0.6094	15.4781
1/8	0.1250	3.1750	5/8	0.6250	15.8750
9/64	0.1406	3.5719	41/64	0.6406	16.2719
5/32	0.1562	3.9688	21/32	0.6562	16.6688
11/64	0.1719	4.3656	43/64	0.6719	17.0656
3/16	0.1875	4.7625	11/16	0.6875	17.4625
13/64	0.2031	5.1594	45/64	0.7031	17.8594
7/32	0.2188	5.5562	23/32	0.7188	18.2562
15/64	0.2344	5.9531	47/64	0.7344	18.6531
1/4	0.2500	6.3500	3/4	0.7500	19.0500
17/64	0.2656	6.7469	49/64	0.7656	19.4469
9/32	0.2812	7.1438	25/32	0.7812	19.8438
19/64	0.2969	7.5406	51/64	0.7969	20.2406
5/16	0.3125	7.9375	13/16	0.8125	20.6375
21/64	0.3281	8.3344	53/64	0.8281	21.0344
11/32	0.3438	8.7312	27/32	0.8438	21.4312
23/64	0.3594	9.1281	55/64	0.8594	21.8281
3/8	0.3750	9.5250	7/8	0.8750	22.2250
25/64	0.3906	9.9219	57/64	0.8906	22.6219
13/32	0.4062	10.3188	29/32	0.9062	23.0188
27/64	0.4219	10.7156	59/64	0.9219	23.4156
7/16	0.4375	11.1125	15/16	0.9375	23.8125
29/64	0.4531	11.5094	61/64	0.9531	24.2094
15/32	0.4688	11.9062	31/32	0.9688	24.6062
31/64	0.4844	12.3031	63/64	0.9844	25.0031
1/2	0.5000	12.7000	1	1.0000	25.4000

Troubleshooting

Contents

This section provides an easy reference guide to the more common problems which may occur during the operation of your vehicle. These problems and their possible causes are grouped under headings denoting various components or systems, such as Engine, Cooling system, etc. They also refer you to the chapter and/or section which deals with the problem.

Remember that successful troubleshooting is not a mysterious "black art" practiced only by professional mechanics. It is simply the result of the right knowledge combined with an intelligent, systematic approach to the problem. Always work by process of elimination, starting with the simplest solution and working through to the most complex - and never overlook the obvious. Anyone can run the gas tank dry or leave the lights on overnight, so don't assume that you are exempt from such oversights.

Finally, always establish a clear idea of why a problem has occurred and take steps to ensure that it doesn't happen again. If the electrical system fails because of a poor connection, check the other connections in the system to make sure that they don't fail as well. If a particular fuse continues to blow, find out why - don't just replace one fuse after another. Remember, failure of a small component can often be indicative of potential failure or incorrect functioning of a more important component or system.

Engine and performance

1 Engine will not rotate when attempting to start

1 Battery terminal connections loose or corroded. Check the cable terminals at the battery; tighten cable clamp and/or clean off corrosion as necessary (Chapter 1).
2 Battery discharged or faulty. If the cable ends are clean and tight on the battery posts, turn the key to the On position and switch on the headlights or windshield wipers. If they won't run, the battery is discharged.
3 Automatic transmission not engaged in Park (P) or Neutral (N).
4 Broken, loose or disconnected wires in the starting circuit. Inspect all wires and connectors at the battery, starter solenoid and ignition switch (on steering column).
5 Starter motor pinion jammed in driveplate ring gear. Remove starter (Chapter 5) and inspect pinion and driveplate (Chapter 2).
6 Starter solenoid faulty (Chapter 5).
7 Starter motor faulty (Chapter 5).
8 Ignition switch faulty (Chapter 12).
9 Engine seized. Try to turn the crankshaft with a large socket and breaker bar on the pulley bolt.

2 Engine rotates but will not start

1 Fuel tank empty.
2 Battery discharged (engine rotates slowly). Check the operation of electrical components as described in previous Section.
3 Battery terminal connections loose or corroded. See previous Section.
4 Fuel not reaching the fuel injectors. Check for clogged fuel filter or lines and defective fuel pump. Also make sure the tank vent lines aren't clogged (Chapter 4).
5 Low cylinder compression. Check as described in Chapter 2.
6 Valve clearances not properly adjusted (Chapter 1).
7 Water in fuel. Drain tank and fill with new fuel.
8 Dirty fuel injector(s) (Chapter 4).
9 Wet or damaged ignition components (Chapters 1 and 5).
10 Worn, faulty or incorrectly gapped spark plugs (Chapter 1).

3 Starter motor operates without turning engine

1 Starter pinion sticking. Remove the starter (Chapter 5) and inspect.
2 Starter pinion or driveplate teeth worn or broken. Remove the inspection cover and inspect.

4 Engine hard to start when cold

1 Battery discharged or low. Check as described in Chapter 1.
2 Fuel not reaching the fuel injectors. Check the fuel filter, lines and fuel pump (Chapters 1 and 4).
3 Defective spark plugs (Chapter 1).
4 Fault with the fuel injection or engine management system (Chapter 4 or 6).

5 Engine hard to start when hot

1 Air filter dirty (Chapter 1).
2 Fault with the fuel injection or engine management system (Chapter 4 or 6).

6 Starter motor noisy or engages roughly

1 Pinion or driveplate teeth worn or broken. Remove the inspection cover and inspect.
2 Starter motor mounting bolts loose or missing.

7 Engine starts but stops immediately

1 Loose or damaged wire harness connections in the ignition system or at the alternator.
2 Intake manifold vacuum leaks. Make sure all mounting bolts/nuts are tight and all vacuum hoses connected to the manifold are attached properly and in good condition (Chapters 2A, 2B and 4).
3 Insufficient fuel flow to fuel injectors (Chapter 4).

8 Engine lopes while idling or idles erratically

1 Vacuum leaks. Check mounting bolts at the intake manifold for tightness. Make sure that all vacuum hoses are connected and in good condition. Use a stethoscope or a length of fuel hose held against your ear to listen for vacuum leaks while the engine is running. A hissing sound will be heard. A soapy water solution will also detect leaks. Check the intake manifold gasket surfaces.
2 Leaking EGR valve or plugged PCV valve (see Chapters 1 and 6).
3 Air filter clogged (Chapter 1).
4 Fuel pump not delivering sufficient fuel (Chapter 4).
5 Leaking head gasket. Perform a cylinder compression check (Chapter 2).
6 Timing chain(s) worn (Chapter 2).
7 Camshaft lobes worn (Chapter 2).
8 Valve clearance out of adjustment (Chapter 1).

9 Valves burned or otherwise leaking (Chapter 2).
10 Ignition system not operating properly (Chapters 1 and 5).
11 Dirty or clogged injector(s) (Chapter 4).

9 Engine misses at idle speed

1 Spark plugs faulty or not gapped properly (Chapter 1).
2 Faulty emissions systems (see Chapter 6).
3 Clogged fuel filter and/or foreign matter in fuel. Remove the fuel filter (Chapter 1) and inspect.
4 Vacuum leaks at intake manifold or hose connections. Check as described in Section 8.
5 Low or uneven cylinder compression. Check as described in Chapter 2.
6 Clogged or dirty fuel injectors (Chapter 4).

10 Excessively high idle speed

1 Intake air leak (Chapters 2 and 4).
2 Malfunction in the engine management system (Chapter 6).

11 Battery will not hold a charge

1 Drivebelt or tensioner defective (Chapter 1).
2 Battery cables loose or corroded (Chapter 1).
3 Alternator not charging properly (Chapter 5).
4 Loose, broken or faulty wires in the charging circuit (Chapter 5).
5 Short circuit causing a continuous drain on the battery.
6 Battery defective internally.

12 Alternator light stays on

1 Fault in alternator or charging circuit (Chapter 5).
2 Drivebelt or tensioner defective (Chapter 1).

13 Alternator light fails to come on when key is turned on

1 Faulty bulb (Chapter 12).
2 Defective alternator (Chapter 5).
3 Fault in the printed circuit, dash wiring or bulb holder (Chapter 12).

14 Engine misses throughout driving speed range

1 Fuel filter clogged and/or impurities in the fuel system. Check fuel filter (Chapter 1)

or clean system (Chapter 4).
2 Faulty or incorrectly gapped spark plugs (Chapter 1).
3 Emissions system components faulty (Chapter 6).
4 Low or uneven cylinder compression pressures. Check as described in Chapter 2.
5 Weak or faulty ignition coil(s) (Chapter 5).
6 Vacuum leaks at intake manifold or vacuum hoses (see Section 8).
7 Dirty or clogged fuel injector (Chapter 4).
8 Leaky EGR valve (Chapter 6).

15 Hesitation or stumble during acceleration

1 Ignition system not operating properly (Chapter 5).
2 Dirty or clogged fuel injector(s) (Chapter 4).
3 Low fuel pressure. Check for proper operation of the fuel pump and for restrictions in the fuel filter and lines (Chapter 4).
4 Fault with the fuel injection or engine management system (Chapter 4 or 6).

16 Engine stalls

1 Fuel filter clogged and/or water and impurities in the fuel system (Chapter 1).
2 Emissions system components faulty (Chapter 6).
3 Faulty or incorrectly gapped spark plugs (Chapter 1).
4 Vacuum leak at the throttle body, the intake manifold or vacuum hoses. Check as described in Section 8.
5 Valve clearances incorrect (Chapter 1).
6 Fault with the fuel injection or engine management system (Chapter 4 or 6).

17 Engine lacks power

1 Faulty or incorrectly gapped spark plugs (Chapter 1).
2 Air filter dirty (Chapter 1).
3 Faulty ignition coil(s) (Chapter 5).
4 Automatic transmission fluid level incorrect, causing slippage (Chapter 1).
5 Fuel filter clogged and/or impurities in the fuel system (Chapters 1 and 4).
6 EGR system not functioning properly (Chapter 6).
7 Use of sub-standard fuel. Fill tank with proper octane fuel.
8 Low or uneven cylinder compression pressures. Check as described in Chapter 2C.
9 Air leak at intake manifold (check as described in Section 8).
10 Fault with the fuel injection or engine management system (Chapter 4 or 6).
11 Brakes binding (Chapters 1 and 9).

18 Engine backfires

1 EGR system not functioning properly (Chapter 6).
2 Thermostatic air cleaner system not operating properly (Chapter 6).
3 Vacuum leak (see Section 8).
4 Valve clearances incorrect (Chapter 1).
5 Damaged valve springs or sticking valves (Chapter 2).
6 Intake air leak (see Section 8).

19 Engine surges while holding accelerator steady

1 Intake air leak (see Section 8).
2 Fuel pump not working properly (Chapter 4).
3 Fault with the fuel injection or engine management system (Chapter 4 or 6).

20 Pinging or knocking engine sounds when engine is under load

1 Incorrect grade of fuel. Fill tank with fuel of the proper octane rating.
2 Carbon build-up in combustion chambers. Remove cylinder heads and clean combustion chambers (Chapter 2).
3 Incorrect spark plugs (Chapter 1).

21 Engine continues to run after being turned off

1 Incorrect spark plug heat range (Chapter 1).
2 Intake air leak (see Section 8).
3 Carbon build-up in combustion chambers. Remove the cylinder heads and clean the combustion chambers (Chapter 2).
4 Valves sticking (Chapter 2).
5 Valve clearances incorrect (Chapter 1).
6 EGR system not operating properly (Chapter 6).
7 Check for causes of overheating (see Section 27).
8 Faulty ignition switch (Chapter 12).

22 Low oil pressure

1 Improper grade of oil.
2 Oil pump worn or damaged (Chapter 2).
3 Engine overheating (refer to Section 27).
4 Clogged oil filter (Chapter 1).
5 Clogged oil strainer (Chapter 2).
6 Oil pressure gauge not working properly (Chapter 2C).

23 Excessive oil consumption

1 Loose oil drain plug.

2 Loose bolts or damaged oil pan gasket (Chapter 2).
3 Loose bolts or damaged front cover gasket (Chapter 2).
4 Front or rear crankshaft oil seal leaking (Chapter 2).
5 Loose bolts or damaged valve cover gasket (Chapter 2).
6 Loose oil filter (Chapter 1).
7 Loose or damaged oil pressure switch (Chapter 2).
8 Pistons and cylinders excessively worn (Chapter 2).
9 Piston rings not installed correctly on pistons (Chapter 2).
10 Worn or damaged piston rings (Chapter 2).
11 Intake and/or exhaust valve oil seals worn or damaged (Chapter 2).
12 Worn valve stems.
13 Worn or damaged valves/guides (Chapter 2).

24 Excessive fuel consumption

1 Dirty or clogged air filter element (Chapter 1).
2 Low tire pressure or incorrect tire size (Chapters 1 and 10).
3 Fuel leakage. Check all connections, lines and components in the fuel system (Chapter 4).
4 Dirty or clogged fuel injectors (Chapter 4).
5 Fault with the fuel injection or engine management system (Chapter 4 or 6).

25 Fuel odor

1 Fuel leakage. Check all connections, lines and components in the fuel system (Chapter 4).
2 Fuel tank overfilled. Fill only to automatic shut-off.
3 Charcoal canister defective (Chapter 6).
4 Vapor leaks from Evaporative Emissions Control system lines (Chapter 6).

26 Miscellaneous engine noises

1 A strong dull noise that becomes more rapid as the engine accelerates indicates worn or damaged crankshaft bearings or an unevenly worn crankshaft. To pinpoint the trouble spot, remove the ignition coil electrical connector from one coil at a time and run the engine. If the noise stops, the cylinder with the removed plug wire indicates the problem area. Replace the bearing and/or service or replace the crankshaft (Chapter 2).
2 A noise similar (yet slightly higher pitched) to the crankshaft knocking described in the previous paragraph that becomes more rapid as the engine accelerates, indicates worn or damaged connecting rod bearings (Chapter 2). The procedure for locating the problem cylinder is the same as described in Paragraph 1.

3   An overlapping metallic noise that increases in intensity as the engine speed increases, yet diminishes as the engine warms up indicates abnormal piston and cylinder wear (Chapter 2). To locate the problem cylinder, use the procedure described in Paragraph 1.

4   A rapid clicking noise that becomes faster as the engine accelerates indicates a worn piston pin or piston pin hole. This sound will happen each time the piston hits the highest and lowest points in the stroke (Chapter 2). The procedure for locating the problem piston is described in Paragraph 1.

5   A metallic clicking noise coming from the water pump indicates worn or damaged water pump bearings or pump. Replace the water pump with a new one (Chapter 3).

6   A rapid tapping sound or clicking sound that becomes faster as the engine speed increases indicates valve tapping or improperly adjusted valve clearances. This can be identified by holding one end of a section of hose to your ear and placing the other end at different spots along the valve cover. The point where the sound is loudest indicates the problem valve. Adjust the valve clearance (Chapter 1). If the problem persists, you likely have a collapsed valve lifter or other damaged valve train component.

7   A steady metallic rattling or rapping sound coming from the area of the timing chain cover indicates a worn, damaged or out-of-adjustment timing chain. Service or replace the chain and related components (Chapter 2).

Cooling system

27   Overheating

1   Insufficient coolant in system (Chapter 1).
2   Drivebelt or tensioner defective (Chapter 1).
3   Radiator core blocked or radiator grille dirty or restricted (Chapter 3).
4   Thermostat faulty (Chapter 3).
5   Fan not functioning properly (Chapter 3).
6   Radiator cap not maintaining proper pressure. Have cap pressure tested by gas station or repair shop.
7   Defective water pump (Chapter 3).
8   Improper grade of engine oil.
9   Inaccurate temperature gauge (Chapter 12).

28   Overcooling

1   Thermostat faulty (Chapter 3).
2   Inaccurate temperature gauge (Chapter 12).

29   External coolant leakage

1   Deteriorated or damaged hoses. Loose clamps at hose connections (Chapter 1).

2   Water pump seals defective. If this is the case, water will drip from the weep hole in the water pump body (Chapter 3).
3   Leakage from radiator core or header tank. This will require the radiator to be professionally repaired (see Chapter 3 for removal procedures).
4   Engine drain plugs or water jacket freeze plugs leaking (see Chapter 1).
5   Leak from coolant temperature switch (Chapter 3).
6   Leak from damaged gaskets or small cracks (Chapter 2).
7   Damaged head gasket. This can be verified by checking the condition of the engine oil (see **Note** in Section 30).

30   Internal coolant leakage

Note: *Internal coolant leaks can usually be detected by examining the oil. Check the dipstick and inside the valve cover for water deposits and an oil consistency like that of a milkshake.*
1   Leaking cylinder head gasket. Have the system pressure tested (Chapter 2).
2   Cracked cylinder bore or cylinder head. Dismantle engine and inspect (Chapter 2).
3   Loose cylinder head bolts (tighten as described in Chapter 2).

31   Abnormal coolant loss

1   Overfilling system (Chapter 1).
2   Coolant boiling away due to overheating (see causes in Section 27).
3   Internal or external leakage (see Sections 29 and 30).
4   Faulty radiator cap. Have the cap pressure tested.
5   Cooling system being pressurized by engine compression. This could be due to a cracked head or block or leaking head gaskets.

32   Poor coolant circulation

1   Inoperative water pump. A quick test is to pinch the top radiator hose closed with your hand while the engine is idling, then release it. You should feel a surge of coolant if the pump is working properly (Chapter 3).
2   Restriction in cooling system. Drain, flush and refill the system (Chapter 1). If necessary, remove the radiator (Chapter 3) and have it reverse flushed or professionally cleaned.
3   Defective drivebelt or tensioner (Chapter 1).
4   Thermostat sticking (Chapter 3).
5   Insufficient coolant (Chapter 1).

33   Corrosion

1   Excessive impurities in the water. Soft, clean water is recommended. Distilled or rain-

water is satisfactory.
2   Insufficient antifreeze solution (refer to Chapter 1 for the proper ratio of water to antifreeze).
3   Infrequent flushing and draining of system. Regular flushing of the cooling system should be carried out at the specified intervals as described in (Chapter 1).

Automatic transmission

Note: *Due to the complexity of the automatic transmission, it's difficult for the home mechanic to properly diagnose and service. For problems other than the following, the vehicle should be taken to a reputable mechanic.*

34   Fluid leakage

1   Automatic transmission fluid is a deep red color, and fluid leaks should not be confused with engine oil which can easily be blown by air flow to the transmission.
2   To pinpoint a leak, first remove all built-up dirt and grime from the transmission. Degreasing agents and/or steam cleaning will achieve this. With the underside clean, drive the vehicle at low speeds so the air flow will not blow the leak far from its source. Raise the vehicle and determine where the leak is located. Common areas of leakage are:

a) *Fluid pan: tighten mounting bolts and/or replace pan gasket as necessary (Chapter 1).*
b) *Rear extension: tighten bolts and/or replace oil seal as necessary.*
c) *Filler pipe: replace the rubber oil seal where pipe enters transmission case.*
d) *Transmission oil lines: tighten fittings where lines enter transmission case and/or replace lines.*
e) *Vent pipe: transmission overfilled and/or water in fluid (see checking procedures, Chapter 1).*
f) *Speedometer connector: replace the O-ring where speedometer cable enters transmission case.*

35   General shift mechanism problems

Chapter 7A deals with checking and adjusting the shift cable on automatic transmissions. Common problems which may be caused by out-of-adjustment linkage are:

a) *Engine starting in gears other than P (park) or N (Neutral).*
b) *Indicator pointing to a gear other than the one actually engaged.*
c) *Vehicle moves with transmission in P (Park) position.*

36 Transmission will not downshift with the accelerator pedal pressed to the floor

Malfunction in the transmission's electronic control system.

37 Engine will start in gears other than Park or Neutral

Chapter 7A deals with adjusting the Park/Neutral Position (PNP) switch/Transmission Range (TR) sensor installed on automatic transmissions.

38 Transmission slips, shifts rough, is noisy or has no drive in forward or Reverse gears

1   There are many probable causes for the above problems, but the home mechanic should concern himself only with one possibility: fluid level.
2   Before taking the vehicle to a shop, check the fluid level and condition as described in Chapter 1. Add fluid, if necessary, or change the fluid and filter if needed. If problems persist, have a professional diagnose the transmission.

Driveshaft
Note: *Refer to Chapter 8, unless otherwise specified, for service information.*

39 Leaks at front of driveshaft

Defective transmission rear seal. See Chapter 7A for replacement procedure. As this is done, check the splined yoke for burrs or roughness that could damage the new seal. Remove burrs with a fine file or whetstone.

40 Knock or clunk when transmission is under initial load (just after transmission is put into gear)

1   Loose or disconnected rear suspension components. Check all mounting bolts and bushings (Chapters 7A and 10).
2   Loose driveshaft bolts. Inspect all bolts and nuts and tighten them securely.
3   Worn or damaged universal joint bearings. Inspect the universal joints.
4   Worn sleeve yoke and mainshaft spline.

41 Metallic grating sound consistent with vehicle speed

Pronounced wear in the universal joint bearings. Replace U-joints or driveshafts, as necessary.

42 Vibration

Note: *Before blaming the driveshaft, make sure the tires are perfectly balanced and perform the following test.*
1   Install a tachometer inside the vehicle to monitor engine speed as the vehicle is driven. Drive the vehicle and note the engine speed at which the vibration (roughness) is most pronounced. Now shift the transmission to a different gear and bring the engine speed to the same point.
2   If the vibration occurs at the same engine speed (rpm) regardless of which gear the transmission is in, the driveshaft is NOT at fault since the driveshaft speed varies.
3   If the vibration decreases or is eliminated when the transmission is in a different gear at the same engine speed, refer to the following probable causes.
4   Bent or dented driveshaft. Inspect and replace as necessary.
5   Undercoating or built-up dirt, etc. on the driveshaft. Clean the shaft thoroughly.
6   Worn universal joint bearings. Replace the U-joints or driveshaft as necessary.
7   Driveshaft and/or companion flange out of balance. Check for missing weights on the shaft. Remove driveshaft and reinstall 180-degrees from original position, then recheck. Have the driveshaft balanced if problem persists.
8   Loose driveshaft mounting bolts/nuts.
9   Defective center bearing, if so equipped.
10 Worn transmission rear bushing (Chapter 7A).

43 Scraping noise

Make sure the dust cover on the sleeve yoke isn't rubbing on the transmission extension housing.

44 Whining or whistling noise

Defective center bearing.

Axles and differential
Note: *For servicing information, refer to Chapter 8, unless otherwise specified.*

45 Noise - same when in drive as when vehicle is coasting

1   Road noise. No corrective action available.

2   Tire noise. Inspect tires and check tire pressures (Chapter 1).
3   Front wheel bearings worn (Chapter 10).
4   Insufficient differential oil (Chapter 1).
5   Defective differential.

46 Knocking sound when starting or shifting gears

Defective or incorrectly adjusted differential.

47 Noise when turning

Defective differential.

48 Vibration

1   See probable causes under Driveshaft. Proceed under the guidelines listed for the driveshaft. If the problem persists, check the rear wheel bearings by raising the rear of the vehicle and spinning the wheels by hand. Listen for evidence of rough (noisy) bearings. Remove and inspect.
2   Worn driveaxle CV joint.

49 Oil leaks

1   Pinion oil seal damaged.
2   Axleshaft oil seals damaged.
3   Differential cover leaking. Tighten mounting bolts or replace the gasket as required.
4   Loose filler or drain plug on differential (Chapter 1).
5   Clogged or damaged breather on differential.

Transfer case
Note: *Unless otherwise specified, refer to Chapter 7B for service and repair information.*

50 Gear jumping out of mesh

1   Interference between the control lever and the console.
2   Internal wear or incorrect adjustments.

51 Difficult shifting

1   Lack of oil.
2   Internal wear, damage or incorrect adjustment.

52 Noise

1   Lack of oil in transfer case.
2   Noise in 4H and 4L, but not in 2H indicates cause is in the front differential or front axle.

3   Noise in 2H, 4H and 4L indicates cause is in rear differential or rear axle.
4   Noise in 2H and 4H but not in 4L, or in 4L only, indicates internal wear or damage in transfer case.

Brakes

Note: *Before assuming a brake problem exists, make sure the tires are in good condition and inflated properly, the front end alignment is correct and the vehicle is not loaded with weight in an unequal manner. All service procedures for the brakes are included in Chapter 9, unless otherwise noted.*

53   Vehicle pulls to one side during braking

1   Defective, damaged or oil contaminated brake pad on one side. Inspect as described in Chapter 1. Refer to Chapter 9 if replacement is required.
2   Excessive wear of brake pad material or disc on one side. Inspect and repair as necessary.
3   Loose or disconnected front suspension components. Inspect and tighten all bolts securely (Chapters 1 and 10).
4   Defective caliper assembly. Remove caliper and inspect for stuck piston or damage.
5   Scored disc.

54   Noise (high-pitched squeal)

1   Brake pads worn out. Replace pads with new ones immediately!
2   Glazed or contaminated pads.
3   Dirty or scored disc.
4   Bent support plate.

55   Excessive brake pedal travel

1   Partial brake system failure. Inspect entire system (Chapter 1) and correct as required.
2   Insufficient fluid in master cylinder. Check (Chapter 1) and add fluid, bleed system and inspect for leaks.
3   Air in system. Bleed system.

56   Brake pedal feels spongy when depressed

1   Air in brake lines. Bleed the brake system.
2   Deteriorated rubber brake hoses. Inspect all system hoses and lines. Replace parts as necessary.
3   Master cylinder mounting nuts loose.
4   Master cylinder faulty.
5   Deformed rubber brake lines.

57   Excessive effort required to stop vehicle

1   Power brake booster not operating properly.
2   Excessively worn pads. Check and replace if necessary.
3   One or more caliper pistons seized or sticking. Inspect and replace caliper as required.
4   Brake pads contaminated with oil or grease. Inspect and replace as required.
5   New pads installed and not yet seated. It'll take a while for the new material to seat against the disc.
6   Worn or damaged master cylinder or caliper assemblies. Check particularly for frozen pistons.

58   Pedal travels to the floor with little resistance

Little or no fluid in the master cylinder reservoir caused by leaking caliper piston(s) or loose, damaged or disconnected brake lines. Inspect entire system and repair as necessary.

59   Brake pedal pulsates during brake application

1   Wheel bearings worn (Chapter 1).
2   Disc not within specifications. Check for excessive lateral runout and parallelism. Have the discs resurfaced or replace them with new ones. Also make sure that all discs are the same thickness.

60   Brakes drag (indicated by sluggish engine performance or wheels being very hot after driving)

1   Push rod adjustment incorrect at the brake pedal.
2   Obstructed master cylinder compensation port.
3   Master cylinder piston seized in bore.
4   Caliper sticking.
5   Piston cups in master cylinder or caliper assembly deformed.
6   Parking brake assembly will not release.
7   Clogged or internally split brake lines.
8   Brake pedal height improperly adjusted.

61   Rear brakes lock up under light brake application

1   Tire pressures too high.
2   Tires excessively worn (Chapter 1).

62   Rear brakes lock up under heavy brake application

1   Tire pressures too high.
2   Tires excessively worn (Chapter 1).
3   Front brake pads contaminated with oil, mud or water. Clean or replace the pads.
4   Front brake pads excessively worn.
5   Defective master cylinder or caliper assembly.

Suspension and steering

Note: *All service procedures for the suspension and steering systems are included in Chapter 10, unless otherwise noted.*

63   Vehicle pulls to one side

1   Tire pressures uneven (Chapter 1).
2   Defective tire (Chapter 1).
3   Excessive wear in suspension or steering components (Chapter 1).
4   Wheel alignment incorrect.
5   Front brakes dragging.

64   Shimmy, shake or vibration

1   Tire or wheel out-of-balance or out-of-round. Have them balanced on the vehicle.
2   Worn wheel bearings.
3   Shock absorbers and/or suspension components worn or damaged. Check for worn bushings in the upper and lower links.
4   Wheel lug nuts loose.
5   Incorrect tire pressures.
6   Excessively worn or damaged tire.
7   Loosely mounted steering gear housing.
8   Steering gear worn.
9   Tie-rod end worn.
10 Worn balljoint.

65   Excessive pitching and/or rolling around corners or during braking

1   Defective shock absorbers. Replace as a set.
2   Broken or weak springs and/or suspension components.
3   Worn or damaged stabilizer bar or bushings.

66   Wandering or general instability

1   Improper tire pressures.
2   Incorrect front end alignment.
3   Worn or damaged steering or suspension components.
4   Worn rear shock absorbers.
5   Fatigued or damaged rear leaf springs.

67 Excessively stiff steering

1   Lack of lubricant in power steering fluid reservoir (Chapter 1).
2   Incorrect tire pressures (Chapter 1).
3   Balljoints worn.
4   Front end out of alignment.
5   Worn or damaged steering gear.
6   Low tire pressures.

68 Excessive play in steering

1   Worn wheel bearings.
2   Excessive wear in suspension bushings (Chapter 1).
3   Steering gear worn.
4   Incorrect wheel alignment.
5   Steering gear mounting bolts loose.
6   Worn tie-rod ends.

69 Lack of power assistance

1   Drivebelt or tensioner faulty (Chapter 1).
2   Fluid level low (Chapter 1).
3   Hoses or pipes restricting the flow. Inspect and replace parts as necessary.
4   Air in power steering system. Bleed system.
5   Defective power steering pump.

70 Steering wheel fails to return to straight-ahead position

1   Incorrect front end alignment.
2   Tire pressures low.

3   Steering gear worn or damaged.
4   Lack of fluid in power steering pump.

71 Steering effort not the same in both directions (power system)

1   Leaks in steering gear.
2   Clogged fluid passage in steering gear.

72 Noisy power steering pump

1   Insufficient oil in pump.
2   Clogged hoses or oil filter in pump.
3   Loose pulley.
4   Drivebelt or tensioner faulty (Chapter 1).
5   Defective pump.

73 Miscellaneous noises

1   Improper tire pressures.
2   Worn balljoint or tie-rod end.
3   Loose or worn steering gear, or suspension components.
4   Defective shock absorber.
5   Defective wheel bearing.
6   Worn or damaged suspension bushings.
7   Damaged leaf spring.
8   Loose wheel lug nuts.
9   Worn or damaged rear axleshaft spline.
10   Worn or damaged rear shock absorber mounting bushing.
11   Excessive rear axle end play.
12   See also causes of noises at the rear axle and driveshaft.

74 Excessive tire wear (not specific to one area)

1   Incorrect tire pressures.
2   Tires out of balance. Have them balanced on the vehicle.
3   Wheels damaged. Inspect and replace as necessary.
4   Suspension or steering components worn (Chapter 1).

75 Excessive tire wear on outside edge

1   Incorrect tire pressure.
2   Excessive speed in turns.
3   Front end alignment incorrect (excessive toe-in).

76 Excessive tire wear on inside edge

1   Incorrect tire pressure.
2   Front end alignment incorrect (toe-out).
3   Loose or damaged steering components (Chapter 1).

77 Tire tread worn in one place

1   Tires out of balance. Have them balanced on the vehicle.
2   Damaged wheel. Inspect and replace if necessary.
3   Defective tire.

Chapter 1
Tune-up and routine maintenance

Contents

Specifications

Recommended lubricants and fluids

Engine oil
Type ... API "certified for gasoline engines"
Viscosity ... See accompanying chart
Coolant ... Toyota Genuine Long Life Coolant (SLLC) pre-mixed coolant or equivalent (do not dilute)

Caution: *In order to avoid damage to the engine cooling system and other problems, only use TOYOTA SLLC or similar high quality ethylene glycol based non-silicate, non-amine, non-nitrite, non-borate coolant with long-life hybrid organic acid technology.*

Brake fluid ... DOT 3 brake fluid
Power steering fluid ... DEXRON III automatic transmission fluid
Automatic transmission fluid ... TOYOTA Genuine ATF WS
Transfer case lubricant
 2013 and earlier Tundra models and all Sequoia models (JF1A and JF3A) ... API GL-4 or GL-5 SAE 75W-90 gear oil
 2014 and later Tundra models (WF1AM) ... TOYOTA Genuine Transfer gear oil LF 75W
Differential lubricant ... Toyota Genuine Differential gear oil LT 75W-85 GL-5 or equivalent

Note: *On models with a limited slip rear differential, use hypoid gear oil.*

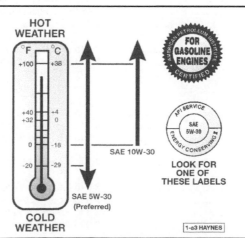

Engine oil viscosity chart - for best fuel economy and cold starting, select the lowest SAE viscosity grade for the expected temperature range

Capacities*

Chassis grease	NLGI No. 2 lithium-base chassis grease

Engine oil (with filter change)
   1GR-FE

2010 and earlier models	5.5 qts (5.2 liters)
2011 and later models	6.4 qts (6.1 liters)

   1UR-FE

2010 through 2017 models	7.9 qts (7.5 liters)
2018 and later models	8.5 qts (8.0 liters)
2UZ-FE	6.6 qts (6.2 liters)

   3UR-FE/3UR-FBE

2009 and earlier models	7.4 qts (7.0 liters)
2010 through 2017 models	7.9 qts (7.5 liters)
2018 and later models	8.5 qts (8.0 liters)

Cooling system
   Tundra models
      1GR-FE engine

2010 and earlier models	10.1 qts (9.6 liters)
2011 and later models	11.3 qts (10.7 liters)

      1UR-FE engines
         2013 and earlier models

w/ Trailer towing package	13.9 qts (13.2 liters)
w/o Trailer towing package	12.8 qts (12.1 liters)

         2014 and later models

w/ Trailer towing package	12.6 qts (11.9 liters)
w/o Trailer towing package	11.4 qts (10.8 liters)
2UZ-FE engine 2010 and earlier models	10.3 qts (9.7 liters)

      3UR-FE/3UR-FBE engines
         2013 and earlier models

w/ Trailer towing package	13.7 qts (13.0 liters)
w/o Trailer towing package	12.8 qts (12.1 liters)

         2014 and later models

w/ Trailer towing package	12.4 qts (11.7 liters)
w/o Trailer towing package	11.4 qts (10.8 liters)

   Sequoia models
      1UR-FE engine

w/ Trailer towing package	15.6 qts (14.8 liters)
w/o Trailer towing package	14.5 qts (13.7 liters)
2UZ-FE engine	13.3 qts (12.6 liters)

      3UR-FE/3UR-FBE engines
         2013 and earlier models

w/ Trailer towing package	15.4 qts (14.6 liters)
w/o Trailer towing package	14.5 qts (13.7 liters)

         2014 and later models

w/ Trailer towing package	14.1 qts (13.3 liters)
w/o Trailer towing package	13.1 qts (12.4 liters)

Automatic transmission (drain and refill) (see Section 6 for refilling procedure)

A750 (E,F)	1.8 qts (1.7 liters)
A760 (E,F)/AB60 (E,F)	2.2 qts (2.1 liters)

Transfer case
   Tundra models

2013 and earlier	1.2 $\pm$ 0.6 qts (1.1 $\pm$ 0.6 liters)
2014 and later	1.48 to 1.58 qts (1.4 to 1.5 liters)
Sequoia models	1.4 $\pm$ 0.6 qts (1.3 $\pm$ 0.6 liters)

Front differential

2009 and earlier models	2.2 to 2.3 qts (2.08 to 2.18 liters)
2010 and later models	2.12 to 2.21 qts (2.0 to 2.10 liters)

Rear differential
   2009 and earlier models
      2UZ-FE (2WD/4WD)

10 ft. 6.8 in. Wheelbase	4.2 to 4.3 qts (3.97 to 4.07 liters)
12 ft. 1.7 in. Wheelbase	4.8 to 4.9 qts (4.55 to 4.65 liters)
13 ft. 8.6 in. Wheelbase	4.8 to 4.9 qts (4.55 to 4.65 liters)

      3UR-FE

10 ft. 6.8 in. Wheelbase (2WD/4WD)	4.2 to 4.3 qts (3.97 to 4.07 liters)

         12 ft. 1.7 in. Wheelbase

Regular/Double cab (2WD/4WD)	3.8 to 3.9 qts (3.59 to 3.69 liters)
Crew cab 2WD	3.9 to 4 qts (3.69 to 3.78 liters)
Crew cab 4WD	3.8 to 3.9 qts (3.59 to 3.69 liters)
13 ft. 8.6 in. Wheelbase, double cab (2WD/4WD)	3.9 to 4 qts (3.69 to 3.78 liters)

Rear differential (continued)
    2010 and later
        3UR-FE
            10 ft. 6.8 in. Wheelbase (2WD/4WD) .... 3.65 to 3.75 qts (3.45 to 3.55 liters)
            12 ft. 1.7 in. Wheelbase
                Regular/Double cab (2WD/4WD)..................................... 3.76 to 3.85 qts (3.55 to 3.65 liters)
                Crew cab 2WD... 3.91 to 4.01 qts (3.70 to 3.80 liters)
                Crew cab 4WD... 3.76 to 3.85 qts (3.55 to 3.65 liters)
            13 ft. 8.6 in. Wheelbase, double cab (2WD/4WD).................. 3.91 to 4.01 qts (3.70 to 3.80 liters)

All capacities approximate. Add as necessary to bring to appropriate level.

Spark plug type and gap

Type.. Denso
    1GR-FE engine
        2010 and earlier models K20HR11
        2011 and 2012 models SK20HR11
        2013 models .... SK20HR11 or SK16HR11
        2014 models SK16HR11
    2UZ-FE engine.. SK20R11
    1UR-FE, 3UR-FE/3UR-FBE engines SK20HR11
Gap (all models).. 0.039 to 0.043 inch (1.0 to 1.1 mm)

Firing order

V6 engine 1-2-3-4-5-6
V8 engine
    2UZ-FE... 1-8-4-3-6-5-7-2
    3UR-FE, 3UR-FBE and 1UR-FE.. 1-8-7-3-6-5-4-2

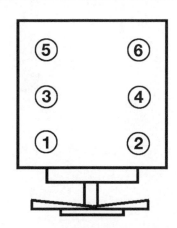

Cylinder location diagram - V6 engine

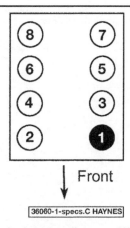

↓ Front

36060-1-specs.C HAYNES

Cylinder location diagram - V8 engines

Valve clearances (engine cold)

    1GR-FE
        Intake... 0.006 to 0.010 inch (0.15 to 0.25 mm)
        Exhaust.. 0.011 to 0.015 inch (0.29 to 0.39 mm)
    2UZ-FE
        Intake... 0.006 to 0.010 inch (0.15 to 0.25 mm)
        Exhaust.. 0.010 to 0.014 inch (0.25 to 0.35 mm)

Note: *Use the information printed on the Vehicle Emissions Control Information label, if different than the Specifications listed here.*

Brakes

Disc brake pad lining thickness (minimum) .... 1/16 inch (1.5 mm)
Brake light switch plunger length.. 0.060 to 0.100 inch (1.5 to 2.5 mm)
Brake pedal height.. 7.08 to 7.48 inches (179.9 to 189.9 mm)
Brake pedal freeplay... 0.0394 to 0.236 inch (1.0 to 6.0 mm)
Brake pedal reserve height.. 4.09 inches (104 mm)
Parking brake pedal adjustment.. 6 to 9 clicks

Suspension and steering

Steering wheel freeplay limit.. 1.2 inches (30 mm)
Balljoint allowable vertical movement... 0.0 inch

Torque specifications

Note: *One foot-pound (ft-lb) of torque is equivalent to 12 inch-pounds (in-lbs) of torque. Torque values below approximately 15 ft-lbs are expressed in inch-pounds, since most foot-pound torque wrenches are not accurate at these smaller values.*

	Ft-lbs (unless otherwise indicated)	Nm
Automatic transmission		
Filter bolts	84 in-lbs	10
Fluid pan bolts	65 in-lbs	7
Drain plug	15	20
Overflow plug	15	20
Refill plug	29	39
Drivebelt tensioner mounting bolt and nuts		
1GR-FE	27	37
2UZ-FE	144 in-lbs	16
3UR-FE, 3UR-FBE and 1UR-FE	17	23
Transfer case filler plug and drain plug		
2013 and earlier Tundra models		
and all Sequoia models (JF1A and JF3A)	27	37
2014 and later Tundra models (WF1AM)	18	25
Front differential		
Drain plug	29	39
Fill plug	29	39
Rear differential fill/drain plug		
Tundra models	36	49
Sequoia models	29	39
Spark plugs		
2UZ-FE	156 in-lbs	18
1GR-FE, 3UR-FE, 3UR-FBE and 1UR-FE	15	20
Engine oil drain plug		
2UZ-FE	29	39
1GR-FE, 3UR-FE, 3UR-FBE and 1UR-FE	30	40
Engine oil filter		
Drain plug	108 in-lbs	13
Cap	18	25
Wheel lug nuts		
Aluminum wheel	97	131
Steel wheel	154	209

1 Maintenance schedule

Every 250 (400 km) miles or weekly, whichever comes first

Check the engine oil level (Section 4)
Check the engine coolant level (Section 4)
Check the windshield washer fluid level (Section 4)
Check the brake fluid level (Section 4)
Check the tires and tire pressures (Section 5)

Every 3000 miles (4800 km) or 3 months, whichever comes first

All items listed above, plus . . .
Check the power steering fluid level (Section 7)

Every 5000 miles (8000 km)

Change the engine oil and filter (Section 8)*

Every 7500 miles (12,000 km) or 6 months, whichever comes first

Check and service the battery (Section 9)
Check the cooling system (Section 10)
Inspect and replace, if necessary, all underhood hoses
  (Section 11)
Inspect and replace, if necessary, the windshield wiper
  blades (Section 12)
Rotate the tires (Section 13)
Inspect the suspension and steering components
  (Section 14)
Lubricate the driveshaft and body components (Section 15)**
Inspect the exhaust system (Section 16)

Every 15,000 miles (24,000 km) or 12 months, whichever comes first

All items listed above, plus . . .
Check the transfer case lubricant level (4WD models)
  (Section 17)
Check the differential lubricant level (Section 18)
Check the seat belts (Section 19)
Check the driveaxle boots (Section 20)
Replace the air filter (Section 22)**
Change cabin air filter (Section 23)**
Check the drivebelt (Section 24)
Inspect the fuel system (Section 25)
Check the brakes (Section 26)**
Check the brake pedal for proper height and
  freeplay (Section 27)

Every 30,000 miles (48,000 km) or 24 months, whichever comes first

All items listed above, plus . . .
Check the automatic transmission fluid level (Section 6)
Replace the spark plugs (non-platinum or iridium type)
  (Section 28)
Service the cooling system (drain, flush and refill)
  (Section 29)
Change the transfer case lubricant (4WD models)
  (Section 31)
Change the differential lubricant (Section 32)**

Every 60,000 miles (96,000 km) or 48 months, whichever comes first

Replace the spark plugs (platinum or iridium type)
  (Section 28)
Change the automatic transmission fluid and filter
  (Section 30)***
Check, and adjust if necessary, the valve clearance
  (Section 33)
Inspect the evaporative emissions control system
  (Section 34)
Replace the timing belt (Chapter 2B [2UZ-FE engines only])

Every 80,000 miles (128,000 km)

Replace the oxygen sensors (see Chapter 6)

*If E85 fuel is used more than 50-percent of the time,
  change the oil and filter at 2500 mile (4000 km) intervals.*

**This item is affected by "severe" operating conditions, as
  described below. If the vehicle is operated under severe
  conditions, perform all maintenance indicated (**) at
  5000 mile/four-month intervals. Severe conditions exist if
  you mainly operate the vehicle . . .*

in dusty areas
towing a trailer
idling for extended periods and/or driving at low speeds
when outside temperatures remain below freezing and
most trips are less than four miles long

****If operated under one or more of the following conditions,
  change the automatic transmission fluid every 30,000
  miles:*

in heavy city traffic where the outside temperature
  regularly reaches 90-degrees F or higher
in hilly or mountainous terrain
frequent trailer pulling

Engine compartment component locations - 4.7L (2UZ-FE) V8 engine

1	Brake fluid reservoir	7	Diagnostic connector	13	Windshield (and back window on
2	PCV valve	8	Battery		Sequoia) washer fluid reservoir
3	Engine oil filler cap	9	Coolant reservoir	14	Air filter housing
4	Main fuse	10	Radiator cap	15	Power steering fluid reservoir
5	Fuse box	11	Upper radiator hose	16	Ignition coils (spark plugs underneath)
6	Engine oil dipstick	12	Drivebelt		– 6 of 8 shown

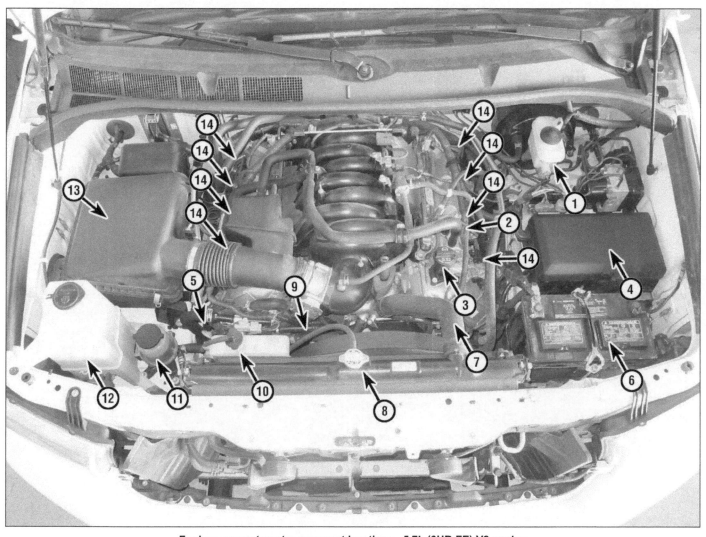

Engine compartment component locations - 5.7L (3UR-FE) V8 engine

1	Brake fluid reservoir	
2	PCV valve	
3	Engine oil filler cap	
4	Engine compartment fuse/relay box	
5	Engine oil dipstick	

6   Battery
7   Upper radiator hose
8   Radiator cap
9   Drivebelt
10 Coolant reservoir

11 Power steering fluid reservoir
12 Windshield (and back window on Sequoia) washer fluid reservoir
13 Air filter housing
14 Ignition coils (spark plugs underneath)

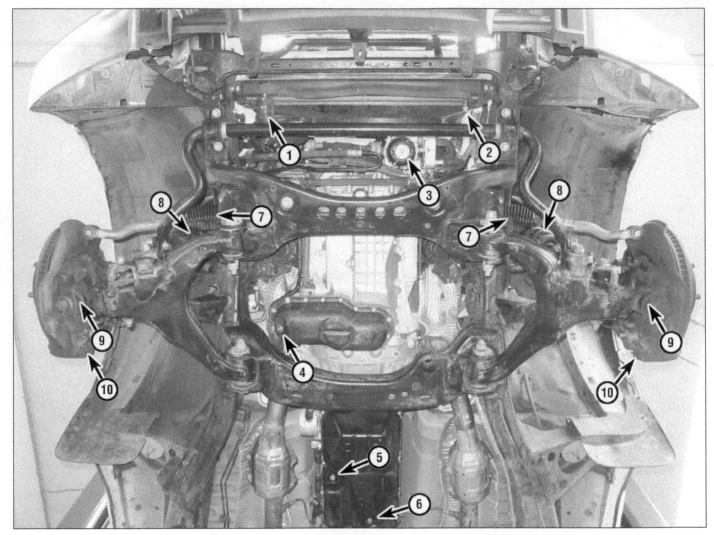

Typical engine compartment underside component locations

1	Lower radiator hose	5	Automatic transmission fluid overflow/	8	Shock absorber/coil spring assembly
2	Radiator drain valve		check plug	9	Lower balljoint
3	Engine oil filter	6	Automatic transmission fluid drain plug	10	Brake caliper
4	Engine oil drain plug	7	Steering gear boot		

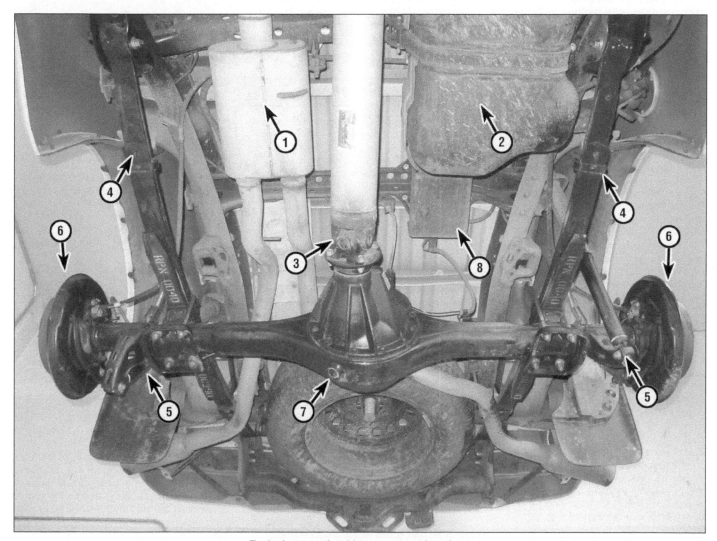

Typical rear underside component locations

1	Muffler	4	Leaf spring	7	Differential drain plug
2	Fuel tank	5	Shock absorber	8	EVAP canister
3	Driveshaft universal joint	6	Brake disc/parking brake drum		

2 Introduction

This Chapter is designed to help the home mechanic maintain the Toyota Tundra and Sequoia for peak performance, economy, safety and long life.

Included is a master maintenance schedule, followed by Sections dealing specifically with each item on the schedule. Visual checks, adjustments, component replacement and other helpful items are included. Refer to the **accompanying illustrations** of the engine compartment and the underside of the vehicle for the location of various components.

Servicing your vehicle in accordance with the mileage/time maintenance schedule and the following Sections will provide it with a planned maintenance program that should result in a long and reliable service life. This is a comprehensive plan, so maintaining some items but not others at the specified service intervals will not produce the same results.

As you service your vehicle, you will discover that many of the procedures can, and should, be grouped together because of the nature of the particular procedure you're performing or because of the close proximity of two otherwise unrelated components to one another.

For example, if the vehicle is raised for any reason, you should inspect the exhaust, suspension, steering and fuel systems while you're under the vehicle. When you're rotating the tires, it makes good sense to check the brakes and wheel bearings since the wheels are already removed.

Finally, let's suppose you have to borrow or rent a torque wrench. Even if you only need to tighten the spark plugs, you might as well check the torque of as many critical fasteners as time allows.

The first step of this maintenance program is to prepare yourself before the actual work begins. Read through all Sections pertinent to the procedures you're planning to do, then make a list of and gather together all the parts and tools you will need to do the job. If it looks as if you might run into problems during a particular segment of some procedure, seek advice from your local auto parts store or dealer service department.

Owner's Manual and VECI label information

Your vehicle Owner's Manual was written for your year and model and contains very specific information on component locations, specifications, fuse ratings, part numbers, etc. The Owner's Manual is an important resource for the do-it-yourselfer to have; if one was not supplied with your vehicle, it can generally be ordered from a dealer parts department.

Among other important information, the Vehicle Emissions Control Information (VECI) label contains specifications and procedures for tune-up adjustments (if applicable) and, in some instances, spark plugs (see Chapter 6

for more information on the VECI label). The information on this label is the *exact* maintenance data recommended by the manufacturer. This data often varies by intended operating altitude, local emissions regulations, month of manufacture, etc.

This Chapter contains procedural details, safety information and more ambitious maintenance intervals than you might find in the manufacturer's literature. However, you may also find procedures or specifications in your Owner's Manual or VECI label that differ with what's printed here. In these cases, the Owner's Manual or VECI label can be considered correct, since it is specific to your particular vehicle.

3 Tune-up general information

The term tune-up is used in this manual to represent a combination of individual operations rather than one specific procedure.

If, from the time the vehicle is new, the routine maintenance schedule is followed closely and frequent checks are made of fluid levels and high wear items, as suggested throughout this manual, the engine will be kept in relatively good running condition and the need for additional work will be minimized.

More likely than not, however, there will be times when the engine is running poorly due to lack of regular maintenance. This is even more likely if a used vehicle, which has not received regular and frequent maintenance checks, is purchased. In such cases, an engine tune-up will be needed outside of the regular routine maintenance intervals.

The first step in any tune-up or diagnostic procedure to help correct a poor running engine is a cylinder compression check. A compression check (see Chapter 2, Part C) will help determine the condition of internal engine components and should be used as a guide for tune-up and repair procedures. If, for instance, the compression check indicates serious internal engine wear, a conventional tune-up won't improve the performance of the engine and would be a waste of time and money. Because of its importance, the compression check should be done by someone with the right equipment and the knowledge to use it properly.

The following procedures are those most often needed to bring a generally poor running engine back into a proper state of tune.

Minor tune-up

Check all engine related fluids (Section 4)
Clean, inspect and test the battery (Section 9)
Check the cooling system (Section 10)
Check all underhood hoses (Section 11)
Check the PCV valve (Section 21)
Check the air filter (Section 22)
Check the drivebelt(s) (Section 24)

4.2a On the 2UZ-FE V8 engines, the oil dipstick is located on the left side of the engine (on the V6 it's at the front of the engine)

Major tune-up

All items listed under Minor tune-up, plus . . .
Replace the PCV valve (Section 21)
Replace the air filter (Section 22)
Check the fuel system (Section 25)
Replace the spark plugs (Section 28)
Check and adjust the valve clearances (Section 33)
Check the ignition system (Chapter 5)
Check the charging system (Chapter 5)

4 Fluid level checks (every 250 miles [400 km] or weekly)

Note: *The following are fluid level checks to be done on a 250 mile or weekly basis. Additional fluid level checks can be found in specific maintenance procedures which follow. Regardless of intervals, be alert to fluid leaks under the vehicle which would indicate a fault to be corrected immediately.*

1   Fluids are an essential part of the lubrication, cooling, brake and windshield washer systems. Because the fluids gradually become depleted and/or contaminated during normal operation of the vehicle, they must be periodically replenished. See *Recommended lubricants and fluids* in this Chapter's Specifications before adding fluid to any of the following components. **Note:** *The vehicle must be on level ground when fluid levels are checked.*

Engine oil

Refer to illustrations 4.2a, 4.2b, 4.4 and 4.6
2   The engine oil level is checked with a dipstick that extends through a tube and into the oil pan at the bottom of the engine **(see illustrations)**.
3   The oil level should be checked before the vehicle has been driven, or about 5 minutes after the engine has been shut off. If the oil is checked immediately after driving the vehicle, some of the oil will remain in the

4.2b On all other V8 engines, the dipstick is located at the front right side of the engine

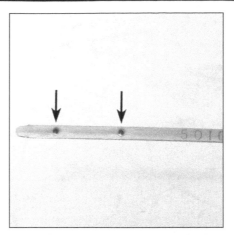

4.4 The oil level must be maintained between the marks at all times - it takes one quart of oil to raise the level from the lower mark to the upper mark

4.6 Oil is added to the engine after removing the twist-off cap located on the valve cover (V8 model shown, V6 models similar)

upper engine components, resulting in an inaccurate reading on the dipstick.

4   Pull the dipstick out of the tube and wipe all the oil from the end with a clean rag or paper towel. Insert the clean dipstick all the way back into the tube, then pull it out again. Note the oil at the end of the dipstick. Add oil as necessary to keep the level between the upper and lower marks on the dipstick **(see illustration)**.

5   Do not overfill the engine by adding too much oil since this may result in oil-fouled spark plugs, oil leaks or oil seal failures.

6   Oil is added to the engine after removing the threaded cap from the valve cover **(see illustration)**.

7   Checking the oil level is an important preventive maintenance step. A consistently low oil level indicates oil leakage through damaged seals, defective gaskets or past worn rings or valve guides. If the oil looks milky or has water droplets in it, the cylinder head gasket(s) may be blown or the head(s) or block may be cracked. The engine should be checked immediately. The condition of the oil should also be checked. Whenever you check the oil level, slide your thumb and index finger up the dipstick before wiping off the oil. If you see small dirt or metal particles clinging to the dipstick, the oil should be changed (see Section 8).

Engine coolant

Refer to illustration 4.8

Warning: *Do not allow antifreeze to come in contact with your skin or painted surfaces of the vehicle. Flush contaminated areas immediately with plenty of water. Don't store new coolant or leave old coolant lying around where it's accessible to children or pets - they're attracted by its sweet smell. Ingestion of even a small amount of coolant can be fatal! Wipe up garage floor and drip pan spills immediately. Keep antifreeze containers cov-*

ered and repair cooling system leaks as soon as they're noticed.

8   All vehicles covered by this manual are equipped with a pressurized coolant recovery system. A coolant reservoir, which is located in the left front of the engine compartment (on the fan shroud), is connected by a hose to the base of the coolant filler cap **(see illustration)**. If the coolant gets too hot during engine operation, coolant can escape through a spring-loaded radiator cap, then through a connecting hose into the reservoir. As the engine cools, the coolant is automatically drawn back into the cooling system to maintain the correct level.

9   The coolant level should be checked regularly. It must be between the FULL and LOW lines on the tank. The level will vary with the temperature of the engine. When the engine is cold, the coolant level should be at or slightly above the LOW mark on the tank. Once the engine has warmed up, the level should be at or near the FULL mark. If it isn't, allow the fluid in the tank to cool, then remove the cap from the reservoir and add coolant to bring the level up to the FULL line. Use only the type of coolant recommended in this Chapter's Specifications. Do not use supplemental inhibitor additives. If only a small amount of coolant is required to bring the system up to the proper level, water can be used. However, repeated additions of water will dilute the recommended antifreeze solution. In order to maintain the proper ratio of antifreeze and water, it is advisable to top up the reservoir level with the correct coolant.

10   If the coolant level drops within a short time after replenishment, there may be a leak in the system. Inspect the radiator, hoses, engine coolant filler cap, drain plugs and water pump. If no leak is evident, have the radiator cap pressure tested.

Warning: *Never remove the radiator cap or the coolant recovery reservoir cap when the engine is running or has just been shut down, because the cooling system is hot. Escaping*

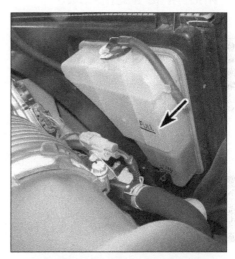

4.8 Check the coolant level in the reservoir with the engine hot - it should be visible through the translucent reservoir

steam and scalding liquid could cause serious injury.

11   If it is necessary to open the radiator cap, wait until the system has cooled completely, then wrap a thick cloth around the cap and turn it to the first stop. If any steam escapes, wait until the system has cooled further, then remove the cap.

12   When checking the coolant level, always note its condition. It should be relatively clear. If it is brown or rust colored, the system should be drained, flushed and refilled. Even if the coolant appears to be normal, the corrosion inhibitors wear out with use, so it must be replaced at the specified intervals.

13   Do not allow antifreeze to come in contact with your skin or painted surfaces of the vehicle. Flush contacted areas immediately with plenty of water.

4.14 The windshield washer fluid reservoir is located in the right front corner of the engine compartment

4.15 On some models the battery electrolyte level can be visually checked through the translucent battery case

4.17 The fluid level inside the brake fluid reservoir can be checked by observing the level from the outside

Windshield washer fluid

Refer to illustration 4.14

14   Fluid for the windshield washer system is located on the right (passenger) side of the engine compartment **(see illustration)**. Sequoia models are equipped with a rear window washer system built into the front washer system. In milder climates, plain water can be used to top up the reservoir, but the reservoir should be kept no more than two-thirds full to allow for expansion should the water freeze. In colder climates, the use of a specially designed windshield washer fluid, available at your dealer and any auto parts store, will help lower the freezing point of the fluid. Mix the solution with water in accordance with the manufacturer's directions on the container. Do not use regular antifreeze. It will damage the vehicle's paint.

Battery electrolyte

Refer to illustration 4.15

15   On models not equipped with a sealed battery, unscrew the filler/vent cap and check the electrolyte level. It must be between the upper and lower levels. On some batteries you can view the level through the translucent case **(see illustration)**. If the level is low, remove the caps and add distilled water. Install and securely retighten the caps.
Caution: *Overfilling the cells may cause electrolyte to spill over during periods of heavy charging, causing corrosion or damage.*

Brake fluid

Refer to illustration 4.17

16   The brake master cylinder is mounted on the front of the power booster unit in the engine compartment.
17   To check the fluid level of the brake master cylinder, simply look at the MAX and MIN marks on the reservoir **(see illustration)**. The level should be 1/4 inch below the maximum fill line.
18   If the level is low, wipe the top of the res-

ervoir cover with a clean rag to prevent contamination of the brake system before lifting the cover off.
19   Add only the specified brake fluid to the brake reservoir (refer to *Recommended lubricants and fluids* in this Chapter's Specifications or to your owner's manual). Mixing different types of brake fluid can damage the system.
Warning: *Use caution when filling the reservoir - brake fluid can harm your eyes and damage painted surfaces. Do not use brake fluid that has been opened for more than one year or has been left open. Brake fluid absorbs moisture from the air. Excess moisture can cause a dangerous loss of braking.*
20   While the reservoir cover is removed, inspect the reservoir for contamination. If deposits, dirt particles or water droplets are present, the system should be drained and refilled.
21   After filling the reservoir to the proper level, make sure the cover is properly seated to prevent fluid leakage.

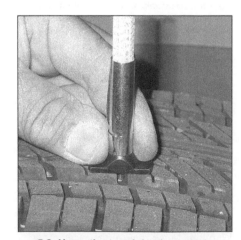

5.2 Use a tire tread depth gauge to monitor tire wear - they are available at auto parts stores and service stations and cost very little

22   The fluid in the brake master cylinder will drop slightly as the brake pads at each wheel wear down during normal operation. If the master cylinder requires repeated replenishing to keep it at the proper level, this is an indication of leakage in the brake system, which should be corrected immediately. If the brake system shows an indication of leakage check all brake lines and connections, along with the calipers, wheel cylinders and booster (see Section 26 for more information).
23   If, upon checking the brake master cylinder fluid level, you discover the reservoir empty or nearly empty, the system should be bled (see Chapter 9).

5   Tire and tire pressure checks (every 250 miles [400 km] or weekly)

Refer to illustrations 5.2, 5.3, 5.4a, 5.4b and 5.8

1   Periodic inspection of the tires may spare you the inconvenience of being stranded with a flat tire. It can also provide you with vital information regarding possible problems in the steering and suspension systems before major damage occurs.
2   The original tires on this vehicle are equipped with 1/2-inch wide wear bands that will appear when tread depth reaches 1/16-inch, at which point the tires can be considered worn out. Tread wear can be monitored with a simple, inexpensive device known as a tread depth gauge **(see illustration)**.
3   Note any abnormal tread wear **(see illustration)**. Tread pattern irregularities such as cupping, flat spots and more wear on one side than the other are indications of front end alignment and/or balance problems. If any of these conditions are noted, take the vehicle to a tire shop or service station to correct the problem.
4   Look closely for cuts, punctures and embedded nails or tacks. Sometimes a tire will hold air pressure for a short time or leak

UNDERINFLATION

CUPPING

Cupping may be caused by:
- Underinflation and/or mechanical irregularities such as out-of-balance condition of wheel and/or tire, and bent or damaged wheel.
- Loose or worn steering tie-rod or steering idler arm.
- Loose, damaged or worn front suspension parts.

OVERINFLATION

INCORRECT TOE-IN OR EXTREME CAMBER

FEATHERING DUE TO MISALIGNMENT

5.3 This chart will help you determine the condition of the tires, the probable cause(s) of abnormal wear and the corrective action necessary

down very slowly after a nail has embedded itself in the tread. If a slow leak persists, check the valve stem core to make sure it's tight **(see illustration)**. Examine the tread for an object that may have embedded itself in the tire or for a plug that may have begun to leak (radial tire punctures are repaired with a rubber plug that's installed in the hole). If a puncture is suspected, it can be easily verified by spraying a solution of soapy water onto the puncture area **(see illustration)**. The soapy solution will bubble if there's a leak. Unless the puncture is unusually large, a tire shop or service station can usually repair the tire.

5   Carefully inspect the inner sidewall of each tire for evidence of brake fluid leakage. If you see any, inspect the brakes immediately.

6   Correct air pressure adds miles to the life of the tires, improves mileage and enhances overall ride quality. Tire pressure cannot be accurately estimated by looking at a tire, especially if it's a radial. A tire pressure gauge is essential. Keep an accurate gauge in the vehicle. The pressure gauges attached to the

5.4a If a tire loses air on a steady basis, check the valve core first to make sure it's snug (special inexpensive wrenches are commonly available at auto parts stores)

5.4b If the valve core is tight, raise the corner of the vehicle with the low tire and spray a soapy water solution onto the tread as the tire is turned slowly - leaks will cause small bubbles to appear

5.8 **To extend the life of the tires, check the air pressure at least once a week with an accurate gauge (don't forget the spare)**

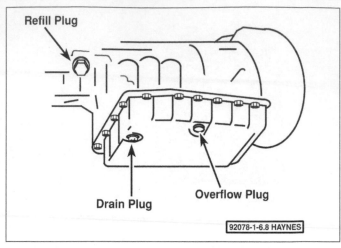

6.7 **Typical transmission plug identifications**

nozzles of air hoses at gas stations are often inaccurate.

7   Always check tire pressure when the tires are cold. Cold, in this case, means the vehicle has not been driven over a mile in the three hours preceding a tire pressure check. A pressure rise of four to eight pounds is not uncommon once the tires are warm.

8   Unscrew the valve cap protruding from the wheel or hubcap and push the gauge firmly onto the valve stem **(see illustration)**. Note the reading on the gauge and compare the figure to the recommended tire pressure shown on the placard on the driver's side door jamb. Reinstall the valve cap to keep dirt and moisture out of the valve stem mechanism. Check all four tires and, if necessary, add enough air to bring them up to the recommended pressure.

9   Don't forget to keep the spare tire inflated to the specified pressure (consult your owner's manual).

---

6   **Automatic transmission fluid level check (every 30,000 miles [48,000 km] or 24 months)**

Refer to illustration 6.7

Warning: *On Sequoia models equipped with rear height control suspension, adjust the height control to the NORMAL mode, turn OFF the height control, then turn off the engine BEFORE raising the vehicle.*
Note: *These models are not equipped with an automatic transmission fluid dipstick.*
Note: *The vehicle must be level for this check. If there is not enough room to crawl under the vehicle, raise the vehicle and support it securely on jackstands.*

1   The level of the automatic transmission fluid should be carefully maintained. Low fluid level can lead to slipping or loss of drive, while overfilling can cause foaming, loss of fluid and transmission damage.

2   The transmission fluid level should only be checked when the transmission is hot (at its normal operating temperature). If the vehicle has just been driven over 10 miles (15 miles in a frigid climate), and the fluid temperature is 160 to 175-degrees F, the transmission is hot.
Caution: *If the vehicle has just been driven for a long time at high speed or in city traffic in hot weather, or if it has been pulling a trailer, an accurate fluid level reading cannot be obtained. Allow the fluid to cool down for about 30 minutes.*

3   If a repair had been done, such as filter replacement or removal and installation, add the correct amount of fluid listed in this Chapter's Specifications. Start the engine and allow the transmission fluid temperature to reach 160 to 175-degrees F.

4   If the vehicle has not been driven, park the vehicle on level ground, set the parking brake, then start the engine and bring it to operating temperature. While the engine is idling, depress the brake pedal and move the selector lever through all the gear ranges, beginning and ending in Park.

5   On models equipped with trailer towing package, clean the area around the transmission thermostat cap located on the side of the transmission. Depress the cap with a screwdriver and insert a pin (3/64 to 5/64-inch [1.0 to 1.8 mm] thick) through the sides of the cap from bottom to top to hold the cap in place, then remove the screwdriver.

6   Turn off the engine. Remove the overflow plug from the bottom of the transmission fluid pan **(see illustration 30.5)**.
Note: *The overflow plug is located along the right side of the fluid pan; the drain plug is located at the back of the pan (don't remove the drain plug).*

7   If the fluid runs out of the hole, allow it to drip until it stops. If no fluid comes out of the hole, remove the refill plug, located on the side of the transmission housing, near the rear **(see illustration)**.
Note: *On 4WD models, the refill plug is*

located on the driver's side towards rear of the transmission.

8   Add the proper type of transmission fluid (see this Chapter's Specifications) until fluid flows from the overflow hole in the bottom of the pan. When the flow of fluid slows to a trickle, install the overflow plug and tighten it to the torque listed in this Chapter's Specifications.

9   Once the fluid is installed into the transmission pan, it will be necessary to warm the transmission fluid and check the fluid level of the transmission (final stage). Lower the vehicle, start the engine and allow it to idle with the air conditioning OFF. Move the shift lever through the entire range (P-R-N-D-2-1) to circulate transmission fluid throughout transmission.

10   Check the temperature of the transmission fluid. Use an infrared temperature gauge to determine transmission fluid/pan temperature. Shut down the engine once it reaches 115 degrees F.

11   Raise the vehicle and support it securely on jackstands, making sure it is level. With the engine idling and the transmission in Park, remove the overflow plug. See if transmission fluid flows slowly out of the transmission. If there is no transmission fluid, add some (engine NOT running).

12   If there was slight amounts of transmission fluid dripping from the overflow plug, wait until the drips cease, install the plug and tighten it to the Specifications listed in this Chapter.

13   On models equipped with trailer towing package, depress the cap with a screwdriver and remove the pin (see Step 5).

14   Tighten the fill plug to the torque listed in this Chapter's Specifications.

15   The condition of the fluid should also be checked along with the level. If the fluid at the end of the dipstick is black or a dark reddish brown color, or if it emits a burned smell, the fluid should be changed (see Section 30). If you are in doubt about the condition of the fluid, purchase some new fluid and compare the two for color and smell.

7 Power steering fluid level check (every 3000 miles [4800] or 3 months)

Refer to illustration 7.4

1 The reservoir for the power steering pump is remotely mounted on the right side of the engine compartment.

2 The power steering fluid level can be checked with the engine either hot or cold.

3 With the engine off, use a rag to clean the reservoir cap and the area around the cap. This will help prevent foreign material from falling into the reservoir when the cap is removed.

4 The power steering fluid reservoir is translucent and the level can be checked without removing the cap **(see illustration)**.

5 If additional fluid is required, pour the specified type fluid (see *Recommended lubricants and fluids* in this Chapter's Specifications or your owner's manual) directly into the reservoir using a funnel to prevent spills.

6 If the reservoir requires frequent topping up, all power steering hoses, hose connections, the power steering pump and the steering gear should be carefully examined for leaks.

8 Engine oil and filter change (see Maintenance schedule for service interval)

Refer to illustrations 8.2, 8.6a, 8.6b and 8.7

Warning: *On Sequoia models equipped with rear height control suspension, adjust the height control to the NORMAL mode, turn OFF the height control, then turn off the engine BEFORE raising the vehicle.*

1 Frequent oil changes are the best preventive maintenance the home mechanic can give the engine, because aging oil becomes diluted and contaminated, which leads to premature engine wear.

2 Make sure that you have all the neces-

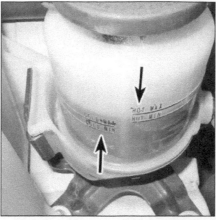

7.4 The power steering fluid level can be viewed through the translucent reservoir

sary tools before you begin this procedure **(see illustration)**. You should also have plenty of rags or newspapers handy for mopping up any spills.

3 Access to the underside of the vehicle is greatly improved if the vehicle can be lifted on a hoist, driven onto ramps or supported by jackstands.

4 If this is your first oil change, get under the vehicle and familiarize yourself with the location of the oil drain plug. The engine and exhaust components will be warm during the actual work, so try to anticipate any potential problems before the engine and accessories are hot.

5 Park the vehicle on a level spot. Start the engine and allow it to reach its normal operating temperature (the needle on the temperature gauge should be at least above the bottom mark). Warm oil and contaminates will flow out more easily. Turn off the engine when it's warmed up. Remove the filler cap on the valve cover.

6 Raise the vehicle and support it securely on jackstands.

Warning: *To avoid personal injury, never get beneath the vehicle when it is supported by only by a jack. The jack provided with your*

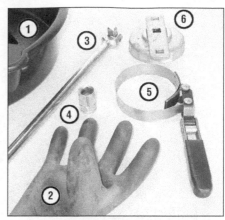

8.2 These tools are required when changing the engine oil and filter

1 **Drain pan** - *It should be fairly shallow in depth, but wide to prevent spills*

2 **Rubber gloves** - *When removing the drain plug and filter, you will get oil on your hands (the gloves will prevent burns)*

3 **Breaker bar** - *Sometimes the oil drain plug is tight, and a long breaker bar is needed to loosen it*

4 **Socket** - *To be used with the breaker bar or a ratchet (must be the correct size to fit the drain plug)*

5 **Filter wrench** - *This is a metal band-type wrench, which requires clearance around the filter to be effective*

6 **Filter wrench** - *This type fits on the bottom of the filter and can be turned with a ratchet or breaker bar (different size wrenches are available for different types of filters)*

vehicle is designed solely for raising the vehicle to remove and replace the wheels. Always use jackstands to support the vehicle when it becomes necessary to place your body underneath the vehicle. Remove the under-vehicle splash shield fasteners and unhook the shield from the body **(see illustrations)**.

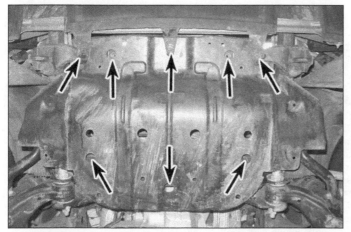

8.6a To gain access to various components under the front of the vehicle, the splash shield must be removed; remove these bolts . . .

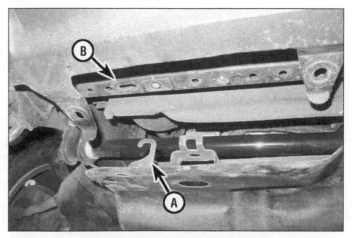

8.6b . . . then, lower the shield, unhook the shield hook (A), from the body (B), and remove the shield

8.7 The engine oil drain plug is located on the bottom of the oil pan - it is usually very tight, so use the proper size box-end wrench or socket to avoid rounding it off

8.13a The oil filter on 2UZ-FE V8 engines shown, is usually on very tight and will require a special wrench for removal - DO NOT use the wrench to tighten the new filter!

8.13b On 2010 and earlier 1GR-FE V6 engines, the oil filter is located on the timing chain cover, so it's easier to access than on other models. But when you unscrew it, some oil will run out before you can tilt the open end up. The shield catches any spilled oil, and by removing a rubber plug from the underside of the shield, you can drain spilled oil through a small drain hole onto a shop rag

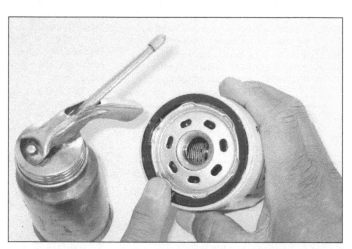

8.15 Lubricate the oil filter gasket with clean engine oil before installing the filter on the engine - 2UZ-FE V8 engines and 2010 and earlier 1GR-FE V6 engines

7   Being careful not to touch the hot exhaust components, place the drain pan under the drain plug in the bottom of the pan and remove the plug **(see illustration)**. You may want to wear gloves while unscrewing the plug the final few turns if the engine is really hot.

8   Allow the old oil to drain into the pan. It may be necessary to move the pan farther under the engine as the oil flow slows to a trickle. Inspect the old oil for the presence of metal shavings and chips.

9   After all the oil has drained, wipe off the drain plug with a clean rag. Even minute metal particles clinging to the plug would immediately contaminate the new oil.

10   Clean the area around the drain plug opening, reinstall the plug and tighten it to the torque listed in this Chapter's Specifications.

All 2UZ-FE V8 engines and 2010 and earlier 1GR-FE V6 engines

Refer to illustrations 8.13a, 8.13b and 8.15

11   Move the drain pan into position under the oil filter.

12   Remove all tools, rags, etc. from under the vehicle, being careful not to spill the oil in the drain pan, then lower the vehicle.

13   Loosen the oil filter by turning it counterclockwise with the filter wrench **(see illustrations)**. Any standard filter wrench should work. Once the filter is loose, use your hands to unscrew it from the block. Just as the filter is detached from the block, immediately tilt the open end up to prevent the oil inside the filter from spilling out.

Warning: *The engine exhaust manifold may still be hot, so be careful.*

14   With a clean rag, wipe off the mounting surface on the block. If a residue of old oil is allowed to remain, it will smoke when the block is heated up. It will also prevent the new filter from seating properly. Also make sure that the none of the old gasket remains stuck to the mounting surface.

15   Compare the old filter with the new one to make sure they are the same type. Smear some engine oil on the rubber gasket of the new filter **(see illustration)**. Attach the filter to the engine, following the tightening directions printed on the filter canister or packing box. Most filter manufacturers recommend against using a filter wrench due to the possibility of overtightening and damage to the seal.

8.16 Use a 3/8 inch extension to remove the oil filter drain plug

8.17a Insert the drain pipe into the filter cap . . .

All other V8 engines and 2011 and later 1GR-FE V6 engines

Refer to illustrations 8.16, 8.17a, 8.17b, 8.18a, 8.18b, 8.19 and 8.21

16   Remove the oil filter drain plug **(see illustration)**.

17   Insert the filter drain pipe into the filter cap and allow the oil to drain out of the cap **(see illustrations)**.

18   Remove the drain pipe and insert the special tool #09228-06501 to the housing and carefully unscrew the filter cap from the housing **(see illustration)**. If the special tool is not available, loosen the oil filter by turning it counterclockwise with the filter wrench on the lowest part of the housing **(see illustration)**, then remove it by hand.

Note: *Do not remove the oil filter clip from the side of the housing.*

19   Remove the old filter element and O-ring from the cap **(see illustration)**.

20   With a clean rag, wipe off the cap and cap mounting surface on the housing.

21   Compare the old filter element with the new one to make sure they are the same type. Smear some engine oil on the O-ring and install it on the filter cap **(see illustra-** tion). Slide the filter back onto the cap, attach the filter assembly to the housing, then hand tighten the housing.

22   Tighten the filter cap to the torque listed

8.17b . . . and drain the oil from the filter

8.18a Insert the special tool #09228-06501 to the housing and carefully unscrew the filter cap

8.18b If the special tool is not available loosen the oil filter by turning it counterclockwise with the filter wrench then remove it by hand

8.19 Remove the filter element (A) then the O-ring (B) from the filter cap

8.21 Smear some engine oil on the O-ring and install it on the filter cap

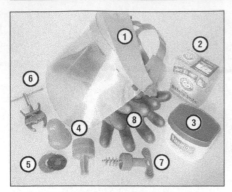

9.1 Tools and materials required for battery maintenance

1   **Face shield/safety goggles** - *When removing corrosion with a brush, the acidic particles can easily fly up into your eyes*
2   **Baking soda** - *A solution of baking soda and water can be used to neutralize corrosion*
3   **Petroleum jelly** - *A layer of this on the battery posts will help prevent corrosion*
4   **Battery post/cable cleaner** - *This wire brush cleaning tool will remove all traces of corrosion from the battery posts and cable clamps*
5   **Treated felt washers** - *Placing one of these on each post, directly under the cable clamps, will help prevent corrosion*
6   **Puller** - *Sometimes the cable clamps are very difficult to pull off the posts, even after the nut/bolt has been completely loosened. This tool pulls the clamp straight up and off the post without damage*
7   **Battery post/cable cleaner** - *Here is another cleaning tool which is a slightly different version of Number 4 above, but it does the same thing*
8   **Rubber gloves** - *Another safety item to consider when servicing the battery; remember that's acid inside the battery!*

in this Chapter's Specifications.
23   Install a new O-ring on the filter drain plug and tighten the plug to the torque listed in this Chapter's Specifications.

All models

24   Add new oil to the engine through the oil filler cap in the valve cover. Use a spout or funnel to prevent oil from spilling onto the top of the engine. Pour five quarts of fresh oil into the engine. Wait a few minutes to allow the oil to drain into the pan, then check the level on the oil dipstick (see Section 4 if necessary). If the oil level is at or near the H mark, install the filler cap hand tight, start the engine and allow the new oil to circulate.
25   Allow the engine to run for about a minute, then turn it off. While the engine is running, look under the vehicle and check for

9.6a Battery terminal corrosion usually appears as light, fluffy powder

leaks at the oil pan drain plug and around the oil filter. If either is leaking, stop the engine and tighten the plug or filter slightly.
26   Wait a few minutes to allow the oil to trickle down into the pan, then recheck the level on the dipstick and, if necessary, add enough oil to bring the level to the upper mark.
27   During the first few trips after an oil change, make it a point to check frequently for leaks and proper oil level.
28   The old oil drained from the engine cannot be reused in its present state and should be disposed of. Check with your local auto parts store, disposal facility or environmental agency to see if they will accept the oil for recycling. After the oil has cooled it can be drained into a container (capped plastic jugs, topped bottles, milk cartons, etc.) for transport to one of these disposal sites. Don't dispose of the oil by pouring it on the ground or down a drain!

9   Battery maintenance and charging (every 7500 miles [12,000 km] or 6 months)

Warning: *Certain precautions must be followed when checking and servicing the battery. Hydrogen gas, which is highly flammable, is always present in the battery cells, so keep lighted tobacco and all other open flames and sparks away from the battery. The electrolyte inside the battery is actually dilute sulfuric acid, which will cause injury if splashed on your skin or in your eyes. It will also ruin clothes and painted surfaces. When removing the battery cables, always detach the negative cable first and hook it up last!*

Maintenance

Refer to illustrations 9.1, 9.6a, 9.6b, 9.7a and 9.7b

1   A routine preventive maintenance program for the battery in your vehicle is the only

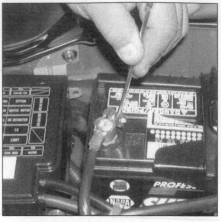

9.6b Removing the cable from a battery post with a wrench - sometimes special battery pliers are required for this procedure if corrosion has caused deterioration of the nut hex (always remove the negative cable first and hook it up last!)

way to ensure quick and reliable starts. But before performing any battery maintenance, make sure that you have the proper equipment necessary to work safely around the battery (**see illustration**).
2   There are also several precautions that should be taken whenever battery maintenance is performed. Before servicing the battery, always turn the engine and all accessories off and disconnect the cable from the negative terminal of the battery.
3   The battery produces hydrogen gas, which is both flammable and explosive. Never create a spark, smoke or light a match around the battery. Always charge the battery in a ventilated area.
4   Electrolyte contains poisonous and corrosive sulfuric acid. Do not allow it to get in your eyes, on your skin on your clothes. Never ingest it. Wear protective safety glasses when working near the battery. Keep children away from the battery.
5   Note the external condition of the battery. If the positive terminal and cable clamp on your vehicle's battery is equipped with a rubber protector, make sure that it's not torn or damaged. It should completely cover the terminal. Look for any corroded or loose connections, cracks in the case or cover or loose hold-down clamps. Also check the entire length of each cable for cracks and frayed conductors.
6   If corrosion, which looks like white, fluffy deposits (**see illustration**) is evident, particularly around the terminals, the battery should be removed for cleaning. Loosen the cable clamp bolts with a wrench, being careful to remove the negative cable first, and slide them off the terminals (**see illustration**). Then disconnect the hold-down clamp bolt and nut, remove the clamp and lift the battery from the engine compartment.
7   Clean the cable clamps thoroughly with a battery brush or a terminal cleaner and a

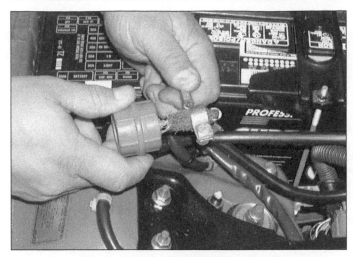

9.7a When cleaning the cable clamps, all corrosion must be removed

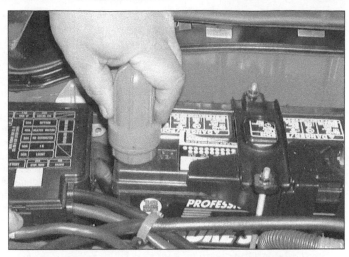

9.7b Regardless of the type of tool used on the battery posts, a clean, shiny surface should be the end result

solution of warm water and baking soda **(see illustration)**. Wash the terminals and the top of the battery case with the same solution but make sure that the solution doesn't get into the battery. When cleaning the cables, terminals and battery top, wear safety goggles and rubber gloves to prevent any solution from coming in contact with your eyes or hands. Wear old clothes too - even diluted, sulfuric acid splashed onto clothes will burn holes in them. If the terminals have been extensively corroded, clean them up with a terminal cleaner **(see illustration)**. Thoroughly wash all cleaned areas with plain water.

8   Make sure that the battery tray is in good condition and the hold-down clamp bolts are tight. If the battery is removed from the tray, make sure no parts remain in the bottom of the tray when the battery is reinstalled. When reinstalling the hold-down clamp bolts, do not overtighten them.

9   Any metal parts of the vehicle damaged by corrosion should be covered with a zinc-based primer, then painted.

10   Information on removing and installing the battery can be found in Chapter 5. Information on jump starting can be found at the front of this manual. For more detailed battery checking procedures, refer to the *Haynes Automotive Electrical Manual.*

Charging

Warning: *When batteries are being charged, hydrogen gas, which is very explosive and flammable, is produced. Do not smoke or allow open flames near a battery. Wear eye protection when near the battery during charging. Also, make sure the charger is unplugged before connecting or disconnecting the battery from the charger.*

Note: *The manufacturer recommends the battery be removed from the vehicle for charging because the gas that escapes during this procedure can damage the paint. Fast charging with the battery cables connected can result in damage to the electrical system.*

11   Slow-rate charging is the best way to restore a battery that's discharged to the point where it will not start the engine. It's also a good way to maintain the battery charge in a vehicle that's only driven a few miles between starts. Maintaining the battery charge is particularly important in the winter when the battery must work harder to start the engine and electrical accessories that drain the battery are in greater use.

12   It's best to use a one or two-amp battery charger (sometimes called a trickle charger). They are the safest and put the least strain on the battery. They are also the least expensive. For a faster charge, you can use a higher amperage charger, but don't use one rated more than 1/10th the amp/hour rating of the battery. Rapid boost charges that claim to restore the power of the battery in one to two hours are hardest on the battery and can damage batteries not in good condition. This type of charging should only be used in emergency situations.

13   The average time necessary to charge a battery should be listed in the instructions that come with the charger. As a general rule, a trickle charger will charge a battery in 12 to 16 hours.

14   Remove all the cell caps (if equipped) and cover the holes with a clean cloth to prevent spattering electrolyte. Disconnect the negative battery cable and hook the battery charger cable clamps up to the battery posts (positive to positive, negative to negative), then plug in the charger. Make sure it is set at 12-volts if it has a selector switch.

15   If you're using a charger with a rate higher than two amps, check the battery regularly during charging to make sure it doesn't overheat. If you're using a trickle charger, you can safely let the battery charge overnight after you've checked it regularly for the first couple of hours.

16   If the battery has removable cell caps, measure the specific gravity with a hydrometer every hour during the last few hours of

the charging cycle. Hydrometers are available inexpensively from auto parts stores - follow the instructions that come with the hydrometer. Consider the battery charged when there's no change in the specific gravity reading for two hours and the electrolyte in the cells is gassing (bubbling) freely. The specific gravity reading from each cell should be very close to the others. If not, the battery probably has a bad cell(s).

17   Some batteries with sealed tops have built-in hydrometers on the top that indicate the state of charge by the color displayed in the hydrometer window. Normally, a bright-colored hydrometer indicates a full charge and a dark hydrometer indicates the battery still needs charging.

18   If the battery has a sealed top and no built-in hydrometer, you can hook up a voltmeter across the battery terminals to check the charge. A fully charged battery should read 12.6 volts or higher after the surface charge has been removed.

19   Further information on the battery and jump starting can be found in Chapter 5 and at the front of this manual.

10   Cooling system check (every 7500 miles [12,000 km] or 6 months)

Refer to illustration 10.4

1   Many major engine failures can be attributed to a faulty cooling system. The cooling system also cools the transmission fluid, and thus plays an important role in prolonging transmission life.

2   The cooling system should be checked with the engine cold. Do this before the vehicle is driven for the day or after it has been shut off for at least three hours.

3   Remove the radiator cap by turning it to the left until it reaches a stop. If you hear a hissing sound (indicating there is still pres-

Check for a chafed area that could fail prematurely.

Check for a soft area indicating the hose has deteriorated inside.

Overtightening the clamp on a hardened hose will damage the hose and cause a leak.

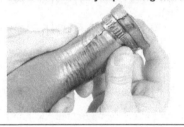

Check each hose for swelling and oil-soaked ends. Cracks and breaks can be located by squeezing the hose.

10.4 Hoses, like drivebelts, have a habit of failing at the worst possible time - to prevent the inconvenience of a blown radiator or heater hose, inspect them carefully as shown here

sure in the system), wait until this stops. Now press down on the cap with the palm of your hand and continue turning to the left until the cap can be removed. Thoroughly clean the cap, inside and out, with clean water. Also clean the filler neck on the radiator. All traces of corrosion should be removed. The coolant inside the radiator should be relatively transparent. If it is rust colored, the system should be drained and refilled (see Section 29). If the coolant level is not up to the top, add additional antifreeze of the proper type (see Section 4).

4   Carefully check the large upper and lower radiator hoses along with the smaller diameter heater hoses which run from the engine to the firewall. Inspect each hose along its entire length, replacing any hose which is cracked, swollen or shows signs of deterioration. Cracks may become more apparent if the

hose is squeezed (see illustration). Regardless of condition, it's a good idea to replace hoses with new ones every two years.

5   Make sure all hose connections are tight. A leak in the cooling system will usually show up as white or rust colored deposits on the areas adjoining the leak. If wire-type clamps are used at the ends of the hoses, it may be a good idea to replace them with more secure screw-type clamps.

6   Use compressed air or a soft brush to remove bugs, leaves, etc. from the front of the radiator or air conditioning condenser. Be careful not to damage the delicate cooling fins or cut yourself on them.

7   Every other inspection, or at the first indication of cooling system problems, have the cap and system pressure tested. If you don't have a pressure tester, most gas stations and repair shops will do this for a minimal charge.

11   Underhood hose check and replacement (every 7500 miles [12,000 km] or 6 months)

General

Warning: *Replacement of air conditioning hoses must be left to a dealer service department or air conditioning shop that has the equipment to depressurize the system safely. Never remove air conditioning components or hoses until the system has been depressurized.*

1   High temperatures in the engine compartment can cause the deterioration of the rubber and plastic hoses used for engine, accessory and emission systems operation. Periodic inspection should be made for cracks, loose clamps, material hardening and leaks. Information specific to the cooling system hoses can be found in Section 10.

2   Some, but not all, hoses are secured to the fittings with clamps. Where clamps are used, check that they haven't lost their tension, allowing the hose to leak. If clamps aren't used, make sure the hose has not expanded and/or hardened where it slips over the fitting, allowing it to leak.

Vacuum hoses

3   It's quite common for vacuum hoses, especially those in the emissions system, to be color coded or identified by colored stripes molded into them. Various systems require hoses with different wall thickness, collapse resistance and temperature resistance. When replacing hoses, be sure the new ones are made of the same material.

4   Often the only effective way to check a hose is to remove it completely from the vehicle. If more than one hose is removed, label the hoses and fittings to ensure correct installation.

5   When checking vacuum hoses, include any plastic T-fittings in the check. Inspect the fittings for cracks and the hose where it

fits over the fitting for distortion, which could cause leakage.

6   A small piece of vacuum hose (1/4-inch inside diameter) can be used as a stethoscope to detect vacuum leaks. Hold one end of the hose to your ear and probe around vacuum hoses and fittings, listening for the hissing sound characteristic of a vacuum leak. **Warning:** *When probing with the vacuum hose stethoscope, be very careful not to come into contact with moving engine components such as the drivebelt, cooling fan, etc.*

Fuel hose

Warning: *Gasoline is extremely flammable, so take extra precautions when you work on any part of the fuel system. Don't smoke or allow open flames or bare light bulbs near the work area, and don't work in a garage where a gas-type appliance (such as a water heater or clothes dryer) is present. Since gasoline is carcinogenic, wear fuel-resistant gloves when there's a possibility of being exposed to fuel, and, if you spill any fuel on your skin, rinse it off immediately with soap and water. Mop up any spills immediately and do not store fuel-soaked rags where they could ignite. The fuel system is under constant pressure, so if any fuel lines are to be disconnected, the fuel pressure in the system must be relieved first (see Chapter 4). When you perform any kind of work on the fuel system, wear safety glasses and have a Class B Type fire extinguisher on hand.*

7   Check all rubber fuel lines for deterioration and chafing. Check especially for cracks in areas where the hose bends and just before fittings, such as where a hose attaches to the fuel filter.

8   Only high quality fuel line, designed for high-pressure fuel injection systems, must be used for fuel line replacement. Never, under any circumstances, use standard fuel hose, un-reinforced vacuum line, clear plastic tubing or water hose for fuel lines.

9   Spring-type clamps are commonly used on fuel lines. These clamps often lose their tension over a period of time, and can be "sprung" during removal. Replace all spring-type clamps with screw clamps whenever a hose is replaced.

Metal lines

10   Sections of metal line are often used in the fuel system. Check carefully that the line has not been bent or crimped and that cracks have not started in the line.

11   If a section of metal fuel line must be replaced, only seamless steel tubing should be used, since copper and aluminum tubing don't have the strength necessary to withstand normal engine vibration.

12   Check the metal brake lines where they enter the master cylinder and brake proportioning unit (if used) for cracks in the lines or loose fittings. Any sign of brake fluid leakage calls for an immediate thorough inspection of the brake system.

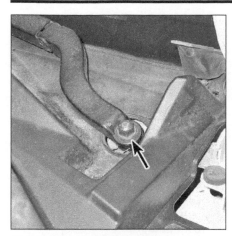

12.3 Check the tightness of the wiper arm retaining nut

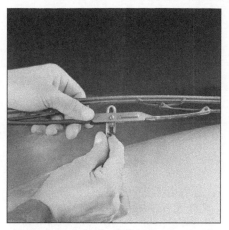

12.5 Press on the release tab, then push the blade assembly down and out of the hook in the arm

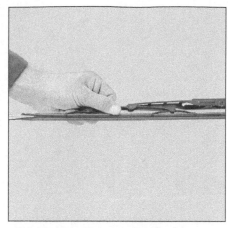

12.6 Use needle-nose pliers to compress the rubber element, then slide the element out - slide the new element in and lock the blade assembly fingers into the notches of the wiper element

12 Wiper blade inspection and replacement (every 7500 miles [12,000 km] or 6 months)

Refer to illustrations 12.3, 12.5 and 12.6

1   The windshield wiper and blade assemblies should be inspected periodically for damage, loose components and cracked or worn blade elements.

2   Road film can build up on the wiper blades and affect their efficiency, so they should be washed regularly with a mild detergent solution.

3   The action of the wiping mechanism can loosen bolts, nuts and fasteners, so they should be checked and tightened, as necessary **(see illustration)**, at the same time the wiper blades are checked.

4   If the wiper blade elements are cracked, worn or warped, or no longer clean adequately, they should be replaced with new ones.

5   Lift the arm assembly away from the glass for clearance, press on the release lever, then slide the wiper blade assembly out of the hook in the end of the arm **(see illustration)**.

6   Use needle-nose pliers to compress the blade element, then slide the element out of the frame and discard it **(see illustration)**.

7   Installation is the reverse of removal.

13 Tire rotation (every 7500 miles [12,000 km] or 6 months)

Refer to illustration 13.2

Warning: *On Sequoia models equipped with rear height control suspension, adjust the height control to the NORMAL mode, turn OFF the height control, then turn off the engine BEFORE raising the vehicle.*

1   The tires should be rotated at the specified intervals and whenever uneven wear is noticed. Since the vehicle will be raised and

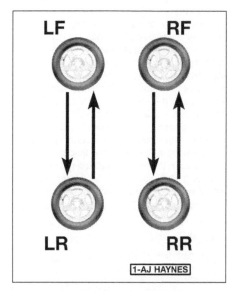

13.2 Tire rotation diagram

the tires removed anyway, check the brakes (see Section 26) at this time.

2   Radial tires must be rotated in a specific pattern **(see illustration)**.

3   Refer to the information in "Jacking and towing" at the front of this manual for the proper procedures to follow when raising the vehicle and changing a tire. If the brakes are to be checked, do not apply the parking brake as stated. Make sure the tires are blocked to prevent the vehicle from rolling.

4   Preferably, the entire vehicle should be raised at the same time. This can be done on a hoist or by jacking up each corner, then lowering the vehicle onto jackstands placed under the frame rails. Always use four jackstands and make sure the vehicle is firmly supported.

5   After rotation, check and adjust the tire pressures as necessary and check the lug nut tightness.

6   For further information on the wheels and tires, refer to Chapter 10.

14.1 Steering wheel freeplay is the amount of travel between an initial steering input and the point at which the front wheels begin to turn (indicated by a slight resistance)

14 Suspension and steering check (every 7500 miles [12,000 km] or 6 months)

Note: *For detailed illustrations of the steering and suspension components, refer to Chapter 10.*

With the wheels on the ground

Refer to illustration 14.1

1   With the vehicle stopped and the front wheels pointed straight ahead, rock the steering wheel gently back and forth. If freeplay **(see illustration)** is excessive, a front wheel bearing, main shaft U-joint, intermediate shaft U-joint or tie rod end is worn or the steering gear is out of adjustment or broken. Refer to Chapter 10 for the appropriate repair procedure.

2   Other symptoms, such as excessive vehicle body movement over rough roads,

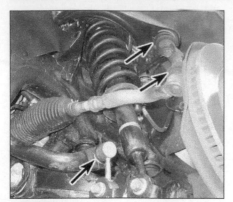

14.6a Inspect the suspension for deteriorated rubber bushings and torn grease seals

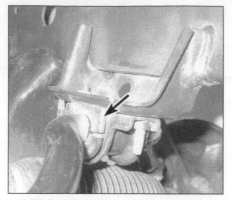

14.6b Check the stabilizer bar bushings for deterioration at the front and the rear of the vehicle

swaying (leaning) around corners and binding as the steering wheel is turned, may indicate faulty steering and/or suspension components.

3   Check the shock absorbers by pushing down and releasing the vehicle several times at each corner. If the vehicle does not come back to a level position within one bounce, the shocks are worn and must be replaced. When bouncing the vehicle up and down, listen for squeaks and noises from the suspension components.

Under the vehicle

Refer to illustrations 14.6a, 14.6b and 14.7

Warning: *On Sequoia models equipped with rear height control suspension, adjust the height control to the NORMAL mode, turn OFF the height control, then turn off the engine BEFORE raising the vehicle.*

4   Raise the vehicle with a floor jack and support it securely on jackstands. See "Jacking and towing" at the front of this book for proper jacking points.

5   Check the tires for irregular wear patterns and proper inflation. See Section 5 in this Chapter for information regarding tire wear.

6   Inspect the universal joint between the

steering shaft and the steering rack. Check the steering rack and driveaxle boots for grease leakage. Check the steering linkage for looseness or damage. Check the tie-rod ends for excessive play. Look for loose bolts, broken or disconnected parts and deteriorated rubber bushings on all suspension and steering components **(see illustrations)**. While an assistant turns the steering wheel from side to side, check the steering components for free movement, chafing and binding. If the steering components do not seem to be reacting with the movement of the steering wheel, try to determine where the slack is located.

7   Check the wheel bearings. Do this by spinning the front wheels. Listen for any abnormal noises and watch to make sure the

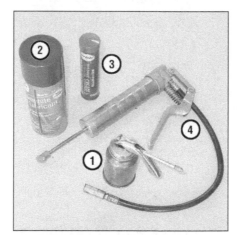

15.1 Materials required for chassis and body lubrication

1   Engine oil - Light engine oil in a can like this can be used for door and hood hinges

2   Graphite spray - Used to lubricate lock cylinders

3   Grease - Grease, in a variety of types and weights, is available for use in a grease gun. Check the Specifications for your requirements

4   Grease gun - A common grease gun, shown here with a detachable hose and nozzle, is needed for chassis lubrication. After use, clean it thoroughly

wheel spins true (doesn't wobble). Grab the top and bottom of the tire and pull in-and-out on it. Notice any movement which would indicate a loose wheel bearing assembly **(see illustration)**. If the bearings are loose, they are in need of replacement. Refer to Chapter 10 for more information.

8   Inspect the driveshafts for worn U-joints and for excessive play in the slip yoke and spline area (see Chapter 8).

9   Check the transfer case and differentials for evidence of fluid leakage.

15   Driveshaft and body lubrication (every 7500 miles [12,000 km or 6 months)

Refer to illustrations 15.1, 15.5, 15.8 and 15.10

Warning: *On Sequoia models equipped with rear height control suspension, adjust the height control to the NORMAL mode, turn OFF the height control, then turn off the engine BEFORE raising the vehicle.*

1   Refer to *Recommended lubricants and fluids* in this Chapter's Specifications to obtain the necessary grease, etc. You will also need a grease gun **(see illustration)**.

2   Look under the vehicle for grease fittings on the driveline components. They are normally found on the driveshaft universal joints and slip yokes.

3   For easier access under the vehicle, raise it with a jack and place jackstands under the frame. Make sure it is safely supported by the stands. If the wheels are to be removed at this interval for tire rotation or brake inspection, loosen the lug nuts slightly while the vehicle is still on the ground.

4   Before beginning, force a little grease out of the nozzle to remove any dirt from the end of the gun. Wipe the nozzle clean with a rag.

5   With the grease gun and plenty of clean rags, crawl under the vehicle and begin lubricating the driveshaft universal joints **(see illustration)**.

Note: *Not all models have grease fittings installed with original Toyota equipment.*

14.7 Grasp the tire as shown and check for endplay at the wheel bearings - if endplay is found the wheel bearings must be replaced

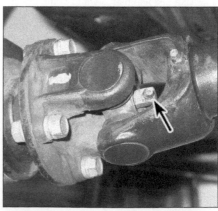

15.5 Pump grease into the universal joints until it can be seen coming out from the seals

15.8 The slip joint grease fitting is located on the yoke - pump grease into it until it comes out of the slip joint seal

15.10 Use lithium base grease to lubricate the contact points on the steering knuckle stops

6   Wipe the area around the grease fitting free of dirt, then squeeze the trigger on the grease gun to force grease into the component. Continue pumping grease into the fitting until it just oozes out of the bearing cup seals. If it escapes around the grease gun nozzle, the fitting is clogged or the nozzle is not completely seated on the fitting. Re-secure the gun nozzle to the fitting and try again. If necessary, replace the fitting with a new one.

7   Wipe the excess grease from the components and the grease fitting. Repeat the procedure for the remaining fittings.

8   Lubricate the driveshaft slip yoke by pumping grease into the fitting until it can be seen coming out of the slip yoke seal **(see illustration)**.

9   While you are under the vehicle, clean and lubricate the parking brake cable along with the cable guides and levers. This can be done by smearing some chassis grease onto the cable and its related parts with your fingers.

10   Lubricate the contact points on the steering knuckle stop **(see illustration)**.

11   Open the hood and smear a little chassis grease on the hood latch mechanism. Have an assistant pull the hood release lever from inside the vehicle as you lubricate the cable at the latch.

12   Lubricate all the hinges (door, hood, etc.) with engine oil to keep them in proper working order.

13   The key lock cylinders can be lubricated with spray-on graphite or silicone lubricant, which is available at auto parts stores.

14   Lubricate the door weather-stripping with silicone spray. This will reduce chafing and retard wear.

---

16   Exhaust system check (every 7500 miles [12,000 km] or 6 months)

Refer to illustrations 16.2a and 16.2b
Warning: *On Sequoia models equipped*

with rear height control suspension, adjust the height control to the NORMAL mode, turn OFF the height control, then turn off the engine BEFORE raising the vehicle.

1   With the engine cold (at least three hours after the vehicle has been driven), check the complete exhaust system from the manifold to the end of the tailpipe. Be careful around the catalytic converter (if equipped), which may be hot even after three hours. The inspection should be done with the vehicle on a hoist to permit unrestricted access. If a hoist isn't available, raise the vehicle and support it securely on jackstands.

2   Check the exhaust pipes and connections for signs of leakage and/or corrosion indicating a potential failure. Make sure that all brackets and hangers are in good condition and tight **(see illustrations)**.

3   Inspect the underside of the body for holes, corrosion, open seams, etc. which may allow exhaust gasses to enter the passenger compartment. Seal all body openings with silicone sealant or body putty.

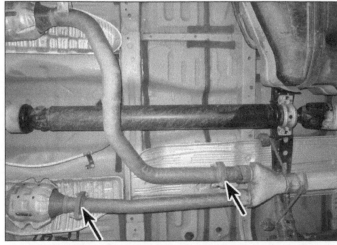

16.2a Check the exhaust pipes and connections for signs of leakage and corrosion

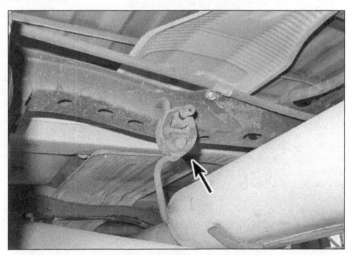

16.2b Check the exhaust system rubber hangers for cracks and damage

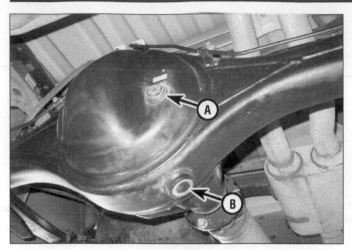

18.2 The rear differential fill plug (A) and drain plug (B) are located on the axle housing - use your finger as a dipstick to check the lubricant level

20.2 Check the condition of the driveaxle boot for signs of cracks and grease leaks

4   Rattles and other noises can often be traced to the exhaust system, especially the hangers, mounts and heat shields. Try to move the pipes, mufflers and catalytic converter. If the components can come in contact with the body or suspension parts, secure the exhaust system with new brackets and hangers.

17 Transfer case lubricant level check (4WD models) (every 15,000 miles [24,000 km] or 12 months)

1   The transfer case lubricant level is checked by removing the upper plug located in the side of the case.
2   Use a finger to reach inside the housing to determine the lubricant level. The lubricant level should be just at the bottom of the hole. If not, add the appropriate lubricant through the opening.
3   Install and tighten the plug and check for leaks after the first few miles of driving.

18 Differential lubricant level check (every 15,000 miles [24,000 km] or 6 months)

Refer to illustration 18.2
Warning: *On Sequoia models equipped with rear height control suspension, adjust the height control to the NORMAL mode, turn OFF the height control, then turn off the engine BEFORE raising the vehicle.*
Note: *4WD models covered by this manual have two differentials; check the lubricant level in both differentials.*
1   The differential has a check/fill plug which must be removed to check the lubricant level. If the vehicle must be raised to gain access to the plug, support it safely on

jackstands - DO NOT crawl under the vehicle when it's supported only by the jack.
2   Remove the oil check/fill plug from the back of the rear differential or the front of the front differential **(see illustration)**. On some models a tag is located in the area of the plug which gives information regarding lubricant type, particularly on models equipped with a limited-slip differential.
3   Use a finger to reach inside the housing to determine the lubricant level. The oil level should be at the bottom of the plug opening. If it isn't, use a hand pump (available at auto parts stores) to add the specified lubricant until it just starts to run out of the opening.
4   Install the plug and tighten it securely.

19 Seat belt check (every 15,000 miles [24,000 km] or 12 months)

1   Check the seat belts, buckles, latch plates and guide loops for any obvious damage or signs of wear.
2   Make sure the seat belt reminder light comes on when the key is turned on.
3   The seat belts are designed to lock up during a sudden stop or impact, yet allow free movement during normal driving. The retractors should hold the belt against your chest while driving and rewind the belt when the buckle is unlatched.
4   If any of the above checks reveal problems with the seat belt system, replace parts as necessary.

20 Driveaxle boot check (every 15,000 miles [24,000 km] or 12 months)

Refer to illustration 20.2
1   The driveaxle boots are important because they prevent dirt, water and foreign

material from entering and damaging the constant velocity joints (CV). Oil and grease can cause the boot material to deteriorate prematurely, so it's a good idea to wash the boots with soap and water.
2   Inspect the boots for tears and cracks as well as loose clamps **(see illustration)**. If there is any evidence of cracks or leaking grease, they must be replaced (see Chapter 8).

21 Positive Crankcase Ventilation (PCV) valve check and replacement (every 15,000 miles or 12 months)

The Positive Crankcase Ventilation valve on these vehicles is not a routine maintenance item. For more information on the PCV system, refer to Chapter 6.

22 Air filter check and replacement (every 15,000 miles [24,000 km] or 12 months)

Caution: *Never drive the vehicle with the air filter removed. Excessive engine wear could result and backfiring could even cause a fire under the hood.*
1   At the specified intervals, the air filter should be replaced with a new one. A thorough preventive maintenance schedule would also require the filter to be inspected between filter changes.

1GR-FE V6 engines

2010 and earlier models

Refer to illustrations 22.3 and 22.4
2   The air filter housing is located on the right side of the intake manifold plenum.
3   Open the two retaining clips and swing

22.3 To remove the air filter housing, open these two clips and swing open the housing (the firewall end of the housing is hinged) – 2010 and earlier models

22.4 Note how the old air filter is installed, then pull it out of the air filter housing. The new filter must be installed in exactly the same way – 2010 and earlier models

open the air filter housing **(see illustration)**.
4   Remove the air filter element from the housing **(see illustration)**, then wipe out the inside of the housing with a clean rag.
5   Inspect the outer surface of the filter element. If it's dirty, replace it. If it's only moderately dusty, blow it out from the back to the front surface with low-pressure compressed air. Because the filter element is a pleated paper type filter, it cannot be oiled or washed. If you can't clean the filter element satisfactorily with compressed air, replace it.
6   Installation is the reverse of removal.

2011 and later models

7   The air filter housing is mounted in the right front corner of the engine compartment.
8   Remove the cover retaining screws and detach all hoses that might interfere with the removal of the air filter cover from the housing. While the top cover is off, be careful not to drop anything down into the housing.
9   Lift the air filter element out of the housing and wipe out the inside of the housing with a clean rag.

10   Inspect the outer surface of the filter element. If it is dirty, replace it. If it is only moderately dusty, it can be reused by blowing it clean from the back to the front surface with compressed air. Because it is a pleated paper type filter, it cannot be washed or oiled. If it cannot be cleaned satisfactorily with compressed air, discard and replace it.
11   Place the new filter in the air filter housing, making sure it seats properly.
12   Installation of the cover is the reverse of removal.

V8 engines
Refer to illustrations 22.14 and 22.16
13   The air filter housing is mounted in the right front corner of the engine compartment.
14   Unlatch the cover retaining clips **(see illustration)**.
15   Detach all hoses that might interfere with the removal of the air filter cover from the housing. While the top cover is off, be careful not to drop anything down into the housing.
16   Lift the air filter element out of the housing **(see illustration)** and wipe out the inside

of the housing with a clean rag.
17   Inspect the outer surface of the filter element. If it is dirty, replace it. If it is only moderately dusty, it can be reused by blowing it clean from the back to the front surface with compressed air. Because it is a pleated paper type filter, it cannot be washed or oiled. If it cannot be cleaned satisfactorily with compressed air, discard and replace it.
18   Place the new filter in the air filter housing, making sure it seats properly.
19   Installation of the cover is the reverse of removal.

23 Cabin air filter replacement (every 15,000 miles [24,000 km] or 12 months)

Refer to illustration 23.2, 23.3, 23.4 and 23.5
1   There is an air filter in the blower housing that cleans the air before it enters the passenger's compartment.
2   To remove the air filter, remove the inset

22.14 The first step in removing the air filter is detaching the clips on the housing – 3UR-FE V8 model shown, other V8 models similar

22.16 After raising the cover, the filter element can be lifted out of the housing

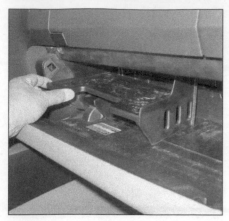

23.2 Unclip and remove the inset from the glove box

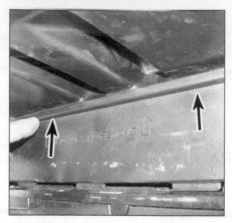

23.3 Slide the filter panel up until the tabs on the bottom of the panel can be seen

23.4 Cabin air filter cover mounting tabs

in the glove box **(see illustration)**.
Note: *On some models it may be necessary to remove the glove box (see Chapter 11).*
3   Slide the filter access panel up until the tabs on the bottom of the panel are visible then remove the panel **(see illustration)**.
4   Depress the filter cover mounting tabs and pull the cover off **(see illustration)**.
5   Remove the air filter element **(see illustration)**.
6   Installation is the reverse of removal.

---

24   Drivebelt check and replacement/ tensioner replacement (every 15,000 miles [24,000 km] or 12 months)

Warning: *Before checking or replacing a drivebelt, make sure the ignition key is not in the ignition lock cylinder.*
Warning: *On Sequoia models equipped with rear height control suspension, adjust the height control to the NORMAL mode, turn OFF the height control, then turn off the engine BEFORE raising the vehicle.*

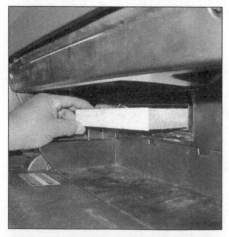

23.5 Slide the filter out from the housing

Check

Refer to illustration 24.2

1   The drivebelt is located at the front of the engine and plays an important role in the operation of the vehicle and its components. Due to its function and material makeup, the belt is prone to failure after a period of time and should be inspected periodically to prevent major damage. All engines covered in this manual use a single serpentine belt; no adjustment is necessary because an automatic tensioner is used.
2   With the engine turned off, open the hood and locate the drivebelt at the front of the engine. Use a flashlight to carefully check the ribs for separation from the adhesive rubber and for cracking or separation of the ribs,

torn or worn ribs or cracks in the inner ridges of the ribs **(see illustration)**.
3   Also check for fraying and glazing, which gives the belt a shiny appearance. Inspect both sides of the belt by twisting the belt to check the underside. Use your fingers to feel the belt where you can't see it. If any of the above conditions are evident, replace the belt.
Note: *On some models, drivebelt inspection can be made easier by removing the under-vehicle splash shield (see illustration 8.6a).*
4   On 2UZ-FE models, the belt should be replaced when the pointer on top of the automatic tensioner falls outside of the acceptable range.

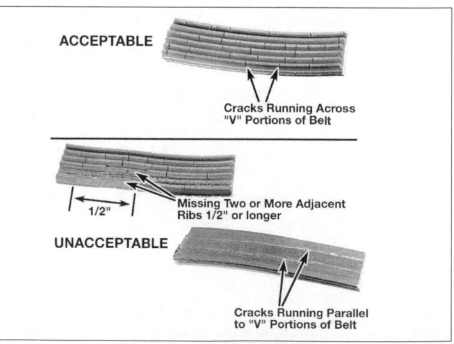

24.2 Small cracks in the underside of the V-ribbed belt are acceptable - lengthwise cracks or missing pieces are cause for replacement

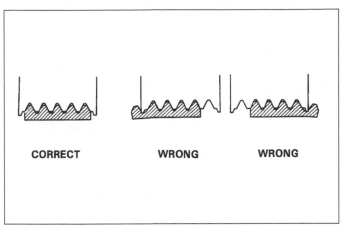

24.5 Make sure the belt is centered on all of the pulleys when installed

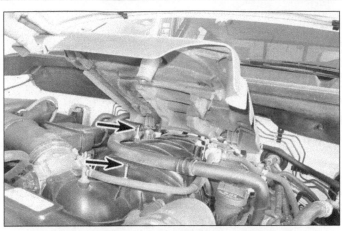

24.12a Lift the engine cover up until the grommets are free of the locking pins . . .

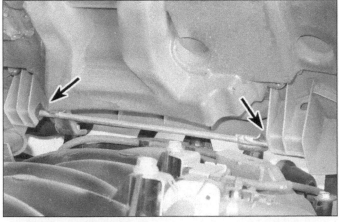

24.12b . . . then pull the cover forward until the hooks at the rear of the cover are free, and remove the cover

24.13 To release tension on the drivebelt, put a box wrench on the tensioner pulley bolt and rotate the tensioner counterclockwise

Replacement

2UZ-FE engine

Drivebelt

Refer to illustration 24.5

5   The automatic tensioner must be released to allow drivebelt replacement. Check to make sure the key has been removed from the ignition lock cylinder, then place a wrench on the bolt in the center of the tensioner pulley and rotate it counterclockwise to release tension on the belt. Remove the belt and slowly release the tensioner. Install the new belt and rotate the tensioner counterclockwise to allow the belt to slip over it, then release the tensioner slowly until it contacts the drivebelt. Make sure the drivebelt is centered on all of the pulleys **(see illustration)**.

Tensioner

6   Be sure the key is not in the ignition lock cylinder, then remove the drivebelt (see Step 5).
7   Remove the under-vehicle splash shield **(see illustrations 8.6a and 8.6b)**.
8   Working from above, remove the two tensioner mounting nuts.

9   Working from below, remove the tensioner mounting bolt, then remove the tensioner and guide it out towards the bottom of the vehicle.
10   Installation is the reverse of the removal procedure. Tighten the fasteners to the torque listed in this Chapter's Specifications.

All other engines

Drivebelt removal

Refer to illustrations 24.12a, 24.12b and 24.13

11   Disconnect the cable from the negative battery terminal.
12   Remove the engine cover **(see illustrations)** and the under-vehicle splash shield **(see illustrations 8.6a and 8.6b)**.
13   Working from below, put a box wrench on the belt tensioner pulley bolt, rotate the tensioner counterclockwise **(see illustration)** to release tension on the drivebelt and remove the belt from the tensioner.

Drivebelt installation (method one)

Refer to illustrations 24.14, 24.15a and 24.15b

14   Before installing the drivebelt, lock the tensioner as follows: turn the tensioner clock-

wise, align the two holes on the tensioner assembly and insert a 0.24-inch dowel pin or drill bit through the two holes **(see illustration)**.

24.14 To lock the belt tensioner while installing the drivebelt, turn the tensioner counterclockwise until the two holes are aligned, then insert a 0.24-inch dowel pin or drill bit through the holes

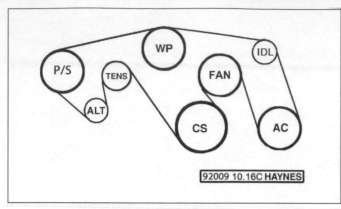

24.15a Serpentine drivebelt routing (1UR-FE, 3UR-FE and 3UR-FBE engines)

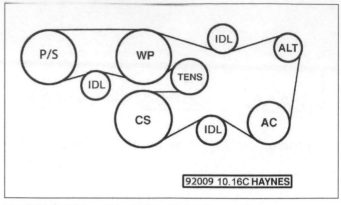

24.15b Serpentine drivebelt routing (1GR-FE engines)

15   Install the drivebelt. Be sure to route it correctly **(see illustrations)**. Remove the dowel pin or drill bit and tension the belt. Make sure that the belt is centered on the all of the pulleys **(see illustration 24.5)**.

Drivebelt installation (method two)

16   If you have difficulty installing the drivebelt using method one, try this method instead.
17   Release tension on the belt tensioner pulley, then route the belt over the pulleys, except for the power steering pump pulley.
18   Release belt tension by rotating the belt tensioner counterclockwise, put the belt on the power steering pump pulley, then tension the belt. Make sure that the belt is centered on all of the pulleys **(see illustration 24.5)**.
19   Installation is otherwise the reverse of removal.

Tensioner (V6 engines)

20   Remove the drivebelt (see Steps 11 through 13).
21   Remove the five tensioner mounting bolts from the side of the engine and remove the tensioner.
22   Installation is the reverse of removal. Tighten the bolts to the torque listed in this Chapter's Specifications.

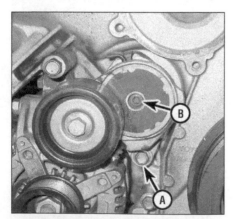

24.25 Remove the standard-head bolt (A), then remove the Allen head bolt (B) and remove the tensioner from the engine (1UR-FE, 3UR-FE and 3UR-FBE engines)

Tensioner (V8 engines)

Refer to illustration 24.25

23   Remove the drivebelt (see Steps 11 through 13).
24   Remove the fan shroud and water pump pulley (see Chapter 3).
25   Remove the two tensioner mounting bolts and remove the tensioner **(see illustration)**.
26   Installation is the reverse of removal. Tighten the bolts to the torque listed in this Chapter's Specifications.

---

25   Fuel system check (every 15,000 miles [24,000 km] or 12 months)

Warning: *Gasoline is extremely flammable, so take extra precautions when you work on any part of the fuel system. Don't smoke or allow open flames or bare light bulbs near the work area, and don't work in a garage where a gas-type appliance (such as a water heater or clothes dryer) is present. Since gasoline is carcinogenic, wear fuel-resistant gloves when there's a possibility of being exposed to fuel, and, if you spill any fuel on your skin, rinse it off immediately with soap and water. Mop up any spills immediately and do not store fuel-soaked rags where they could ignite. The fuel system is under constant pressure, so, if any fuel lines are to be disconnected, the fuel pressure in the system must be relieved first (see Chapter 4 for more information). When you perform any kind of work on the fuel system, wear safety glasses and have a Class B type fire extinguisher on hand.*
Warning: *On Sequoia models equipped with rear height control suspension, adjust the height control to the NORMAL mode, turn OFF the height control, then turn off the engine BEFORE raising the vehicle.*
1   The fuel system is most easily checked with the vehicle raised on a hoist so the components underneath the vehicle are readily visible and accessible.
2   If the smell of gasoline is noticed while driving or after the vehicle has been in the sun, the system should be thoroughly inspected immediately.

3   Remove the fuel tank cap and check for damage, corrosion and an unbroken sealing imprint on the gasket. Replace the cap with a new one if necessary.
4   With the vehicle raised and safely supported, inspect the gas tank and filler neck for punctures, cracks and other damage. The connection between the filler neck and the tank is particularly critical. Sometimes a rubber filler neck will leak because of loose clamps or deteriorated rubber. These are problems a home mechanic can usually rectify. **Warning:** *Do not, under any circumstances, try to repair a fuel tank (except rubber components). A welding torch or any open flame can easily cause fuel vapors inside the tank to explode.*
5   Carefully check all rubber hoses and metal lines leading away from the fuel tank. Check for loose connections, deteriorated hoses, crimped lines and other damage. Follow the lines to the front of the vehicle, carefully inspecting them all the way to the carburetor or fuel injection system. Repair or replace damaged sections as necessary.
6   If a fuel odor is still evident after the inspection, check the EVAP system (see Chapter 6).

---

26   Brake check (every 15,000 miles [24,000 km] or 12 months)

Warning: *The dust created by the brake system is harmful to your health. Never blow it out with compressed air and don't inhale any of it. An approved filtering mask should be worn when working on the brakes. Do not, under any circumstances, use petroleum-based solvents to clean brake parts. Use brake system cleaner only! Try to use non-asbestos replacement parts whenever possible.*
Warning: *On Sequoia models equipped with rear height control suspension, adjust the height control to the NORMAL mode, turn OFF the height control, then turn off the engine BEFORE raising the vehicle.*
Note: *For detailed photographs of the brake system, refer to Chapter 9.*
1   In addition to the specified intervals, the

brakes should be inspected every time the wheels are removed or whenever a defect is suspected. Any of the following symptoms could indicate a potential brake system defect: The vehicle pulls to one side when the brake pedal is depressed; the brakes make squealing or dragging noises when applied; brake pedal travel is excessive; the pedal pulsates; brake fluid leaks, usually onto the inside of the tire or wheel.

2   Loosen the wheel lug nuts.

3   Raise the vehicle and place it securely on jackstands.

4   Remove the wheels.

Disc brakes

Refer to illustrations 26.5 and 26.10

5   There are two pads (an outer and an inner) in each caliper. The pads are visible after the wheels are removed **(see illustration)**.

6   Measure the pad thickness. If the lining material is less than the minimum thickness listed in this Chapter's Specifications, replace the pads. **Note:** *Keep in mind that the lining material is riveted or bonded to a metal backing plate and the metal portion is not included in this measurement.*

7   If it is difficult to determine the exact thickness of the remaining pad material by the above method, or if you are at all concerned about the condition of the pads, remove the pads for further inspection (see Chapter 9).

8   Once the pads are removed, clean them with brake cleaner and re-measure them.

9   Measure the disc thickness with a micrometer to make sure that it still has service life remaining. If any disc is thinner than the specified minimum thickness, replace it (see Chapter 9). Even if the disc has service life remaining, check its condition. Look for scoring, gouging and burned spots. If these conditions exist, remove the disc and have it resurfaced (see Chapter 9).

10   Before installing the wheels, check all brake lines and hoses for damage, wear, deformation, cracks, corrosion, leakage, bends and twists, particularly in the vicinity of the rubber hoses at the calipers **(see illustration)**. Check the clamps for tightness and the connections for leakage. Make sure that all hoses and lines are clear of sharp edges, moving parts and the

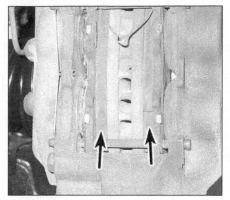

26.5 With the wheels removed, the brake pad lining can be inspected

exhaust system. If any of the above conditions are noted, repair, reroute or replace the lines and/or fittings as necessary (see Chapter 9). Tighten the lug nuts to the torque listed in this Chapter's Specifications after the vehicle has been lowered.

Brake booster check

11   Sit in the driver's seat and perform the following sequence of tests.

12   With the brake fully depressed, start the engine - the pedal should move down a little when the engine starts.

13   With the engine running, depress the brake pedal several times - the travel distance should not change.

14   Depress the brake, stop the engine and hold the pedal in for about 30 seconds - the pedal should neither sink nor rise.

15   Restart the engine, run it for about a minute and turn it off. Then firmly depress the brake several times - the pedal travel should decrease with each application.

16   If your brakes do not operate as described above when the preceding tests are performed, the brake booster has failed. Refer to Chapter 9 for the replacement procedure.

Parking brake

17   Slowly depress the parking brake pedal and count the number of clicks you hear until maximum travel has been reached. The adjustment should be within the specified number of

26.10 Check for any sign of brake fluid leakage at the brake line fittings and the brake hoses

clicks listed in this Chapter's Specifications. If you hear more or fewer clicks, it's time to adjust the parking brake (see Chapter 9).

18   An alternative method of checking the parking brake is to park the vehicle on a steep hill with the parking brake set and the transmission in Neutral (stay in the vehicle during this check!). If the parking brake cannot prevent the vehicle from rolling, it is in need of adjustment (see Chapter 9).

---

27   **Brake pedal height, freeplay and reserve height check, and brake light switch adjustment (every 15,000 miles [24,000 km] or 12 months)**

Pedal height

Refer to illustration 27.1

1   The height of the brake pedal is the distance the pedal sits off the floor **(see illustration)**. If the pedal height is not within the specified range, check for damaged to the brake pedal, pedal arm, the pedal mounting bracket and the dash panel. If the pedal height is incorrect, adjust the brake light switch, Do not try to adjust the pushrod length, this will alter the pedal force ratio.

Note: *If the pedal reserve height is OK then no changes are necessary.*

2   Before measuring the brake pedal height make sure the pedal is in the fully returned position, then start the engine and depress the accelerator several times to activate the power brake booster. Measure the pedal height and adjust the brake light switch if necessary.

Note: *All measurements for the pedal height should be done with the carpet pulled back and the brake pedal pad removed.*

Pedal freeplay

Refer to illustration 27.3

3   The freeplay is the pedal slack, or the distance the pedal can be depressed before it begins to have any effect on the brake sys-

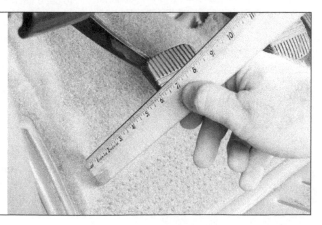

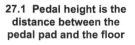

27.1 Pedal height is the distance between the pedal pad and the floor

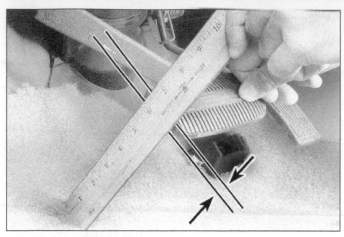

27.3 Pedal freeplay is the distance from the natural resting point of the pedal to the point at which resistance is felt

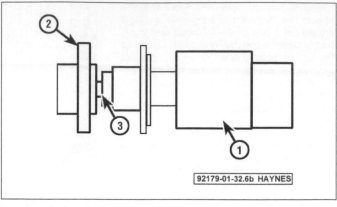

92179-01-32.6b HAYNES

27.8 Brake light switch details

1   *Brake light switch*   3   *Plunger extension*
2   *Brake pedal support*

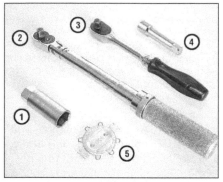

28.1 Tools required for changing spark plugs

1   **Spark plug socket** - *This will have special padding inside to protect the spark plug's porcelain insulator*
2   **Torque wrench** - *Although not mandatory, using this tool is the best way to ensure the plugs are tightened properly*
3   **Ratchet** - *Standard hand tool to fit the spark plug socket*
4   **Extension** - *Depending on model and accessories, you may need special extensions and universal joints to reach one or more of the plugs*
5   **Spark plug gap gauge** - *This gauge for checking the gap comes in a variety of styles. Make sure the gap for your engine is included*

tem **(see illustration)**. If the pedal height is not within the specified range, check for damaged to the brake pedal, pedal arm, the pedal mounting bracket and the dash panel. If the pedal height is incorrect, adjust the brake light switch. Do not try to adjust the pushrod length, this will alter the pedal force ratio.
Note: *If the pedal reserve height is OK, no changes are necessary.*
4   Before measuring the brake pedal freeplay, turn off the engine and depress the brake pedal a half dozen times. Measure the pedal

freeplay; if it isn't correct, adjust the brake light switch (see Step 6). If the freeplay is still incorrect, troubleshoot the brake system.

Pedal reserve height
5   The pedal reserve is the distance the pedal sits off the floor when it is fully depressed, measured from the center of the brake pedal pad to the dash panel. If the pedal reserve height is not within the specified range, check for damaged to the brake pedal, pedal arm, the pedal mounting bracket and the dash panel. If the reserve height is still incorrect, troubleshoot the brake system (see Chapter 9).

Brake light switch adjustment
Refer to illustration 27.8
6   Disconnect the electrical connector to the switch (see Chapter 9), and rotate the brake light switch counterclockwise and remove it.
7   Reinsert the brake light switch until the body of the switch contacts the cushion, then turn the switch 1/4-turn clockwise to lock it into position.
Note: *Support the brake pedal from below so it does not push the pedal inwards.*
8   Measure the how far the plunger on the brake light switch is extended **(see illustration)**. If the plunger length is not within the range listed in this Chapter's Specifications, remove the switch and pull the brake pedal out slightly and reinsert the switch and measure the plunger. Repeat this procedure until the correct distance is obtained.

28   Spark plug replacement (see Maintenance schedule for service interval)

Refer to illustrations 28.1, 28.4a, 28.4b and 28.4c
1   Spark plug replacement requires a spark plug socket, extension and ratchet. This socket is lined with a rubber grommet to protect the porcelain insulator of the spark plug and to

hold the plug while you insert it into the spark plug hole. You will also need a wire-type feeler gauge to check and adjust the spark plug gap and a torque wrench to tighten the new plugs to the specified torque **(see illustration)**.
2   If you are replacing the plugs, purchase the new plugs, adjust them to the proper gap and replace each plug one at a time. **Note:** *When buying new spark plugs, it's essential that you obtain the correct plugs for your specific vehicle. This information can be found in the Specifications Section in this Chapter's Specifications, on the Vehicle Emissions Control Information (VECI) label located on the underside of the hood or in the owner's manual. If these sources specify different plugs, purchase the spark plug type specified on the VECI label because that information is provided specifically for your engine.*
3   Inspect each of the new plugs for defects. If there are any signs of cracks in the porcelain insulator of a plug, don't use it.
4   Check the electrode gaps of the new plugs. Check the gap by inserting the wire gauge of the proper thickness between the electrodes at the tip of the plug **(see illustrations)**. The gap between the electrodes should be identical to that listed in this Chapter's Specifications. If the gap is incorrect, use the notched adjuster on the feeler gauge body to bend the curved side electrode slightly **(see illustration)**.
Caution: *Platinum and iridium spark plugs generally come pre-gapped. If you check the gap, treat them very gently and do not scratch the coating on the electrodes. Also, don't attempt to adjust the gap on used platinum or iridium spark plugs.*
5   If the side electrode is not exactly over the center electrode, use the notched adjuster to align them.

Removal
Refer to illustrations 28.7a, 28.7b and 28.8
6   If compressed air is available, blow any dirt or foreign material away from the coil/spark plug area before proceeding.

28.4a Spark plug manufacturers recommend using a wire type gauge when checking the gap - if the wire does not slide between the electrodes with a slight drag, adjustment is required

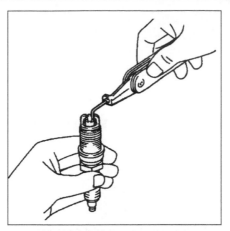

28.4b Some spark plugs have dual ground electrodes; on these plugs, check the gap between each ground electrode and the center electrode

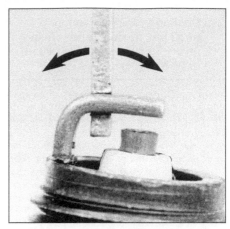

28.4c To change the gap, bend the side electrode only, as indicated by the arrows, and be very careful not to crack or chip the porcelain insulator surrounding the center electrode

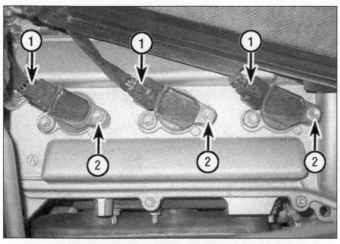

28.7a Ignition coil removal details (V6 engines)

1   *Depress the release tab and pull off the electrical connector*
2   *Remove the coil mounting bolt and pull the coil off the spark plug*

28.7b To remove an ignition coil from a V8, disconnect the electrical connector (A), then unscrew the mounting bolt (B) and twist the coil back-and-forth while pulling it up and out of the valve cover

7   The ignition coils are mounted directly over each spark plug. Remove the ignition coil **(see illustrations)**.
Note: *Most models require multiple components to be removed before the ignition coil(s) can be removed (see Chapter 5).*
8   Remove the spark plug **(see illustration)**.
9   Whether you are replacing the plugs at this time or intend to reuse the old plugs, compare each old spark plug with the chart shown on the inside back cover of this manual to determine the overall running condition of the engine.

Installation

Refer to illustrations 28.10a and 28.10b
10   Prior to installation, apply a coat of anti-seize compound to the plug threads. It's often difficult to insert spark plugs into their holes without cross-threading them. To avoid this possibility, fit a short piece of rubber hose over the end of the spark plug **(see illustrations)**. The flexible hose acts as a universal joint to help align the plug with the plug hole. Should the plug begin to cross-thread, the hose will slip on the spark plug, preventing thread damage. Tighten the plug to the torque listed in this Chapter's Specifications.
11   Attach the coil to the new spark plug, again using a twisting motion until it is firmly seated on the end of the spark plug then tighten the mounting bolt(s) securely.
12   Follow the above procedure for the remaining spark plugs, replacing them one at a time to prevent mixing up the ignition coils.

28.8 Use a socket with a long extension to unscrew the spark plugs

29 Cooling system servicing (draining, flushing and refilling) (every 30,000 miles [48,000 km] or 24 months)

Warning: *Do not allow antifreeze to come in contact with your skin or painted surfaces of the vehicle. Rinse off spills immediately with plenty of water. Antifreeze is highly toxic if ingested. Never leave antifreeze lying around in an open container or in puddles on the floor; children and pets are attracted by it's sweet smell and may drink it. Check with local authorities about disposing of used antifreeze. Many communities have collection centers which will see that antifreeze is disposed of safely.*

Warning: *On Sequoia models equipped with rear height control suspension, adjust the height control to the NORMAL mode, turn OFF the height control, then turn off the engine BEFORE raising the vehicle.*

1   Periodically, the cooling system should be drained, flushed and refilled to replenish the antifreeze mixture and prevent formation of rust and corrosion, which can impair the performance of the cooling system and cause engine damage. When the cooling system is serviced, all hoses and the radiator cap should be checked and replaced if necessary.

Draining

Refer to illustrations 29.3a and 29.3b

2   Apply the parking brake and block the wheels.

Warning: *If the vehicle has just been driven, wait several hours to allow the engine to cool down before beginning this procedure.* Remove the under-vehicle splash shield **(see illustrations 8.6a and 8.6b).**

3   Move a large container under the radiator drain to catch the coolant. The radiator drain plug is located on the left (driver's) side lower corner of the radiator **(see illustrations).** Unscrew the drain plug until coolant starts flowing from the drain hole (a pair of pliers may be required to turn it).

4   Remove the radiator cap and allow the radiator to drain, then, move the container

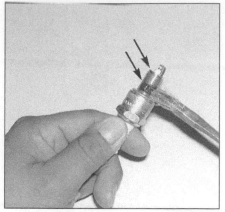

28.10a Apply a thin coat of anti-seize compound to the spark plug threads (but not close to the electrode end)

under the engine. Loosen the engine block drain plug(s) and allow the coolant in the block to drain **(see illustration 29.3b).**

Note: *Frequently, the coolant will not drain from the block after the plug is removed. This is due to a rust layer that has built up behind the plug. Insert a thin screwdriver or punch into the hole to break the rust barrier.*

5   While the coolant is draining, check the condition of the radiator hoses, heater hoses and clamps (refer to Section 10 if necessary).

6   Replace any damaged clamps or hoses. Close the drain plugs.

Flushing

7   Close the radiator and engine block drain plugs. Fill the cooling system with clean water, following the *Refilling* procedure (see Step 13).

8   Start the engine and allow it to reach normal operating temperature, then rev up the engine a few times.

9   Turn the engine off and allow it to cool completely, then drain the system as described earlier.

10   Repeat Steps 8 and 9 until the water being drained is free of contaminants.

11   In severe cases of contamination or clogging of the radiator, remove the radiator (see

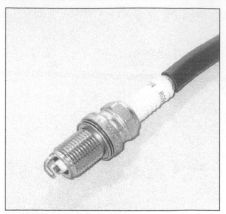

28.10b A length of rubber hose will save time and prevent damaged threads when installing the spark plugs

Chapter 3) and have a radiator repair facility clean and repair it if necessary.

12   Many deposits can be removed by the chemical action of a cleaner available at auto parts stores. Follow the procedure outlined in the manufacturer's instructions.

Note: *When the coolant is regularly drained and the system refilled with the correct antifreeze/water mixture, there should be no need to use chemical cleaners or descalers.*

Refilling

13   Reconnect the upper radiator hose and reinstall the thermostat.

14   Place the heater temperature control in the maximum heat position, if not already done.

15   Slowly add new coolant (a 50/50 mixture of water and antifreeze) to the radiator until it's full. Add coolant to the reservoir up to the lower mark.

16   Install the radiator cap and run the engine at approximately 2,000 to 2,500 rpm in a well-ventilated area until the thermostat opens (coolant will begin flowing through the radiator and the upper radiator hose will become hot).

17   Turn the engine off and let it cool. Add more coolant mixture to bring the level back up to the lip on the radiator filler neck.

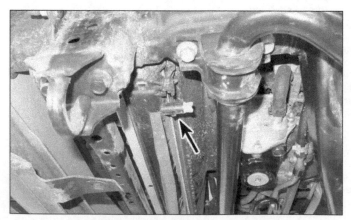

29.3a Radiator drain plug location

29.3b Typical coolant drain plug location - 3UZ-FE V8 engine shown, others similar

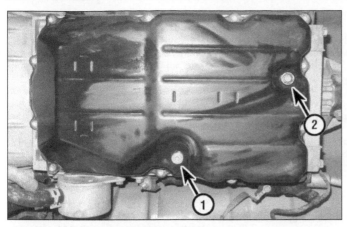

30.5 The transmission drain plug is located on the bottom of the transmission pan – the overflow plug is also on the bottom of the pan. The overflow plug is round with an Allen head

1   *Overflow or check plug*   2   *Drain plug*

30.8 Use a rubber mallet to carefully tap on the pan to break the seal between the pan and the transmission case, then carefully lower the pan from the vehicle

18   Squeeze the upper radiator hose to expel air, then add more coolant mixture if necessary. Reinstall the radiator cap and the under-vehicle splash shield. Add coolant to the reservoir, if necessary.
19   Start the engine, allow it to reach normal operating temperature and check for leaks.

30   Automatic transmission fluid and filter change (every 60,000 miles [96,000 km] or 48 months)

Refer to illustrations 30.5, 30.8, 30.9a, 30.9b, 30.11 and 30.12

Warning: *On Sequoia models equipped with rear height control suspension, adjust the height control to the NORMAL mode, turn OFF the height control, then turn off the engine BEFORE raising the vehicle.*
1   At the specified intervals, the transmission fluid should be drained and replaced. Since the fluid will remain hot long after driving, perform this procedure only after the engine has cooled down completely.
2   Before beginning work, purchase the specified transmission fluid (see *Recom-*

mended lubricants and fluids in this Chapter's Specifications) and a new filter.
3   Other tools necessary for this job include a floor jack, jackstands to support the vehicle in a raised position, a drain pan capable of holding at least eight quarts, newspapers and clean rags.
4   Raise the vehicle and support it securely on jackstands.
5   Place the drain pan underneath the transmission pan and remove the drain plug **(see illustration)**. Allow the fluid to completely drain from the transmission, then reinstall the drain plug.
6   Detach the transmission pan rock shield (if equipped) and remove the pan mounting bolts from the outer edges of the pan.
7   Using a rubber mallet, carefully tap on the pan to break the layer of gasket sealant between the pan and the transmission case.
Note: *Prying between the pan and the transmission case with a screwdriver or similar tool may result in damage to the sealing surface on the transmission case.*
8   Lower the pan **(see illustration)**. Once the pan is free, drain any remaining transmission fluid from the pan.
9   Remove the filter retaining bolts from the valve body and remove the filter **(see illustra-**

tion**)**, then remove the O-ring from the valve body **(see illustration)**.
Note: *On some models different length bolts are used; note the length and locations of the bolts as you remove them.*
10   Thoroughly inspect the bottom of the pan, the filter and the fluid. Although normally bright red, transmission fluid may turn dark red or brown during normal use. If you find the fluid very dark colored, or if it smells burned, it usually indicates the transmission has been overheated. If you find small pieces of metal or clutch material in the pan or filter, it indicates wear or damage have occurred to the internal parts or clutches. If you have any concerns about the condition of your transmission based on what you find in the fluid, pan and filter, it's a good idea to take your vehicle to your dealer or a transmission shop for further evaluation.
11   Clean the pan with solvent and dry it. Use a gasket scraper to remove any traces of old gasket material remaining on the transmission case or valve body.
Note: *Be very careful not to gouge the delicate gasket surfaces.*
Install a new O-ring on the filter, then install the filter, tightening the bolts to the torque listed in this Chapter's Specifications **(see illustration)**.

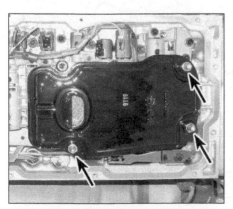

30.9a Remove the filter bolts and lower the filter

30.9b Remove the O-ring from the valve body

30.11 Install a new O-ring around the filter

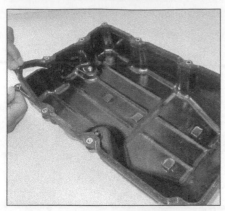

30.12 Place the gasket onto the pan, making sure the metal inserts in the gasket are aligned with the holes in the pan

12   Be sure the gasket surface on the transmission pan is clean, then install the gasket to the oil pan **(see illustration)**. Place the pan against the transmission case and, working around the pan, tighten each bolt a little at a time to the torque listed in this Chapter's Specifications.

13   Refer to Section 6 for the refilling procedure.

14   Check under the vehicle for leaks during the first few trips. Check the fluid level again when the transmission is hot (see Section 6).

31 Transfer case lubricant change (4WD models) (every 30,000 miles [48,000 km] or 24 months)

Warning: *On Sequoia models equipped with rear height control suspension, adjust the height control to the NORMAL mode, turn OFF the height control, then turn off the engine BEFORE raising the vehicle.*

1   Drive the vehicle for at least 15 minutes to warm the lubricant in the case. Perform this warm-up procedure with 4WD engaged, if possible. Use all gears, including Reverse, to ensure the lubricant is sufficiently warm to drain completely.

2   Raise the vehicle and support it securely on jackstands.

3   Remove the transfer case shield bolts and shield to expose the drain plug.

4   Remove the drain plug from the lower part of the case and allow the old lubricant to drain completely.

5   After the lubricant has drained completely, reinstall the plug and tighten it securely.

6   Remove the filler plug from the case.

7   Fill the case with the specified lubricant until it is level with the lower edge of the filler hole.

8   Install the filler plug and tighten it securely.

9   Drive the vehicle for a short distance and recheck the lubricant level. In some instances a small amount of additional lubricant will have to be added.

32 Differential lubricant change (every 30,000 miles [48,000 km] or 24 months)

Warning: *On Sequoia models equipped with rear height control suspension, adjust the height control to the NORMAL mode, turn OFF the height control, then turn off the engine BEFORE raising the vehicle.*

Note: *The following procedure applies to the front and rear differentials.*

1   Drive the vehicle for several miles to warm up the differential oil, then raise the vehicle and support it securely on jackstands.

2   Move a drain pan, rags, newspapers and the proper tools under the vehicle.

3   Remove the differential check/fill plug (see Section 18). With the drain pan under the differential, use a socket and ratchet to loosen the drain plug. It's the lower of the two plugs **(see illustration 18.2)**.

4   Once loosened, carefully unscrew it with your fingers until you can remove it from the case.

5   Allow all of the oil to drain into the pan, then replace the drain plug and tighten it securely.

6   Feel with your hands along the bottom of the drain pan for any metal bits that may have come out with the oil. If there are any, it's a sign of excessive wear, indicating that the internal components should be carefully inspected in the near future.

7   Using a hand pump, syringe or funnel, fill the differential with the correct amount and grade of oil (see this Chapter's Specifications) until the level is just at the bottom of the plug hole.

8   Reinstall the plug and tighten it securely.

9   Lower the vehicle. Check for leaks at the drain plug after the first few miles of driving.

33 Valve clearance check and adjustment (every 60,000 miles [96,000 km] or 48 months)

Note: *This procedure applies to the 4.0L V6 engine (1GR-FE) and the 4.7L V8 engine (2UZ-FE). The 4.6L V8 engine (1UR-FE) and the 5.7L V8 engine (3UR-FE and 3UR-FBE) utilize hydraulic valve lash adjusters.*

1   Position the number 1 piston at TDC on the compression stroke (see Chapter 2A or 2B).

2   Disconnect the cable from the negative terminal of the battery.

3   Remove the coils (see Chapter 5) and any other components that will interfere with valve cover removal.

4   Blow out the recessed area around the spark plug openings with compressed air, if available, to remove any debris that might fall into the cylinders, then remove the spark plugs (see Section 28).

5   Remove the valve cover (see Chapter 2A or 2B).

Check

V6 engine

Refer to illustrations 33.6a, 33.6b, 33.7a and 33.7b

6   Measure the clearances of the indicated valves with feeler gauges **(see illustrations)**. Record the measurements which are out of specification. They will be used

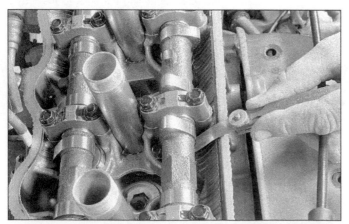

33.6a Check the clearance of each valve with a feeler gauge of the specified thickness - if the clearance is correct, you should feel a slight drag on the gauge as you pull it out

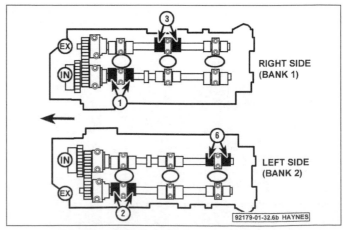

33.6b With the No. 1 piston at TDC on the compression stroke, check the indicated valves (1GR-FE V6 engine)

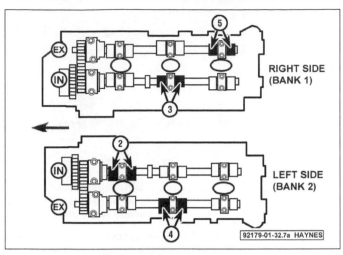

33.7a Rotate the crankshaft 2/3 turn (240 degrees) and check the indicated valves (1GR-FE V6 engine)

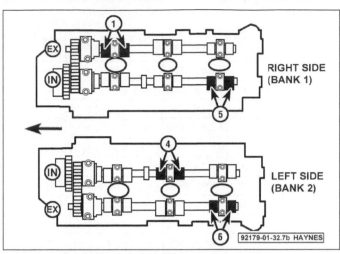

33.7b Rotate the crankshaft 2/3 turn (240 degrees) and check the remaining valves indicated (1GR-FE V6 engine)

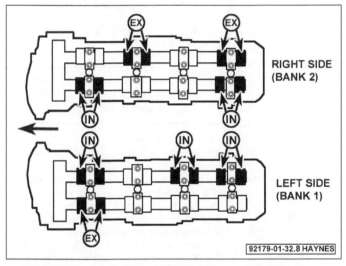

33.8 With the No. 1 piston at TDC on the compression stroke, check the valves indicated by the blackened cam lobes (2UZ-FE V8 engine)

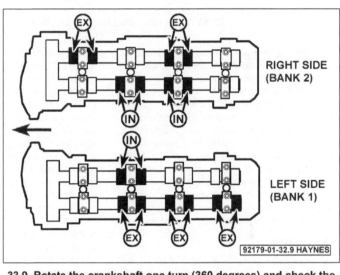

33.9 Rotate the crankshaft one turn (360 degrees) and check the remaining valves indicated by the blackened cam lobes (2UZ-FE V8 engine)

later to determine the required replacement shims.

7   Turn the crankshaft 2/3 revolution (240 degrees) and check the indicated valves **(see illustration)**. Turn the crankshaft 2/3 revolution again and check the next group of valves **(see illustration)**. Make notes of the cylinder number and the valve type (intake or exhaust) that need to be adjusted.

V8 engine

Refer to illustrations 33.8 and 33.9

8   Measure the clearances of the indicated valves with feeler gauges **(see illustration)**. Record the measurements which are out of specification. They will be used later to determine the required replacement shims.

9   Turn the crankshaft one revolution (360 degrees) and check the remaining valves **(see illustration)**.

Adjustment

V6 engine

Refer to illustrations 33.11 and 33.13

10   Remove the camshaft(s) for the valve(s) that you intend to adjust (see Chapter 2A).

11   Remove and measure each lifter (whose clearance is not correct) with a micrometer **(see illustration)**. Put each lifter back into its bore in the cylinder head before moving on to the next lifter. Record the measurement for each lifter.

12   To calculate the correct thickness of a replacement lifter that will put the valve clearance within the specified range, use the following formula:

$N = T + (A - V)$, where:
N = *thickness of the new lifter*
T = *thickness of the old lifter*
A = *measured valve clearance*
V = *specified valve clearance (see this Chapter's Specifications)*

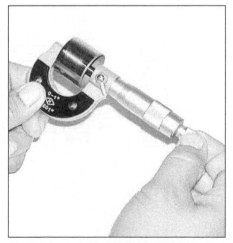

33.11 On V6 engines, measure the thickness of each lifter head with a micrometer

New lifter thickness mm (in.)					
Lifter No.	Thickness	Lifter No.	Thickness	Lifter No.	Thickness
06	5.060 (0.1992)	30	5.300 (0.2087)	54	5.540 (0.2181)
08	5.080 (0.2000)	32	5.320 (0.2094)	56	5.560 (0.2189)
10	5.100 (0.2008)	34	5.340 (0.2102)	58	5.580 (0.2197)
12	5.120 (0.2016)	36	5.360 (0.2110)	60	5.600 (0.2205)
14	5.140 (0.2024)	38	5.380 (0.2118)	62	5.620 (0.2213)
16	5.160 (0.2031)	40	5.400 (0.2126)	64	5.640 (0.2220)
18	5.180 (0.2039)	42	5.420 (0.2134)	66	5.660 (0.2228)
20	5.200 (0.2047)	44	5.440 (0.2142)	68	5.680 (0.2236)
22	5.220 (0.2055)	46	5.460 (0.2150)	70	5.700 (0.2244)
24	5.240 (0.2063)	48	5.480 (0.2157)	72	5.720 (0.2252)
26	5.260 (0.2071)	50	5.500 (0.2165)	74	5.740 (0.2260)
28	5.280 (0.2079)	52	5.520 (0.2173)		

33.13 Valve lifter thickness chart (V6 models)

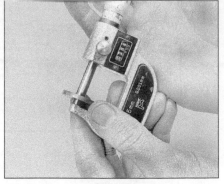

33.19 Measure the shim thickness with a micrometer (V8 engine)

13   Select a lifter with a thickness as close as possible to the calculated valve clearance. Lifters for this V6 engine are available in 35 sizes in increments of 0.008-inch (0.020 mm), and range in size from 0.1992-inch (5.060 mm) to 0.2260-inch (5.740 mm) **(see illustration)**.

14   Install the camshaft(s) (see Chapter 2A).

15   After the camshafts are reinstalled, check the valve clearances again to verify that they're now within the range of clearance listed in this Chapter's Specifications.

16   Installation of the valve cover(s), spark plugs, ignition coils, air filter housing and/or intake manifold is the reverse of removal.

V8 engines

Refer to illustrations 33.19 and 33.21

17   Remove the camshafts (see Chapter 2B). **Note:** *It's only necessary to remove the camshaft(s) over any valve(s) whose clearance is incorrect.*

18   Remove the valve lifter(s) from the valves whose clearances were out of specification. Keep the lifters in order; they must be returned to the same bore they were removed from.

19   Remove the shim from the underside of the lifter. Clean the shim then measure its thickness with a micrometer **(see illustration)**.

20   Calculate the required thickness of the new shim by using the formula given in Step 12.

21   Select a shim with a thickness as close as possible to the valve clearance calculated. Shims, which are available in 41 sizes in increments of 0.0008-inch (0.020 mm), range in size from 0.0787-inch (2.000 mm) to 0.1102-inch (2.800 mm) **(see illustration)**. **Note:** *Through careful analysis of the shim sizes needed to bring the out-of-specification valve clearance within specification, it is often possible to simply move a shim that has to come out anyway to another valve lifter requiring a shim of that particular size, thereby reducing the number of new shims that must be purchased.*

22   Once the proper shims have been selected, apply a thin coat of engine assembly lube to the shim and stick it in place on the underside of the lifter.

23   Repeat this procedure until all the valves which are out of clearance have been corrected.

24   Reinstall the lifters and camshafts (see Chapter 2B).

34   Evaporative emissions control system check (every 60,000 miles [98,000 km] or 48 months)

Refer to illustration 34.2

1   The function of the evaporative emissions control system is to draw fuel vapors from the gas tank and fuel system, store them in a charcoal canister and route them to the intake manifold during normal engine operation.

2   The most common symptom of a fault in the evaporative emissions system is a strong fuel odor. If a fuel odor is detected, inspect the charcoal canister, located in the front of the engine compartment **(see illustration)**. Check the canister and all hoses for damage and deterioration.

3   The evaporative emissions control system is explained in more detail in Chapter 6.

New shim thickness					mm (in.)
Shim No.	Thickness	Shim No.	Thickness	Shim No.	Thickness
00	2.000 (0.0787)	28	2.280 (0.0898)	56	2.560 (0.1008)
02	2.020 (0.0795)	30	2.300 (0.0906)	58	2.580 (0.1016)
04	2.040 (0.0803)	32	2.320 (0.0913)	60	2.600 (0.1024)
06	2.060 (0.0811)	34	2.340 (0.0921)	62	2.620 (0.1031)
08	2.080 (0.0819)	36	2.360 (0.0929)	64	2.640 (0.1039)
10	2.100 (0.0827)	38	2.380 (0.0937)	66	2.660 (0.1047)
12	2.120 (0.0835)	40	2.400 (0.0945)	68	2.680 (0.1055)
14	2.140 (0.0843)	42	2.420 (0.0953)	70	2.700 (0.1063)
16	2.160 (0.0850)	44	2.440 (0.0961)	72	2.720 (0.1071)
18	2.180 (0.0858)	46	2.460 (0.0969)	74	2.740 (0.1079)
20	2.200 (0.0866)	48	2.480 (0.0976)	76	2.760 (0.1087)
22	2.220 (0.0874)	50	2.500 (0.0984)	78	2.780 (0.1094)
24	2.240 (0.0882)	52	2.520 (0.0992)	80	2.800 (0.1102)
26	2.260 (0.0890)	54	2.540 (0.1000)		

33.21 Valve adjusting shim thickness chart (V8 engine)

34.2 Check the evaporative emissions control canister and the hose connections for cracks and damage. The EVAP canister is mounted underneath the vehicle, near the fuel tank

Chapter 2 Part A
V6 engine

Contents

Specifications

General

Engine designation	1GR-FE
Displacement	4.0 liters
Cylinder numbers (timing chain end-to-transmission end)	
Right (passenger) side	1-3-5
Left (driver) side	2-4-6
Firing order	1-2-3-4-5-6

Warpage limits

Cylinder head	0.0031 inch (0.08 mm)
Intake manifold	
Intake plenum side	0.031 inch (0.79 mm)
Cylinder head side	0.008 inch (0.20 mm)
Exhaust manifolds	0.031 inch (0.79 mm)

```
┌──────────────────┐
│  ⑤         ⑥    │
│                  │
│  ③         ④    │
│                  │
│  ①         ②    │
└──────────────────┘
       92076-2B- HAYNES
```

Cylinder location diagram

Camshaft and related components

Valve clearance (engine cold)	See Chapter 1
Bearing journal diameter	
No. 1 journal	1.416 to 1.417 inches (35.966 to 35.991 mm)
Other journals	0.904 to 0.905 inch (22.962 to 22.987 mm)
Bearing oil clearance	
Standard	
Right side No. 1 (intake)	0.0003 to 0.0015 inch (0.008 to 0.038 mm)
Right side No. 1 (exhaust)	0.0016 to 0.0031 inch (0.041 to 0.079 mm)
Left side (intake)	0.0016 to 0.0031 inch (0.041 to 0.079 mm)
Others	0.0010 to 0.0024 inch (0.025 to 0.061 mm)
Service limit	
Right side No. 1 (intake)	0.0028 inch (0.07 mm)
Others	0.0039 inch (0.10 mm)
Lobe height	
Intake	
Standard	1.739 to 1.742 inches (44.172 to 44.247 mm)
Service limit	1.73 inches (43.942 mm)
Exhaust	
Standard	1.755 to 1.759 inches (44.577 to 44.679 mm)
Service limit	1.75 inches (44.45 mm)
Thrust clearance (endplay)	
Standard	0.016 to 0.035 inch (0.41 to 0.89 mm)
Service limit	0.0043 inch (0.11 mm)
Runout limit (total indicator reading)	0.0024 inch (0.06 mm)
Lifters	
Outside diameter	1.2191 to 1.220 inches (30.966 to 30.988 mm)
Bore diameter	1.220 to 1.221 inches (30.988 to 31.013 mm)
Lifter-to-bore (oil) clearance	
Standard	0.0013 to 0.0023 inch (0.033 to 0.059 mm)
Service limit	0.0031 inch (0.08 mm)

Timing chain

Timing chain stretch limit (15 pins) (No. 1 and both No. 2 chains)	5.780 inches (146.8 mm)
Timing gear assembly wear limits	
Larger gear (No. 1 chain installed)	4.55 inches (115.5 mm)
Smaller (No. 2 chain installed)	2.88 inches (73.1 mm)
Timing sprocket wear limits	
Camshaft sprocket (No. 2 chain installed)	2.88 inches (73.1 mm)
Crankshaft sprocket (No. 1 chain installed)	2.40 inches (61.0 mm)
Idle gear wear limits	
With No. 1 chain installed	2.40 inches (61.0 mm)
Shaft diameter	0.905 to 0.906 inch (22.987 to 23.01 mm)
Inside diameter	0.906 to 0.907 inch (23.01 to 23.04 mm)
Oil clearance	
Standard	0.000787 to 0.00169 inch (0.020 to 0.043 mm)
Maximum	0.00366 inch (0.093 mm)
Chain tensioner No. 2 wear limit	0.039 inch (1.0 mm)
Chain tensioner slipper wear limit	0.039 inch (1.0 mm)
Vibration damper No. 1 and No. 2 wear limit	0.039 inch (1.0 mm)

Oil pump

Driven rotor-to-pump body clearance	
Standard	0.0098 to 0.0128 inch (0.250 mm to 0.325 mm)
Service limit	0.0128 inch (0.325 mm)
Rotor tip clearance	
Standard	0.0024 to 0.0063 inch (0.06 mm to 0.16 mm)
Service limit	0.0063 inch (0.16 mm)
Rotor side clearance	
Standard	0.0012 to 0.0035 inch (0.030 to 0.089 mm)
Service limit	0.0035 inch (0.089 mm)

Torque specifications

Note: *One foot-pound (ft-lb) of torque is equivalent to 12 inch-pounds (in-lbs) of torque. Torque values below approximately 15 foot-pounds are expressed in inch-pounds, because most foot-pound torque wrenches are not accurate at these smaller values.*

	Ft-lbs (unless otherwise indicated)	Nm
Intake manifold assembly		
Upper intake manifold bolts/nuts	21	28
Lower intake manifold bolts/nuts		
2010 and earlier models	19	26
2011 and later models	15	21
Valve cover bolts/nuts		
2009 and earlier models		
Bolt A	84 in-lbs	10
Bolt B and nut	80 in-lbs	9
2010 and later models		
Bolts A and D	84 in-lbs	10
Bolts B and C	15	21
Exhaust manifold nuts	15	21
Crankshaft pulley bolt		
2010 and earlier models	184	250
2011 and later models	192	260
Timing chain cover bolts/nuts	17	23
Idler pulley bolts		
Idler pulley No. 1 bolt	40	54
Idler pulley No. 2 bolt	29	39
Drivebelt tensioner mounting bolts	27	36
Camshaft timing gear bolt (applies to timing gear and timing sprocket)	74	100
Camshaft bearing cap bolts		
10 mm bolts	80 in-lbs	9
12 mm bolts	18	24
Cylinder head bolts (in sequence - **see illustrations 10.24a and 10.24b**)		
Step 1	27	36
Step 2	Tighten an additional 90 degrees	
Step 3	Tighten an additional 90 degrees	
Left cylinder head (two front bolts)	22	30
Oil filter assembly		
Oil filter bracket mounting bolts	168 in-lbs	14
Oil cooler-to-oil filter bracket union bolt	50	68
Oil pan bolts/nuts		
Oil pan No. 1		
All bolts & nuts except 2 bolts at flywheel/driveplate end of pan	15	21
2 bolts at flywheel/driveplate end of pan	84 in-lbs	10
Oil pan No. 2		
Bolts	80 in-lbs	9
Nuts	84 in-lbs	10
Oil pick-up tube nuts	80 in-lbs	9
Oil pump assembly		
Oil pump cover bolts	80 in-lbs	9
Oil pipe flange bolts	80 in-lbs	9
Oil pump relief valve plug	36	49
Driveplate bolts	61	83
Engine rear oil seal retainer		
Bolts	84 in-lbs	10
Nuts	80 in-lbs	9
Timing chain assembly		
Timing chain tensioner bolts		
Chain tensioner No. 1	84 in-lbs	10
Chain tensioner No. 2 and No. 3	15	21
Timing chain (guide) vibration damper No. 1 mounting bolts	168 in-lbs	19

1   General information

This Part of Chapter 2 is devoted to in-vehicle repair procedures for the 4.0L (1GR-FE) V6 engine. All information concerning engine removal and installation and engine overhaul can be found in Part C of this Chapter.

The following repair procedures are based on the assumption that the engine is installed in the vehicle. If the engine has been removed from the vehicle and mounted on a stand, many of the steps outlined in this Part of Chapter 2 will not apply.

2   Repair operations possible with the engine in the vehicle

1   Many major repair operations can be accomplished without removing the engine from the vehicle. Clean the engine compartment and the exterior of the engine with some type of degreaser before any work is done. It will make the job easier and help keep dirt out of the internal areas of the engine.

2   Depending on the components involved, it may be helpful to remove the hood to improve access to the engine as repairs are performed (refer to Chapter 11 if necessary). Cover the fenders to prevent damage to the paint. Special pads are available, but an old bedspread or blanket will also work.

3   If vacuum, exhaust, oil or coolant leaks develop, indicating a need for gasket or seal replacement, the repairs can generally be made with the engine in the vehicle. The intake and exhaust manifold gaskets, oil pan gasket, crankshaft oil seals and cylinder head gaskets are all accessible with the engine in place.

4   Exterior engine components, such as the intake and exhaust manifolds, the oil pan, the oil pump, the water pump, the starter motor, the alternator and the fuel system components can be removed for repair with the engine in place. **Note:** *On 4WD models, it will be necessary to either remove the front axle/differential assembly (see Chapter 8) or remove the engine from the vehicle (see Chapter 2C) in order to access the oil pan and the oil pump.*

5   Since the cylinder heads can be removed without pulling the engine, valve component servicing can also be accomplished with the engine in the vehicle. Replacement of the camshafts, timing belt and pulleys is also possible with the engine in the vehicle.

6   In extreme cases caused by a lack of necessary equipment, repair or replacement of piston rings, pistons, connecting rods and rod bearings is possible with the engine in the vehicle. However, this practice is not recommended because of the cleaning and preparation work that must be done to the components involved.

3.7 Turn the crankshaft until the notch in the pulley aligns with the zero on the timing plate

3   Top Dead Center (TDC) for number one piston - locating

Refer to illustration 3.7

1   Top Dead Center (TDC) is the highest point in the cylinder that each piston reaches as it travels up the cylinder bore. Each piston reaches TDC on the compression stroke and again on the exhaust stroke, but TDC generally refers to piston position on the compression stroke.

2   Positioning the piston(s) at TDC is an essential part of certain procedures such as valve adjustment and timing chain/camshaft removal and installation.

3   Before beginning this procedure, remove the Number One spark plug (see Chapter 1, if necessary). Place the transmission in Park and apply the parking brake or block the rear wheels. Disable the ignition system by disconnecting the electrical connector from each ignition coil. If method b) or c) will be used in the next step, disable the fuel pump (see Chapter 4, Section 3).

4   In order to bring any piston to TDC, the crankshaft must be turned using one of the methods outlined below. When looking at the front of the engine, normal crankshaft rotation is clockwise.

a)   *The preferred method is to turn the crankshaft with a socket and ratchet attached to the bolt threaded into the front of the crankshaft. Turn the crankshaft in a clockwise direction only.*

b)   *A remote starter switch, which may save some time, can also be used. Follow the instructions included with the switch. Once the piston is close to TDC, use a socket and ratchet as described in the previous paragraph.*

c)   *If an assistant is available to turn the ignition switch to the Start position in short bursts, you can get the piston close to TDC without a remote starter switch. Make sure your assistant is out of the vehicle, away from the ignition switch, then use a socket and ratchet as described in Paragraph (a) to complete the procedure.*

5   This engine does not have a distributor, but rather a separate coil for each cylinder. If not already done, disconnect the wires from the coils and remove the Number One spark plug (see Chapter 1).

6   Install a compression pressure gauge in the number one spark plug hole (see Chapter 2C). It should be a gauge with a screw-in fitting and a hose at least six inches long.

7   Rotate the crankshaft using one of the methods described above while observing the compression gauge. When TDC for the compression stroke of number one cylinder is reached, compression pressure will show on the gauge as the marks on the crankshaft pulley are beginning to line up **(see illustration)**. If you go past the marks, release the gauge pressure and rotate the crankshaft around two more revolutions.

8   After the number one piston has been positioned at TDC on the compression stroke, TDC for the next cylinder in the firing order can be located by turning the crankshaft another 120-degrees (refer to the firing order in this Chapter's Specifications).

4   Valve covers - removal and installation

Refer to illustrations 4.7, 4.9, 4.10 and 4.11

1   Disconnect the cable from the negative terminal of the battery (see Chapter 5).

2   Remove the engine cover.

3   Remove the air intake duct and the air filter housing (see *Air filter housing - removal and installation* in Chapter 4).

4   If you're going to remove the left valve cover, remove the upper intake manifold (see Section 5).

5   Disconnect the PCV ventilation hose from the valve cover.

6   Remove the ignition coils.

7   Remove the valve cover retaining bolts and nuts and remove the valve cover **(see illustration)**.

8   Remove and discard the valve cover gasket.

4.7 To detach the valve cover from the cylinder head, remove the ignition coils before removing the valve cover bolts

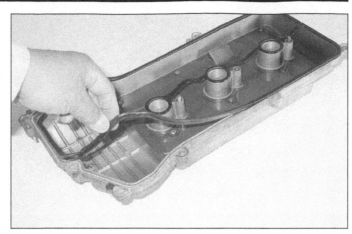

4.9 Make sure that the gasket is fully seated into the valve cover groove, and that the spark plug tube seals are seated as well

9   Installation is the reverse of removal. Inspect the main cover gasket and spark plug tube seals (which are part of the gasket), replacing them if necessary **(see illustration)**.
10   Put new seals on the three center bolts **(see illustration)**.
11   Apply dabs of RTV sealant to the joints where the timing chain cover meets the engine block **(see illustration)**.
12   Evenly tighten the valve cover nuts and bolts to the torque listed in this Chapter's Specifications.
13   Start the engine and check for oil leaks around the edges of the valve cover.

5   Intake manifold - removal and installation

Warning: *Wait until the engine is completely cool before beginning this procedure.*

Upper intake manifold

Refer to illustrations 5.11 and 5.12
Note: *Toyota refers to the upper intake mani-*

fold as the intake air surge tank. If you're buying a gasket for the upper intake manifold at a dealer parts department, use the Toyota terminology.
1   Disconnect the cable from the negative terminal of the battery (see Chapter 5).
2   Remove the engine cover.
3   Remove the air intake duct and the air filter housing (see *Air filter housing - removal and installation* in Chapter 4).
4   Clamp-off and disconnect the coolant hoses from the throttle body.
5   Disconnect the (EVAP system) fuel vapor feed hose.
6   Disconnect the ventilation hose. Disconnect the throttle body wiring.
7   Disconnect the two Vacuum Switching Valve (VSV) electrical connectors.
8   Remove the two throttle body bracket bolts and remove the bracket.
9   Remove the baffle plate.
10   Remove both bolts from each intake manifold support bracket (Toyota refers to these two support brackets as surge tank stays) and remove both brackets.
11   Remove the two intake manifold mount-

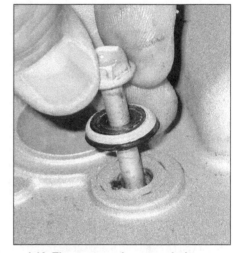

4.10 The center valve cover bolts use sealing washers

ing nuts and four mounting bolts and remove the upper intake manifold **(see illustration)**.
12   Check the condition of the upper intake

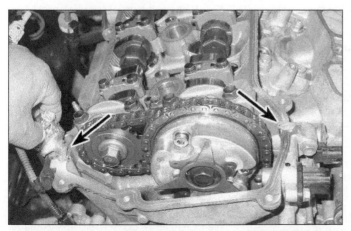

4.11 After removing the old sealant, apply a fresh dab of RTV sealant to the joints at the front of the valve cover where the cylinder head meets the timing chain cover

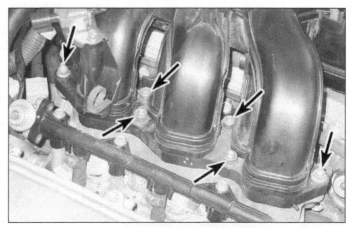

5.11 Upper intake manifold fasteners

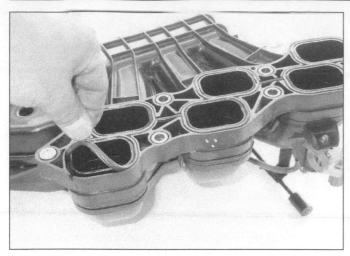

5.12 Check the condition of the upper intake manifold gasket, replacing it if necessary

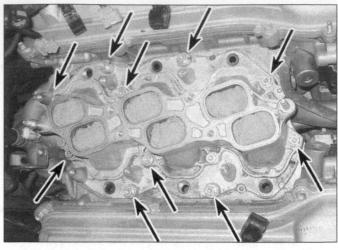

5.17 To detach the lower intake manifold, remove these ten mounting bolts (fuel rail removed for clarity)

manifold gasket **(see illustration)**. If it isn't cracked, hardened or flattened-out, it can be reused.

13   Installation is the reverse of removal. Be sure to tighten the upper intake manifold mounting bolts and nuts to the torque listed in this Chapter's Specifications.

Lower intake manifold

Refer to illustration 5.17

Note: *Toyota refers to the lower intake manifold as the intake manifold. If you're buying a gasket for the lower intake manifold at a dealer parts department, use the Toyota terminology.*

14   Remove the upper intake manifold (see Steps 1 through 11).

15   If you're removing the lower intake manifold to replace it, remove the fuel rail now (see Chapter 4).

16   If you're just removing the lower intake manifold to replace the gaskets, it's not necessary to remove the fuel rail, but you'll have to disconnect the fuel supply and return line connections (see Chapter 4).

17   Remove the lower intake manifold bolts and remove the lower intake manifold **(see illustration)**.

18   Remove and discard the old lower intake manifold gaskets. Clean off all traces of old gasket material from the mating surfaces of the manifold and cylinder heads, then wipe the surfaces with brake system cleaner.

19   Installation is the reverse of removal. Be sure to use new gaskets and tighten the lower intake manifold bolts to the torque listed in this Chapter's Specifications.

6   Exhaust manifolds - removal and installation

Warning: *The engine must be completely cool before beginning this procedure.*
Note: *The following procedure applies to either exhaust manifold.*

1   Disconnect the cable from the negative terminal of the battery (see Chapter 5.

2   Raise the front of the vehicle and place it securely on jackstands.

Warning: *On Sequoia models equipped with rear height control suspension, adjust the height control to the NORMAL mode, turn OFF the height control, then turn off the engine BEFORE raising the vehicle.*

3   Unbolt and disconnect the front exhaust pipes from the exhaust manifolds.

4   Disconnect the electrical connectors from the upstream oxygen sensors.

5   Remove the three bolts that secure the exhaust manifold support bracket and remove it.

6   Remove the six nuts that secure the exhaust manifold and remove the exhaust manifold.

7   Remove and discard the old exhaust manifold gasket.

8   Installation is the reverse of removal. Be sure to use a new gasket and install it with the oval-shaped protruding tip facing in the correct direction. For the left (driver's side) manifold, the tip must face to the rear; on the right manifold it must face to the front.

9   Tighten the exhaust manifold nuts to the torque listed in this Chapter's Specifications.

7   Timing chain and sprockets - removal, inspection and installation

Removal

Refer to illustrations 7.17a, 7.17b, 7.17c, 7.23a, 7.23b, 7.23c, 7.24a, 7.24b, 7.24c, 7.25, 7.26, 7.27, 7.31 and 7.32

Warning: *Wait until the engine is completely cool before beginning this procedure.*
Caution: *The timing system is complex, and severe engine damage will occur if you make any mistakes. Do not attempt this procedure*

unless you are highly experienced with this type of repair. If you are at all unsure of your abilities, be sure to consult an expert. Double-check all your work and be sure everything is correct before you attempt to start the engine.

Note: *There is an access plate for the timing chain tensioner in the right side of the timing chain cover.*

Note: *If you're working on a 4WD model, the front axle/differential assembly will have to be removed (see Chapter 8).*

1   Disconnect the cable from the negative terminal of the battery (see Chapter 5).

2   Drain the engine oil and coolant (see Chapter 1).

3   Remove the battery (see Chapter 5).

4   Remove the engine cover.

5   Remove the air filter housing (see Chapter 1).

6   Remove the radiator (see Chapter 3). Remove the cooling fan mounting nuts and remove the cooling fan.

7   Remove the serpentine drivebelt (see Chapter 1).

8   Remove the upper intake manifold (see Section 5).

9   Remove the ignition coils (see Chapter 5) and the valve covers (see Section 4).

10   Remove the Variable Valve Timing (VVT) sensor (see Chapter 6).

11   Remove the dipstick, then remove the dipstick tube retaining bolt and the dipstick tube. Remove and discard the old dipstick tube O-ring.

12   Disconnect the electrical connector from the power steering pressure switch, then detach the power steering pump (see Chapter 10) and set it aside. Don't disconnect the power steering fluid hoses from the pump!

13   Remove the alternator (see Chapter 5).

14   Detach the air conditioning compressor (see Chapter 3). Do NOT disconnect the air conditioning refrigerant hoses!

15   Unbolt the drivebelt tensioner (see Chapter 1).

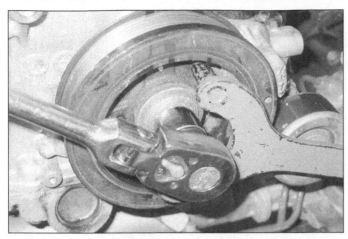

7.17a Hold the crankshaft pulley with a pin spanner while removing the bolt

7.17b A chain wrench can also be used to hold the pulley, but only if special precautions are taken; note the sockets used as spacers to raise the chain over the timing scale, and the piece of old drivebelt used to pad the pulley

7.17c If the crankshaft pulley can't be removed by hand, use a puller that bolts to the hub of the pulley - not a jaw-type puller. Also, be sure to use the correct adapter between the nose of the crankshaft and the puller screw, so as not to damage the threads in the crankshaft

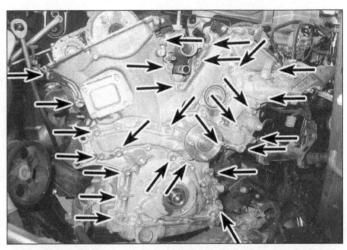

7.23a Remove the timing chain cover fasteners - make sure you keep track of the locations of the long and short bolts

16   Remove the idler pulley bolts and remove the two idler pulleys.

17   Remove the crankshaft pulley (see illustrations).

18   Remove the four bolts from the front of the upper oil pan.

19   Remove the two oil cooler hoses.

20   Disconnect the two radiator hoses from the water inlet.

21   Disconnect all five water bypass hoses.

22   Remove the water inlet (see Chapter 3, illustration 7.9). Remove the O-ring and gasket from the water inlet and discard them.

23   Remove the timing chain cover mounting fasteners and remove the timing chain cover (see illustration). There are only four spots where you can pry between the timing chain cover and the engine (see illustrations). Do NOT pry the timing chain cover loose at any other spot or you will damage the sealing surface of the cover. After removing the timing chain cover, carefully pry out the old crankshaft seal with a screwdriver (see Sec-

tion 8). Make sure that you don't scratch the seal bore. If you want to inspect or replace any oil pump parts, refer to Section 12.

Note: *Keep track of the locations of all of the bolts. They are of two different lengths and can't be interchanged.*

7.23b The timing chain cover can be pried loose at the lower corners . . .

7.23c . . . and at the upper corners; prying anywhere else may damage the cover

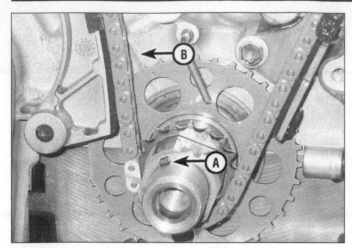

7.24a To bring the No. 1 cylinder to TDC on the compression stroke, rotate the crankshaft until the crankshaft key (A) is aligned with the timing line on the cylinder block (B) . . .

7.24b . . . the cast dot on the left cylinder head should line up with the center of the marks on the sprocket assembly . . .

24   Bring the piston in the No. 1 cylinder to TDC on its compression stroke: Install the crankshaft pulley bolt, then rotate the crankshaft until the crankshaft set key is aligned with the timing line on the cylinder block **(see illustration)**. Verify that the timing marks on the camshaft timing gear assemblies and the marks on the timing sprockets are aligned with their corresponding marks on top of the front camshaft bearing caps **(see illustrations)**. If the marks are not aligned, rotate the crankshaft another 360-degrees and recheck the marks.

25   Turn the stopper plate on the No. 1 tensioner clockwise and push in the tensioner plunger **(see illustration)**. To lock the plunger in this position, turn the stopper plate counterclockwise and insert a drill or punch (0.138-inch diameter) through the holes in the stopper plate and the tensioner body. Remove the two tensioner mounting bolts and remove the No. 1 tensioner.

26   Remove the chain tensioner slipper **(see illustration)**.
27   Using a 10 mm hex bit, unscrew the idler sprocket shaft and remove the idler shaft, sprocket and collar **(see illustration)**. Note which side of the sprocket faces out.
28   Remove the two chain vibration dampers.
29   Remove the No. 1 timing chain.
Caution: *While the No. 1 timing chain is removed, DO NOT ROTATE THE CRANKSHAFT!*
30   Remove the crankshaft timing chain sprocket.
31   Compress chain tensioner No. 2 and insert a drill or punch (0.039-inch diameter) into the hole **(see illustration)**.
Note: *Timing chain No. 2 and the No. 2 chain tensioner are on the right (passenger's side) cylinder head. Timing chain No. 3 and the No. 3 tensioner are on the left (driver's side) cylinder head.*

32   Hold the hex on the exhaust camshaft with a wrench and unscrew the two bolts that secure the camshaft timing sprockets to the camshafts **(see illustration)**. Remove the sprockets and chain as an assembly. Keep the components in a resealable plastic bag to ensure that none of these components is mixed with the other timing chain set.
Caution: *Don't attempt to disassemble the adjustable intake sprocket assembly. If disassembled, it will have to be replaced.*
33   Remove the chain tensioner No. 2 mounting bolt and remove chain tensioner No. 2. Store the tensioner in the plastic bag with the other No. 2 timing chain components.
34   To remove the No. 3 timing chain and tensioner, repeat Steps 31 through 33. Again, store the components in a resealable plastic bag.
Caution: *While the timing chains are removed, DO NOT ROTATE THE CRANKSHAFT!*

7.24c . . . and the cast dots on the right cylinder head should line up with the mark on the sprocket assembly

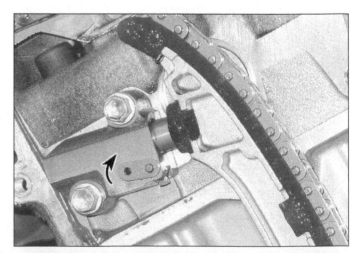

7.25 To lock the tensioner in the retracted position, rotate the stopper plate clockwise and push the plunger in, then rotate the stopper plate counterclockwise and insert a pin through the hole in the stopper plate and the tensioner body

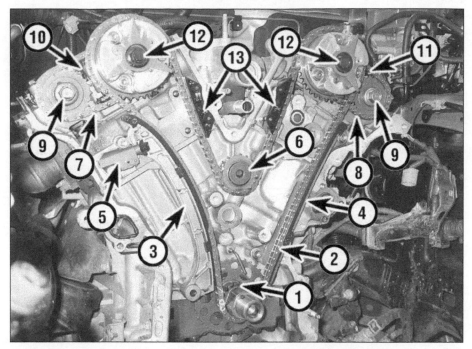

7.27 Use a 10 mm hex bit to remove the idler shaft and sprocket

7.26 Details of the timing chains and related components

1	Crankshaft sprocket	8	No. 3 timing chain
2	No. 1 timing chain	9	Exhaust camshaft sprocket bolt
3	Chain tensioner slipper	10	No. 2 timing chain tensioner
4	Chain guide	11	No. 3 timing chain tensioner
5	No. 1 timing chain tensioner	12	Intake camshaft sprocket bolt
6	Idler sprocket and shaft	13	Chain vibration dampers
7	No. 2 timing chain		

7.31 Insert a pin through the small tensioners to lock them in the retracted position

Inspection

Refer to illustrations 7.36, 7.37 and 7.39

35   Inspect all parts of the timing chain assembly for wear and damage. Inspect the three timing chains for loose pins, cracks, and worn rollers and side plates. Inspect the sprockets for hook-shaped, chipped and/or broken teeth.

36   Inspect timing chain No. 1 for stretching. To measure timing chain stretch, measure the distance between 15 pins at three or more places around the length of the chain (see illustration). Compare your measurements with the distance listed in this Chapter's Specifications.

37   Measure the diameter of each tim-

ing chain sprocket and idler sprocket with the appropriate timing chain installed on the

7.32 Hold the camshafts with a wrench to remove the sprocket bolts; never try to disassemble the intake camshaft sprocket assembly - only remove or install it as an assembly

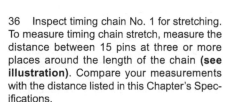

(15 pins)

92078-2A-8.36 HAYNES

7.36 Measure timing chain stretch by measuring the distance between 15 pins at three or more places around the length of the chain

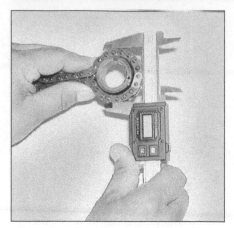

7.37 Wrap the chain around each of the timing chain sprockets and measure the diameter of the sprockets across the chain rollers. If the measurement is less than the minimum sprocket diameter, replace the chain and the timing sprockets

sprocket **(see illustration)**. The sprocket diameter, with the chain in place, should not exceed the dimensions listed in this Chapter's Specifications.

38   Measure the idler sprocket oil clearance as follows. First, measure the diameter of the idler sprocket collar with a micrometer and record your measurement. Then measure the inside diameter of the idler sprocket and record that measurement as well. Subtract the idler sprocket collar diameter from the inside diameter of the idler sprocket and compare the result with the clearance listed in this Chapter's Specifications. If the clearance is excessive, replace the idler sprocket and/or collar, as necessary.

39   Some scoring and wear of the timing chain tensioners and vibration dampers is normal, but excessive wear will increase

7.39 When inspecting the chain tensioners, the chain tensioner slipper and the chain vibration dampers, measure timing chain wear from the top of the chain contact surface to the bottom of the wear grooves

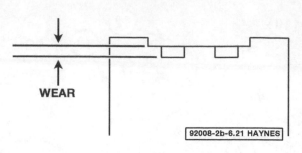

chain noise, will accelerate chain, gear and sprocket wear and could damage the engine if a chain jumps timing. Inspect chain tensioners No. 2 and 3, the timing chain tensioner slipper and the timing chain vibration dampers for excessive wear **(see illustration)**. If the measured chain wear for any of these components exceeds the depth listed in this Chapter's Specifications, replace the component.

40   Check the chain tensioners for correct operation. On the No. 1 tensioner, raise the ratchet pawl and verify that the plunger moves smoothly in and out of the tensioner, then release the ratchet pawl and verify that it prevents the plunger from sliding back into the tensioner. Also verify that the plungers on the No. 2 and No. 3 tensioners move in and out smoothly.

Installation

Refer to illustrations 7.41, 7.43, 7.44, 7.52a, 7.52b, 7.56 and 7.57

41   Install the crankshaft pulley bolt, then rotate the crankshaft in a counterclockwise direction until the crankshaft set key is at the 270-degree (9 o'clock) position (on the left and aligned with an imaginary horizontal line),

as you're looking at the front of the engine **(see illustration)**.

42   Push in the tensioner plunger on chain tensioner No. 2 and insert a drill bit or punch (0.039 inch diameter) into the hole of each tensioner to lock the plunger in the retracted position **(see illustration 7.31)**. Install the tensioners and tighten the mounting bolts to the torque listed in this Chapter's Specifications.

43   Install the No. 2 timing chain on the camshaft timing sprocket and camshaft timing gear assembly. Make sure that the yellow mark links on the chain are aligned with the timing marks (dots) on the camshaft timing sprockets **(see illustration)**.

44   Align the yellow links on the No. 2 timing chain with the timing marks on the bearing caps **(see illustration)** and install timing chain No. 2, the timing sprocket and the timing gear as an assembly. Install the two bolts that secure the timing sprockets to the camshafts. Immobilize the hex on the exhaust camshaft with an adjustable wrench and tighten these two bolts to the torque listed in this Chapter's Specifications. Remove the drill bit or punch that you used to lock the tensioner plunger in its retracted position and verify that the plunger tensions the chain.

7.41 Install the crankshaft pulley bolt, then rotate the crankshaft in a counterclockwise direction until the crankshaft set key is at the 270-degree position (on the left and aligned with an imaginary horizontal line), as you're looking at the front of the engine

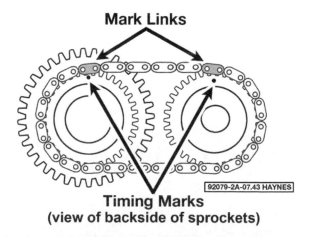

Mark Links

Timing Marks
(view of backside of sprockets)

7.43 When installing the short timing chain for the right bank (number 2 chain) between the camshaft sprockets, make sure that the yellow mark links on the chain are aligned with the timing marks (dots) on the sprockets

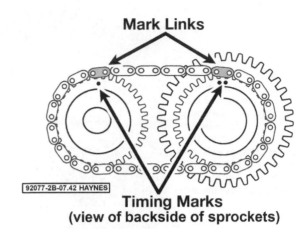

7.44 When installing the No. 2 (or No. 3) timing chain and sprocket, be sure to align the yellow links on the chain with the timing marks on the bearing caps

7.45 When installing the short timing chain for the left bank (number 3 chain), align the yellow colored links with the timing marks on the sprockets

45 Install the No. 3 timing chain (repeat Steps 42 through 44) **(see illustration)**.

46 Install the chain guide and tighten the bolts to the torque listed in this Chapter's Specifications.

47 Install the crankshaft timing sprocket on the crankshaft. Be sure to align the timing sprocket keyway with the key on the crankshaft, and make sure that the sprocket end of the crank timing sprocket faces in, toward the engine.

48 Install the chain tensioner slipper.

49 Turn the stopper plate on the No. 1 tensioner clockwise and push in the tensioner plunger. To lock the plunger in this position, turn the stopper plate counterclockwise and insert a drill bit or punch (0.138-inch diameter) through the holes in the stopper plate and the tensioner. Install the tensioner and tighten the two tensioner mounting bolts to the torque listed in this Chapter's Specifications.

50 Verify that the timing marks on the camshaft timing gear assemblies and the marks on the camshaft timing sprockets are aligned with their corresponding marks on top of the front camshaft bearing caps **(see illustrations 7.24b and 7.24c)**.

51 Using the crankshaft pulley bolt, rotate the crankshaft clockwise until the crankshaft set key is aligned with the timing line on the cylinder block **(see illustration 7.24a)**.

52 Install the long timing chain (No. 1) on the camshaft timing gear assemblies and on the crankshaft timing sprocket. Make sure that the yellow mark link is aligned with the timing mark (dot) on the crankshaft timing sprocket **(see illustration)** and that the orange mark links are aligned with the timing marks on the intake camshaft sprockets **(see illustration)**.

53 Apply a light coat of engine oil to the bearing surface of the idler sprocket collar. Install the idler sprocket collar, sprocket and shaft. Make sure that the sprocket teeth on the idle gear are facing forward.

Tighten the idler sprocket shaft to the torque listed in this Chapter's Specifications. Remove the drill bit or punch that you inserted into chain tensioner No. 1 (in Step 49) and verify that it tensions timing chain No. 1.

Caution: *Carefully rotate the crankshaft by hand through at least two full revolutions (use a socket and breaker bar on the crankshaft pulley center bolt). If you feel any resistance, STOP! There is something wrong - most likely valves are contacting the pistons. You must find the problem before proceeding. Check your work to make sure all timing marks line-up properly, and see if any updated repair information is available.*

54 Remove all old RTV sealant from the gasket mating surfaces of the timing chain cover and from the front of the cylinder heads and engine block.

55 Install a new crankshaft oil seal in the timing chain cover (see Section 8).

56 Install a new O-ring on the left cylinder head **(see illustration)**.

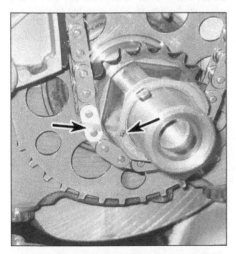

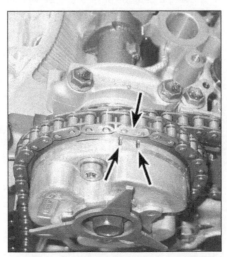

7.52a When installing timing chain No. 1 (the long timing chain) on the camshaft timing gear assemblies and the crankshaft timing sprocket, make sure that the yellow mark link is aligned with the timing mark (dot) on the crankshaft timing sprocket . . .

7.52b . . . and the orange mark links are aligned with mark(s) on the intake camshaft timing sprocket

7.56 Be sure to install a new O-ring in the front of the left cylinder head

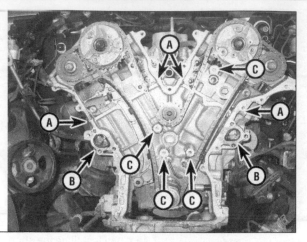

7.57 Apply a continuous bead of gray RTV sealant along the edges of the timing chain cover and across the top of the oil pan, then apply dabs at the cylinder head-to-block joints (A), around the water pump passages (B) and at the bosses (C) - but don't get any on the O-ring seal below the topmost boss

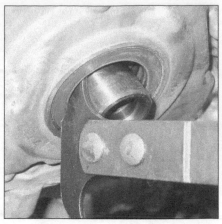

8.2 Cut away the crankshaft seal lip and pry out the seal with a seal removal tool or a screwdriver wrapped with tape

57   Apply gray RTV sealant on the timing chain cover and to the indicated locations of the mating surfaces on the engine **(see illustration)**. These beads should be about 1/8 to 3/16-inch (3 to 4.5 mm) wide.

Caution: *Once you have installed the sealant on the engine and timing chain cover, you have three minutes to install the cover. If you take longer than that, the sealant might not set up properly, so you'll have to remove the sealant and re-apply it.*

Make sure to get sealant into the corners where the oil pan meets the engine block and avoid getting any on the O-rings.

58   Rotate the flats on the oil drive rotor about 15-degrees to the right of vertical to align it with the square part of the crankshaft timing sprocket and slide the timing chain cover into place.

59   Install the timing chain cover mounting bolts and nuts and tighten all fasteners gradually and evenly, in a criss-cross fashion until they're all snug.

Caution: *Do not put long bolts in short holes or vice versa.*

When all of the fasteners are snug, tighten them to the torque listed in this Chapter's Specifications.

60   Install and tighten the four oil pan bolts

that go into the timing chain cover.

61   The remainder of installation is the reverse of the removal procedure.

62   Refill the engine with oil and coolant (see Chapter 1).

63   Reconnect the cable to the negative battery terminal.

64   Start the engine check for leaks.

---

8   Crankshaft front oil seal - replacement

Refer to illustrations 8.2 and 8.4

1   Remove the crankshaft pulley (see Section 7).

2   Carefully pry the seal out of the timing chain cover with a screwdriver or seal removal tool **(see illustration)**. If you use a screwdriver, wrap tape around the tip - don't scratch the housing bore or damage the crankshaft (if the crankshaft is damaged, the new seal will end up leaking).

3   Clean the bore in the timing chain cover and coat the lip and the outer edge of the new seal with engine oil or multi-purpose grease.

4   Using a seal driver or a socket with an outside diameter slightly smaller than the out-

side diameter of the seal, carefully drive the new seal into place with a hammer **(see illustration)**. Make sure it's installed squarely and driven in flush with the surface of the timing chain cover. Check the seal after installation to make sure the spring didn't pop out of place.

5   Reinstall the crankshaft pulley, tightening the bolt to the torque listed in this Chapter's Specifications.

6   Run the engine and check for oil leaks at the front seal.

---

9   Camshafts and lifters - removal, inspection and installation

Removal

Refer to illustration 9.4

Note: *The following procedure is not for beginners. Please read the entire procedure carefully before deciding whether this is a job that you want to tackle at home.*

1   Disconnect the cable from the negative terminal of the battery (see Chapter 5).

2   Drain the engine oil and coolant (see Chapter 1).

3   Remove the timing chains (see Section 7).

4   To remove the camshafts from the *right* (passenger's side) cylinder head, rotate the camshafts counterclockwise so that the nose of the intake cam lobe for the No. 1 cylinder is facing toward 7 o'clock, and the nose of the exhaust cam lobe is facing 12 o'clock, or straight up **(see illustration)**. Use an open-end wrench on the integral hex cast into each camshaft to rotate the cams. This Step is NOT necessary for the camshafts on the left cylinder head.

5   Gradually and evenly loosen all 16 camshaft bearing cap bolts in the opposite order of the tightening sequence **(see illustrations 9.15 and 9.16)**.

6   Remove all eight bearing caps and remove the intake and exhaust camshafts.

7   Store the bearing caps in the correct order. One way to do this is to put them in a

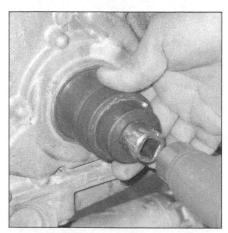

8.4 Lubricate the seal lip and drive the new crankshaft seal into place with a large socket or piece of pipe and a hammer

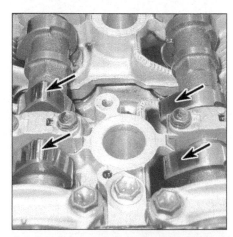

9.4 Correct camshaft position for removal of the right cylinder head camshafts - the camshafts on the left cylinder head can be removed in any position

box and label the cap numbers with a utility marker pen.

8   Remove the lifters with a magnet and put them in the same box with the cam bearing caps. Again, be sure to label each lifter with a marker. You can also write the lifter number directly on top of the lifter.

9   If you're removing the camshafts from the *other* cylinder head, repeat Steps 5 through 8. Don't forget that Step 4 applies *only* to camshaft removal for the *right* cylinder.

Caution: *While the timing chains and camshafts are removed, DO NOT ROTATE THE CRANKSHAFT!*

Inspection

Refer to illustrations 9.10, 9.11, 9.12, 9.13a and 9.13b

10   Inspect each lifter for scuffing and score marks **(see illustration)**.

11   Visually examine the cam lobes and bearing journals for score marks, pitting, galling and evidence of overheating (blue, discolored areas). Look for flaking away of the hardened surface layer of each lobe. Using a micrometer, measure the height of each camshaft lobe **(see illustration)**. Compare your measurements with this Chapter's Specifications. If the height for any one lobe is less than

the specified minimum, replace the camshaft.

12   Using a micrometer, measure the diameter of each journal at several points **(see illustration)**. Compare your measurements with this Chapter's Specifications. If the diameter of any one journal is less than specified, replace the camshaft.

13   Check the oil clearance for each camshaft journal as follows:

a)   Clean the bearing caps and the camshaft journals with brake system cleaner.

b)   Carefully lay the camshaft(s) in place in the cylinder head. Don't install the lifters or intake camshaft sub-gear and don't use any lubrication.

c)   Lay a strip of Plastigage on each journal.

d)   Install the bearing caps with the arrows pointing toward the front (timing chain end) of the engine **(see illustration)**.

e)   Tighten the bolts to the torque listed in this Chapter's Specifications in 1/4-turn increments.
   Note: *Don't turn the camshaft while the Plastigage is in place.*

f)   Remove the bolts and detach the caps.

g)   Compare the width of the crushed Plastigage (at its widest point) to the scale on the Plastigage envelope **(see illustration)**.

9.10 Inspect each lifter for wear and scuffing

h)   If the clearance is greater than specified, replace the camshaft and/or cylinder head.

i)   Scrape off the Plastigage with your fingernail or the edge of a credit card - don't scratch or nick the journals or bearing caps.

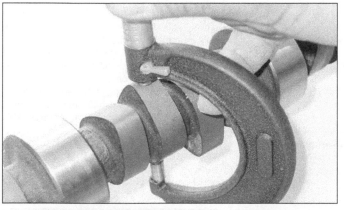

9.11 Measure the lobe heights on each camshaft - if any lobe height is less than the specified allowable minimum, replace that camshaft

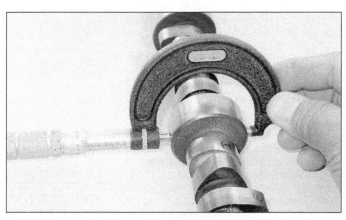

9.12 Measure each journal diameter with a micrometer - if any journal measures less than the specified limit, replace the camshaft

9.13a The camshaft bearing caps are numbered with an arrow facing the front of the engine

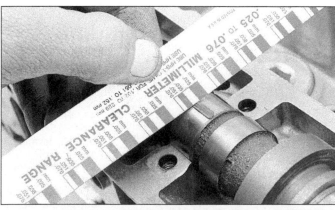

9.13b Compare the width of the crushed Plastigage to the scale on the envelope to determine the oil clearance

9.15 Camshaft bearing cap bolt tightening sequence (right cylinder head)

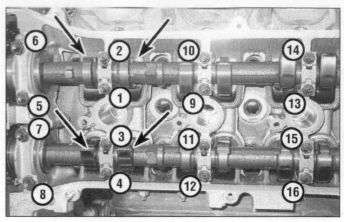

9.16 Tighten the camshaft bearing caps on the left cylinder head in this sequence; the front lobes of the exhaust camshaft must be facing upward, and the front lobes of the intake camshaft must be in the 7 o'clock position

Installation

Refer to illustrations 9.15 and 9.16

14   Lightly lubricate the lifter bores and the lifters with clean engine oil, then install the lifters in the same bores from which they were removed. Verify that each lifter rotates smoothly in its bore.

15   Right side: Lightly lubricate the camshaft journals with clean engine oil. Install the camshafts on the right cylinder head so that the nose of the intake cam lobe for the No. 1 cylinder is facing toward 7 o'clock, and the nose of the exhaust cam lobe for No. 1 is facing 12 o'clock, or straight up **(see illustration 9.4)**. Apply a light coat of engine oil to the upper bearings, then install the eight bearing caps in their correct locations. Apply a light coat of engine oil to the threads of the bearing cap bolts and install the bolts. Gradually and evenly tighten the bearing cap bolts, in the indicated sequence **(see illustration),** to the torque listed in this Chapter's Specifications. Rotate the camshafts 90-degrees clockwise until the knock pins on the front end of the camshafts are at a 90-degree position in relation to the cylinder head mating surface.

16   Left side: Lightly lubricate the camshaft journals with clean engine oil. Install the camshafts on the left cylinder head so that the nose of the intake cam lobe for the No. 2 cylinder is facing toward 7 o'clock, and the nose of the exhaust cam lobe for No. 2 is facing 12 o'clock, or straight up **(see illustration)**. Apply a light coat of engine oil to the upper bearings, then install the eight bearing caps in their correct locations. Apply a light coat of engine oil to the threads of the bearing cap bolts and install the bolts. Then gradually and evenly tighten the bearing cap bolts, in the indicated sequence, to the torque listed in this Chapter's Specifications.

17   The remainder of installation is the reverse of removal. Install the timing chains, the timing chain cover and all of the components attached to the timing cover (see Section 7).

Caution: *Carefully rotate the crankshaft by hand through at least two full revolutions (use a socket and breaker bar on the crankshaft pulley center bolt). If you feel any resistance, STOP! There is something wrong - most likely valves are contacting the pistons. You must find the problem before proceeding. Check your work to make sure all timing marks line-up properly, and see if any updated repair information is available.*

18   Refill the engine with oil and coolant (see Chapter 1), reconnect the cable to the negative battery terminal, start the engine and check for leaks.

Note: *It may take a few minutes for lifter clatter to disappear.*

10 Cylinder heads - removal and installation

Warning: *The engine must be completely cool before starting this procedure.*
Note: *This procedure applies to either cylinder head.*

Removal

Refer to illustration 10.13

1   Position the engine at TDC compression for cylinder no. 1 (see Section 3). Disconnect the cable from the negative terminal of the battery (see Chapter 5).

2   Drain the engine oil and coolant (see Chapter 1).

3   Remove the engine cover.

4   Remove the air filter housing (see Chapter 4).

5   Remove the upper intake manifold (see Section 5).

6   Remove the valve cover(s) (see Section 4).

7   Disconnect the fuel supply and return lines from the fuel rail (see Chapter 4). Remove the lower intake manifold and the fuel rail as a single assembly (see Section 5).

8   Locate the coolant passage on the back of the engine. It is bolted to the upper rear part of the block and is attached to both cylinder heads by a pair of nuts at each end. Disconnect the Engine Coolant Temperature (ECT) sensor (see Chapter 6). Disconnect the heater hose from the coolant passage. Remove the two bolts and four nuts and remove the coolant passage. Remove and discard the old gaskets. Remove the old O-ring from the coolant outlet pipe and discard it.

9   Disconnect the front exhaust pipe from the exhaust manifold and remove the exhaust manifold (see Section 6).

10   Remove the timing chain (see Section 7).

11   Remove the camshaft timing oil control valve, the oil control valve filter and the VVT-i sensor (see Chapter 6).

12   Remove the camshafts (see Section 9).

13   If you're removing the left cylinder head, remove the two front bolts in the indicated sequence **(see illustration)**.

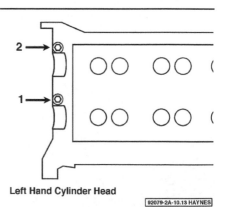

Left Hand Cylinder Head

`92079-2A-10.13 HAYNES`

10.13 If you're removing the left cylinder head, remove these two bolts first, in this order. When installing the left head, install these two bolts, in reverse order, AFTER you have installed the rest of the cylinder head bolts

14   Remove the cylinder head bolts gradually and evenly, in the order opposite that of the tightening sequence **(see illustrations 10.24a and 10.24b)**, then remove the cylinder head from the engine block.

15   Remove and discard the old cylinder head gasket.

16   If you're removing both cylinder heads, repeat this procedure for the other cylinder head.

Installation

Refer to illustrations 10.22, 10.24a and 10.24b

17   The mating surfaces of the cylinder heads and the block must be perfectly clean as the heads are installed.

18   Use a gasket scraper to remove all traces of carbon and old gasket material, then clean the mating surfaces with brake system cleaner. If there's oil on the mating surfaces when the head is installed, the gasket may not seal correctly and leaks could develop. When working on the block, stuff the cylinders with clean shop rags to keep out debris. Use a vacuum cleaner to remove material that falls into the cylinders.

19   Check the block and head mating surfaces for nicks, deep scratches and other damage. If damage is slight it can be removed with a file; if it's excessive, machining may be the only alternative.

20   Use a tap of the correct size to chase the threads in the cylinder head bolt holes, then clean them with compressed air - make sure that nothing remains in the holes.

Warning: *Wear eye protection when using compressed air!*

21   Mount each bolt in a vise and run a die down the threads to remove corrosion and restore the threads. Dirt, corrosion and damaged threads affect torque readings. Measure the outside diameter of each cylinder head bolt in several places, comparing your measurements to the minimum allowable diameter listed in this Chapter's Specifications.

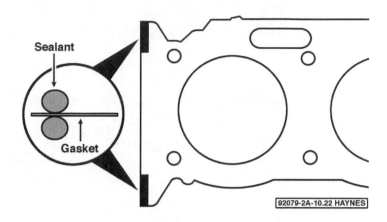

10.22   Apply small beads of sealant to the top and bottom of the head gasket at these two points

Replace any bolt with a diameter less than the allowable minimum.

22   Apply two small beads of RTV sealant to the two indicated areas of the new cylinder head gasket **(see illustration)**. Apply the sealant to the top and the bottom of these two spots.

Caution: *Once you have applied the RTV sealant to the cylinder head gasket, you must install the cylinder head within three minutes, and you must tighten and torque the cylinder head bolts within 15 minutes. If you don't, you must remove the RTV sealant and reapply it.*

23   Position the cylinder head gasket on the engine block so that the lot number stamp is on the center upper edge of the gasket (next to the intake manifold), facing up. Carefully place the cylinder head on the head gasket.

24   Apply a light coat of oil to the cylinder head bolts, then install and tighten them (don't forget the washers!), gradually and evenly, in

the proper sequence **(see illustrations)**, to the *initial* torque listed in this Chapter's Specifications.

25   After tightening all eight bolts to the initial torque, put a paint mark on the front edge of each bolt (the edge facing toward the front of the engine), then retighten each bolt, in the same sequence, another 180-degrees.

26   If you're installing the left cylinder head, install the two front head bolts and tighten them to the torque listed in this Chapter's Specifications, in the opposite order of the sequence indicated in **illustration 10.13**.

27   If you removed both cylinder heads, install the other cylinder head now (see Steps 17 through 25).

28   Install the intake and exhaust camshafts (see Section 9).

29   Install the camshaft timing oil control valve, the oil control valve filter and the VVT-i sensor (see Chapter 6).

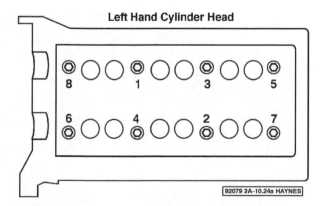

10.24a   Cylinder head bolt tightening sequence - left cylinder head

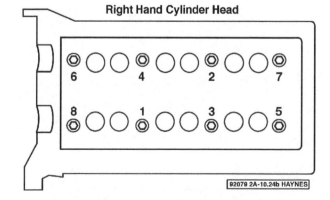

10.24b   Cylinder head bolt tightening sequence - right cylinder head

11.3 Locations of the engine splash shield fasteners

30   Install the timing chains, the timing chain cover and all components attached to the cover (see Section 7).
31   The remainder of installation is the reverse of the removal procedure.
32   Refill the engine with oil and coolant (see Chapter 1).
33   Reconnect the cable to the negative battery terminal, start the engine and check for leaks.

11   Oil pan - removal and installation

Warning: *On Sequoia models equipped with rear height control suspension, adjust the height control to the NORMAL mode, turn OFF the height control, then turn off the engine BEFORE raising the vehicle.*
Note: *On 4WD models, it will be necessary to either remove the front axle/differential assembly (see Chapter 8) or remove the engine from the vehicle (see Chapter 2C) in order to remove the oil pan.*

Removal

Refer to illustration 11.3

1   Disconnect the cable from the negative terminal of the battery (see Chapter 5).
2   Raise the front of the vehicle and place it securely on jackstands.
3   Drain the engine oil (see Chapter 1). Remove the splash shield from underneath the engine **(see illustration)**.
4   Remove the bolts and nuts that secure oil pan No. 2 (the smaller stamped steel pan) to the larger cast aluminum oil pan.
5   Oil pan No. 2 will probably be stuck to the oil pan with RTV sealant. Try tapping it loose with a rubber-tipped mallet. If you're unable to knock it loose, carefully cut the sealant with a putty knife and a hammer. Make sure that you don't damage the mating surfaces of the two pans.
6   Remove the two oil pump pickup tube/strainer mounting nuts and remove the pickup/strainer assembly.
7   Remove the four bolts that attach the flywheel housing cover and remove the cover.
8   Remove the 17 bolts and 2 nuts that secure the oil pan to the engine block. Remove the 4 stud bolts.
9   Carefully pry the oil pan loose from the engine block.
Caution: *Only pry in the small cutout areas along the side of the pan.*

Installation

10   Use a scraper to remove all traces of old sealant from the block and oil pan. Clean the mating surfaces with brake system cleaner.
11   Make sure the threaded holes in the block are clean.
12   Check the flange of the steel oil pan for distortion around the bolt holes. If necessary, place it on a wood block and use a hammer to flatten and restore the gasket surface.
13   Inspect the strainer for cracks or blockage. Clean it with solvent and install it using a new gasket. Tighten the fasteners to the torque listed in this Chapter's Specifications.
14   Apply a 1/8 inch bead of RTV sealant to the upper oil pan flange.
15   Position the pan onto the block and install the fasteners. Working from the center out, tighten the fasteners to the torque listed in this Chapter's Specifications in several steps.
16   After you have installed the aluminum portion of the oil pan, apply a bead of RTV sealant to the flange of the No. 2 oil pan, carefully position it on the upper oil pan and install the bolts. Working from the center out, tighten them to the torque listed in this Chapter's Specifications in several steps.
17   The remainder of installation is the reverse of removal. Add oil and install a new filter (see Chapter 1). Run the engine and check for leaks.

12   Oil pump - removal, inspection and installation

Removal

Refer to illustrations 12.2 and 12.3

1   Remove the timing chain cover (see Section 7).
2   Remove the oil pipe mounting bolts **(see illustration)** and remove the oil pipe. Remove and discard the two old oil pipe O-rings.
3   Remove the oil pump cover bolts and remove the oil pump cover **(see illustration)**. Remove the oil pump drive rotor and driven rotor.

12.2 Oil pump tube bolts

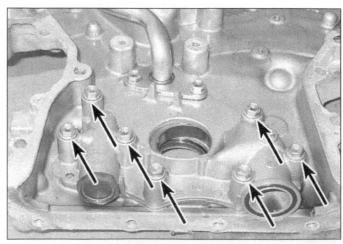

12.3 Oil pump cover bolts

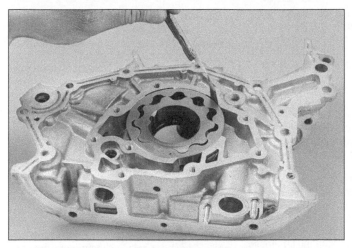

12.7a Measure the driven rotor-to-body clearance with a feeler gauge

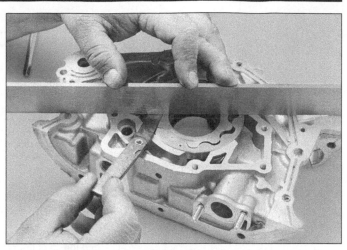

12.7b Measure the rotor side clearance with a precision straightedge and feeler gauge

4   Remove the oil pressure relief plug, valve spring and valve from the oil pump cover.

Inspection

Refer to illustrations 12.7a, 12.7b and 12.7c

5   Clean all components with solvent, then inspect them for wear and damage. Check that the oiled relief valve falls easily through its bore without sticking.

6   Check the oil pressure relief valve sliding surface and valve spring. If either the spring or the valve is damaged, they must be replaced as a set.

7   Check the clearance of the following components with a feeler gauge and compare the measurements to this Chapter's Specifications **(see illustrations)**:

a) *Driven rotor-to-oil pump body*
b) *Rotor side clearance*
c) *Rotor tip clearance*

Installation

8   Pry the old crankshaft seal out of the timing chain cover with a screwdriver.

9   Apply multi-purpose grease or engine oil to the outer edge of the new crank seal and carefully drive it into place with a deep socket and a hammer. Apply multi-purpose grease or engine oil to the seal lip.

10   Apply a coat of petroleum jelly to the pump drive and driven rotors, then place the two rotors into position in the timing chain cover. Make sure that the pump marks (dimples) are facing out, toward the pump cover and away from the timing chain cover.

11   Pack the pump cavity with petroleum jelly (this will help to prime the pump) and install the cover. Tighten the cover bolts to the torque listed in this Chapter's Specifications.

12   Lubricate the oil pressure relief valve with clean engine oil and insert the valve, then the spring, into the pump cover. Screw in the plug and tighten it to the torque listed in this Chapter's Specifications.

13   Install the timing chain cover (see Section 7).

14   The remainder of installation is the reverse of removal. Add oil and install a new filter (see Chapter 1). When you're done, run the engine and check for leaks.

13   Driveplate - removal and installation

Warning: *On Sequoia models equipped with rear height control suspension, adjust the height control to the NORMAL mode, turn OFF the height control, then turn off the engine BEFORE raising the vehicle.*

Removal

1   Disconnect the cable from the negative terminal of the battery (see Chapter 5, Section 3).

2   Raise the vehicle and support it securely on jackstands, then refer to Chapter 7 and remove the transmission.

3   Make alignment marks on the driveplate and crankshaft to ensure correct alignment during reinstallation.

4   Remove the bolts securing the driveplate to the crankshaft. If the crankshaft turns, wedge a screwdriver in the ring gear teeth to hold the driveplate.

5   Remove the driveplate from the crankshaft. Be sure to support it while removing the last bolt. Automatic transmission equipped vehicles have spacers on both sides of the driveplate. Keep them with the driveplate.

Installation

6   Clean the driveplate to remove grease and oil. Inspect the surface for cracks. Check for cracked or broken ring gear teeth.

7   Clean and inspect the mating surfaces of the driveplate and the crankshaft. If the crankshaft rear seal is leaking, replace it before reinstalling the driveplate (see Section 14).

8   Position the driveplate against the crankshaft. Be sure to align the marks made during removal. Note that some engines have an alignment dowel or staggered bolt holes to ensure correct installation. Before installing the bolts, apply thread-locking compound to the threads.

9   Wedge a screwdriver in the ring gear teeth to keep the driveplate from turning and tighten the bolts to the torque listed in this Chapter's Specifications. Follow a criss-cross pattern and work up to the final torque in three or four steps.

10   The remainder of installation is the reverse of the removal procedure.

12.7c Measure the rotor tip clearance with a feeler gauge - note the rotor marks are facing out (when the pump body cover is installed, the marks will be against the cover)

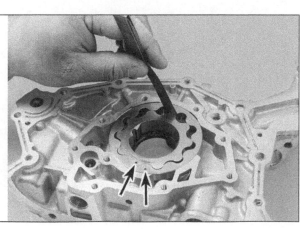

14.6 Drive the new seal into the retainer with a wood block or a section of pipe - make sure that you don't cock the seal in the retainer bore

14 Rear main oil seal - replacement

Refer to illustration 14.6

1   Remove the transmission (see Chapter 7). Remove the rear end plate.

2   The seal can be replaced without removing the oil pan or seal retainer. However, this method is not recommended because the lip of the seal is quite stiff and it's possible to cock the seal in the retainer bore or damage it during installation. If you want to take the chance, pry out the old seal with a screwdriver. Apply multi-purpose grease to the crankshaft seal journal and the lip of the new seal and carefully press the new seal into place. The lip is stiff, so carefully work it onto the seal journal of the crankshaft with a smooth object like the end of an extension as you tap the seal into place. Don't rush it, or you may damage the seal.

3   The following method is recommended but requires removing the seal retainer and resealing the rear of the oil pan.

4   After removing the two rearmost oil pan-to-seal retainer bolts, break the seal between the rear of the oil pan and the bottom of the seal retainer with a putty knife. Remove the retainer-to-engine block bolts, detach the seal retainer and remove all the old gasket material. remove the sealant from the top of the oil pan flange.

Note: *Cover the open area of the oil pan with clean rags to keep debris out while bracing the pan flange.*

5   Position the seal and retainer assembly between two wood blocks on a workbench and drive the old seal out from the back side with a screwdriver.

6   Drive the new seal into the retainer with a wood block **(see illustration)** or a section of pipe slightly smaller in diameter than the outside diameter of the seal.

7   Lubricate the crankshaft seal journal and the lip of the new seal with multi-purpose grease. Note that the engine doesn't have a gasket between the seal retainer and the engine block. Instead, apply a 2 to 3 mm wide bead of RTV sealant to the retainer flange before attaching the retainer to the block.

8   Slowly and carefully push the seal and retainer onto the crankshaft. The seal lip is stiff, so work it onto the crankshaft with a smooth object such as the end of an extension as you push the retainer against the block.

9   Install and tighten the retainer bolts to the torque listed in this Chapter's Specifications.

10   The remainder of installation is the reverse of removal.

15 Engine mounts - check and replacement

Warning: *On Sequoia models equipped with rear height control suspension, adjust the height control to the NORMAL mode, turn OFF the height control, then turn off the engine BEFORE raising the vehicle.*

1   Engine mounts seldom require attention, but broken or deteriorated mounts should be replaced immediately or the added strain placed on the driveline components may cause damage or wear.

Check

2   During the check, the engine must be raised slightly to remove the weight from the mounts.

3   Raise the vehicle and support it securely on jackstands, then position a jack under the engine oil pan. Place a large wood block between the jack head and the oil pan, then carefully raise the engine just enough to take the weight off the mounts. Do not position the wood block under the drain plug.

Warning: *DO NOT place any part of your body under the engine when it's supported only by a jack!*

4   Check the mounts to see if the rubber is cracked, hardened or separated from the metal plates. Sometimes the rubber will split down the center.

5   Check for relative movement between the mount plates and the engine or frame (use a large screwdriver or pry bar to attempt to move the mounts).

6   If movement is noted, lower the engine and tighten the mount fasteners.

Replacement

Refer to illustrations 15.8a and 15.8b

7   Disconnect the cable from the negative terminal of the battery (see Chapter 5), then raise the vehicle and support it securely on jackstands (if not already done). Support the engine as described in Step 3.

8   To remove an engine mount, remove the fasteners, raise the engine and detach the mount **(see illustration)**. The engine can be raised with an engine hoist, or with a floor jack and wood block placed under the oil pan.

Note: *Even if only one mount is being replaced, remove the mount-to-engine bracket nut from the other mount (this will allow the engine to be raised far enough for mount removal).*

9   Installation is the reverse of removal. Use non-hardening thread locking compound on the mount bolts/nuts and be sure to tighten them securely.

10   See Chapter 7 for transmission mount replacement.

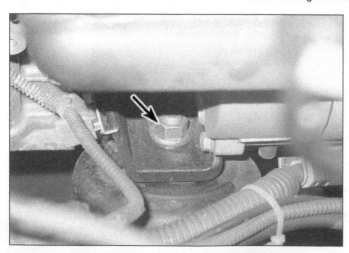

15.8a Engine mount-to-engine bracket nut

15.8b Engine mount-to-frame bracket bolts (A) and nuts (B, on the other side of the bracket)

Chapter 2 Part B
V8 engines

Contents

Specifications

General

Engine designation
Tundra
2007 through 2009.. 2UZ-FE
2007 and later... 3UR-FE
2009 and later... 3UR-FBE
2010 and later... 1UR-FE
Sequoia
2008 and 2009... 2UZ-FE
2008 and later... 3UR-FE
2009 and later... 3UR-FBE
2010 through 2012... 1UR-FE
Displacement
1UR-FE .... 4.6 liters
2UZ-FE... 4.7 liters
3UR-FBE .... 5.7 liters
Cylinder numbers (drivebelt end-to-transmission end)
Right (passenger) side 2-4-6-8
Left (driver) side 1-3-5-7
Firing order
2UZ-FE... 1-8-4-3-6-5-7-2
1UR-FE, 3UR-FE and 3UR-FBE... 1-8-7-3-6-5-4-2

↓ Front

36060-1-specs.C HAYNES

Cylinder numbering diagram

Warpage limits
Cylinder head
  2UZ-FE
    Block side .... 0.00394 inch (0.10 mm)
    Intake manifold .... 0.00394 inch (0.10 mm)
    Exhaust manifolds .... 0.00394 inch (0.10 mm)
  1UR-FE, 3UR-FE and 3UR-FBE
    Block side .... 0.00197 inch (0.05 mm)
    Intake manifold .... 0.00315 inch (0.08 mm)
    Exhaust manifold .... 0.00197 inch (0.05 mm)

Camshaft and related components
Valve clearance (engine cold) 2UZ-FE
  Intake .... 0.006 to 0.010 inch (0.15 to 0.25 mm)
  Exhaust .... 0.010 to 0.0138 inch (0.25 to 0.35 mm)
Bearing journal diameter
  2UZ-FE.. 1.061 to 1.062 inches (26.950 to 26.975 mm)
  1UR-FE, 3UR-FE and 3UR-FBE
    Bearing journal No. 1 diameter.. 1.1793 to 1.1799 inches (29.956 to 29.970 mm)
    Other journals .... 1.0220 to 1.0226 inches (25.959 to 25.975 mm)
Bearing oil clearance
  2UZ-FE
    Standard .... 0.0012 to 0.0028 inch (0.030 to 0.071 mm)
    Service limit .... 0.00394 inch (0.10 mm)
  1UR-FE, 3UR-FE and 3UR-FBE
    No. 1 journal
      Standard.. 0.00118 to 0.00256 inch (0.030 to 0.065 mm)
      Service limit .... 0.00394 inch (0.10 mm)
    Other journals
      Standard.. 0.000984 to 0.00244 inch (0.025 to 0.062 mm)
      Service limit .... 0.00354 inch (0.09 mm)
Lobe height
  2UZ-FE
    Intake
      Standard.. 1.6776 to 1.6815 inches (42.61 to 42.71 mm)
      Service Limit .... 1.6717 inches (42.46 mm)
    Exhaust
      Standard.. 1.6783 to 1.6823 inches (42.63 to 42.73 mm)
      Service Limit .... 1.6724 inches (42.48 mm)
  1UR-FE, 3UR-FE and 3UR-FBE
    Intake
      Standard.. 1.744 to 1.750 inches (44.30 to 44.45 mm)
      Service Limit .... 1.742 inches (44.25 mm)
    Exhaust
      Standard.. 1.740 to 1.746 inches (44.196 to 44.348 mm)
      Service Limit .... 1.738 inches (44.25 mm)
Thrust clearance (endplay)
  2UZ-FE
    Intake
      Standard.. 0 to 0.00157 inch (0 to 0.040 mm)
      Service limit .... 0.00472 inch (0.12 mm)
    Exhaust
      Standard.. 0.00118 to 0.00276 inch (0.030 to 0.070)
      Service limit .... 0.00394 inch (0.10 mm)
  1UR-FE, 3UR-FE and 3UR-FBE
    Standard .... 0.00315 to 0.00531 inch (0.08 to 0.135 mm)
    Service limit .... 0.00591 inch (0.15 mm)
Runout limit (total indicator reading)
  2UZ-FE.. 0.00118 inch (0.03 mm)
  1UR-FE, 3UR-FE and 3UR-FBE... 0.00157 inch (0.04 mm)
Camshaft gear backlash (2UZ-FE)
  Standard.. 0.0008 to 0.0079 inch (0.020 to 0.200 mm)
  Service limit.. 0.0118 inch (0.30 mm)
Camshaft gear spring (gap distance) (2UZ-FE) .... 0.717 to 0.740 inch (18.2 to 18.8 mm)
Timing belt tensioner protrusion (2UZ-FE) .... 0.374 to 0.413 inch (9.5 to 10.5 mm)
Lifters (2UZ-FE)
  Outside diameter.. 1.2192 to 1.2195 inches (30.968 to 30.976 mm)
  Bore diameter... 1.220 to 1.221 inches (30.989 to 31.013 mm)

Lifter-to-bore (oil) clearance
    Standard .... 0.000945 to 0.00196 inch (0.024 to 0.050 mm)
    Service limit .... 0.00276 inch (0.07 mm)

Timing chain

Timing chain (No. 1) stretch limit (between 15 pins) .... 5.390 inches (136.9 mm)
Timing chain (No. 2) stretch limit (between 15 pins) .... 5.420 inches (137.6 mm)
Timing chain sprocket wear limit (with No. 1 chain installed) .... 2.420 inches (61.4 mm)
Chain tensioner pad/guide/vibration damper wear limit........................... 0.0354 inch (0.9 mm)

Oil pump

Driven rotor-to-pump body clearance
  2UZ-FE
    Standard .... 0.00984 to 0.0128 inch (0.250 to 0.325 mm)
    Service limit .... 0.0128 inch (0.325 mm)
  1UR-FE, 3UR-FE and 3UR-FBE
    Standard .... 0.00689 to 0.00984 inch (0.175 to 0.250 mm)
    Service limit .... 0.00984 inch (0.250 mm)
Rotor tip clearance
  2UZ-FE
    Standard .... 0.00236 to 0.00709 inch (0.060 to 0.180 mm)
    Service limit .... 0.00709 inch (0.18 mm)
  1UR-FE, 3UR-FE and 3UR-FBE
    Standard .... 0.00709 to 0.0118 inch (0.180 to 0.300 mm)
    Service limit .... 0.0118 inch (0.300 mm)
Rotor side clearance
    Standard.. 0.00118 to 0.00354 inch (0.030 to 0.090 mm)
    Service limit.. 0.00354 inch (0.090 mm)

Torque specifications

Note: *One foot-pound (ft-lb) of torque is equivalent to 12 inch-pounds (in-lbs) of torque. Torque values below approximately 15 foot-pounds are expressed in inch-pounds, because most foot-pound torque wrenches are not accurate at these smaller values.*

	Ft-lbs (unless otherwise indicated)	Nm
Intake manifold bolts/nuts		
2UZ-FE	156 in-lbs	17.5
1UR-FE, 3UR-FE and 3UR-FBE	15	21
Exhaust manifold nuts		
2UZ-FE	32	43
1UR-FE, 3UR-FE and 3UR-FBE	15	21
Crankshaft pulley bolt		
2UZ-FE	181	245
1UR-FE, 3UR-FE and 3UR-FBE	221	300
Drivebelt idler pulley bolt(s)		
2UZ-FE	26	35
1UR-FE, 3UR-FE and 3UR-FBE	32	43
Drivebelt tensioner bolts		
2UZ-FE	144 in-lbs	16
1UR-FE, 3UR-FE and 3UR-FBE	17	23
Fan bracket bolts (2UZ-FE)		
12 mm bolt head	144 in-lbs	16
14 mm bolt head	24	32
Driveplate-to-crankshaft bolts		
Step 1	22	30
Step 2	Tighten an additional 90 degrees	
Driveplate-to-torque converter bolts		
2UZ-FE	35	47
1UR- FE, 3UR-FE and 3UR-FBE	39	53
Timing belt cover (2UZ-FE)		
Timing belt cover number 2 bolts	144 in-lbs	16
Timing belt cover (left and right) number 3 bolts	66 in-lbs	7.5
Timing belt idler pulley bolts (2UZ-FE)*	26	35
Timing belt tensioner bolts (2UZ-FE)		
Oil pump	19	26
Block side	144 in-lbs	16
Timing chain cover fasteners (1UR-FE, 3UR-FE and 3UR-FBE) – in sequence **(see illustrations 8.48a and 8.48b)**		
Bolts 1 through 11	35	47
Bolts 12 through 35 and the nut	17	23
Timing chain tensioner bolts	84 in-lbs	10
Timing chain guide/vibration damper bolts	15	21

* Apply thread locking compound to the threads prior to installation

Torque specifications (continued)

Note: *One foot-pound (ft-lb) of torque is equivalent to 12 inch-pounds (in-lbs) of torque. Torque values below approximately 15 foot-pounds are expressed in inch-pounds, because most foot-pound torque wrenches are not accurate at these smaller values.*

	Ft-lbs (unless otherwise indicated)	Nm
Valve cover bolts		
2UZ-FE	84 in-lbs	10
1UR-FE, 3UR-FE and 3UR-FBE		
Center bolt A **(see illustration 4.6b)**	15	21
All other bolts	108 in-lbs	12
Camshaft sprocket bolts		
2UZ-FE	72 in-lbs	8
1UR-FE, 3UR-FE and 3UR-FBE	74	100
Camshaft timing tube (2UZ-FE)		
Center bolt	58	79
Screw plug	132 in-lbs	15
Drive gear bolts	66 in-lbs	7.5
Pulley	72 in-lbs	8
Camshaft bearing cap bolts – in sequence		
2UZ-FE		
Bolt A **(see illustrations 10.25 and 10.33)**	66 in-lbs	7.5
All others	144 in-lbs	16
1UR-FE, 3UR-FE/3UR-FBE **(see illustrations 11.16a and 11.16b)**	144 in-lbs	16
Camshaft housing bolts – in sequence (1UR-FE, 3UR-FE and 3UR-FBE)		
(see illustrations 11.18a and 11.18b)		
Bolt 18	84 in-lbs	10
All others	22	30
Cylinder head bolts **(in sequence - see illustration 12.28)**		
2UZ-FE		
Step 1	24	32
Step 2	Turn an additional 90-degrees (1/4-turn)	
Step 3	Turn an additional 90-degrees (1/4-turn)	
1UR-FE, 3UR-FE and 3UR-FBE		
Step 1	27	36
Step 2	Turn an additional 90-degrees (1/4-turn)	
Step 3	Turn an additional 90-degrees (1/4-turn)	
For 12 mm bolt head	15	21
Oil pump body cover bolts		
2UZ-FE	84 in-lbs	10
1UR-FE, 3UR-FE and 3UR-FBE	96 in-lbs	11
Oil pump mounting bolts		
2UZ-FE		
14 mm bolt head - 1.10 inch (1.73 mm) long	23	31
All others	144 in-lbs	16
1UR-FE, 3UR-FE and 3UR-FBE	84 in-lbs	10
Oil pan		
2UZ-FE		
Upper pan bolts		
10 mm bolt head	66 in-lbs	7.5
12 mm bolt head	21	28
Lower pan bolts	66 in-lbs	7.5
1UR-FE, 3UR-FE and 3UR-FBE		
Upper pan bolts		
Long bolts	84 in-lbs	10
Short bolts	26	35
Lower pan nuts/bolts	84 in-lbs	10
Baffle plate	84 in-lbs	10
Oil pick-up tube mounting bolts		
2UZ-FE	71 in-lbs	8
1UR-FE, 3UR-FE and 3UR-FBE	108 in-lbs	12
Rear crankshaft oil seal retainer mounting bolts	84 in-lbs	10

4.6a Typical valve cover bolt locations on the right bank valve cover (2UZ-FE engines)

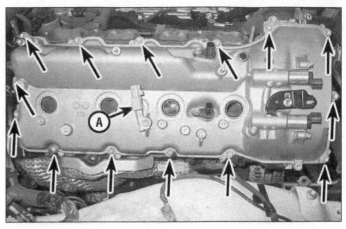

4.6b Typical valve cover bolt locations on the right bank valve cover - don't forget the bolt hidden under the capacitor (A) (1UR-FE, 3UR-FE and 3UR-FBE engines)

1 General information

This Part of Chapter 2 is devoted to in-vehicle repair procedures for the 4.6L (1UR-FE), 4.7L (2UZ-FE) and 5.7L (3UR-FE and 3UR-FBE) V8 engines. All these engines are designed with a Double-OverHead-Cam (DOHC) arrangement with four valves per cylinder (32 in all). 2UZ-FE engines use a Variable Valve Timing-intelligent (VVT-i) system and 1UR-FE, 3UR-FE and 3UR-FBE engines use a Dual Variable Valve Timing-intelligent (VVT-i) system. All models have Direct Ignition System (DIS), Acoustic Control Induction System (ACIS) and Electronic Throttle Control System-intelligent (ETCS-i) system. All these systems help the engine with fuel economy, reliable engine performance and cleaner emissions.

All information concerning engine removal and installation and engine block and cylinder head overhaul can be found in Part C of this Chapter.

The following repair procedures are based on the assumption that the engine is installed in the vehicle. If the engine has been removed from the vehicle and mounted on a stand, many of the steps outlined in this Part of Chapter 2 will not apply.

Note: *Some electrical systems need to be re-booted once the electrical connector is reconnected. Also there are some systems that need time for the memory features to set before they are disconnected. Wait approximately 90 seconds or more before disconnecting the cable from the negative battery terminal once the key has been turned to the OFF position.*

2 Repair operations possible with the engine in the vehicle

1   Many major repair operations can be accomplished without removing the engine from the vehicle. Clean the engine compartment and the exterior of the engine with some type of degreaser before any work is done. It will make the job easier and help keep dirt out of the internal areas of the engine.

2   Depending on the components involved, it may be helpful to remove the hood to improve access to the engine as repairs are performed (refer to Chapter 11 if necessary). Cover the fenders to prevent damage to the paint. Special pads are available, but an old bedspread or blanket will also work.

3   If vacuum, exhaust, oil or coolant leaks develop, indicating a need for gasket or seal replacement, the repairs can generally be made with the engine in the vehicle. The intake and exhaust manifold gaskets, crankshaft oil seals and cylinder head gaskets are all accessible with the engine in place.

Note: *It will be necessary to remove the engine from the vehicle in order to access the oil pump and oil pan.*

4   Exterior engine components, such as the intake and exhaust manifolds, the oil pan, the oil pump, the water pump, the starter motor, the alternator and the fuel system components can be removed for repair with the engine in place.

5   Since the cylinder heads can be removed without pulling the engine, valve component servicing can also be accomplished with the engine in the vehicle. Replacement of the camshafts, timing belt and pulleys is also possible with the engine in the vehicle.

6   In extreme cases caused by a lack of necessary equipment, repair or replacement of piston rings, pistons, connecting rods and rod bearings is possible with the engine in the vehicle. However, this practice is not recommended because of the cleaning and preparation work that must be done to the components involved.

3 Top Dead Center (TDC) for number one piston - locating

Note: *The most positive method for finding TDC is to examine the crankshaft timing marks and camshaft sprocket timing marks (see Sections 7 and 8).*

1   Refer to Chapter 2, Part A for this procedure.

2   TDC for the remaining cylinders can be found by turning the crankshaft 90-degrees (1/4-turn) at a time (in the normal direction of rotation) and following the firing order. Turning the crankshaft 90-degrees past TDC for cylinder number one would bring cylinder number 8 to TDC compression, and so on.

4 Valve covers - removal and installation

Removal

Refer to illustrations 4.6a and 4.6b

1   Disconnect the cable from the negative battery terminal (see Chapter 5).

2   Disconnect the electrical connectors to the valve sensors and actuator, remove the wire clamps from the valve covers and set the harness off to the side.

3   Remove the igniter/coil assemblies from the spark plugs. Mark each igniter/coil assembly using tape or other marking device to insure proper reassembly. See Chapter 5 for additional details.

4   Disconnect any vacuum lines, PCV hose or wires that may interfere with the removal process: Mark them carefully to insure proper reassembly.

5   Remove the bypass hose bracket mounting bolts and move the hoses out of the way. On 1UR-FE, 3UR-FE and 3UR-FBE engines, remove the dipstick tube bolts and move the tube out of the way.

6   Remove the retaining bolts **(see illustration)**, then detach the cover(s). If the cover is stuck to the head, bump the end with a wood block and a hammer to jar it loose. If that doesn't work, try to slip a flexible putty knife between the head and cover to break the seal. **Caution:** *Don't pry at the cover-to-head joint or damage to the sealing surfaces may occur, leading to oil leaks after the cover is reinstalled.*

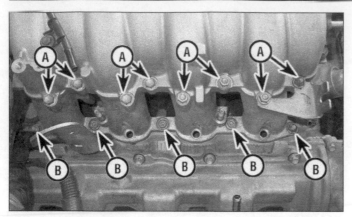

5.7 Upper (A) and lower (B) intake manifold fasteners

5.8 The upper and lower intake manifolds can be removed as a complete assembly

Installation

Note: *On 1UR-FE, 3UR-FE and 3UR-FBE engines, an oil control valve filter located in the valve cover is covered by a spacer. The filter should be removed and be cleaned if the valve cover is removed.*

7   The mating surfaces of the cylinder head and cover must be clean when the cover is installed. Use a gasket scraper to remove all traces of sealant and old gasket material, then clean the mating surfaces with lacquer thinner or acetone. If there's residue or oil on the mating surfaces when the cover is installed, oil leaks may develop.

8   On 2UZ-FE engines, position the semicircular plugs in the cylinder head cutouts, sealing them with RTV sealant, then apply a thin, uniform layer of RTV sealant to the gasket/plug joints.

9   Position a new gasket on the valve cover, then install the valve cover, sealing washers and bolts.

Note: *Install new spark plug tube seals into the valve cover.*

10   Tighten the bolts to the torque listed in this Chapter's Specifications in three or four equal steps.

11   Reinstall the remaining parts, run the engine and check for oil leaks.

5   Intake manifold(s) - removal and installation

Warning: *The engine must be completely cool before beginning this procedure.*

1   On 2UZ-FE engines, if you're going to remove the lower intake manifold, relieve the fuel system pressure (see Chapter 4).

2   Disconnect the cable from the negative battery terminal (see Chapter 5).

Removal

2UZ-FE engines

Upper intake manifold

Refer to illustration 5.7

3   Remove the throttle body (see Chapter 4).

4   Clearly label, then detach all remaining wires, hoses and brackets still attached to the upper intake manifold.

5   Remove the mounting bolt for the EVAP vacuum switching valve and separate it from the upper intake manifold.

6   Evenly loosen the intake manifold fasteners; make several passes to avoid warping it.

7   Remove the mounting nuts and bolts, then detach the upper intake manifold from the lower intake manifold **(see illustration)**. If it's stuck, don't pry between the gasket mating surfaces or damage may result.

Lower intake manifold

Refer to illustration 5.8

8   Remove the upper intake manifold from the lower intake manifold (see Steps 1 through 7).

Note: *If you're removing the lower intake manifold for access to other components, the upper intake manifold can remain attached to the lower manifold* **(see illustration)**. *The throttle body can remain attached as well, but the coolant hoses must be clamped-off and disconnected from it.*

9   Remove the fuel rail and injectors (see Chapter 4).

10   Clearly label, then detach all remaining wires, hoses and brackets still attached to the lower intake manifold, coolant outlets and rear coolant bypass casting.

11   Remove the mounting nuts and bolts, then detach the intake manifold from the cylinder heads. Start with the outer bolts first and work your way to the inner bolts on the manifold. Make several passes to ensure that the manifold is separated from the cylinder heads evenly to avoid damage to the cylinder head and manifold surfaces. If it's stuck, don't pry between the gasket mating surfaces or damage may result.

1UR-FE, 3UR-FE and 3UR-FBE engines

Refer to illustration 5.22

12   Remove the engine cover (see Chapter 1, **illustration 24.12a**).

13   Remove the windshield wiper motor and linkage (see Chapter 12).

14   Remove the eight cowl panel bolts and remove the cowl.

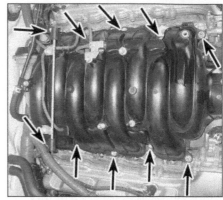

5.22 Intake manifold mounting fastener locations

15   Drain the cooling system (see Chapter 1).

16   Remove the air intake duct (see Chapter 4).

17   Remove the throttle body (see Chapter 4).

18   Clearly label, then detach all wires, hoses and brackets still attached to the intake manifold.

19   Disconnect the water bypass hoses and move the hoses out of the way.

20   Remove the fuel rail insulators from both sides of the engine (see Chapter 4, Section 12).

21   Disconnect the wiring harness clamps at the rear of the intake manifold.

22   Remove the mounting nuts and bolts **(see illustration)**, then detach the intake manifold from the cylinder heads. Start with the outer bolts first and work your way to the inner bolts on the manifold. Make several passes to ensure that the manifold is separated from the cylinder heads evenly to avoid damage to the cylinder head and manifold surfaces. If it's stuck, don't pry between the gasket mating surfaces or damage may result.

Installation

2UZ-FE engines

Refer to illustration 5.23

23   Remove all traces of old gasket material and sealant from the upper and lower intake manifolds and cylinder heads, then

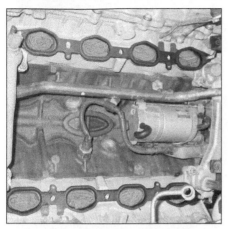

5.23 Plug the cylinder head ports with rags while cleaning the mating surfaces, but don't forget to remove them later

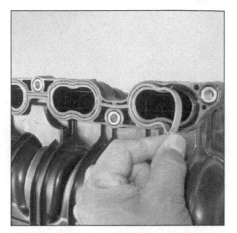

5.30 Install the new gasket into the groves on the manifold

6.4 Disconnect the oxygen sensor, then unbolt the exhaust pipe from the ends of the exhaust manifold

clean the mating surfaces with brake system cleaner **(see illustration)**.

24   Install new gaskets. Position the lower intake manifold on the engine, then install the fasteners.

25   Tighten the fasteners, in three or four equal steps, to the torque listed in this Chapter's Specifications. Work from the center out towards the ends to avoid warping the manifold.

26   Install a new gasket and the upper intake manifold onto the lower intake manifold. Install the nuts and bolts and tighten them in three or four equal steps to the torque listed in this Chapter's Specifications.

27   Install the remaining parts in the reverse order of removal.

28   Check the coolant level (see Chapter 1). Run the engine and check for fuel, vacuum and coolant leaks around the throttle body hoses.

1UR-FE, 3UR-FE and 3UR-FBE engines

Refer to illustration 5.30

29   Remove all traces of old gasket from the intake manifolds then clean the mating surfaces with brake system cleaner.

30   Install new gaskets to the manifold **(see illustration)**. Position the intake manifold on the engine, then install the fasteners.

31   Tighten the fasteners, in three or four equal steps, to the torque listed in this Chapter's Specifications. Work from the center out towards the ends to avoid warping the manifold.

32   Install the remaining parts in the reverse order of removal.

33   Refill the cooling system (see Chapter 1). Run the engine and check for fuel, vacuum and coolant leaks around the throttle body hoses.

---

6   Exhaust manifolds - removal and installation

Warning: *The engine must be completely cool before beginning this procedure.*
Warning: *On Sequoia models equipped with rear height control suspension, adjust the height control to the NORMAL mode,*

turn OFF the height control, then turn off the engine BEFORE raising the vehicle.

2UZ-FE engines

Refer to illustration 6.4

1   Disconnect the cable from the negative battery terminal (see Chapter 5).

2   Remove the heat shields from the exhaust manifolds.

3   Disconnect the oxygen sensor wiring.

4   Spray penetrating oil on the exhaust manifold and exhaust flange fasteners and allow it to soak in. Remove the exhaust pipe flange nuts from the exhaust manifolds **(see illustration)**.

5   Unbolt the exhaust manifolds from the cylinder heads.

1UR-FE, 3UR-FE and 3UR-FBE engines

6   Remove the exhaust pipe flange nuts from the exhaust manifolds and lower the exhaust pipe.

7   Remove the lower engine splash shield (see Chapter 1, **illustrations 8.6a and 8.6b**).

8   Drain the engine coolant (see Chapter 1).

9   Remove the engine cover (see Chapter 1, **illustration 24.12a**).

10   Remove the air filter housing (see Chapter 4).

11   Remove the drivebelt (see Chapter 1).

12   Remove the upper and lower radiator hoses and fan shroud (see Chapter 3).

13   Remove the inner fender seal and wheel housing splash shield (see Chapter 11, Section 14) from the side the manifold is being removed from.

Right side manifold

Refer to illustration 6.18

14   Remove the engine oil dipstick tube mounting bolts and remove the tube.

15   Remove the power steering pump (see Chapter 10).

16   Remove the alternator (see Chapter 5).

17   Remove the heat shield fasteners and heat shield.

6.18 Remove all of the gaskets used with the exhaust manifold

18   Remove the exhaust manifold nuts and remove the manifold and gaskets **(see illustration)**.

Left side manifold

19   Remove the A/C compressor (see Chapter 3) and place it out of the way.

20   On 4WD models, remove the front driveshaft (see Chapter 8) then remove the driveshaft heat shield fasteners and shield.

21   Make aligning marks, then disconnect the steering coupler (see Chapter 10).

22   Remove the heat shield fasteners and heat shield.

23   Remove the exhaust manifold nuts and remove the manifold.

All models

24   Carefully inspect the manifolds and fasteners for cracks and damage.

25   Use a scraper to remove all traces of old gasket material and carbon deposits from the manifolds and cylinder head mating surfaces. If the gasket was leaking, use a precision straight-edge to check the manifolds for warpage and compare your readings with those listed in this Chapter's Specifications. They can be resurfaced by a machine shop, if necessary.

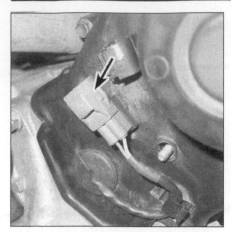

7.9a Disconnect the camshaft sensor from the left side number 3 timing belt cover

7.9b Remove the bolts from the left side number 3 timing belt cover (right side similar)

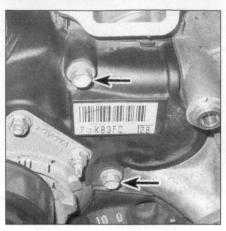

7.11 Remove the number 2 (center) timing belt cover mounting bolts

26   Position new gaskets over the cylinder head studs.
27   Install the manifolds and thread the mounting nuts into place.
28   Working from the center out, tighten the nuts to the torque listed in this Chapter's Specifications in three or four equal steps.
29   Reinstall the remaining parts in the reverse order of removal. Use new gaskets when connecting the exhaust pipes.
30   Run the engine and check for exhaust leaks.

7   Timing belt and sprockets (2UZ-FE engines) - removal, inspection and installation

Warning: *Wait until the engine is completely cool before beginning this procedure.*
Warning: *On Sequoia models equipped with rear height control suspension, adjust the height control to the NORMAL mode, turn OFF the height control, then turn off the engine BEFORE raising the vehicle.*
Caution: *After the timing belt is installed,*

be sure to check all the camshaft sprockets and the crankshaft sprocket alignment marks before starting the engine. If the procedure is not followed exactly, damage to the valves and pistons may occur.

Removal

Refer to illustrations 7.9a, 7.9b, 7.11, 7.15a, 7.15b, 7.15c, 7.16a, 7.16b, 7.17a, 7.17b, 7.18, 7.19, 7.20, 7.21, 7.24 and 7.25

1   Disconnect the cable from the negative battery terminal (see Chapter 5).
2   Remove the under-vehicle splash shield (see Chapter 1). Drain the engine coolant (see Chapter 1).
3   Remove the fan shroud and the fan assembly (see Chapter 3).
4   Remove the drivebelt (see Chapter 1).
5   Remove the power steering pump and set it aside, without detaching the hoses (see Chapter 10). Remove the alternator (see Chapter 5) and the air conditioning compressor, without disconnecting the refrigerant lines (see Chapter 3).
6   Remove the throttle body cover.
7   Remove the drivebelt idler pulley.

8   On models so equipped, disconnect the hoses from the oil cooler pipe mounted on the left side of the engine (plug the hoses to prevent leakage). Remove the fasteners and detach the oil cooler pipe.
9   Remove the left side number 3 timing belt cover **(see illustrations)**.
10   Remove the right side number 3 timing belt cover.
11   Remove the number 2 timing belt cover **(see illustration)**.
12   Remove the fan bracket.
13   Remove the drivebelt tensioner mounting bolts and tensioner from the engine block.
14   Position the number one piston at TDC (see Section 3).
15   Check to make sure the marks on the camshaft and crankshaft sprockets are lined up properly **(see illustrations)**. Also check to see if there are installation marks on the timing belt - if you intend to re-use the belt and the marks have been obscured, make new ones. Place a new mark on the belt at the exact location of each timing mark on each camshaft sprocket. This will allow easy and accurate timing belt alignment if the old timing belt is reused.

7.15a The mark on the left camshaft sprocket must be aligned with the line on the rear timing belt cover

7.15b The mark on the right camshaft sprocket must also be aligned with the line on the rear timing belt cover

7.15c Make sure the engine is still set at TDC

7.16a Use a chain wrench to lock the crankshaft pulley into place - note the piece of old drivebelt used to prevent damage to the pulley

7.16b Use a puller to remove the crankshaft pulley

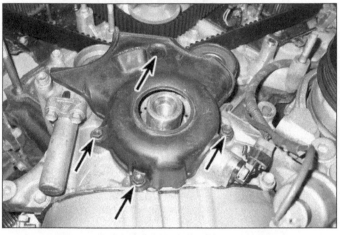

7.17a Locations of the fasteners securing the number 1 timing belt cover

7.17b Note which side faces out, then remove the crankshaft sensor timing plate from the crankshaft

16   Remove the crankshaft pulley. If air tools are not available, lock the crankshaft pulley with a special tool to prevent the engine from rotating and remove the crankshaft pulley bolt with a breaker bar and socket **(see illustration)**. A puller is usually required for removing the pulley **(see illustration)**.

17   Remove the number one timing belt cover and the crankshaft sensor timing plate **(see illustrations)**.

18   Double-check to make sure the camshaft sprockets and the crankshaft timing sprocket timing marks are properly aligned after removal of the crankshaft pulley. If the engine was rotated slightly, re-align the timing marks **(see illustrations 7.15a, 7.15b and 7.15c)**. If you're re-using the belt, check for a mark on the belt adjacent to the drilled mark on the crankshaft sprocket. If the original mark is gone, make a new one, then slip the belt off the sprocket **(see illustration)**. Remove the timing belt from the engine.

19   Temporarily reinstall the crankshaft pulley and bolt. Immobilize the pulley and tighten

the bolt, then turn the crankshaft approximately 45-degrees counterclockwise (1/8-turn), from TDC **(see illustration)**.

7.18 Mark the timing belt where it aligns with the mark on the crankshaft sprocket before you remove it - if you plan to reuse the belt

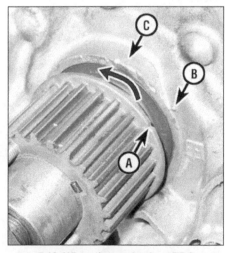

7.19 When the engine is at TDC compression for cylinder no. 1, the mark on the crankshaft sprocket (A) will be aligned with the mark on the oil pump housing (B). From this point, the engine must be rotated about 45-degrees counterclockwise (C)

7.20 Remove the tensioner mounting bolts

7.21 Use a special tool to turn the right camshaft sprocket slightly to relieve belt tension

20   Remove the timing belt tensioner **(see illustration)**. Remove the rubber boot as well; it may stick in the tensioner recess.

21   Using a special camshaft sprocket tool or a pin spanner **(see illustration)**, release the tension between the right side camshaft sprocket and the crankshaft sprocket by slightly rotating the right side camshaft sprocket. This will allow slack on the timing belt.

22   Remove the timing belt from the sprockets.

23   The camshaft sprockets can be removed at this point if they are worn or damaged.

Note: *DO NOT remove the four smaller, recessed bolts; only remove the four larger bolts that retain the sprocket to the timing tube. If the four smaller, recessed bolts are removed, the inner gear will lose its adjustment and the entire timing tube will have to be replaced.*

24   If necessary, it is now possible to remove the number 1 and number 2 idler pulleys **(see illustration)**.

25   If it's worn or damaged, or if you're replacing the crankshaft front oil seal, the crankshaft sprocket can now be removed.

If it won't come off by hand, lever it off with two screwdrivers. A steering wheel type puller may be needed to remove the sprocket **(see illustration)**.

Inspection

26   Inspect the timing belt for cracks, tears, torn belt strands, cut edges or broken belt teeth. Replace the timing belt if there are any signs of damage or prolonged wear (high mileage). Also check the water pump for wear (see Chapter 3).

Note: *It's very easy to replace the water pump at this time.*

27   Check the belt tensioner for visible oil leakage. If there's only a faint trace of oil on the pushrod side, the tensioner seal is in satisfactory condition.

28   Hold the tensioner in both hands and push it forcefully against an immovable object. If the pushrod moves, replace the tensioner.

29   Measure the protrusion of the pushrod past the housing end. Compare your measurement to this Chapter's Specifications. If the protrusion is not as specified, replace the tensioner.

30   Check that both idler pulleys turn smoothly.

Installation

Refer to illustrations 7.36a, 7.36b, 7.36c and 7.37

31   Remove all dirt, oil and grease from the timing belt area at the front of the engine.

32   Install the number 1 and number 2 idler pulleys if they were removed, tightening the bolts to the torque listed in this Chapter's Specifications.

Note: *Apply a non-hardening thread locking compound to the threads of the bolts before installing them.*

33   Align the crankshaft timing sprocket keyway with the crankshaft key and install the sprocket with the flange side up against the engine.

34   Align the camshaft sprocket marks with their marks on the rear timing belt cover **(see illustrations 7.15a and 7.15b)**.

Note: *To make belt installation easier, set the right camshaft sprocket clockwise one tooth and the left camshaft sprocket clockwise 1/2 tooth.*

35   Turn the crankshaft 45-degrees back to TDC (in the opposite direction that it was turned in Step 19).

36   Install the timing belt onto the sprockets and idler pulleys, aligning the marks on the belt with the marks on the pulleys **(see illustrations)**, in the following order:

a) *Crankshaft sprocket*
b) *No. 2 idler pulley*
c) *Left-side camshaft sprocket*
d) *Water pump pulley*
e) *Right-side camshaft sprocket*
f) *No. 1 idler pulley*

37   Using a vise, slowly compress the timing belt tensioner pushrod. Insert a metal pin, drill or Allen wrench through the holes in the pushrod and housing **(see illustration)**. Release the tensioner from the vise.

38   Install the timing belt tensioner and tighten the bolts to the torque listed in this Chapter's Specifications. Remove the retaining pin.

7.24 The number 1 and number 2 idler pulleys can be removed if they show signs of wear or damage

7.25 Use a puller to remove the crankshaft sprocket from the crankshaft

7.36a Align the marks on the timing belt with the marks on the left camshaft sprocket, which in turn should line up with the mark on the rear timing belt cover

7.36b Right camshaft sprocket and timing belt detail

7.36c Install the new timing belt with the crankshaft designation (CR) on the timing belt aligned with the mark on the crankshaft sprocket

39   Using a socket and breaker bar on the crankshaft pulley bolt, turn the crankshaft slowly through two complete revolutions (720-degrees). Recheck the timing marks.
Caution: *If you feel any resistance at any point, STOP! Something is wrong; most likely the valves are contacting the pistons. Find out what's wrong before proceeding. Also, if the timing marks are not aligned exactly as shown, repeat the timing belt installation procedure. DO NOT attempt to start the engine until you're absolutely certain that the timing belt is installed correctly. Serious and costly engine damage will occur if the belt is installed incorrectly.*
40   Slip the crankshaft sensor timing plate over the crankshaft with the cupped side facing out.
41   Install the number 1 timing belt cover and gasket.

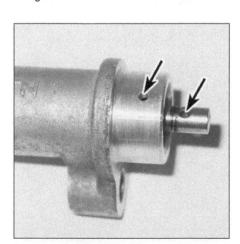

7.37 Collapse the tensioner pushrod by compressing it in a vise and inserting a pin approximately 1.5 mm in diameter through the holes when they are aligned - when installing the tensioner, make sure the rubber boot is in place

42   Slip the crankshaft (drivebelt) pulley onto the crankshaft, aligning the pulley keyway with the crankshaft key. Install the bolt and tighten it to the torque listed in this Chapter's Specifications. Use the method described in Step 16 to keep the crankshaft from turning.
43   Install the drivebelt tensioner. Tighten the bolts to the torque listed in this Chapter's Specifications. Make sure the pulley turns smoothly.
44   Install the fan bracket.
45   Install the air conditioning compressor.
46   Install the upper (number 2) timing belt cover and gasket.
47   Install the right and left number 3 timing belt covers.
48   Reinstall the remaining parts in the reverse order of removal.
49   Refill the cooling system (see Chapter 1).
50   Start the engine and check for leaks and proper operation.

8   Timing chains and sprockets (1UR-FE, 3UR-FE and 3UR-FBE engines) - removal, inspection and installation

Warning: *On Sequoia models equipped with rear height control suspension, adjust the height control to the NORMAL mode, turn OFF the height control, then turn off the engine BEFORE raising the vehicle.*

Removal
Refer to illustrations 8.14, 8.17a, 8.17b, 8.18, 8.20a, 8.20b, 8.21 and 8.26
Warning: *Wait until the engine is completely cool before beginning this procedure.*
Caution: *The timing system is complex, and severe engine damage will occur if you make*

any mistakes. Do not attempt this procedure unless you are highly experienced with this type of repair. If you are at all unsure of your abilities, be sure to consult an expert. Double-check all your work and be sure everything is correct before you attempt to start the engine.
Note: *There are access plugs for the timing chain tensioner on the left and right sides of the timing chain cover.*
1   Disconnect the cable from the negative terminal of the battery (see Chapter 5).
2   Relieve the fuel pressure (see Chapter 4), then disconnect the fuel lines.
3   Drain the engine oil and engine coolant (see Chapter 1) and remove the engine cover (see Chapter 1, **illustration 24.12a**).
4   Remove the fan shroud and radiator (see Chapter 3).
5   Remove the air inlet duct and air filter housing (see Chapter 4).
6   Disconnect all the electrical connectors from the engine compartment fuse/relay box, air injection, fuel injector harness, ignition coils, VVT sensors, engine coolant temperature sensor, camshaft timing oil control valves and camshaft position sensor.
7   Remove the intake manifold (see Section 5).
8   Disconnect the heater hoses and pipe, the thermostat housing, water pump and all the by-pass hoses (see Chapter 3).
9   Remove the A/C compressor (see Chapter 3).
Note: *It is not necessary to completely remove the compressor. With the hoses connected to the compressor, hang the compressor out of the way using a piece of wire.*
10   Remove the oil filter and oil cooler bracket (see Chapter 3, Section 16).
11   Unbolt the oil dipstick tube and remove it from the pan.
12   Remove the power steering pump (see Chapter 10).
13   Remove the alternator (see Chapter 5).

8.14 Remove the fan bracket mounting bolts

8.17a Hold the crankshaft pulley with a pin spanner while removing the bolts; if this is not available, a chain wrench can also be used to hold the pulley, but only if special precautions are taken. Use a piece of old drivebelt to pad the pulley before installing the chain wrench

8.17b If the crankshaft pulley can't be removed by hand, use a puller that bolts to the hub of the pulley - not a jaw-type puller. Also, be sure to use the correct adapter between the nose of the crankshaft and the puller screw, so as not to damage the threads in the crankshaft

14   Remove the fan bracket assembly bolts and bracket (see illustration).
15   Remove the ignition coils and spark plugs (see Chapter 1).
16   Remove both valve covers (see Section 4).
17   Set the engine to TDC (see Section 3). Remove the crankshaft pulley (see illustrations).
18   Remove the timing chain cover mounting fasteners and remove the timing chain cover.
Note: There are six different sized bolts and one nut used on the timing cover (see illustration 8.48a).
There are only a few spots where you can safely pry the cover off without damaging it (see illustration). Do NOT pry the timing chain cover loose at any other spot or you will damage the sealing surface of the cover.
19   After removing the timing chain cover, carefully pry out the old crankshaft seal with a screwdriver (see Section 9). Make sure that you don't scratch the seal bore. If you want to inspect or replace any oil pump parts, refer to Section 14.

Note: Keep track of the locations of all of the bolts. They are of different lengths and can't be interchanged.
20   Verify that the piston in the No. 1 cylinder is near TDC on its compression stroke. If not, install the crankshaft pulley bolt, then rotate the crankshaft clockwise until the dot on the crankshaft timing sprocket is at 6 o'clock and the key way is at 2 o'clock. You can also temporarily install the front cover and the crankshaft pulley and set the pulley at the 0 degree mark. Verify that the timing marks on the camshaft timing sprockets are aligned with their corresponding marks on chains (see illustrations). If the marks are not aligned, rotate the crankshaft another 360-degrees and recheck the marks.

8.18 The timing chain cover can be pried loose at the lower corners and at the upper corners; prying anywhere else may damage the cover

8.20a Verify that the orange link (A) on the No.1 chain is aligned with the mark on the camshaft sprocket and the yellow link (B) on the No.2 chain is also aligned with the mark on the camshaft sprocket – right side shown

8.20b Verify that the orange link (A) on the No.1 chain is aligned with the mark on the camshaft sprocket and the yellow link (B) on the No.2 chain is also aligned with the mark on the camshaft sprocket – left side shown

8.21 To lock the tensioner in the retracted position, rotate the stopper plate clockwise and push the plunger in, then rotate the stopper plate counterclockwise and insert a pin through the hole in the stopper plate and the tensioner body

8.26 Hold the hex on the camshaft with a wrench and unscrew the bolt that secures the camshaft timing sprocket

Caution: *The upper timing marks on this engine do NOT align perfectly when the timing chains are correctly installed. Because of this, the only positive way to proceed is to carefully mark the positions of each timing chain on each camshaft sprocket as well as the crankshaft sprocket. Use a marker or paint that can't be rubbed off easily. It helps to use different colors to avoid confusion later.*

21 Turn the stopper plate on the No. 1 tensioners clockwise and push in the tensioner plunger **(see illustration)**. To lock the plunger in this position, turn the stopper plate counterclockwise and insert a drill or punch (0.39-inch [1.0 mm] diameter) through the holes in the stopper plate and the tensioner body. Remove the tensioner mounting bolts and remove both No. 1 tensioner.

22 Remove the chain vibration damper fastener and dampers from each chain.

23 Note the positions of the timing marks on each camshaft sprocket and the crankshaft sprocket. Make sketches or take pictures if you have a camera. Remove the No. 1 timing chain.

Caution: *While the No. 1 timing chain is removed, DO NOT ROTATE THE CRANK-SHAFT!*

24 Remove the crankshaft timing chain sprocket.

25 Compress chain tensioner No. 2 and insert a drill or punch (0.039-inch diameter) into the hole.

Note: *Timing chain No. 2 and the No. 2 chain tensioner are on the right cylinder head. Timing chain No. 3 and the No. 3 tensioner are on the left cylinder head.*

26 Hold the hex on the exhaust camshaft with a wrench and unscrew the bolt that secure the camshaft timing sprockets to the camshafts **(see illustration)**. Remove the sprockets and chain as an assembly. Keep the components in a resalable plastic bag to ensure that none of these components is mixed with the other timing chain set.

Caution: *Don't attempt to disassemble the adjustable intake sprocket assembly. If disassembled, it will have to be replaced.*

27 Remove the chain tensioner No. 2 mounting bolt and remove chain tensioner No. 2. Store the tensioner in the plastic bag with the other No. 2 timing chain components.

28 To remove the No. 3 timing chain and tensioner, repeat these Steps. Again, store the components in a resealable plastic bag.

Caution: *While the timing chains are removed, DO NOT ROTATE THE CRANKSHAFT!*

Inspection

Refer to illustrations 8.30 and 8.31

29 Inspect all parts of the timing chain assembly for wear and damage. Inspect the three timing chains for loose pins, cracks, and worn rollers and side plates. Inspect the sprockets for hook-shaped, chipped and/or broken teeth.

30 Inspect timing chain No. 1 for stretching. To measure timing chain stretch, measure the distance between 15 pins at three or more places around the length of the chain **(see illustration)**. Measure between the inside of the rollers and compare your measurements with the distance listed in this Chapter's Specifications.

31 Measure the diameter of each timing chain sprocket with the appropriate timing chain installed on the sprocket **(see illustration)**. The sprocket diameter, with the chain in place, should not be less than the dimensions listed in this Chapter's Specifications.

32 Some scoring and wear of the timing chain tensioners and vibration dampers is normal, but excessive wear will increase chain

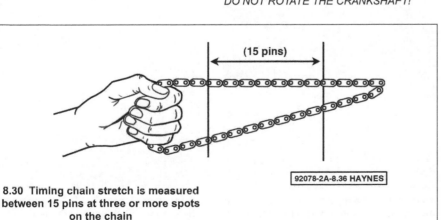

8.30 Timing chain stretch is measured between 15 pins at three or more spots on the chain

(15 pins)

92078-2A-8.36 HAYNES

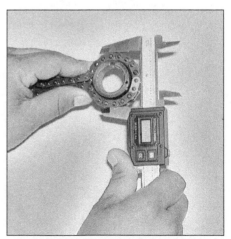

8.31 Measure the timing chain sprocket diameter with the chain installed

8.36 The small No. 2 timing chain maintains the alignment of the intake and exhaust camshaft sprockets of the right cylinder head; set the yellow links (A) on the single lines (B) of the camshaft sprockets

8.38 Set the yellow links (A) of the No. 2 timing chain of the left cylinder head with the single lines (B) of the camshaft sprockets

noise, accelerate chain and sprocket wear and could damage the engine if a chain jumps timing. Inspect chain tensioners No. 2 and 3, the timing chain tensioner slipper and the timing chain vibration dampers for excessive wear.

33　If the measured chain wear for any of these components exceeds the depth listed in this Chapter's Specifications, replace the component.

34　Check the chain tensioners for correct operation. On the No. 1 tensioner, raise the ratchet pawl and verify that the plunger moves smoothly in and out of the tensioner, then release the ratchet pawl and verify that it prevents the plunger from sliding back into the tensioner. Also verify that the plungers on the No. 2 and No. 3 tensioners move in and out smoothly.

Installation

Refer to illustrations 8.36, 8.38, 8.41a, 8.41b, 8.41c, 8.48a and 8.48b

Caution: *Before starting the engine, carefully rotate the crankshaft by hand through at least two full revolutions (use a socket and breaker bar on the crankshaft pulley center bolt). If*

you feel any resistance, STOP! There is something wrong - most likely, valves are contacting the pistons. You must find the problem before proceeding. Check your work and see if any updated repair information is available.

35　Push in the tensioner plunger on chain tensioner No. 2 and insert a drill or punch (0.039 inch diameter) into the hole of the tensioner to lock the plunger in the retracted position. Install the tensioners and tighten the mounting bolts to the torque listed in this Chapter's Specifications.

36　Install the No. 2 (right bank inner) timing chain on the camshaft sprockets. Make sure that the yellow mark links on the chain are aligned with the single timing lines on the camshaft sprockets **(see illustration)**.

37　Align the yellow links on the No. 2 timing chain with the timing marks on the bearing caps and install timing chain No. 2 and the camshaft sprockets as an assembly. Install the two bolts that secure the timing sprockets to the camshafts. Immobilize the hex on the exhaust camshaft with an adjustable wrench and tighten these two bolts to the torque listed in this Chapter's Specifications. Remove

the drill or punch that was used to lock the tensioner plunger in its retracted position and verify that the plunger tensions the chain.

38　Install the other inner timing chain by repeating Steps 35 through 37. Align the yellow links with the lines on the sprockets **(see illustration)**.

39　Install the chain guides and tighten the bolts to the torque listed in this Chapter's Specifications.

40　Install the crankshaft timing sprockets on the crankshaft (sprocket for right side first).

Note: *It may be necessary to install the crankshaft and camshaft sprockets as a unit with all the chains installed.*

41　Install the long timing chain (No. 1) on the camshaft timing sprockets. Turn the camshaft sprockets to remove the slack in the upper part of the chain, then install the chain over the crankshaft sprocket. Make sure that the orange link on each chain is aligned with the timing dot on the crankshaft timing sprockets (it's near the 6 o'clock position) and that the orange link is aligned with the timing mark on the intake camshaft sprockets **(see illustrations)**.

8.41a Align the timing marks (dot) on the sprockets with the orange plated links of the chains

8.41b Install the right side No.1 chain with the orange link aligned with the dot on the camshaft sprocket

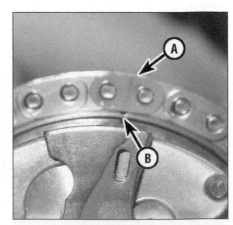

8.41c Install the left side No.1 chain with the orange link aligned with the dot on the camshaft sprocket

8.48a Timing chain cover bolt length locations – 3UR-FE engine shown, 1UR-FE similar

A	Bolt A 0.98-inches (25 mm) length		D	Bolt D 1.38-inches (35 mm) length
B	Bolt B 2.17-inches (55 mm) length		E	Bolt E 2.17-inches (55 mm) length
C	Bolt C 2.76-inches (70 mm) length		F	Bolt F 3.15-inches (80 mm) length

42 Turn the stopper plate on the No. 1 tensioner clockwise and push in the tensioner plunger. To lock the plunger in this position, turn the stopper plate counterclockwise and insert a drill or punch (0.138-inch diameter) through the holes in the stopper plate and the tensioner. Install the tensioner and tighten the tensioner mounting bolts to the torque listed in this Chapter's Specifications. Remove the drill or punch that you inserted into chain tensioner No. 1 and verify that it tensions timing chain No. 1.

Caution: *Carefully rotate the crankshaft by hand through at least two full revolutions (use a socket and breaker bar on the crankshaft pulley center bolt). If you feel any resistance, STOP! There is something wrong - most likely, valves are contacting the pistons. You must find the problem before proceeding. Check your work to make sure all timing marks line-up properly, and see if any updated repair information is available.*

43 Remove all old RTV sealant from the gasket mating surfaces of the timing chain cover and from the front of the cylinder heads and engine block.

44 Install a new crankshaft oil seal in the timing chain cover (see Section 9).

45 Install a new O-ring on the left cylinder head.

46 Apply gray RTV sealant on the timing chain cover in all areas where the cover seals to the engine block and oil pan. These beads should be about 1/8 to 3/16-inch (3 to 4.5 mm) wide.

Caution: *Once you have installed the seal-*

ant on the engine and timing chain cover you have three minutes to install the cover. If you take longer than that, the sealant might not set up properly, so you'll have to remove the sealant and re-apply it.

Be sure to get sealant into the corners where the oil pan meets the engine block and avoid

getting any on the O-rings.

47 Rotate the flats on the oil pump drive rotor to align it with the square part of the crankshaft timing sprocket and slide the timing chain cover into place.

48 Install all of the cover fasteners in the same locations they were removed from, tightening them as you go.

Caution: *Do not put long bolts in short holes or vice versa.*

Tighten all of the fasteners in the proper sequence to the torque listed in this Chapter's Specifications **(see illustrations).**

Note: *To properly complete the tightening sequence, the water pump and fan pulley bracket must be installed after the first three bolts are tightened.*

49 The remainder of installation is the reverse of the removal procedure.

50 Refill the engine with oil and coolant (see Chapter 1).

51 Reconnect the cable to the negative terminal of the battery (see Chapter 5).

52 Start the engine and check for leaks.

9 Crankshaft front oil seal - replacement

Refer to illustrations 9.4 and 9.5

1 On 2UZ-FE engines, remove the timing belt and crankshaft timing belt sprocket (see Section 7).

2 On 1UR-FE, 3UR-FE and 3UR-FBE engines, remove the crankshaft pulley (see Section 8).

3 Note how far the seal is recessed in the bore, then cut away the seal lip with a razor knife.

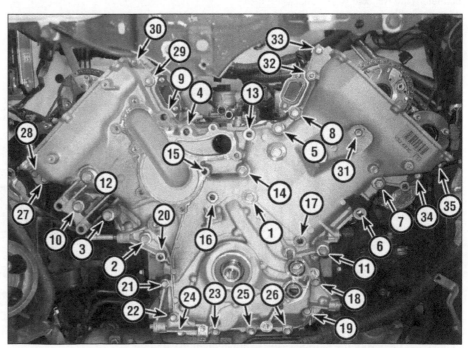

8.48b Timing chain cover tightening sequence – 3UR-FE engine shown, 3UR-FBE and 1UR-FE similar

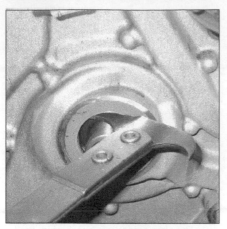

9.4 Cut away the crankshaft seal lip, wrap a screwdriver tip with tape or use a seal removal tool and pry out the seal – 3UR-FE engine shown, all others similar

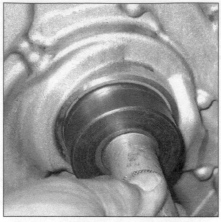

9.5 Lubricate the seal lip and drive the new crankshaft seal into place with a seal driver or a large socket and a hammer – 3UR-FE engine shown, all others similar

4   Carefully pry the seal out of the engine with a screwdriver or seal removal tool **(see illustration)**. If you use a screwdriver, wrap tape around the tip - don't scratch the housing bore or damage the crankshaft (if the crankshaft is damaged, the new seal will end up leaking).

5   Clean the bore in the engine and coat the outer edge of the new seal with engine oil or multi-purpose grease. Apply the same grease to the seal lip.

6   Using a seal driver or a socket with an outside diameter slightly smaller than the outside diameter of the seal, carefully drive the new seal into place with a hammer **(see illustration)**. Make sure it's installed squarely and driven in to the same depth as the original. If a socket isn't available, a short section of large diameter pipe will also work. Check the seal after installation to make sure the spring didn't pop out of place.

7   On 2UZ-FE engines, reinstall the crankshaft timing sprocket and timing belt (see Section 7).

8   On 1UR-FE, 3UR-FE and 3UR-FBE engines reinstall the crankshaft pulley (see Section 8).

9   Run the engine and check for oil leaks at the front seal.

---

10 Camshafts and lifters (2UZ-FE engines) - removal, inspection and installation

---

Warning: *The engine must be completely cool before beginning this procedure.*

Note: *Before beginning this procedure, obtain two 6 x 1.0 mm bolts 16 to 20 mm long. They will be referred to as service bolts in the text.*

Removal

Refer to illustrations 10.3, 10.6, 10.8, 10.12, 10.13 and 10.14

1   Remove the valve covers (see Section 4). Measure the valve clearances (see Chapter 1), then remove the timing belt.

Note: *If cylinder head work involving the valves is to be done, don't measure the valve clearances at this time. Otherwise, the valve clearances should be measured now, and replacement shim thicknesses can be calculated. This will minimize the chances of having to remove the camshafts after this job has been completed, due to incorrect valve clearances.*

Caution: *Make sure the crankshaft is positioned 45-degrees (1/8-turn) counterclockwise from the TDC position* **(see illustration 7.19)**.

2   Remove the cam sprockets (see Section 7, Step 23). Also remove the camshaft position sensor (see Chapter 6).

3   The following steps apply to the removal of the camshafts on each cylinder head. Start the removal process on the right bank cylinder head. Secure the exhaust camshaft sub-gear to the driven gear with a service bolt installed in the threaded hole **(see illustration)**. Turn the camshaft with a wrench if necessary, using the hexagonal portion of the exhaust camshaft.

Note: *For reference purposes, the outside camshafts are the exhaust camshafts, while the inner camshafts are the intake camshafts. The right side cylinder head (passenger's side) is called the right bank, while the left side cylinder head (driver's side) is called the left bank.*

4   Align the cam timing marks on the drive and driven gears at an upward 10-degree angle (right side cylinder head only).

5   Loosen the camshaft bearing cap bolts in 1/4-turn increments until they can be removed by hand. Follow the reverse of the recommended tightening sequence **(see illustrations 10.25 and 10.33)**.

6   Remove the bearing caps and gently lift out the oil feed pipe and the camshafts **(see illustration)**. Be sure to keep it level.

Note: *The camshaft cap bolts vary in length. Be sure to mark each bolt carefully to avoid reassembly problems.*

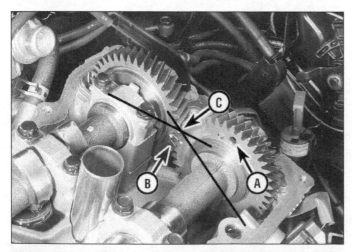

10.3 Insert a service bolt into the threaded hole (A) in the exhaust cam gear, then, with the marks on the gears aligned (B), turn the camshafts (use the hex on the exhaust cam) until the marks are approximately 10-degrees (C) from where they were

10.6 Use care when removing the oil feed pipe

10.8 Mark up a cardboard box to store the lifters/shims and camshaft bearing caps - use a separate box for each set to avoid mix-ups and mark the FRONT, INTAKE and EXHAUST orientation

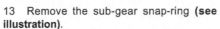

10.12 With the hex portion of the camshaft held in a vise, use a two-pin spanner to remove the tension from the sub-gear and remove the service bolt, then release the sub-gear

10.13 Remove the snap-ring with snap-ring pliers

Caution: *Since the camshaft thrust clearance is minimal, the camshafts must be held level as they are being removed. If they aren't, the portion of the cylinder head next to the cam gears may crack or be damaged. Before lifting a camshaft out of the head, make certain that the torsional spring force of the sub-gear has been eliminated by the service bolt in the exhaust camshaft.*

7   Repeat Steps 3 through 6 for the left-bank cylinder head, including using another service bolt.

Note: *Align the camshaft timing gears together on the left bank cylinder head camshafts - don't set them at a 10-degree angle as you did with the other cylinder head* **(see illustration 10.3)**.

8   Store the bearing caps in the correct order.

Note: *The camshaft cap bolts vary in length. Be sure to mark each bolt carefully to avoid reassembly problems.*

If necessary, the valve lifters and shims can now be removed with a magnetic tool. Be sure to store them separately so they can be reinstalled in their original locations **(see illustration)**.

9   Mount the intake camshaft in a vise with the jaws gripping on the large hex on the shaft. Remove the screw plug and sealing washer from the end of the shaft, unscrew the hex bolt underneath and remove the timing tube assembly from the camshaft.

Caution: *Don't remove the four bolts from the perimeter of the timing tube.*

10   The drive gear that turns the exhaust camshaft can be removed from the timing tube assembly. Hold the drive gear with a pin spanner and unscrew the four bolts securing the drive gear. Pull off the drive gear and oil seal.

11   To disassemble an exhaust camshaft gear, mount the cam in a vise with the jaws gripping the large hex on the shaft.

12   Using a pin spanner, rotate the sub-gear clockwise and remove the service bolt **(see illustration)**.

13   Remove the sub-gear snap-ring **(see illustration)**.

14   The wave washer, sub-gear and camshaft gear spring can now be removed from the camshaft **(see illustration)**. Be sure to keep the parts from the left side camshaft separate from the right side.

Inspection

15   Refer to Chapter 2, Part A for camshaft, lifter and related component inspection procedures. Be sure to use the Specifications in this Part of Chapter 2.

Installation

Refer to illustrations 10.22, 10.25 and 10.33

16   Insert new camshaft plugs into the cylinder head. Apply a small amount of RTV sealant to the grooves.

17   Install a new oil seal, then mate the drive gear to the timing tube by aligning the knock pin with its corresponding groove. Install the four drive gear bolts, tightening them to the torque listed in this Chapter's Specifications. Insert the timing tube into the camshaft, aligning the knock pin with its corresponding

groove, then install the center bolt, tightening it to the torque listed in this Chapter's Specifications. Finally, install the sealing washer and screw plug, tightening it to the torque listed in this Chapter's Specifications.

18   Reassemble the exhaust camshaft gear(s) by installing the camshaft gear spring, sub-gear, wave washer and snap-ring. Mount the camshaft in a padded vise. Using a pin spanner, align the holes of the camshaft driven gear and sub-gear by turning the camshaft sub-gear clockwise. Install a service bolt in the threaded hole, tightening it to clamp the gears together.

19   Apply moly-base grease or engine assembly lube to the lifters, then install them in their original locations in the cylinder heads. Make sure the valve adjustment shims are in place in the lifters, and that all lifters are installed in their original bores.

Right-bank cylinder head

20   Apply moly-base grease or engine assembly lube to the camshaft lobes, bearing journals and gear thrust faces.

21   Set the intake camshaft and exhaust camshaft in place in the cylinder head with the timing marks (two dots) at an 10 degree angle

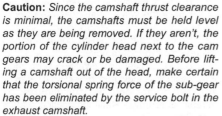

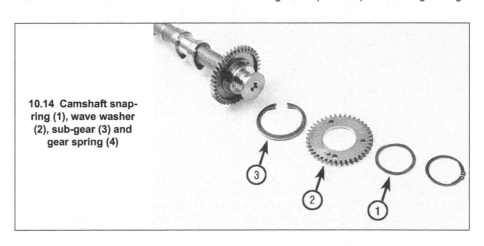

10.14 Camshaft snap-ring (1), wave washer (2), sub-gear (3) and gear spring (4)

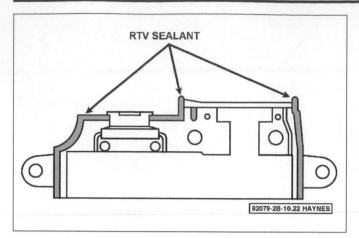

10.22 Apply 1/16-inch (1.5 mm) bead of RTV sealant to the shaded areas on the bearing cap (right bank shown, left bank similar)

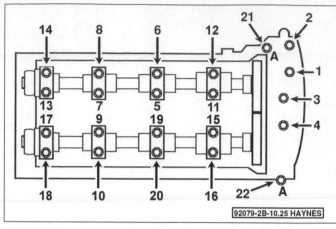

10.25 Tightening sequence for the camshaft bearing caps (right bank). Locate Bolts A (0.98-inch [25 mm] length) and note that they're tightened to a different (lesser) torque

facing each camshaft **(see illustration 10.3)**.

22 Apply a bead of RTV sealant to the edges of the front bearing cap mating surfaces **(see illustration)**.

23 Install the bearing caps in numerical order with the arrows pointing toward the front (timing belt end) of the engine.

Note: *The "I" caps go on the intake side and the "E" caps go on the exhaust side. The lower numbers go toward the timing belt end of the engine.*

24 Before tightening the bearing cap bolts, push inward on the intake camshaft to seat the oil seal.

25 Tighten the bearing cap bolts in 1/4-turn increments to the torque listed in this Chapter's Specifications. Follow the recommended sequence **(see illustration)**.

26 Remove the service bolt from the exhaust cam gear.

27 Install a new camshaft oil seal.

Left-bank cylinder head

28 Apply moly-base grease or engine assembly lube to the camshaft lobes, bearing

journals and gear thrust faces.

29 Set the intake camshaft and exhaust camshaft in place in the cylinder head with the timing marks aligned next to each other on each camshaft **(see illustration 10.3)**.

30 Apply a bead of RTV sealant to the edges of the front bearing cap mating surfaces **(see illustration 10.22)**.

31 Install the bearing caps in numerical order with the arrows pointing toward the front (timing belt end) of the engine.

Note: *The "I" caps go on the intake side and the "E" caps go on the exhaust side. The lower numbers go toward the timing belt end of the engine.*

32 Before tightening the bearing cap bolts, push inward on the intake camshaft to seat the oil seal.

33 Tighten the bearing cap bolts in 1/4-turn increments to the torque listed in this Chapter's Specifications. Follow the recommended sequence **(see illustrations)**.

34 Remove the service bolt from the exhaust cam gear.

35 Install a new camshaft oil seal.

Both cylinder heads

36 Reinstall the timing belt (see Section 7).

37 Check the valve clearances (see Chapter 1).

38 Reinstall the remaining components in the reverse order of removal.

39 Before reinstalling the valve covers, apply RTV sealant as indicated in Section 4, Step 8.

40 The remainder of installation is the reverse of removal. Refill the cooling system (see Chapter 1).

41 Run the engine, then check for leaks and proper operation.

---

11 Camshafts, rocker arms and valve lash adjusters (1UR-FE, 3UR-FE and 3UR-FBE engines) - removal, inspection and installation

---

Warning: *On Sequoia models equipped with rear height control suspension, adjust the height control to the NORMAL mode, turn OFF the height control, then turn off the engine BEFORE raising the vehicle.*

Removal

Refer to illustrations 11.4 and 11.10

Note: *The following procedure is not for beginners. Please read the entire procedure carefully before deciding whether this is a job that you want to tackle at home.*

1 Disconnect the cable from the negative terminal of the battery (see Chapter 5).

2 Drain the engine oil and coolant (see Chapter 1).

3 Set the engine to No. 1 cylinder TDC, then remove the timing chains and sprockets (see Section 8).

4 To remove the camshafts from the right (passenger's side) cylinder head, rotate the camshafts until the knock pins at the front of the camshaft are at 11 o'clock for the exhaust cam and 9 to almost 10 o'clock for the intake cam **(see illustration)**. Use an open-end wrench on the integral hex cast into each camshaft to rotate the cams if necessary.

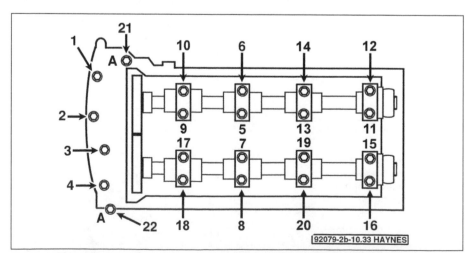

10.33 Tightening sequence for the camshaft bearing caps (left bank). Locate Bolts A (0.98-inch [25 mm] length) and note that they're tightened to a different (lesser) torque

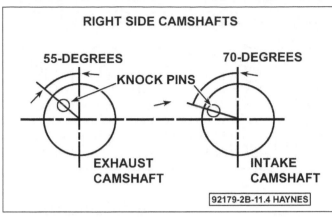

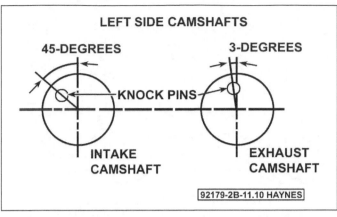

11.4 On the right side cylinder head, make sure the knock pins at the front of the camshaft are at 11 o'clock for the exhaust cam and 9 to almost 10 o'clock for the intake cam before removing the bearing caps

11.10 On the left side cylinder head, make sure that the knock pins at the front of the camshafts are at 10 o'clock for the intake camshaft and 11 o'clock for the exhaust cams before removing the bearing caps

5   Gradually loosen and remove the 10 bearing cap bolts in the reverse of the tightening sequence **(see illustrations 11.16a and 11.16b)**.

6   Gradually loosen and remove the remaining 18 housing bolts in the reverse of the tightening sequence **(see illustrations 11.18a and 11.18b)**.

7   Remove all 6 bearing caps and remove the intake and exhaust camshafts. Be sure to keep all of the components in the correct order. One way to do this is to put them in a box and label the cap numbers with a utility marker pen.

8   Carefully pry the camshaft housing from the top of the cylinder head.

Caution: *Use a screwdriver wrapped with tape to avoid scratching the parts.*

9   Remove all of the rocker arms and lash adjusters and put them in the same box with the cam bearing caps. Every component should be marked or labeled so that it can be returned to its original location.

10   If you're removing the camshafts from the other cylinder head, repeat Steps 5 through 9.

Rotate the camshafts until the knock pins at the front of the camshafts are at 10 o'clock for the intake camshaft and 11 o'clock for the exhaust cams **(see illustration)**.

Caution: *While the timing chains and camshafts are removed, DO NOT ROTATE THE CRANKSHAFT!*

Inspection

Refer to illustrations 11.12, 11.13 and 11.14

11   Inspect each rocker and lash adjuster arm for wear.

12   Visually examine the cam lobes and bearing journals for score marks, pitting, galling and evidence of overheating (blue, discolored areas). Look for flaking away of the hardened surface layer of each lobe. Using a micrometer, measure the height of each camshaft lobe **(see illustration)**. Compare your measurements with this Chapter's Specifications. If the height for any one lobe is less than the specified minimum, replace the camshaft.

13   Using a micrometer, measure the diameter of each journal at several points **(see**

illustration). Compare your measurements with this Chapter's Specifications. If the diameter of any one journal is less than specified, replace the camshaft.

14   Check the oil clearance for each camshaft journal as follows:

a) *Clean the bearing caps and the camshaft journals with brake system cleaner.*

b) *Carefully lay the camshaft(s) in place in the cylinder head. Don't install the lifters and don't use any lubrication.*

c) *Lay a strip of Plastigage on each journal.*

d) *Install the bearing caps with the arrows pointing toward the front (timing chain end) of the engine.*

e) *Tighten the bolts to the torque listed in this Chapter's Specifications in 1/4-turn increments.*

Note: *Don't turn the camshaft while the Plastigage is in place.*

f) *Remove the bolts and detach the caps.*

g) *Compare the width of the crushed Plastigage (at its widest point) to the scale on the Plastigage envelope **(see illustration)**.*

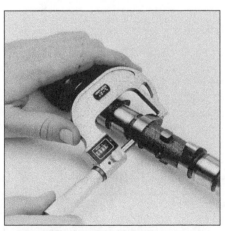

11.12 Measure the lobe heights on each camshaft - if any lobe height is less than the specified allowable minimum, replace that camshaft

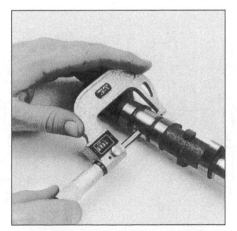

11.13 Measure each journal diameter with a micrometer - if any journal measures less than the specified limit, replace the camshaft

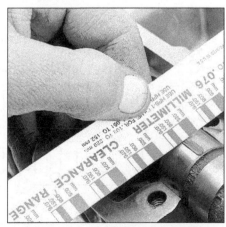

11.14 Compare the width of the crushed Plastigage to the scale on the envelope to determine the oil clearance

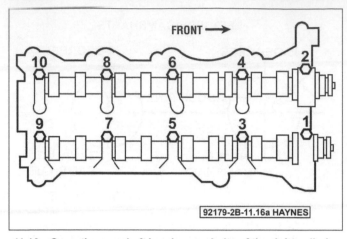

11.16a Snug the camshaft bearing cap bolts of the right cylinder head in this sequence

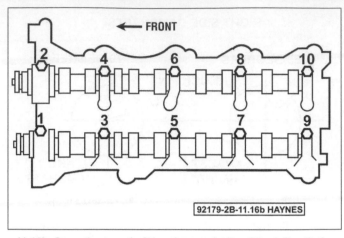

11.16b Snug the camshaft bearing cap bolts of the left cylinder head in this sequence

h) *If the clearance is greater than specified, replace the camshaft and/or cylinder head.*

i) *Scrape off the Plastigage with your fingernail or the edge of a credit card - don't scratch or nick the journals or bearing caps.*

Installation

Refer to illustrations 11.16a, 11.16b, 11.18a and 11.18b

15   Lightly lubricate the lash adjuster bores and the adjusters themselves with clean engine oil, then install them in the same bores from which they were removed. Install the rocker arms in their original positions, oiling all wear points as you do so.

16   Right side: Lightly lubricate the camshaft journals with clean engine oil. Install the camshafts on the right camshaft housing so that the knock pins at the front of the camshaft are at 10 o'clock for the exhaust cam and 9 to almost 10 o'clock for the intake cam **(see illustration 11.4)**. Apply a light coat of engine oil to the upper bearing caps, then install them in their correct locations. Tighten the bolts snug at this time in the correct sequence **(see illustrations)**.

17   Thoroughly clean the sealing surfaces of the bottom of the camshaft bearing support and the top of the cylinder head. Apply a continuous 1/8-inch to 3/16-inch (3.5 to 4.0 mm) bead of RTV silicone sealer to the surface of the top of the cylinder head that mates with the camshaft bearing housing.

18   Set the camshaft housing assembly into place. Install the 18 mounting bolts and tighten them in the correct sequence to the torque listed in this Chapter's Specifications **(see illustrations)**.

19   Loosen the 10 bearing cap bolts you previously tightened, then tighten them in the correct sequence to the torque listed in this Chapter's Specifications **(see illustration 11.16a and 11.16b)**.

20   Left side: Lightly lubricate the camshaft journals with clean engine oil. Install the camshafts on the left cylinder head so that the knock pin of each camshaft is in the 12 o'clock position **(see illustration 11.10)**. Proceed as with the right cylinder head.

21   The remainder of installation is the reverse of removal. Install the timing chains, the timing chain cover and all of the components attached to the timing chain cover (see Section 8).

Caution: *Carefully rotate the crankshaft by hand through at least two full revolutions (use a socket and breaker bar on the crankshaft pulley center bolt). If you feel any resistance, STOP! There is something wrong - most likely valves are contacting the pistons. You must find the problem before proceeding. Check your work to make sure all timing marks line-up properly and see if any updated repair information is available.*

22   Refill the engine with oil and coolant (see Chapter 1), reconnect the cable to the negative battery terminal, start the engine and check for leaks.

Note: *It may take a few minutes for lifter clatter to disappear.*

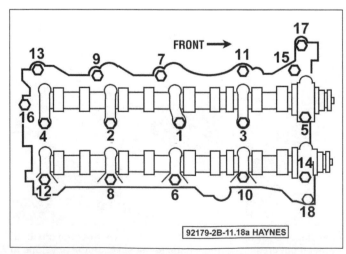

11.18a Camshaft housing bolt tightening sequence for the right cylinder head

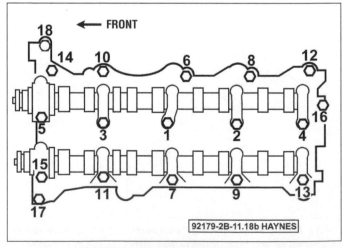

11.18b Camshaft housing bolt tightening sequence for the left cylinder head

12 Cylinder heads - removal and installation

Warning: *The engine must be completely cool before beginning this procedure.*
Note: *Before beginning this procedure, obtain a set of new cylinder head bolts.*

Removal

1   Disconnect the cable from the negative battery terminal (see Chapter 5).
2   Drain the cooling system, including the block (see Chapter 1).
3   Remove the throttle body, fuel rails and injectors (see Chapter 4).
4   Remove the engine dipstick tube.
5   Remove the thermostat housing (see Chapter 3).
6   Remove the front and rear coolant passages. Remove the interfering front coolant pipe assembly.
7   Remove the exhaust manifold(s) (see Section 6).
Note: *The exhaust pipes can be disconnected from the exhaust manifolds to allow the manifolds to be removed along with the cylinder heads if desired.*
8   Remove the alternator (see Chapter 5).
9   Remove the intake manifold(s) (see Section 5).
10 Remove the valve covers (see Section 4).

2UZ-FE engines

11 Remove the timing belt, camshaft sprockets, the drivebelt idler pulley and the drivebelt tensioner (see Section 7).
12 Remove upper timing belt cover number 3.
13 Remove the camshaft(s) from the cylinder heads (see Section 10).

1UR-FE, 3UR-FE and 3UR-FBE engines

14 Set the No. 1 cylinder to TDC on the compression stroke (see Section 3).
15 Remove the timing chain cover, timing chains and sprocket (see Section 8).
16 Remove the camshaft housing and camshafts (see Section 11).

All engines

17 Loosen the cylinder head bolts in 1/4-turn increments until they can be removed by hand. Follow the reverse order of the factory recommended tightening sequence **(see illustration 12.27)**.
Note: *Stuff a rag into the oil drain hole at the end of the cylinder head to avoid dropping a head bolt washer into it. Pieces dropped into this hole end up in the oil pan.*
18 Lift the cylinder head off the engine block. If the head is stuck, place a wood block against it and strike the wood with a hammer.
Caution: *Don't pry between the head and block. The gasket surfaces may be damaged and leaks could result.*
19 Repeat the procedure for the other head.

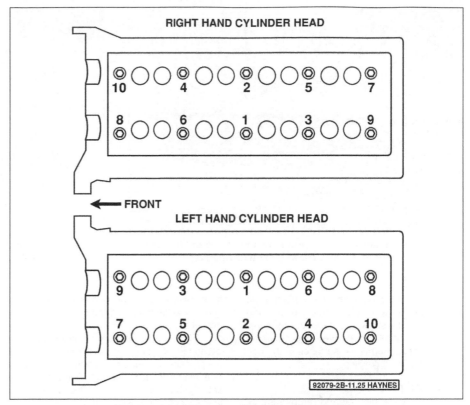

12.27 Cylinder head bolt TIGHTENING sequence

Installation

Refer to illustration 12.27

20 The mating surfaces of the cylinder heads and block must be perfectly clean when the heads are installed.
21 Use a gasket scraper to remove all traces of carbon and old gasket material, then clean the mating surfaces with brake system cleaner. If there's oil on the mating surfaces when the head is installed, the gasket may not seal correctly and leaks could develop. When working on the block, stuff the cylinders with clean shop rags to keep out debris. Use a vacuum cleaner to remove material that falls into the cylinders.
22 Check the block and head mating surfaces for nicks, deep scratches and other damage. If damage is slight, it can be removed with a file; if it's excessive, machining may be the only alternative.
23 Use a tap of the correct size to chase the threads in the cylinder head bolt holes, then clean the holes with compressed air - make sure that nothing remains in the holes.
Warning: *Wear eye protection when using compressed air!*
24 Position the new gaskets over the dowel pins in the block.
25 Carefully set the head on the block without disturbing the gasket.
26 Before installing the **NEW** head bolts, apply a small amount of clean engine oil to the threads and the washers.
27 Install the bolts and tighten them finger tight. Following the recommended sequence, tighten the bolts to the torque listed for Step 1 in this Chapter's Specifications **(see illustration)**.
28 Mark the front of each bolt head with paint. You can also mark the socket you are using. Place the socket over the 12-point bolt so that you can observe the mark.
29 Following the same sequence, tighten each bolt an additional 1/4-turn (90-degrees).
30 Tighten each bolt another 1/4-turn (90-degrees) following the same sequence. The paint marks should now all be 180-degrees from the starting point.
31 Repeat the entire procedure to install the other cylinder head.
32 The remaining installation steps are the reverse of removal.
33 Refill the cooling system, change the oil and filter (see Chapter 1), run the engine and check for leaks.

13 Oil pan - removal and installation

Warning: *On Sequoia models equipped with rear height control suspension, adjust the height control to the NORMAL mode, turn OFF the height control, then turn off the engine BEFORE raising the vehicle.*

Removal

Lower oil pan

Refer to illustration 13.3

1   Remove the lower engine splash shield (see Chapter 1, **illustrations 8.6a and 8.6b**).
2   Drain the engine oil (see Chapter 1).

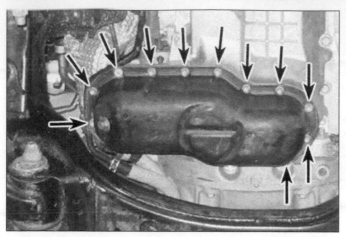

13.3 Lower steel oil pan mounting fasteners - 11 of 16 shown (3UR-FE engine shown, all others similar)

14.9 Remove the pump cover fasteners and remove the cover - 3UR-FE engine shown, other engines similar

3   Remove the bolts and nut securing the lower steel oil pan **(see illustration)** and detach it. If it's stuck, pry it loose very carefully with a small screwdriver or putty knife. Don't damage the mating surfaces of the pan or oil leaks could develop.

Upper oil pan

4   Drain the engine oil and remove the oil filter.
5   Remove the engine from the vehicle (see Chapter 2C).
6   On 2UZ-FE engines, remove the timing belt, the number 1 and number 2 idler pulleys, the crankshaft pulley (see Section 7) and crankshaft sensor (see Chapter 6).
Note: *This step won't be necessary if the oil pump isn't going to be removed.*
7   Remove the oil dipstick tube from the engine.
8   Remove the oil filter, the oil cooler and oil filter bracket assembly from the engine.
9   Remove the bolts and nuts securing the lower steel oil pan **(see illustration 13.3)** and detach it. If it's stuck, pry it loose very carefully with a small screwdriver or putty knife. Don't damage the mating surfaces of the pan or oil leaks could develop.
10   Remove the baffle plate from the bottom of the engine.
11   Remove the bolts and nuts securing the upper oil pan to the engine block and detach it. If it's stuck, pry it loose very carefully with a small screwdriver or putty knife. Don't damage the mating surfaces of the pan or oil leaks could develop.
12   Remove the oil pump strainer.

Installation

13   Use a scraper to remove all traces of old sealant from the block and both oil pans. Clean the mating surfaces with brake system cleaner.
14   Make sure the threaded bolt holes in the engine block and the upper oil pan are clean.
15   Check the flange of the lower oil pan for distortion, particularly around the bolt holes.

If necessary, place the pan on a wood block and use a hammer to flatten and restore the gasket surface.
16   Inspect the oil pump pick-up/strainer assembly for cracks and a blocked strainer. If the pick-up was removed, clean it with solvent or thinner and install it now, using a new gasket. Tighten the fasteners to the torque listed in this Chapter's Specifications.
17   Apply a 3/16-inch wide bead of RTV sealant to the flange of the upper oil pan.
18   Carefully position the upper oil pan on the engine block and install the bolts. Working from the center out, tighten them to the torque listed in this Chapter's Specifications in three or four steps.
19   Install the baffle plate.
20   Apply a 3/16-inch wide bead of RTV sealant to the flange of the lower oil pan.
21   Carefully position the lower oil pan onto the upper oil pan and install the bolts. Working from the center out, tighten them to the torque listed in this Chapter's Specifications in three or four steps.
22   The remainder of installation is the reverse of removal. Be sure to add oil and install a new oil filter, and refill the cooling system (see Chapter 1).
23   Run the engine and check for oil pressure and leaks.

14 Oil pump - removal, inspection and installation

Warning: *On Sequoia models equipped with rear height control suspension, adjust the height control to the NORMAL mode, turn OFF the height control, then turn off the engine BEFORE raising the vehicle.*

Removal

Note: *On 2UZ-FE engines, the oil pump is mounted to the front of the crankshaft. On 1UR-FE, 3UR-FE and 3UR-FBE engines, the oil pump is mounted to the back side of the timing chain cover.*

2UZ-FE engines

1   Remove the engine (see Chapter 2C).
2   Remove the oil pan, oil pick-up tube and strainer (see Section 13).
3   Remove the timing belt (see Section 7) and number 1 and number 2 idler pulleys.
4   Remove the crankshaft timing sprocket and the crankshaft position sensor (see Section 7 and Chapter 6). The oil pump body is behind the crankshaft sprocket.
5   Remove the oil filter/cooler assembly.
6   Remove the eight mounting bolts and detach the oil pump from the front of the engine. If the oil pump is stuck to the engine block, it may be pried off using a screwdriver, but take care to only pry at the center bottom or near the driver-side upper mounting bolt holes. Other areas of the oil pump body are prone to sealing surface damage.
7   Remove the O-ring from the engine block, then remove the plug from the bottom of the oil pump. This will allow the retainer, spring and valve to slide out.
Warning: *The spring is tightly compressed - be careful and wear eye protection.*

1UR-FE, 3UR-FE and 3UR-FBE engines

8   Remove the timing chain cover (see Section 8).

All models

Refer to illustration 14.9

9   Use a large Phillips screwdriver to remove the screws retaining the body cover to the rear of the oil pump housing **(see illustration)**.
10   Lift the cover off and remove the pump rotors.
11   Use a scraper to remove all traces of sealant and old gasket material from the pump body and engine block, then clean the mating surfaces with lacquer thinner or acetone.

Inspection

12   Refer to Chapter 2, Part A for this procedure, but be sure to use the clearance specifications in this Part of Chapter 2.

14.25a Install a new oil pump gasket . . .

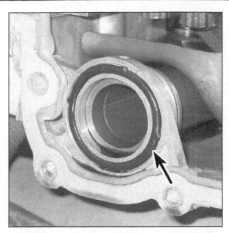

14.25b . . . and O-ring to the engine block

16.4 Drive the new seal into the retainer with a wood block or a section of pipe - make sure that you don't cock the seal in the retainer bore

Installation

13   Place the drive and driven rotors into the pump body with the marks facing out.
14   Pack the pump cavity with petroleum jelly and install the cover.
15   Tighten the cover screws securely following a criss-cross pattern.
16   Lubricate the oil pressure relief valve with engine oil and install the valve components in the pump body.
17   Use brake system cleaner and a clean rag to remove all traces of oil from the gasket surfaces.

2UZ-FE engines

18   Apply a 2 to 3 mm wide bead of RTV sealant to the oil pump. Avoid using an excessive amount of sealant, especially around oil passages and bolt holes. Run the bead of sealant on the inside edge of the bolt holes. Assembly must be completed within five minutes of sealant application, otherwise the material must be removed and reapplied.
19   Position a new O-ring on the block at the top of the oil pump.
20   Engage the spline teeth on the oil pump drive rotor with the large teeth on the crankshaft and slide the pump into place.
21   Install the oil pump mounting bolts in their original locations and tighten them to the torque listed in this Chapter's Specifications in a criss-cross pattern.
22   Using a new gasket, install the oil pick-up tube and tighten the fasteners to the torque listed in this Chapter's Specifications.
23   Reinstall the remaining parts in the reverse order of removal. Install the engine
24   Refill the engine oil and coolant (see Chapter 1). Start the engine and check for oil leaks.

1UR-FE, 3UR-FE and 3UR-FBE engines

Refer to illustrations 14.25a and 14.25b

25   Install new O-rings to the inlet pipe and new oil pump O-rings to the engine block **(see illustrations)**.

26   Install the oil pump mounting bolts in their original locations and tighten them to the torque listed in this Chapter's Specifications in a criss-cross pattern.
27   Reinstall the remaining parts in the reverse order of removal (see Section 8).
28   Refill the engine oil and coolant (see Chapter 1). Start the engine and check for oil leaks.

15   Driveplate - removal and installation

Removal

Warning: *On Sequoia models equipped with rear height control suspension, adjust the height control to the NORMAL mode, turn OFF the height control, then turn off the engine BEFORE raising the vehicle.*
1   Disconnect the negative cable from the battery.
2   Raise the vehicle and support it securely on jackstands, then remove the transmission (see Chapter 7A).
3   Make alignment marks on the driveplate and crankshaft to ensure correct alignment during reinstallation.
4   Remove the bolts securing the driveplate to the crankshaft. If the crankshaft turns, wedge a screwdriver in the ring gear teeth to hold the driveplate.
5   Remove the driveplate from the crankshaft. Retrieve the spacers on both sides of the driveplate. Keep them with the driveplate.
Warning: *The ring-gear teeth may be sharp, wear gloves to protect your hands.*

Installation

6   Clean the driveplate to remove grease and oil. Inspect the surface for cracks. Check for cracked or broken ring gear teeth.
7   Clean and inspect the mating surfaces of the driveplate and the crankshaft. If the crankshaft rear seal is leaking, replace it before reinstalling the driveplate (see Section 16).
8   Position the driveplate against the crankshaft. Align the marks made during removal.

Note that some engines have an alignment dowel or staggered bolt holes to ensure correct installation. Before installing the bolts, apply thread-locking compound to the threads.
9   Wedge a screwdriver in the ring gear teeth to keep the driveplate from turning and tighten the bolts to the torque listed in this Chapter's Specifications. Follow a criss-cross pattern and work up to the final torque in three or four steps.
10   The remainder of installation is the reverse of the removal procedure.

16   Rear main oil seal - replacement

Refer to illustration 16.4

Warning: *On Sequoia models equipped with rear height control suspension, adjust the height control to the NORMAL mode, turn OFF the height control, then turn off the engine BEFORE raising the vehicle.*
Note: *This procedure assumes that the engine has been removed from the vehicle.*
1   Remove the transmission (see Chapter 7A). Refer to Section 15 and remove the driveplate.
2   The seal can be replaced without removing the oil pan or seal retainer. The easiest method involves using a seal-removal tool. If this tool isn't available, you can use a sharp knife to cut the lip off the old seal while carefully avoiding scratching the crankshaft. With the lip gone, use a screwdriver wrapped with tape to pry the seal out.
3   Lubricate the crankshaft seal journal and the lip of the new seal with multi-purpose grease.
4   Evenly drive the new seal into the retainer with a wood block or a section of pipe slightly smaller in diameter than the outside diameter of the seal **(see illustration)**. The new seal should be approximately flush with the surface of the retainer.
5   The remainder of installation is the reverse of removal.

17   Engine mounts - check and replacement

Warning: *On Sequoia models equipped with rear height control suspension, adjust the height control to the NORMAL mode, turn OFF the height control, then turn off the engine BEFORE raising the vehicle.*

1   Engine mounts seldom require attention, but broken or deteriorated mounts should be replaced immediately or the added strain placed on the driveline components may cause damage or wear.

Check

2   During the check, the engine must be raised slightly to remove the weight from the mounts.

3   Raise the vehicle and support it securely on jackstands, then position a jack under the engine oil pan. Place a large wood block between the jack head and the oil pan, then carefully raise the engine just enough to take the weight off the mounts. Do not position the wood block under the drain plug.
Warning: *DO NOT place any part of your body under the engine when it's supported only by a jack!*

4   Check the mounts to see if the rubber is cracked, hardened or separated from the metal plates. Sometimes the rubber will split down the center.

5   Check for relative movement between the mount plates and the engine or frame (use a large screwdriver or pry bar to attempt to move the mounts). If movement is noted, lower the engine and tighten the mount fasteners.

6   Rubber preservative should be applied to the mounts to slow deterioration.

Replacement

7   Disconnect the negative battery cable from the battery, then raise the vehicle and support it securely on jackstands (if not already done). Support the engine as described in Step 3.

8   To remove an engine mount, remove the fasteners, raise the engine and detach the mount.

9   Installation is the reverse of removal. Use thread locking compound on the mount bolts/nuts and tighten them securely.

10   See Chapter 7A for transmission mount replacement.

Chapter 2 Part C
General engine overhaul procedures

Contents

Specifications

General

Displacement
1GR-FE	241.4 cubic inches (4.0 liters)
1UR-FE	281.2 cubic inches (4.6 liters)
2UZ-FE	284.5 cubic inches (4.7 liters)
3UR-FE	345.6 cubic inches (5.7 liters)
3UR-FBE	345.6 cubic inches (5.7 liters)

Bore and stroke
1GR-FE	3.70 x 3.74 inches (94.0 x 95.0 mm)
1UR-FE	3.70 x 3.27 inches (94.0 x 83.0 mm)
2UZ-FE	3.70 x 3.31 inches (94.0 x 84.0 mm)
3UR-FE	3.70 x 4.02 inches (94.0 x 102.0 mm)
3UR-FBE	3.70 x 4.02 inches (94.0 x 102.0 mm)

Cylinder compression pressure

1GR-FE

    Standard
2010 and earlier models	189 psi or more (1300 kPa)
2011 and later models	203 psi or more (1400 kPa)

    Minimum pressure
2010 and earlier models	145 psi (1000 kPa)
2011 and later models	160 psi (1100 kPa)
Difference between each cylinder	15 psi or less (100 kPa)

2UZ-FE
Standard	199 psi or more (1373 kPa)
Minimum pressure	149 psi (1030 kPa)
Difference between each cylinder	14.2 psi or less (98 kPa)

1UR-FE, 3UR-FE and 3UR-FBE
Standard	189 psi or more (1300 kPa)
Minimum pressure	145 psi (1000 kPa)
Difference between each cylinder	15 psi or less (100 kPa)

Oil pressure

1GR-FE and 2UZ-FE
At curb idle	4.3 psi or more (29 kPa)
At 3000 rpm	43 to 85 psi (294 to 588 kPa)

1UR-FE, 3UR-FE and 3UR-FBE
At curb idle	10.1 psi or more (70 kPa)
At 2500 rpm	32 psi or more (220 kPa)

Torque specifications

	Ft-lbs (unless otherwise indicated)	Nm

Note: *One foot-pound (ft-lb) of torque is equivalent to 12 inch-pounds (in-lbs) of torque. Torque values below approximately 15 ft-lbs are expressed in inch-pounds, since most foot-pound torque wrenches are not accurate at these smaller values.*

Driveplate-to-crankshaft bolts

1GR-FE engines	61	83

2UZ-FE, 1UR- FE, 3UR-FE and 3UR-FBE engines

Step 1	22	30
Step 2	Tighten an additional 90 degrees	

Driveplate-to-torque converter bolts

1GR-FE and 2UZ-FE engines	35	48
1UR- FE, 3UR-FE and 3UR-FBE engines	39	53

Connecting rod bearing cap bolts*

1GR-FE and 2UZ-FE engines

Step 1	18	25
Step 2	Tighten an additional 90 degrees	

1UR- FE, 3UR-FE and 3UR-FBE engines

Step 1	30	40
Step 2	Tighten an additional 90 degrees	

Main bearing cap bolts*

1GR-FE models **(see illustration 10.19a and 10.19b)**

Step 1	45	61
Step 2	Tighten an additional 90 degrees	
Side bolts	33	45

2UZ-FE models **(see illustration 10.19e)**

Step 1	20	27
Step 2	Tighten an additional 90 degrees	

1UR-FE, 3UR-FE and 3UR-FBE V8 models **(see illustrations 10.19c and 10.19d)**

Step 1

Inside cap bolts	45	61
Outside cap bolts	20	27
Step 2 (all cap bolts)	Tighten an additional 90 degrees	
Side bolts	33	45

* Use new bolts

1.1 An engine block being bored. An engine rebuilder will use special machinery to recondition the cylinder bores

1.2 If the cylinders are bored, the machine shop will normally hone the engine on a machine like this

1 General information - engine overhaul

Refer to illustrations 1.1, 1.2, 1.3, 1.4, 1.5 and 1.6

Included in this portion of Chapter 2 are general information and diagnostic testing procedures for determining the overall mechanical condition of your engine.

The information ranges from advice concerning preparation for an overhaul and the purchase of replacement parts and/or components to detailed, step-by-step procedures covering removal and installation.

The following Sections have been written to help you determine whether your engine needs to be overhauled and how to remove and install it once you've determined it needs to be rebuilt. For information concerning in-vehicle engine repair, see Chapter 2A or 2B.

The Specifications included in this Part are general in nature and include only those necessary for testing the oil pressure and checking the engine compression. Refer to Chapter 2A or 2B for additional engine Specifications.

It's not always easy to determine when, or if, an engine should be completely overhauled, because a number of factors must be considered.

High mileage is not necessarily an indication that an overhaul is needed, while low mileage doesn't preclude the need for an overhaul. Frequency of servicing is probably the most important consideration. An engine that's had regular and frequent oil and filter changes, as well as other required maintenance, will most likely give many thousands of miles of reliable service. Conversely, a neglected engine may require an overhaul very early in its service life.

Excessive oil consumption is an indication that piston rings, valve seals and/or valve guides are in need of attention. Make sure

1.3 A crankshaft having a main bearing journal ground

that oil leaks aren't responsible before deciding that the rings and/or guides are bad. Perform a cylinder compression check to determine the extent of the work required (see Section 3). Also check the vacuum readings under various conditions (see Section 4).

Check the oil pressure with a gauge installed in place of the oil pressure sending unit and compare it to this Chapter's Specifications (see Section 2). If it's extremely low, the bearings and/or oil pump are probably worn out.

Loss of power, rough running, knocking or metallic engine noises, excessive valve train noise and high fuel consumption rates may also point to the need for an overhaul, especially if they're all present at the same time. If a complete tune-up doesn't remedy the situation, major mechanical work is the only solution.

An engine overhaul involves restoring the internal parts to the specifications of a new engine. During an overhaul, the piston rings are replaced and the cylinder walls are reconditioned (rebored and/or honed) **(see illustrations 1.1 and 1.2)**. If a rebore is done

by an automotive machine shop, new oversize pistons will also be installed. The main bearings, connecting rod bearings and camshaft bearings are generally replaced with new ones and, if necessary, the crankshaft may be reground to restore the journals **(see illustration 1.3)**. Generally, the valves are serviced as well, since they're usually in less-than-perfect condition at this point. While the engine is being overhauled, other components, such as the distributor, starter and alternator, can be rebuilt as well. The end result should be a like-new engine that will give many trouble-free miles.

Note: *Critical cooling system components such as the hoses, drivebelts, thermostat and water pump should be replaced with new parts when an engine is overhauled. The radiator should be checked carefully to ensure that it isn't clogged or leaking (see Chapter 3). If you purchase a rebuilt engine or short block, some rebuilders will not warranty their engines unless the radiator has been professionally flushed. Also, we don't recommend overhauling the oil pump - always install a new one when an engine is rebuilt.*

1.4 A machinist checks for a bent connecting rod, using specialized equipment

1.5 A bore gauge being used to check a cylinder bore

Overhauling the internal components on today's engines is a difficult and time-consuming task which requires a significant amount of specialty tools and is best left to a professional engine rebuilder **(see illustrations 1.4, 1.5 and 1.6)**. A competent engine rebuilder will handle the inspection of your old parts and offer advice concerning the reconditioning or replacement of the original engine. Never purchase parts or have machine work done on other components until the block has been thoroughly inspected by a professional machine shop. As a general rule, time is the primary cost of an overhaul, especially since the vehicle may be tied up for a minimum of two weeks or more. Be aware that some engine builders only have the capability to rebuild the engine you bring them while other rebuilders have a large inventory of rebuilt exchange engines in stock. Also be aware that many machine shops could take as much

as two weeks time to completely rebuild your engine depending on shop workload. Sometimes it makes more sense to simply exchange your engine for another engine that's already rebuilt to save time.

2   Oil pressure check

Refer to illustrations 2.2b, 2.2d and 2.3

1   Low engine oil pressure can be a sign of an engine in need of rebuilding. A low oil pressure indicator (often called an "idiot light") is not a test of the oiling system. Such indicators only come on when the oil pressure is dangerously low. Even a factory oil pressure gauge in the instrument panel is only a relative indication, although much better for driver information than a warning light. A better test is with a mechanical (not electrical) oil pres-

sure gauge.

2   Locate the oil pressure indicator sending unit on the engine block.

a) On all 2UZ-FE V8 engines, the oil pressure sending unit is located in the side of the oil filter bracket next to the oil filter assembly.

b) On 2010 and earlier 1GR-FE V6 engines, the oil pressure sending unit is located on the timing chain cover, next to the oil filter **(see illustration)**.

c) On 2011 and later 1GR-FE V6 engines, the oil pressure sending unit is located on the right hand side of the engine block.

d) On 1UR-FE, 3UR-FE and 3UR-FBE V8 models, the oil pressure sending unit is located on the side of the oil filter bracket **(see illustration)**.

3   Unscrew and remove the oil pressure

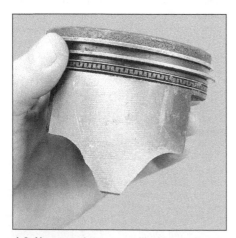

1.6 Uneven piston wear like this indicates a bent connecting rod

2.2b On 2010 and earlier 1GR-FE V6 engines, the oil pressure sending unit is located on the timing chain cover, next to the oil filter

2.2d On 1UR-FE, 3UR-FE and 3UR-FBE V8 models, the oil pressure sending unit is located on the oil filter assembly next to the oil cooler

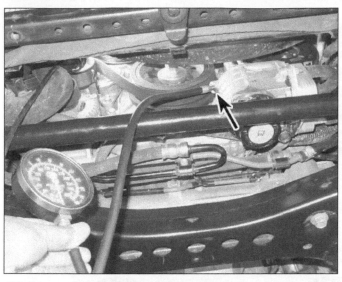

2.3 Remove the oil pressure sending unit then screw in the hose for your oil pressure gauge

3.6 Use a compression gauge with a threaded fitting for the spark plug hole, not the type that requires hand pressure to maintain the seal

sending unit and then screw in the hose for your oil pressure gauge **(see illustration)**. If necessary, install an adapter fitting. Use Teflon tape or thread sealant on the threads of the adapter and/or the fitting on the end of your gauge's hose.

4   Connect an accurate tachometer to the engine, according to the tachometer manufacturer's instructions.

5   Check the oil pressure with the engine running (normal operating temperature) at the specified engine speed, and compare it to this Chapter's Specifications. If it's extremely low, the bearings and/or oil pump are probably worn out.

Caution: *Make sure the hose from the gauge is clear from any moving parts.*

3   Cylinder compression check

Refer to illustration 3.6

1   A compression check will tell you what mechanical condition the upper end of your engine (pistons, rings, valves, head gaskets) is in. Specifically, it can tell you if the compression is down due to leakage caused by worn piston rings, defective valves and seats or a blown head gasket.

Note: *The engine must be at normal operating temperature and the battery must be fully charged for this check.*

2   Begin by cleaning the area around the spark plugs before you remove them (compressed air should be used, if available). The idea is to prevent dirt from getting into the cylinders as the compression check is being done.

3   Remove all of the spark plugs from the engine (see Chapter 1).

4   Block the throttle wide open.

5   Disable the ignition system by discon-

necting the electrical connector(s) from the igniter/ignition coil assemblies (see Chapter 5). The fuel pump circuit should also be disabled by removing the circuit opening relay from the fuse/relay center in the engine compartment (see Chapter 4).

6   Install a compression gauge in the spark plug hole **(see illustration)**.

7   Crank the engine over at least seven compression strokes and watch the gauge. The compression should build up quickly in a healthy engine. Low compression on the first stroke, followed by gradually increasing pressure on successive strokes, indicates worn piston rings. A low compression reading on the first stroke, which doesn't build up during successive strokes, indicates leaking valves or a blown head gasket (a cracked head could also be the cause). Deposits on the undersides of the valve heads can also cause low compression. Record the highest gauge reading obtained.

8   Repeat the procedure for the remaining cylinders and compare the results to this Chapter's Specifications.

9   Add some engine oil (about three squirts from a plunger-type oil can) to each cylinder, through the spark plug hole, and repeat the test.

10   If the compression increases after the oil is added, the piston rings are definitely worn. If the compression doesn't increase significantly, the leakage is occurring at the valves or head gasket. Leakage past the valves may be caused by burned valve seats and/or faces or warped, cracked or bent valves.

11   If two adjacent cylinders have equally low compression, there's a strong possibility that the head gasket between them is blown. The appearance of coolant in the combustion chambers or the crankcase would verify this condition.

12   If one cylinder is slightly lower than the others, and the engine has a slightly rough idle, a worn lobe on the camshaft could be the cause.

13   If the compression is unusually high, the combustion chambers are probably coated with carbon deposits. If that's the case, the cylinder head(s) should be removed and decarbonized.

14   If compression is way down or varies greatly between cylinders, it would be a good idea to have a leak-down test performed by an automotive repair shop. This test will pinpoint exactly where the leakage is occurring and how severe it is.

4   Vacuum gauge diagnostic checks

Refer to illustrations 4.4 and 4.6

1   A vacuum gauge provides inexpensive but valuable information about what is going on in the engine. You can check for worn rings or cylinder walls, leaking head or intake manifold gaskets, incorrect carburetor adjustments, restricted exhaust, stuck or burned valves, weak valve springs, improper ignition or valve timing and ignition problems.

2   Unfortunately, vacuum gauge readings are easy to misinterpret, so they should be used in conjunction with other tests to confirm the diagnosis.

3   Both the absolute readings and the rate of needle movement are important for accurate interpretation. Most gauges measure vacuum in inches of mercury (in-Hg). The following references to vacuum assume the diagnosis is being performed at sea level. As elevation increases (or atmospheric pressure decreases), the reading will decrease. For every 1,000 foot increase in elevation above approximately 2000 feet, the gauge readings

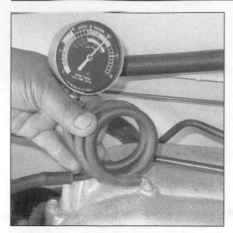

4.4 A simple vacuum gauge can be handy in diagnosing engine condition and performance

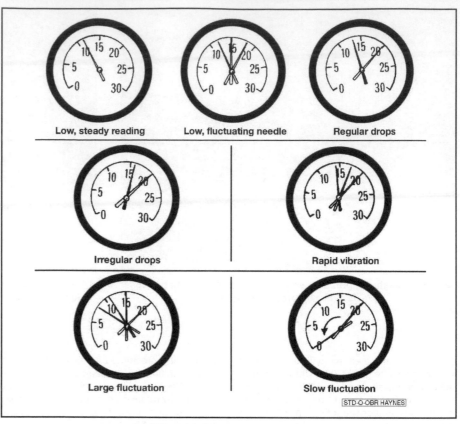

Low, steady reading   Low, fluctuating needle   Regular drops

Irregular drops   Rapid vibration

Large fluctuation   Slow fluctuation

4.6 Typical vacuum gauge readings

will decrease about one inch of mercury.

4   Connect the vacuum gauge directly to the intake manifold vacuum, not to ported (throttle body) vacuum **(see illustration)**. Be sure no hoses are left disconnected during the test or false readings will result.

5   Before you begin the test, allow the engine to warm up completely. Block the wheels and set the parking brake. With the transmission in Park, start the engine and allow it to run at normal idle speed. **Warning:** *Keep your hands and the vacuum gauge clear of the fans.*

6   Read the vacuum gauge; an average, healthy engine should normally produce about 17 to 22 in-Hg with a fairly steady needle **(see illustration)**. Refer to the following vacuum gauge readings and what they indicate about the engine's condition:

7   A low steady reading usually indicates a leaking gasket between the intake manifold and cylinder head(s) or throttle body, a leaky vacuum hose, late ignition timing or incorrect camshaft timing. Check ignition timing with a timing light and eliminate all other possible causes, utilizing the tests provided in this Chapter before you remove the timing chain cover to check the timing marks.

8   If the reading is three to eight inches below normal and it fluctuates at that low reading, suspect an intake manifold gasket leak at an intake port or a faulty fuel injector.

9   If the needle has regular drops of about two-to-four inches at a steady rate, the valves are probably leaking. Perform a compression check or leak-down test to confirm this.

10   An irregular drop or down-flick of the needle can be caused by a sticking valve or an ignition misfire. Perform a compression check or leak-down test and read the spark plugs.

11   A rapid vibration of about four in-Hg vibration at idle combined with exhaust smoke indicates worn valve guides. Perform a leak-down test to confirm this. If the rapid vibration occurs with an increase in engine speed, check for a leaking intake manifold gasket

or head gasket, weak valve springs, burned valves or ignition misfire.

12   A slight fluctuation, say one inch up and down, may mean ignition problems. Check all the usual tune-up items and, if necessary, run the engine on an ignition analyzer.

13   If there is a large fluctuation, perform a compression or leak-down test to look for a weak or dead cylinder or a blown head gasket.

14   If the needle moves slowly through a wide range, check for a clogged PCV system, incorrect idle fuel mixture, throttle body or intake manifold gasket leaks.

15   Check for a slow return after revving the engine by quickly snapping the throttle open until the engine reaches about 2,500 rpm and let it shut. Normally the reading should drop to near zero, rise above normal idle reading (about 5 in-Hg over) and then return to the previous idle reading. If the vacuum returns slowly and doesn't peak when the throttle is snapped shut, the rings may be worn. If there is a long delay, look for a restricted exhaust system (often the muffler or catalytic converter). An easy way to check this is to temporarily disconnect the exhaust ahead of the suspected part and redo the test.

5   Engine rebuilding alternatives

The do-it-yourselfer is faced with a number of options when purchasing a rebuilt

engine. The major considerations are cost, warranty, parts availability and the time required for the rebuilder to complete the project. The decision to replace the engine block, piston/connecting rod assemblies and crankshaft depends on the final inspection results of your engine. Only then can you make a cost effective decision whether to have your engine overhauled or simply purchase an exchange engine for your vehicle.

Some of the rebuilding alternatives include:

Individual parts - If the inspection procedures reveal that the engine block and most engine components are in reusable condition, purchasing individual parts and having a rebuilder rebuild your engine may be the most economical alternative. The block, crankshaft and piston/connecting rod assemblies should all be inspected carefully by a machine shop first.

Short block - A short block consists of an engine block with a crankshaft and piston/connecting rod assemblies already installed. All new bearings are incorporated and all clearances will be correct. The existing camshafts, valve train components, cylinder head and external parts can be bolted to the short block with little or no machine shop work necessary.

Long block - A long block consists of a short block plus an oil pump, oil pan, cylinder head, valve cover, camshaft and valve train components, timing sprockets and chain or

6.1 After tightly wrapping water-vulnerable components, use a spray cleaner on everything, with particular concentration on the greasiest areas, usually around the valve cover and lower edges of the block. If one section dries out, apply more cleaner

6.2 Depending on how dirty the engine is, let the cleaner soak in according to the directions and then hose off the grime and cleaner. Get the rinse water down into every area you can get at; then dry important components with a hair dryer or paper towels

gears and timing cover. All components are installed with new bearings, seals and gaskets incorporated throughout. The installation of manifolds and external parts is all that's necessary.

Low mileage used engines - Some companies now offer low mileage used engines which is a very cost effective way to get your vehicle up and running again. These engines often come from vehicles which have been in totaled in accidents or come from other countries which have a higher vehicle turn over rate. A low mileage used engine also usually has a similar warranty like the newly remanufactured engines.

Give careful thought to which alternative is best for you and discuss the situation with local automotive machine shops, auto parts dealers and experienced rebuilders before ordering or purchasing replacement parts.

6   Engine removal - methods and precautions

Refer to illustrations 6.1, 6.2, 6.3 and 6.4

If you've decided that an engine must be removed for overhaul or major repair work, several preliminary steps should be taken. Read all removal and installation procedures carefully prior to committing this job. Some engines are removed by lowering to the floor and then raising the vehicle sufficiently to slide it out; this will require a vehicle hoist.

Locating a suitable place to work is extremely important. Adequate work space, along with storage space for the vehicle, will be needed. If a shop or garage isn't available, at the very least a flat, level, clean work surface made of concrete or asphalt is required.

Cleaning the engine compartment and engine before beginning the removal procedure will help keep tools clean and organized

6.3 Get an engine hoist that's strong enough to easily lift your engine in and out of the engine compartment; an adapter, like the one shown here (arrow), can be used to change the angle of the engine as it's being removed or installed

(see illustrations 6.1 and 6.2).

An engine hoist or A-frame will also be necessary. Make sure the equipment is rated in excess of the combined weight of the engine and transmission. Safety is of primary importance, considering the potential hazards involved in lifting the engine out of the vehicle.

If you're a novice at engine removal, get at least one helper. One person cannot easily do all the things you need to do to lift a big heavy engine out of the engine compartment. Also helpful is to seek advice and assistance from someone who's experienced in engine removal.

Plan the operation ahead of time.

6.4 Get an engine stand sturdy enough to firmly support the engine while you're working on it. Stay away from three-wheeled models: they have a tendency to tip over more easily, so get a four-wheeled unit

Arrange for or obtain all of the tools and equipment you'll need prior to beginning the job **(see illustrations 6.3 and 6.4).** some of the equipment necessary to perform engine removal and installation safely and with relative ease are (in addition to an engine hoist) a heavy duty floor jack, complete sets of wrenches and sockets as described in the front of this manual, wooden blocks, plenty of rags and cleaning solvent for mopping up spilled oil, coolant and gasoline. If the hoist must be rented, make sure that you arrange for it in advance and have everything disconnected and/or removed before bringing the hoist home. This will save you money and time.

Plan for the vehicle to be out of use for quite a while. A machine shop can do the

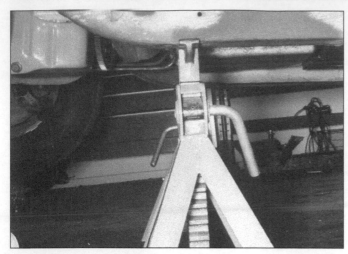

7.5 Place sturdy jackstands under the frame of the vehicle and set them both at uniform height

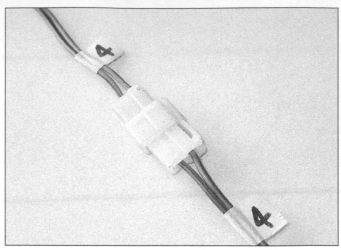

7.9 Label both ends of each wire and hose before disconnecting it - do the same for vacuum hoses

work that is beyond the scope of the home mechanic. Machine shops often have a busy schedule, so before removing the engine, consult the shop for an estimate of how long it will take to rebuild or repair the components that may need work.

7   Engine - removal and installation

Warning: *DO NOT use a cheap engine hoist designed for lifting four-cylinder engines. Obtain a heavy-duty hoist designed for lifting heavy engines. And always be extremely careful when removing and installing the engine. Serious injury can result from careless actions.*

Warning: *The models covered by this manual are equipped with Supplemental Restraint systems (SRS), more commonly known as airbags. Always disable the airbag system before working in the vicinity of any airbag system component to avoid the possibility of accidental deployment of the airbag, which could cause personal injury (see Chapter 12).*

Warning: *Gasoline is extremely flammable, so take extra precautions when you work on any part of the fuel system. Don't smoke or allow open flames or bare light bulbs near the work area, and don't work in a garage where a gas-type appliance (such as a water heater or a clothes dryer) is present. Since gasoline is carcinogenic, wear fuel-resistant gloves when there's a possibility of being exposed to fuel, and, if you spill any fuel on your skin, rinse it off immediately with soap and water. Mop up any spills immediately and do not store fuel-soaked rags where they could ignite. The fuel system is under constant pressure, so, if any fuel lines are to be disconnected, the fuel pressure in the system must be relieved first (see Chapter 4 for more information). When you perform any kind of work on the fuel system, wear safety glasses and have a Class B type fire extinguisher on hand.*

Warning: *The air conditioning system is under high pressure, and refrigerant is expensive. Have a dealer service department or an automotive air conditioning shop discharge the system before beginning this procedure.*

Warning: *On Sequoia models equipped with rear height control suspension, adjust the height control to the NORMAL mode, turn OFF the height control, then turn off the engine BEFORE raising the vehicle.*

Removal

Refer to illustrations 7.5 and 7.9

1   On vehicles with air conditioning, have the air conditioning system discharged.
2   Relieve the fuel system pressure (see Chapter 4).
3   Disconnect the battery cables and remove the battery (see Chapter 5).
4   Remove the hood (see Chapter 11) and cover the fenders and cowl. Special pads are available to protect the fenders, but an old bedspread or blanket will also work.
5   Raise the vehicle and place it securely on jackstands **(see illustration)**.
Note: *On 4WD models, and on models with large tires, this step may not be necessary, because some models already have sufficient ground clearance to allow disconnection of the exhaust system, the engine mounts, etc. from underneath the vehicle. Raising these vehicles any higher might even make engine removal more difficult because it might position the vehicle too high to lift the engine out of the engine compartment with a hoist.*
6   Drain the engine oil and remove the oil filter (see Chapter 1). Detach the engine oil dipstick tube bracket and remove the engine oil dipstick tube.
7   Drain the cooling system (see Chapter 1).
8   Remove the air filter housing and the air intake duct (see Chapter 4).
9   Label all vacuum lines, emissions sys-

tem hoses, wiring harness electrical connectors and ground straps to ensure correct reinstallation, then disconnect them. Pieces of masking tape with numbers or letters written on them work well **(see illustration)**. So does colored electrical tape. If there's any possibility of confusion, make a sketch of the engine compartment and clearly label the lines, hoses and wires. You can also use an inexpensive disposable or digital camera to take photos of connectors, grounds, harness routing, etc.
10   Disconnect the fuel lines running from the engine to the chassis (see Chapter 4). Plug or cap all open fittings/lines.

4WD models

11   Remove the front exhaust pipes (see Chapter 4).
12   Remove the front and rear driveshafts (see Chapter 8).
13   Remove the front stabilizer bar (see Chapter 10).
14   Remove the transmission (see Chapter 7A).

All models

Refer to illustrations 7.28, 7.29a, 7.29b and 7.31

15   Disconnect the VVT sensors, CKP sensor, CMP sensor, camshaft oil control valves and MAF sensor electrical connector (see Chapter 6).
16   Disconnect the electrical connectors to the throttle body, then remove the throttle body (see Chapter 4). Disconnect the fuel lines from the fuel rail, then remove the fuel rail and the injectors (see Chapter 4). Remove the intake manifold (see Chapter 2A or 2B). Also remove all fuel and/or emission control components that might be damaged during engine removal (see Chapters 4 and 6).
17   Clearly label and disconnect all coolant and heater hoses. Remove the cooling fan, shroud and radiator (see Chapter 3).
18   Remove the accessory drivebelt(s) (see

7.28 Secure the engine hangers to the lifting device with heavy chain

7.29a Remove the right side engine mount fasteners

Chapter 1), then remove the alternator (see Chapter 5).

19 On vehicles with air conditioning, remove the compressor (see Chapter 3). Look carefully at the air conditioning system lines. If any section(s) of the air conditioning lines - particularly any section consisting of rigid metal lines - looks like it's going to impede engine removal and installation, detach it from the engine and/or vehicle and set it aside. Secure it with wire if necessary to make sure that it won't be damaged by the engine when the engine is lifted out of the engine compartment.

20 Disconnect the air ducts near the timing belt cover and remove them from the engine compartment (see Chapter 4).

21 Disconnect the vacuum lines from the intake manifold (see Chapter 2A or 2B).

22 Remove the transmission oil cooler pipes from the radiator and the transmission cooler (see Chapter 3).

23 Remove the radiator (see Chapter 3).

24 Remove the power steering pump from its mounting bracket (see Chapter 10) and secure it with wire so that it won't interfere with engine removal.

25 If you're going to be replacing the block, now is a good time to remove all large brackets such as the alternator, air conditioning compressor and power steering pump brackets (see Chapter 9).

26 Remove the part of the exhaust system that's routed underneath the engine, between the exhaust manifolds and the downstream catalytic converter (see Chapter 4). It's not absolutely necessary to remove the exhaust manifolds in order to remove the engine, but removing the manifolds will shave a little weight off the engine.

27 Remove the starter motor (see Chapter 5).

28 Install engine hangers, Toyota special tools #12281-31060 and #12282-31040, or an equivalent to each corner of the engine. Roll a heavy-duty hoist into position and attach it to the lifting brackets with a couple pieces of heavy-duty chain **(see illustration)**. Take up the slack in the sling or chain, but don't lift the engine.

Warning: *DO NOT place any part of your body under the engine when it's supported only by a hoist or other lifting device.*

29 Remove the engine mount fasteners **(see illustrations)**.

30 Recheck to be sure nothing is still connecting the engine to the transmission or vehicle. Disconnect anything still remaining. Raise the engine slightly and inspect it thoroughly once more to make sure that *nothing* is still attached, then slowly lift the engine out of the engine compartment. Check carefully to make sure nothing is hanging up.

31 Remove the driveplate (see Chapter 2A or 2B) and mount the engine on an engine stand **(see illustration)**.

32 Inspect the engine and transmission mounts (see Chapter 2A). If they're worn or damaged, replace them.

7.29b Remove the left side engine mount fasteners

7.31 Use long high-strength bolts to hold the engine block on the engine stand - make sure they are tight before resting all the weight on the stand

9.1 Before you try to remove the pistons, use a ridge reamer to remove the raised material (carbon) from the top of the cylinders

9.3 Checking the connecting rod endplay (side clearance)

Installation

33   Install the driveplate (see Chapter 2A or 2B).

34   Carefully lower the engine into the engine compartment, then reattach it to the engine mounts (see Chapter 2A).

4WD models

35   Install the transmission (see Chapter 7A). Guide the torque converter into the crankshaft following the procedure outlined in Chapter 7A. Install the transmission-to-engine bolts and tighten them securely.

Caution: *DO NOT use the bolts to force the transmission and engine together!*

36   Install the front exhaust pipes (see Chapter 4).

37   Install the front and rear driveshafts (see Chapter 8).

38   Install the front stabilizer bar (see Chapter 10).

All models

39   Reinstall the remaining components in the reverse order of removal.

40   Add coolant, oil and transmission fluid as needed.

41   Run the engine and check for leaks and proper operation of all accessories, then install the hood and test drive the vehicle.

42   Have the air conditioning system recharged and leak tested, if it was discharged.

---

8   Engine overhaul - disassembly sequence

1   It's much easier to remove the external components if it's mounted on a portable engine stand. A stand can often be rented quite cheaply from an equipment rental yard. Before the engine is mounted on a stand, the driveplate should be removed from the engine.

2   If a stand isn't available, it's possible to

remove the external engine components with it blocked up on the floor. Be extra careful not to tip or drop the engine when working without a stand.

3   If you're going to obtain a rebuilt engine, all external components must come off first, to be transferred to the replacement engine. These components include:

 Driveplate
 Ignition system components
 Emissions-related components
 Engine mounts and mount brackets
 Engine rear cover (spacer plate between
  driveplate and engine block)
 Fuel injection components
 Intake/exhaust manifolds
 Oil filter
 Ignition coils/spark plugs
 Thermostat and housing assembly
 Water pump

Note: *When removing the external components from the engine, pay close attention to details that may be helpful or important during installation. Note the installed position of gaskets, seals, spacers, pins, brackets, washers, bolts and other small items.*

4   If you're going to obtain a short block (assembled engine block, crankshaft, pistons and connecting rods), then remove the timing belt, cylinder head, oil pan, oil pump pick-up tube, oil pump and water pump from your engine so that you can turn in your old short block to the rebuilder as a core. See *Engine rebuilding alternatives* for additional information regarding the different possibilities to be considered.

---

9   Pistons and connecting rods - removal and installation

Removal

Refer to illustrations 9.1, 9.3, 9.4 and 9.6

Note: *Prior to removing the piston/connecting rod assemblies, remove the cylinder head*

and oil pan (see Chapter 2A or 2B).

1   Use your fingernail to feel if a ridge has formed at the upper limit of ring travel (about 1/4-inch down from the top of each cylinder). If carbon deposits or cylinder wear have produced ridges, they must be completely removed with a special tool **(see illustration)**. Follow the manufacturer's instructions provided with the tool. Failure to remove the ridges before attempting to remove the piston/ connecting rod assemblies may result in piston breakage.

2   After the cylinder ridges have been removed, turn the engine so the crankshaft is facing up.

3   Before the main bearing cap assembly and connecting rods are removed, check the connecting rod endplay with feeler gauges. Slide them between the first connecting rod and the crankshaft throw until the play is removed **(see illustration)**. Repeat this procedure for each connecting rod. The endplay is equal to the thickness of the feeler gauge(s). Check with an automotive machine shop for the endplay service limit. If the play exceeds the service limit, new connecting rods will be required. If new rods (or a new crankshaft) are installed, the endplay may fall under the minimum allowable clearance. If it does, the rods will have to be machined to restore it. If necessary, consult an automotive machine shop for advice.

4   Check the connecting rods and caps for identification marks **(see illustration)**. If they aren't plainly marked, use a small center-punch to make the appropriate number of indentations on each rod and cap (1, 2, 3, etc., depending on the cylinder they're associated with).

5   Loosen each of the connecting rod cap nuts or bolts 1/2-turn at a time until they can be removed by hand. Remove the number one connecting rod cap and bearing insert. Don't drop the bearing insert out of the cap.

6   Slip a short length of plastic or rubber hose over each connecting rod bolt to protect

9.4 If the connecting rods and caps are not marked, use a center punch or numbered impression stamps to mark the caps to the rods by cylinder number (for example, this would be the No. 4 connecting rod)

9.6 Push a short section of plastic or rubber hose over the connecting rod bolts to prevent damage to the crankshaft journals during piston/rod removal

the crankshaft journal and cylinder wall as the rod is removed **(see illustration)**.

7   Remove the bearing insert and push the connecting rod/piston assembly out through the top of the engine. Use a wooden or plastic hammer handle to push on the upper bearing surface in the connecting rod. If resistance is felt, double-check to make sure that all of the ridge was removed from the cylinder.

8   Repeat the procedure for the remaining cylinders.

Note: *If the connecting rod caps are secured by bolts (instead of nuts), discard the old rod cap bolts. Use new bolts when reassembling the engine.*

9   After removal, reassemble the connecting rod caps and bearing inserts in their respective connecting rods and install the cap bolts finger tight. Leaving the old bearing inserts in place until reassembly will help prevent the connecting rod bearing surfaces from being accidentally nicked or gouged.

10   The pistons and connecting rods are now ready for inspection and overhaul at an automotive machine shop.

Piston ring installation

Refer to illustrations 9.13, 9.14, 9.15, 9.19a, 9.19b, 9.21 and 9.22

11   Before installing the new piston rings, the ring end gaps must be checked. It's assumed that the piston ring side clearance has been checked and verified correct.

12   Lay out the piston/connecting rod assemblies and the new ring sets so the ring sets will be matched with the same piston and cylinder during the end gap measurement and engine assembly.

13   Insert the top (number one) ring into the first cylinder and square it up with the cylinder walls by pushing it in with the top of the piston **(see illustration)**. The ring should be near the bottom of the cylinder, at the lower limit of ring travel.

14   To measure the end gap, slip feeler gauges between the ends of the ring until a

gauge equal to the gap width is found **(see illustration)**. The feeler gauge should slide between the ring ends with a slight amount of drag. Check with an automotive machine shop for the correct end gap for your engine. If the gap is larger or smaller than specified, double-check to make sure you have the correct rings before proceeding.

15   If the gap is too small, it must be enlarged or the ring ends may come in contact with each other during engine operation, which can cause serious damage to the engine. The end gap can be increased by filing the ring ends very carefully with a fine file. Mount the file in a vise equipped with soft jaws, slip the ring over the file with the ends contacting the file face and slowly move the ring to remove material from the ends. When performing this operation, file only by pushing the ring from the outside end of the file towards the vise **(see illustration)**.

16   Excess end gap isn't critical unless it's greater than approximately 0.040-inch. Again,

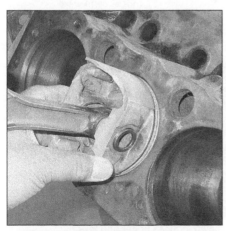

9.13 Install the piston ring into the cylinder then push it down into position using a piston so the ring will be square in the cylinder

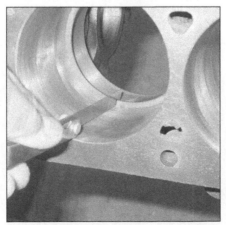

9.14 With the ring square in the cylinder, measure the ring end gap with a feeler gauge

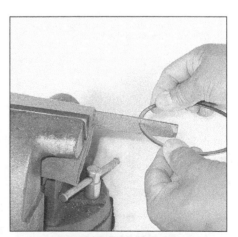

9.15 If the ring end gap is too small, clamp a file in a vise and file the piston ring ends - be sure to remove all raised material

ENGINE BEARING ANALYSIS

Debris

Babbitt bearing embedded with debris from machinings

Microscopic detail of debris

Microscopic detail of gouges

Overplated copper alloy bearing gouged by cast iron debris

Aluminum bearing embedded with glass beads

Microscopic detail of glass beads

Damaged lining caused by dirt left on the bearing back

Misassembly

Result of a lower half assembled as an upper - blocking the oil flow

Excessive oil clearance is indicated by a short contact arc

Polished and oil-stained backs are a result of a poor fit in the housing bore

Result of a wrong, reversed, or shifted cap

Overloading

Damage from excessive idling which resulted in an oil film unable to support the load imposed

Damaged upper connecting rod bearings caused by engine lugging; the lower main bearings (not shown) were similarly affected

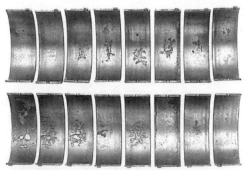

The damage shown in these upper and lower connecting rod bearings was caused by engine operation at a higher-than-rated speed under load

Misalignment

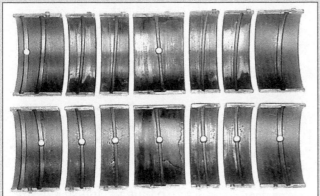

A warped crankshaft caused this pattern of severe wear in the center, diminishing toward the ends

A poorly finished crankshaft caused the equally spaced scoring shown

A bent connecting rod led to the damage in the "V" pattern

A tapered housing bore caused the damage along one edge of this pair

Lubrication

Result of dry start: The bearings on the left, farthest from the oil pump, show more damage

Result of a low oil supply or oil starvation

Severe wear as a result of inadequate oil clearance

Corrosion

Microscopic detail of corrosion

Corrosion is an acid attack on the bearing lining generally caused by inadequate maintenance, extremely hot or cold operation, or inferior oils or fuels

Microscopic detail of cavitation

Example of cavitation - a surface erosion caused by pressure changes in the oil film

Damage from excessive thrust or insufficient axial clearance

Bearing affected by oil dilution caused by excessive blow-by or a rich mixture

9.19a Installing the spacer/expander in the oil ring groove

9.19b DO NOT use a piston ring installation tool when installing the oil control side rails

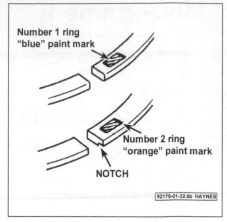

9.21 On 1UR-FE, 3UR-FE AND 3UR-FBE engines, the number 1 piston ring (top) is painted with a "BLUE" mark and the number 2 piston ring (second) is painted with an "ORANGE" mark. Install the piston ring with the mark UP

double-check to make sure you have the correct ring type and that you are referencing the correct section and category of specifications.

17   Repeat the procedure for each ring that will be installed in the first cylinder and for each ring in the remaining cylinders. Remember to keep rings, pistons and cylinders matched up.

18   Once the ring end gaps have been checked/corrected, the rings can be installed on the pistons.

19   The oil control ring (lowest one on the piston) is usually installed first. It's composed of three separate components. Slip the spacer/expander into the groove (see illustration). If an anti-rotation tang is used, make sure it's inserted into the drilled hole in the ring groove.

9.22 Use a piston ring installation tool to install the number 2 and the number 1 (top) rings - be sure the directional mark on the piston ring(s) is facing toward the top of the piston

Next, install the upper side rail in the same manner (see illustration). Don't use a piston ring installation tool on the oil ring side rails, as they may be damaged. Instead, place one end of the side rail into the groove between the spacer/expander and the ring land, hold it firmly in place and slide a finger around the piston while pushing the rail into the groove. Finally, install the lower side rail.

20   After the three oil ring components have been installed, check to make sure that both the upper and lower side rails can be rotated smoothly inside the ring grooves.

21   The number two (middle) ring is installed next. On 2UZ-FE engines it is stamped "2R", on 2010 and earlier 1GR-FE engines the mark is painted on the outside edge of the ring, on 2011 and later 1GR-FE engines there is no mark on the ring. On 1UR-FE, 3UR-FE and 3UR-FBE engines, it is painted It's usually colored with a "BLUE" paint mark which must face up (see illustration), toward the top of the piston. Do not mix up the top and middle rings, as they have different cross-sections.

Note: Always follow the instructions printed on the ring package or box - different manufacturers may require different approaches.

22   Use a piston ring installation tool and make sure the identification mark is facing the top of the piston, then slip the ring into the middle groove on the piston (see illustration). Don't expand the ring any more than necessary to slide it over the piston.

23   Install the number one (top) ring in the same manner. On 2UZ-FE engines there is no mark, on 2010 and earlier 1GR-FE engines there is no mark, on 2011 and later 1GR-FE engines the mark is painted on the outside the ring. On 1UR-FE, 3UR-FE and 3UR-FBE engines, make sure the "ORANGE" paint mark is facing up. Be careful not to confuse

the number one and number two rings (see illustration 9.21).

24   Repeat the procedure for the remaining pistons and rings.

Installation

25   Before installing the piston/connecting rod assemblies, the cylinder walls must be perfectly clean, the top edge of each cylinder bore must be chamfered, and the crankshaft must be in place.

26   Remove the cap from the end of the number one connecting rod (refer to the marks made during removal). Remove the original bearing inserts and wipe the bearing surfaces of the connecting rod and cap with a clean, lint-free cloth. They must be kept spotlessly clean.

Connecting rod bearing oil clearance check

Refer to illustrations 9.30, 9.35, 9.37, 9.38 and 9.41

27   Clean the back side of the new upper bearing insert, then lay it in place in the connecting rod. Make sure the tab on the bearing fits into the recess in the rod. Don't hammer the bearing insert into place and be very careful not to nick or gouge the bearing face. Don't lubricate the bearing at this time.

28   Clean the back side of the other bearing insert and install it in the rod cap. Again, make sure the tab on the bearing fits into the recess in the cap, and don't apply any lubricant. It's critically important that the mating surfaces of the bearing and connecting rod are perfectly clean and oil free when they're assembled.

29   Install a short section of plastic or rubber hose over the connecting rod bolts to avoid damaging the cylinder wall or crankshaft journal (see illustration 9.6).

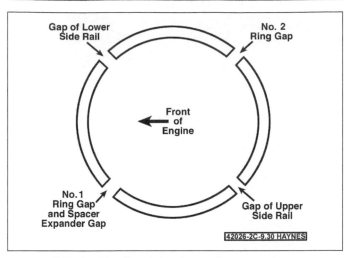

9.30 Position the piston ring end gaps as shown

9.35 Use a plastic or wooden hammer handle to push the piston into the cylinder

30   Position the piston ring gaps at 90-degree intervals around the piston as shown **(see illustration)**.

31   Lubricate the piston and rings with clean engine oil and attach a piston ring compressor to the piston. Leave the skirt protruding about 1/4-inch to guide the piston into the cylinder. The rings must be compressed until they're flush with the piston.

32   Rotate the crankshaft until the number one connecting rod journal is at BDC (bottom dead center) and apply a liberal coat of engine oil to the cylinder walls.

33   With the mark on top of the piston and the connecting rods facing the front (drivebelt end) of the engine, gently insert the piston/connecting rod assembly into the number one cylinder bore and rest the bottom edge of the ring compressor on the engine block. Install the pistons with the cavity mark(s) facing toward the drive belt **(see illustration)**.

Note: *On all models, the connecting rod also has a mark on it that must face the correct*

direction. The marks on the connecting rods face the front (drivebelt) of the engine just like he pistons.

34   Tap the top edge of the ring compressor to make sure it's contacting the block around its entire circumference.

35   Gently tap on the top of the piston with the end of a wooden or plastic hammer handle **(see illustration)** while guiding the end of the connecting rod into place on the crankshaft journal. The piston rings may try to pop out of the ring compressor just before entering the cylinder bore, so keep some downward pressure on the ring compressor. Work slowly, and if any resistance is felt as the piston enters the cylinder, stop immediately. Find out what's hanging up and fix it before proceeding. Do not, for any reason, force the piston into the cylinder - you might break a ring and/or the piston.

36   Once the piston/connecting rod assembly is installed, the connecting rod bearing oil clearance must be checked before the rod

cap is permanently installed.

37   Cut a piece of the appropriate size Plastigage slightly shorter than the width of the connecting rod bearing and lay it in place on the number one connecting rod journal, parallel with the journal axis **(see illustration)**.

38   Clean the connecting rod cap bearing face and install the rod cap. Make sure the mating mark on the cap is on the same side as the mark on the connecting rod **(see illustration)**.

39   Install the old rod bolts or nuts, at this time, and tighten them to the torque listed in this Chapter's Specifications, working up to it in three steps.

Note: *Use a thin-wall socket to avoid erroneous torque readings that can result if the socket is wedged between the rod cap and the bolt or nut. If the socket tends to wedge itself between the fastener and the cap, lift up on it slightly until it no longer contacts the cap. DO NOT rotate the crankshaft at any time during this operation.*

9.37 Place Plastigage on each connecting rod bearing journal parallel to the crankshaft centerline

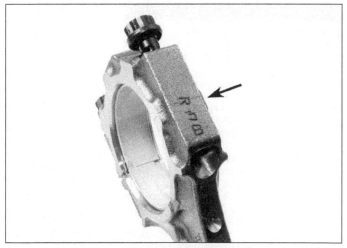

9.38 Install the connecting rod cap making sure the cap and rod identification numbers match

9.41 Use the scale on the Plastigage package to determine the bearing oil clearance - be sure to measure the widest part of the Plastigage and use the correct scale; it comes with both standard and metric scales

10.1 Checking crankshaft endplay with a dial indicator

40   Remove the fasteners and detach the rod cap, being very careful not to disturb the Plastigage. Discard the cap bolts at this time as they cannot be reused.

Note: *You MUST use new connecting rod bolts.*

41   Compare the width of the crushed Plastigage to the scale printed on the Plastigage envelope to obtain the oil clearance **(see illustration)**. The connecting rod oil clearance is usually about 0.001 to 0.002 inch. Consult an automotive machine shop for the clearance specified for the rod bearings on your engine.

42   If the clearance is not as specified, the bearing inserts may be the wrong size (which means different ones will be required). Before deciding that different inserts are needed, make sure that no dirt or oil was between the bearing inserts and the connecting rod or cap when the clearance was measured. Also, recheck the journal diameter. If the Plastigage was wider at one end than the other, the journal may be tapered. If the clearance still exceeds the limit specified, the bearing will have to be replaced with an undersize bearing.

Caution: *When installing a new crankshaft always use a standard size bearing.*

Final installation

43   Carefully scrape all traces of the Plastigage material off the rod journal and/or bearing face. Be very careful not to scratch the bearing - use your fingernail or the edge of a plastic card.

44   Make sure the bearing faces are perfectly clean, then apply a uniform layer of clean moly-base grease or engine assembly lube to both of them. You'll have to push the piston into the cylinder to expose the face of the bearing insert in the connecting rod.

45   **Caution:** *On V8 engines, carefully inspect the connecting rod bolts for distortion or signs of stretching. Replace the bolts if any undesirable conditions exist.* Slide the connecting rod back into place on the journal,

install the rod cap, install the nuts or bolts and tighten them to the torque listed in this Chapter's Specifications. Again, work up to the torque in three steps.

46   Repeat the entire procedure for the remaining pistons/connecting rods.

47   The important points to remember are:

a) *Keep the back sides of the bearing inserts and the insides of the connecting rods and caps perfectly clean when assembling them.*

b) *Make sure you have the correct piston/rod assembly for each cylinder.*

c) *The mark on the piston must face the front (timing belt end) of the engine.*

d) *Lubricate the cylinder walls liberally with clean oil.*

e) *Lubricate the bearing faces when installing the rod caps after the oil clearance has been checked.*

48   After all the piston/connecting rod assemblies have been correctly installed, rotate the crankshaft a number of times by hand to check for any obvious binding.

49   As a final step, check the connecting rod endplay again. If it was correct before disassembly and the original crankshaft and rods were reinstalled, it should still be correct. If new rods or a new crankshaft were installed, the endplay may be inadequate. If so, the rods will have to be removed and taken to an automotive machine shop for resizing.

---

10   Crankshaft - removal and installation

Removal

Refer to illustrations 10.1 and 10.3

Note: *The crankshaft can be removed only after the engine has been removed from the vehicle. It's assumed that the driveplate, crankshaft pulley, timing belt, oil pan, oil*

pump body, oil filter and piston/connecting rod assemblies have already been removed. The rear main oil seal retainer must be unbolted and separated from the block before proceeding with crankshaft removal.

1   Before the crankshaft is removed, measure the endplay. Mount a dial indicator with the indicator in line with the crankshaft and touching the end of the crankshaft **(see illustration)**.

2   Pry the crankshaft all the way to the rear and zero the dial indicator. Next, pry the crankshaft to the front as far as possible and check the reading on the dial indicator. The distance traveled is the endplay. A typical crankshaft endplay will fall between 0.003 to 0.010-inch. If it's greater than that, check the crankshaft thrust surfaces for wear after its removed. If no wear is evident, new main bearings should correct the endplay.

3   If a dial indicator isn't available, feeler gauges can be used. Gently pry the crankshaft all the way to the front of the engine. Slip feeler gauges between the crankshaft and the front face of the thrust bearing or washer to determine the clearance **(see illustration)**.

10.3 Checking crankshaft endplay with feeler gauges at the thrust bearing journal

10.17 Place the Plastigage onto the crankshaft bearing journal as shown

4   Loosen the main bearing cap bolts 1/4-turn at a time each, until they can be removed by hand.

Note: *On all models except 2UZ-FE V8, first remove the main bearing cap side bolts in the reverse of the tightening sequence (see illustration 10.19a, 10.19b, 10.19c, 10.19d or 10.19e).*

5   Gently tap the main bearing cap(s) with a soft-face hammer. Pull the main bearing cap(s) straight up and off the cylinder block. Try not to drop the bearing inserts if they come out with the assembly.

6   Carefully lift the crankshaft out of the engine. It may be a good idea to have an assistant available, since the crankshaft is quite heavy and awkward to handle. With the bearing inserts in place inside the engine block and main bearing caps, reinstall the main bearing cap assembly onto the engine block and tighten the bolts finger tight. Make sure you install the main bearing cap(s) with the arrow facing the front end of the engine.

Installation

7   Crankshaft installation is the first step in engine reassembly. It's assumed at this point that the engine block and crankshaft have been cleaned, inspected and repaired or reconditioned.

8   Position the engine block with the bottom facing up.

9   Remove the mounting bolts and lift off the main bearing cap(s).

10   If they're still in place, remove the original bearing inserts from the block and from the main bearing cap(s). Wipe the bearing surfaces of the block and main bearing cap(s) with a clean, lint-free cloth. They must be kept spotlessly clean. This is critical for determining the correct bearing oil clearance.

Main bearing oil clearance check

Refer to illustrations 10.17, 10.19a, 10.19b, 10.19c, 10.19d, 10.19e and 10.21

11   Without mixing them up, clean the back sides of the new upper main bearing inserts (with grooves and oil holes) and lay one in each main bearing saddle in the block. Each upper bearing has an oil groove and oil hole in it.

Caution: *The oil holes in the block must line up with the oil holes in the upper bearing inserts.*

The thrust washer on V6 models is located on the number 2 crankshaft journal. The thrust washer set on V8 models is located on the number 3 crankshaft journal. Install the thrust washers with the grooved side facing out. Install the thrust washers so that one set is located in the block and the other set is with the main bearing cap assembly. Clean the back sides of the lower main bearing inserts (without grooves) and lay them in the corresponding location in the main bearing cap assembly. Make sure the tab on the bearing

insert fits into the recess in the block or main bearing cap assembly.

Caution: *Do not hammer the bearing insert into place and don't nick or gouge the bearing faces. DO NOT apply any lubrication at this time.*

12   Clean the faces of the bearing inserts in the block and the crankshaft main bearing journals with a clean, lint-free cloth.

13   Check or clean the oil holes in the crankshaft, as any dirt here can go only one way - straight through the new bearings.

14   Once you're certain the crankshaft is clean, carefully lay it in position in the cylinder block.

15   Before the crankshaft can be permanently installed, the main bearing oil clearance must be checked.

16   Cut several strips of the appropriate size of Plastigage (they must be slightly shorter than the width of the main bearing journal).

17   Place one piece on each crankshaft main bearing journal, parallel with the journal axis **(see illustration)**.

18   Clean the faces of the bearing inserts in the main bearing cap assembly. Hold the bearing inserts in place and install the assembly (or caps) onto the crankshaft and cylinder block. DO NOT disturb the Plastigage. Make sure you install the main bearing cap assembly with the arrow facing the front of the engine.

19   Apply clean engine oil to all bolt threads prior to installation, then install all bolts finger-tight. Tighten main bearing cap bolts in the sequence shown **(see illustrations)** progressing in two steps, to the torque listed in this Chapter's Specifications. DO NOT rotate the crankshaft at any time during this operation.

Note: *On all models except 2UZ-FE engines, tighten the main bearing cap bolts first, then the main bearing cap side bolts.*

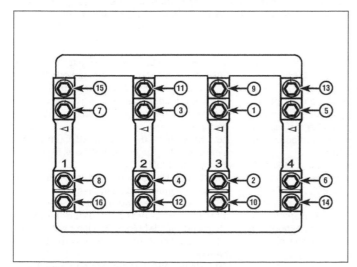

10.19a Main bearing cap bolt tightening sequence (1GR-FE V6 engine)

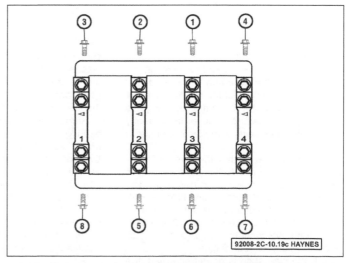

10.19b Main bearing cap side bolt tightening sequence (1GR-FE V6 engine)

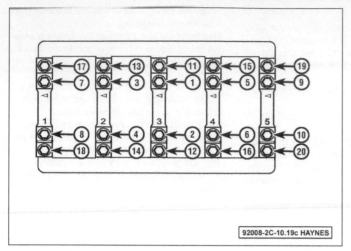

10.19c Main bearing cap bolt tightening sequence (1UR-FE, 3UR-FE AND 3UR-FBE V8 engines)

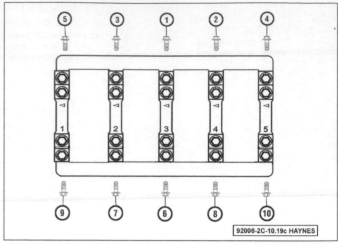

10.19d Main bearing cap side bolt tightening sequence (1UR-FE, 3UR-FE AND 3UR-FBE V8 engines)

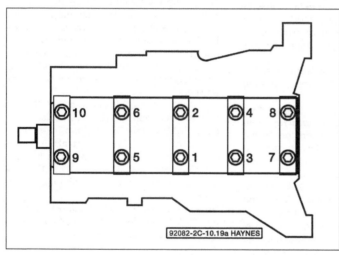

10.19e Main bearing cap bolt tightening sequence (2UZ-FE V8 models)

10.21 Use the scale on the Plastigage package to determine the bearing oil clearance - be sure to measure the widest part of the Plastigage and use the correct scale; it comes with both standard and metric scales

20   Remove the bolts in the *reverse* order of the tightening sequence and carefully lift the main bearing cap assembly straight up and off the block. Do not disturb the Plastigage or rotate the crankshaft. If the main bearing cap assembly is difficult to remove, tap it gently from side-to-side with a soft-face hammer to loosen it.

21   Compare the width of the crushed Plastigage on each journal to the scale printed on the Plastigage envelope to determine the main bearing oil clearance **(see illustration)**. A typical main bearing oil clearance should fall between 0.0015 to 0.0023-inch. Check with an automotive machine shop for the clearance specified for your engine.

22   If the clearance is not as specified, the bearing inserts may be the wrong size (which means different ones will be required). Before deciding if different inserts are needed, make sure that no dirt or oil was between the bearing inserts and the cap assembly or block when the clearance was measured. If the Plastigage was wider at one end than the other, the crankshaft journal may be tapered. If the clearance still exceeds the limit specified, the bearing insert(s) will have to be replaced with an undersize bearing insert(s).

Caution: *When installing a new crankshaft always install a standard bearing insert set.*

23   Carefully scrape all traces of the Plastigage material off the main bearing journals and/or the bearing insert faces. Remove all residue from the oil holes. Use your fingernail or the edge of a plastic card - don't nick or scratch the bearing faces.

Final installation

24   Carefully lift the crankshaft out of the cylinder block.

25   Clean the bearing insert faces in the cylinder block, then apply a thin, uniform layer of moly-base grease or engine assembly lube to each of the bearing surfaces. Coat the thrust faces as well as the journal face of the thrust bearing.

26   Make sure the crankshaft journals are clean, then lay the crankshaft back in place in the cylinder block.

27   Clean the bearing insert faces and then apply the same lubricant to them.

28   Hold the bearing inserts in place and install the main bearing caps on the crankshaft and cylinder block. Tap the bearing caps into place with a brass punch or a soft-face hammer.

29   Using NEW main bearing cap bolts, apply clean engine oil to the bolt threads, wipe off any excess oil and then install the bolts finger-tight.

30 Tighten the main bearing cap bolts in the indicated sequence **(see illustration 10.19a, 10.19c or 10.19e)** to 120 to 144 inch-pounds.
31 Push the crankshaft forward with a screwdriver or prybar to seat the thrust bearing. Once the crankshaft is pushed fully forward, leave the screwdriver in position so that pressure stays on the crankshaft until after all main bearing cap bolts have been tightened.
32 Tighten the main bearing cap bolts in two steps, in the indicated sequence **(see illustration 10.19a, 10.19c, or 10.19e)**, to the torque and angle listed in this Chapter's Specifications.
33 On all models except 2UZ-FE engines, install and tighten the main bearing cap side bolts in the correct sequence **(see illustration 10.19b or 10.19d)**.
34 Recheck crankshaft endplay with a feeler gauge or a dial indicator. The endplay should be correct if the crankshaft thrust faces aren't worn or damaged and if new bearings have been installed.
35 Rotate the crankshaft a number of times by hand to check for any obvious binding. It should rotate with a running torque of 50 in-lbs or less. If the running torque is too high, correct the problem at this time.
36 Install the new rear main oil seal (see Chapter 2A or 2B).

11 Engine overhaul - reassembly sequence

1 Before beginning engine reassembly, make sure you have all the necessary new parts, gaskets and seals as well as the following items on hand:

Common hand tools
A 1/2-inch drive torque wrench
New engine oil
Gasket sealant
Thread locking compound

2 If you obtained a short block, it will be necessary to install the cylinder head, the oil pump and pick-up tube, the oil pan, the water pump, the timing belt and timing cover, and the valve cover (see Chapter 2A or 2B). In order to save time and avoid problems, the external components must be installed in the following general order:

Thermostat and housing cover
Water pump
Intake and exhaust manifolds
Fuel injection components
Emission control components
Spark plugs
Ignition coils
Oil filter
Engine mounts and mount brackets
Driveplate (automatic transmission)

12 Initial start-up and break-in after overhaul

Warning: *Have a fire extinguisher handy when starting the engine for the first time.*
1 Once the engine has been installed in the vehicle, double-check the engine oil and coolant levels.
2 With the spark plugs out of the engine and the ignition system and fuel pump disabled, crank the engine until oil pressure registers on the gauge or the light goes out.
3 Install the spark plugs, hook up the plug wires and/or ignition coils and restore the ignition system and fuel pump functions.
4 Start the engine. It may take a few moments for the fuel system to build up pressure, but the engine should start without a great deal of effort.
5 After the engine starts, it should be allowed to warm up to normal operating temperature. While the engine is warming up, make a thorough check for fuel, oil and coolant leaks.
6 Shut the engine off and recheck the engine oil and coolant levels.
7 Drive the vehicle to an area with minimum traffic, accelerate from 30 to 50 mph, then allow the vehicle to slow to 30 mph with the throttle closed. Repeat the procedure 10 or 12 times. This will load the piston rings and cause them to seat properly against the cylinder walls. Check again for oil and coolant leaks.
8 Drive the vehicle gently for the first 500 miles (no sustained high speeds) and keep a constant check on the oil level. It is not unusual for an engine to use oil during the break-in period.
9 At approximately 500 to 600 miles, change the oil and filter.
10 For the next few hundred miles, drive the vehicle normally. Do not pamper it or abuse it.
11 After 2000 miles, change the oil and filter again and consider the engine broken in.

Notes

Chapter 3
Cooling, heating and air conditioning systems

Contents

Specifications

General

Radiator cap pressure rating	
2UZ-FE engines	10.7 to 14.9 psi
All others engines	13.5 to 17.8 psi
Thermostat rating	176 to 183-degrees F
Refrigerant type	R-134a

Torque specifications

Note: *One foot-pound (ft-lb) of torque is equivalent to 12 inch-pounds (in-lbs) of torque. Torque values below approximately 15 ft-lbs are expressed in inch-pounds, since most foot-pound torque wrenches are not accurate at these smaller values.*

	Ft-lbs (unless otherwise indicated)	Nm
Oil cooler		
Union bolt		
1GR-FE	50	68
2UZ-FE	51	69
Nuts (1UR-FE, 3UR-FE/3UR-FBE)	15	20
Oil filter bracket bolts/nuts (1UR-FE, 3UR-FE/3UR-FBE)	26	35
Pressure cycling switch	80 in-lbs	9
Thermostat housing mounting nuts		
1GR-FE	80 in-lbs	9
1UR-FE	84 in-lbs	10
2UZ-FE	168 in-lbs	19
3UR-FE/3UR-FBE	84 in-lbs	10
Water inlet housing fasteners		
1GR-FE	80 in-lbs	9
2UZ-FE	156 in-lbs	18
1UR-FE, 3UR-FE	15	20
Water pump-to-block bolts		
1GR-FE		
2010 and earlier models		
10 mm head bolt	80 in-lbs	9
12 mm head bolt	17	23
2011 and later models **(see illustration 7.15)**		
Bolt A	35	47
Bolt B	96 in-lbs	11
Bolt C	17	23
2UZ-FE		
Bolt	15	20
Stud bolt and nut	156 in-lbs	18
1UR-FE, 3UR-FE and 3UR-FBE		
14 mm head bolt (A)	35	47
12 mm head bolt (Long) (B)	17	23
12 mm head bolt (Short) (C)	15	20

1   General information

Engine cooling system

All vehicles covered by this manual employ a pressurized engine cooling system with thermostatically controlled coolant circulation. An impeller type water pump mounted on the front of the block pumps coolant through the engine. The coolant flows around each cylinder and toward the rear of the engine. Cast-in coolant passages direct coolant around the intake and exhaust ports, near the spark plug areas and in proximity to the exhaust valve guides. The water pump is mounted on the block and requires timing belt removal.

A wax-pellet type thermostat is located in the thermostat housing at the radiator end of the engine. During warm up, the closed thermostat prevents coolant from circulating through the radiator. When the engine reaches normal operating temperature, the thermostat opens and allows hot coolant to travel through the radiator, where it is cooled before returning to the engine.

The cooling system is sealed by a pressure-type radiator cap. This raises the boiling point of the coolant, and the higher boiling point of the coolant increases the cooling efficiency of the radiator. If the system pressure exceeds the cap pressure-relief value, the excess pressure in the system forces the spring-loaded valve inside the cap off its seat and allows the coolant to escape through the overflow tube into a coolant overflow reservoir. When the system cools, the excess coolant is automatically drawn from the reservoir back into the radiator.

The coolant reservoir serves as both the point at which fresh coolant is added to the cooling system to maintain the proper fluid level and as a holding tank for overheated coolant.

This type of cooling system is known as a closed design because coolant that escapes past the pressure cap is saved and reused.

Transmission cooling systems

All automatic transmissions are equipped with a transmission cooler (located on the side of the transmission) which cools the transmission fluid. The transmission is connected to the cooler by a pair of hoses: one delivers hot transmission fluid to the radiator and the other brings the cooled fluid back to the transmission. Models equipped with the trailer tow system are also equipped with an auxiliary external cooler, located in front of the air conditioning condenser and the radiator.

For more information on transmission oil coolers, refer to Chapter 7A.

Engine oil cooling system

Besides the engine and transmission cooling systems described above, engine heat is also dissipated through an external oil cooler that's integrated into the lubrication system. The oil cooler helps keep engine and oil temperatures within design limits under extreme load conditions.

The oil cooling system consists of a housing mounted between the oil filter and the filter adapter. Hoses connected to the oil cooler circulate coolant from the radiator through the oil cooler housing.

Heating system

The heating system consists of a blower fan and heater core located within the heater box under the dashboard, the inlet and outlet hoses connecting the heater core to the engine cooling system and the heater/air conditioning control head on the dashboard. Hot engine coolant is circulated through the heater core. When the heater mode is activated, a flap opens to expose the heater box to the passenger compartment. A fan switch on the control head activates the blower motor, which forces air through the core, heating the air.

Some Sequoia models are equipped with a rear heating and cooling system. This separate A/C and heating unit is mounted behind the right quarter trim panel.

Air conditioning system

The air conditioning system consists of a condenser mounted in front of the radiator, an evaporator mounted adjacent to the heater core, a compressor mounted on the engine, a filter-drier which contains a high pressure relief valve and the plumbing connecting all of the above.

A blower fan forces the warmer air of the passenger compartment through the evaporator core (sort of a radiator-in-reverse), transferring the heat from the air to the refrigerant. The liquid refrigerant boils off into low pressure vapor, taking the heat with it when it leaves the evaporator. The compressor keeps refrigerant circulating through the system, pumping the warmed coolant through the condenser where it is cooled and then circulated back to the evaporator.

Some Sequoia models are equipped with a rear heating and cooling system. This separate A/C and heating unit is mounted behind the right quarter panel.

2   Troubleshooting

Coolant leaks

Refer to illustration 2.2

1   A coolant leak can develop anywhere in the cooling system, but the most common causes are:

a) *A loose or weak hose clamp*
b) *A defective hose*
c) *A faulty pressure cap*
d) *A damaged radiator*
e) *A bad heater core*
f) *A faulty water pump*
g) *A leaking gasket at any joint that carries coolant*

2.2 The cooling system pressure tester is connected in place of the pressure cap, then pumped up to pressurize the system

2   Coolant leaks aren't always easy to find. Sometimes they can only be detected when the cooling system is under pressure. Here's where a cooling system pressure tester comes in handy. After the engine has cooled completely, the tester is attached in place of the pressure cap, then pumped up to the pressure value equal to that of the pressure cap rating **(see illustration)**. Now, leaks that only exist when the engine is fully warmed up will become apparent. The tester can be left connected to locate a nagging slow leak.

Coolant level drops, but no external leaks

Refer to illustrations 2.5a and 2.5b

3   If you find it necessary to keep adding coolant, but there are no external leaks, the probable causes include:

a) *A blown head gasket*
b) *A leaking intake manifold gasket (only on engines that have coolant passages in the manifold)*
c) *A cracked cylinder head or cylinder block*

4   Any of the above problems will also usually result in contamination of the engine oil, which will cause it to take on a milkshake-like appearance. A bad head gasket or cracked head or block can also result in engine oil contaminating the cooling system.

5   Combustion leak detectors (also known as block testers) are available at most auto parts stores. These work by detecting exhaust gases in the cooling system, which indicates a compression leak from a cylinder into the coolant. The tester consists of a large bulb-type syringe and bottle of test fluid **(see illustration)**. A measured amount of the fluid is added to the syringe. The syringe is placed over the cooling system filler neck and, with the engine running, the bulb is squeezed and a sample of the gases present in the cooling system are drawn up through the test fluid **(see illustration)**. If any combustion gases are present in the sample taken, the test fluid will change color.

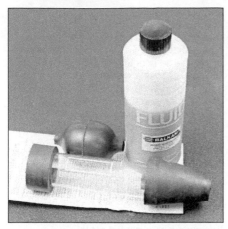

2.5a The combustion leak detector consists of a bulb, syringe and test fluid

2.5b Place the tester over the cooling system filler neck and use the bulb to draw a sample into the tester

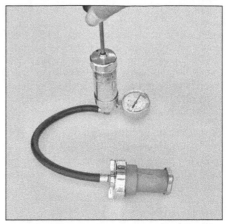

2.8 Checking the cooling system pressure cap with a cooling system pressure tester

6   If the test indicates combustion gas is present in the cooling system, you can be sure that the engine has a blown head gasket or a crack in the cylinder head or block, and will require disassembly to repair.

Pressure cap

Refer to illustration 2.8

Warning: *Wait until the engine is completely cool before beginning this check.*

7   The cooling system is sealed by a spring-loaded cap, which raises the boiling point of the coolant. If the cap's seal or spring are worn out, the coolant can boil and escape past the cap. With the engine completely cool, remove the cap and check the seal; if it's cracked, hardened or deteriorated in any way, replace it with a new one.

8   Even if the seal is good, the spring might not be; this can be checked with a cooling system pressure tester **(see illustration)**. If the cap can't hold a pressure within approximately 1-1/2 lbs of its rated pressure (which is marked on the cap), replace it with a new one.

9   The cap is also equipped with a vacuum relief spring. When the engine cools off, a vacuum is created in the cooling system. The vacuum relief spring allows air back into the system, which will equalize the pressure and prevent damage to the radiator (the radiator tanks could collapse if the vacuum is great enough). If, after turning the engine off and allowing it to cool down you notice any of the cooling system hoses collapsing, replace the pressure cap with a new one.

Thermostat

Refer to illustration 2.10

10   Before assuming the thermostat **(see illustration)** is responsible for a cooling system problem, check the coolant level (see Chapter 1), drivebelt tension (see Chapter 1) and temperature gauge (or light) operation.

11   If the engine takes a long time to warm

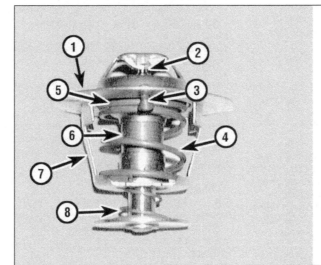

2.10 Typical thermostat:

1   Flange
2   Piston
3   Jiggle valve
4   Main coil spring
5   Valve seat
6   Valve
7   Frame
8   Secondary coil spring

up (as indicated by the temperature gauge or heater operation), the thermostat is probably stuck open. Replace the thermostat with a new one.

12   If the engine runs hot or overheats, a thorough test of the thermostat should be performed.

13   Definitive testing of the thermostat can only be made when it is removed from the vehicle. If the thermostat is stuck in the open position at room temperature, it is faulty and must be replaced.

Caution: *Do not drive the vehicle without a thermostat. The computer may stay in open loop and emissions and fuel economy will suffer.*

14   To test a thermostat, suspend the (closed) thermostat on a length of string or wire in a pot of cold water.

15   Heat the water on a stove while observing thermostat. The thermostat should fully open before the water boils.

16   If the thermostat doesn't open and close as specified, or sticks in any position, replace it.

Cooling fan

Electric cooling fan

17   If the engine is overheating and the cooling fan is not coming on when the engine temperature rises to an excessive level, unplug the fan motor electrical connector(s) and connect the motor directly to the battery with fused jumper wires. If the fan motor doesn't come on, replace the motor.

18   If the radiator fan motor is okay, but it isn't coming on when the engine gets hot, the fan relay might be defective. A relay is used to control a circuit by turning it on and off in response to a control decision by the Powertrain Control Module (PCM). These control circuits are fairly complex, and checking them should be left to a qualified automotive technician. Sometimes, the control system can be fixed by simply identifying and replacing a bad relay.

19   Locate the fan relays in the engine compartment fuse/relay box.

20   Test the relay (see Chapter 12).

21   If the relay is okay, check all wiring and

connections to the fan motor. Refer to the wiring diagrams at the end of Chapter 12. If no obvious problems are found, the problem could be the Engine Coolant Temperature (ECT) sensor or the Powertrain Control Module (PCM). Have the cooling fan system and circuit diagnosed by a dealer service department or repair shop with the proper diagnostic equipment.

Belt-driven cooling fan
22   Disconnect the cable from the negative terminal of the battery and rock the fan back and forth by hand to check for excessive bearing play.
23   With the engine cold (and not running), turn the fan blades by hand. The fan should turn freely.
24   Visually inspect for substantial fluid leakage from the clutch assembly. If problems are noted, replace the clutch assembly.
25   With the engine completely warmed up, turn off the ignition switch and disconnect the negative battery cable from the battery. Turn the fan by hand. Some drag should be evident. If the fan turns easily, replace the fan clutch.

Water pump
26   A failure in the water pump can cause serious engine damage due to overheating.

Drivebelt-driven water pump
Refer to illustration 2.28
27   There are two ways to check the operation of the water pump while it's installed on the engine. If the pump is found to be defective, it should be replaced with a new or rebuilt unit.
28   Water pumps are equipped with weep (or vent) holes **(see illustration)**. If a failure occurs in the pump seal, coolant will leak from the hole.
29   If the water pump shaft bearings fail, there may be a howling sound at the pump while it's running. Shaft wear can be felt with the drivebelt removed if the water pump pulley is rocked up and down (with the engine

2.28 The water pump weep hole is generally located on the underside of the pump

off). Don't mistake drivebelt slippage, which causes a squealing sound, for water pump bearing failure.

Timing chain or timing belt-driven water pump
30   Water pumps driven by the timing chain or timing belt are located underneath the timing chain or timing belt cover.
31   Checking the water pump is limited because of where it is located. However, some basic checks can be made before deciding to remove the water pump. If the pump is found to be defective, it should be replaced with a new or rebuilt unit.
32   One sign that the water pump may be failing is that the heater (climate control) may not work well. Warm the engine to normal operating temperature, confirm that the coolant level is correct, then run the heater and check for hot air coming from the ducts.
33   Check for noises coming from the water pump area. If the water pump impeller shaft or bearings are failing, there may be a howling sound at the pump while the engine is running.
Note: *Be careful not to mistake drivebelt noise (squealing) for water pump bearing or shaft failure.*
34   It you suspect water pump failure due to noise, wear can be confirmed by feeling for play at the pump shaft. This can be done by rocking the drive sprocket on the pump shaft up and down. To do this you will need to remove the tension on the timing chain or belt as well as access the water pump.

All water pumps
35   In rare cases or on high-mileage vehicles, another sign of water pump failure may be the presence of coolant in the engine oil. This condition will adversely affect the engine in varying degrees.
Note: *Finding coolant in the engine oil could indicate other serious issues besides a failed water pump, such as a blown head gasket or a cracked cylinder head or block.*
36   Even a pump that exhibits no outward signs of a problem, such as noise or leakage, can still be due for replacement. Removal for close examination is the only sure way to tell. Sometimes the fins on the back of the impeller can corrode to the point that cooling efficiency is diminished significantly.

Heater system
37   Little can go wrong with a heater. If the fan motor will run at all speeds, the electrical part of the system is okay. The three basic heater problems fall into the following general categories:
  a) *Not enough heat*
  b) *Heat all the time*
  c) *No heat*
38   If there's not enough heat, the control valve or door is stuck in a partially open position, the coolant coming from the engine isn't hot enough, or the heater core is restricted.

If the coolant isn't hot enough, the thermostat in the engine cooling system is stuck open, allowing coolant to pass through the engine so rapidly that it doesn't heat up quickly enough. If the vehicle is equipped with a temperature gauge instead of a warning light, watch to see if the engine temperature rises to the normal operating range after driving for a reasonable distance.
39   If there's heat all the time, the control valve or the door is stuck wide open.
40   If there's no heat, coolant is probably not reaching the heater core, or the heater core is plugged. The likely cause is a collapsed or plugged hose, core, or a frozen heater control valve. If the heater is the type that flows coolant all the time, the cause is a stuck door or a broken or kinked control cable.

Air conditioning system
41   If the cool air output is inadequate:
  a) *Inspect the condenser coils and fins to make sure they're clear*
  b) *Check the compressor clutch for slippage.*
  c) *Check the blower motor for proper operation.*
  d) *Inspect the blower discharge passage for obstructions.*
  e) *Check the system air intake filter for clogging.*
42   If the system provides intermittent cooling air:
  a) *Check the circuit breaker, blower switch and blower motor for a malfunction.*
  b) *Make sure the compressor clutch isn't slipping.*
  c) *Inspect the plenum door to make sure it's operating properly.*
  d) *Inspect the evaporator to make sure it isn't clogged.*
  e) *If the unit is icing up, it may be caused by excessive moisture in the system, incorrect super heat switch adjustment or low thermostat adjustment.*
43   If the system provides no cooling air:
  a) *Inspect the compressor drivebelt. Make sure it's not loose or broken.*
  b) *Make sure the compressor clutch engages. If it doesn't, check for a blown fuse.*
  c) *Inspect the wire harness for broken or disconnected wires.*
  d) *If the compressor clutch doesn't engage, bridge the terminals of the A/C pressure switch(es) with a jumper wire; if the clutch now engages, and the system is properly charged, the pressure switch is bad.*
  e) *Make sure the blower motor is not disconnected or burned out.*
  f) *Make sure the compressor isn't partially or completely seized.*
  g) *Inspect the refrigerant lines for leaks.*
  h) *Check the components for leaks.*
  i) *Inspect the receiver-drier/accumulator or expansion valve/tube for clogged screens.*

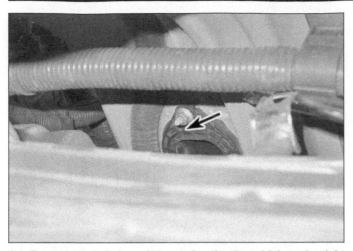

3.1 Evaporator drain hose is located under the vehicle to the right of the transmission

3.9 Insert a thermometer in the center vent, turn on the air conditioning system and wait for it to cool down; depending on the humidity, the output air should be 35 to 40 degrees cooler than the ambient air temperature

44 If the system is noisy:

a) *Look for loose panels in the passenger compartment.*

b) *Inspect the compressor drivebelt. It may be loose or worn.*

c) *Check the compressor mounting bolts. They should be tight.*

d) *Listen carefully to the compressor. It may be worn out.*

e) *Listen to the idler pulley and bearing and the clutch. Either may be defective.*

f) *The winding in the compressor clutch coil or solenoid may be defective.*

g) *The compressor oil level may be low.*

h) *The blower motor fan bushing or the motor itself may be worn out.*

i) *If there is an excessive charge in the system, you'll hear a rumbling noise in the high pressure line, a thumping noise in the compressor, or see bubbles or cloudiness in the sight glass.*

j) *If there's a low charge in the system, you might hear hissing in the evaporator case at the expansion valve, or see bubbles or cloudiness in the sight glass.*

3   Air conditioning and heating system - check and maintenance

Air conditioning system

Refer to illustration 3.1

Warning: *The air conditioning system is under high pressure. Do not loosen any hose fittings or remove any components until after the system has been discharged. Air conditioning refrigerant should be properly discharged into an EPA-approved recovery/recycling unit at a dealer service department or an automotive air conditioning repair facility. Always wear eye protection when disconnecting air conditioning system fittings.*

Caution: *All models covered by this manual use environmentally friendly R-134a. This*

refrigerant (and its appropriate refrigerant oils) are not compatible with R-12 refrigerant system components and must never be mixed or the components will be damaged.

Caution: *When replacing entire components, additional refrigerant oil should be added equal to the amount that is removed with the component being replaced. Be sure to read the can before adding any oil to the system, to make sure it is compatible with the R-134a system.*

1   The following maintenance checks should be performed on a regular basis to ensure that the air conditioning continues to operate at peak efficiency.

a) *Inspect the condition of the compressor drivebelt. If it is worn or deteriorated, replace it (see Chapter 1).*

b) *Check the drivebelt tension (see Chapter 1).*

c) *Inspect the system hoses. Look for cracks, bubbles, hardening and deterioration. Inspect the hoses and all fittings for oil bubbles or seepage. If there is any evidence of wear, damage or leakage, replace the hose(s).*

d) *Inspect the condenser fins for leaves, bugs and any other foreign material that may have embedded itself in the fins. Use a fin comb or compressed air to remove debris from the condenser.*

e) *Make sure the system has the correct refrigerant charge.*

f) *If you hear water sloshing around in the dash area or have water dripping on the carpet, check the evaporator housing drain tube* **(see illustration)** *and insert a piece of wire into the opening to check for blockage.*

2   It's a good idea to operate the system for about ten minutes at least once a month. This is particularly important during the winter months because long term non-use can cause hardening, and subsequent failure, of the seals. Note that using the Defrost function operates the compressor.

3   If the air conditioning system is not working properly, proceed to Step 6 and perform the general checks outlined below.

4   Because of the complexity of the air conditioning system and the special equipment necessary to service it, in-depth troubleshooting and repairs beyond checking the refrigerant charge and the compressor clutch operation are not included in this manual. However, simple checks and component replacement procedures are provided in this Chapter. For more complete information on the air conditioning system, refer to the *Haynes Automotive Heating and Air Conditioning Manual.*

5   The most common cause of poor cooling is simply a low system refrigerant charge. If a noticeable drop in system cooling ability occurs, one of the following quick checks will help you determine if the refrigerant level is low.

Checking the refrigerant charge

Refer to illustration 3.9

6   Warm the engine up to normal operating temperature.

7   Place the air conditioning temperature selector at the coldest setting and put the blower at the highest setting.

8   After the system reaches operating temperature, feel the larger pipe exiting the evaporator at the firewall. The outlet pipe should be cold (the tubing that leads back to the compressor). If the evaporator outlet pipe is warm, the system probably needs a charge.

9   Insert a thermometer in the center air distribution duct **(see illustration)** while operating the air conditioning system at its maximum setting - the temperature of the output air should be 35 to 40 degrees F below the ambient air temperature (down to approximately 40 degrees F). If the ambient (outside) air temperature is very high, say 110 degrees F, the duct air temperature may be as high as 60 degrees F, but generally the air conditioning is 35 to 40 degrees F cooler than the ambient air.

10   Further inspection or testing of the sys-

3.11 R-134a automotive air conditioning charging kit

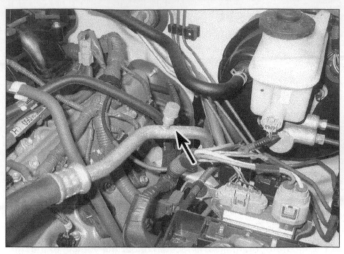

3.13 Location of the low-side charging port

tem requires special tools and techniques and is beyond the scope of the home mechanic.

Adding refrigerant

Refer to illustrations 3.11 and 3.13

Caution: *Make sure any refrigerant, refrigerant oil or replacement component you purchase is designated as compatible with R-134a systems.*

11   Purchase an R-134a automotive charging kit at an auto parts store **(see illustration)**. A charging kit includes a can of refrigerant, a tap valve and a short section of hose that can be attached between the tap valve and the system low side service valve.

Caution: *Never add more than one can of refrigerant to the system. If more refrigerant than that is required, the system should be evacuated and leak tested.*

12   Back off the valve handle on the charging kit and screw the kit onto the refrigerant can, making sure first that the O-ring or rubber seal inside the threaded portion of the kit is in place.

Warning: *Wear protective eyewear when dealing with pressurized refrigerant cans.*

13   Remove the dust cap from the low-side charging port and attach the hose's quick-connect fitting to the port **(see illustration)**.

Warning: *DO NOT hook the charging kit hose to the system high side! The fittings on the charging kit are designed to fit **only** on the low side of the system.*

14   Warm up the engine and turn On the air conditioning. Keep the charging kit hose away from the fan and other moving parts.

Note: *The charging process requires the compressor to be running. If the clutch cycles off, you can put the air conditioning switch on High and leave the car doors open to keep the clutch on and compressor working. The compressor can be kept on during the charging by removing the connector from the pressure switch and bridging it with a paper clip or jumper wire during the procedure.*

15   Turn the valve handle on the kit until the

stem pierces the can, then back the handle out to release the refrigerant. You should be able to hear the rush of gas. Keep the can upright at all times, but shake it occasionally. Allow stabilization time between each addition.

Note: *The charging process will go faster if you wrap the can with a hot-water-soaked rag to keep the can from freezing up.*

16   If you have an accurate thermometer, you can place it in the center air conditioning duct inside the vehicle and keep track of the output air temperature. A charged system that is working properly should cool down to approximately 40 degrees F. If the ambient (outside) air temperature is very high, say 110 degrees F, the duct air temperature may be as high as 60 degrees F, but generally the air conditioning is 35 to 40 degrees F cooler than the ambient air.

17   When the can is empty, turn the valve handle to the closed position and release the connection from the low-side port. Reinstall the dust cap.

18   Remove the charging kit from the can and store the kit for future use with the piercing valve in the UP position, to prevent inadvertently piercing the can on the next use.

Heating systems

19   If the carpet under the heater core is damp, or if antifreeze vapor or steam is coming through the vents, the heater core is leaking. Remove it (see Section 10) and install a new unit (most radiator shops will not repair a leaking heater core).

20   If the air coming out of the heater vents isn't hot, the problem could stem from any of the following causes:

a) *The thermostat is stuck open, preventing the engine coolant from warming up enough to carry heat to the heater core. Replace the thermostat (see Section 4).*

b) *There is a blockage in the system, preventing the flow of coolant through the heater core. Feel both heater hoses at the firewall. They should be hot. If one*

of them is cold, there is an obstruction in one of the hoses or in the heater core, or the heater control valve is shut. Detach the hoses and back flush the heater core with a water hose. If the heater core is clear but circulation is impeded, remove the two hoses and flush them out with a water hose.

c) *If flushing fails to remove the blockage from the heater core, the core must be replaced (see Section 10).*

Eliminating air conditioning odors

Refer to illustration 3.24

21   Unpleasant odors that often develop in air conditioning systems are caused by the growth of a fungus, usually on the surface of the evaporator core. The warm, humid environment there is a perfect breeding ground for mildew to develop.

22   The evaporator core on most vehicles is difficult to access, and factory dealerships have a lengthy, expensive process for eliminating the fungus by opening up the evaporator case and using a powerful disinfectant and rinse on the core until the fungus is gone. You can service your own system at home, but it takes something much stronger than basic household germ-killers or deodorizers.

23   Aerosol disinfectants for automotive air conditioning systems are available in most auto parts stores, but remember when shopping for them that the most effective treatments are also the most expensive. The basic procedure for using these sprays is to start by running the system in the RECIRC mode for ten minutes with the blower on its highest speed. Use the highest heat mode to dry out the system and keep the compressor from engaging by disconnecting the wiring connector at the compressor.

24   The disinfectant can usually comes with a long spray hose. Insert the nozzle into an intake port inside the cabin, and spray according to

3.24 Remove the interior ventilation filter and spray the disinfectant into the blower housing with the blower on high (be sure to wear safety goggles)

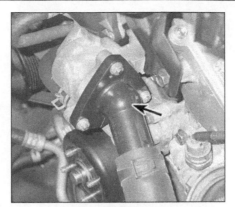

4.11a Thermostat housing details on V6 models

4.11b Remove the thermostat mounting nuts (3UR-FE engine shown)

the manufacturer's recommendations **(see illustration)**. Try to cover the whole surface of the evaporator core, by aiming the spray up, down and sideways. Follow the manufacturer's recommendations for the length of spray and waiting time between applications.

25   Once the evaporator has been cleaned, the best way to prevent the mildew from coming back again is to make sure your evaporator housing drain tube is clear **(see illustration 3.1)**.

Automatic heating and air conditioning systems

26   Some vehicles are equipped with an optional automatic climate control system. This system has its own computer that receives inputs from various sensors in the heating and air conditioning system. This computer, like the PCM, has self-diagnostic capabilities to help pinpoint problems or faults within the system. Vehicles equipped with automatic heating and air conditioning systems are very complex and considered beyond the scope of the home mechanic. Vehicles equipped with automatic heating and air conditioning systems should be taken to dealer service department or other qualified facility for repair.

4   Thermostat - check and replacement

Warning: *Do not remove the radiator cap, coolant or thermostat until the engine has cooled completely.*

Check

1   Before proceeding, check the coolant level and the drivebelt tension (see Chapter 1), then check the operation of the temperature gauge or temperature warning light circuit (see Chapter 12).

2   If the engine takes a long time to warm up, the thermostat is probably stuck open. Replace the thermostat.

3   If the engine runs hot, check the temper-

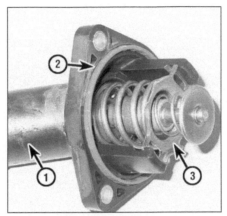

4.12 Thermostat assembly details

1   *Thermostat housing*
2   *O-ring gasket*
3   *Thermostat*

ature of the upper radiator hose. If the hose isn't hot, the thermostat is probably stuck shut. Replace the thermostat.

4   If the upper radiator hose is hot, then the coolant is circulating and the thermostat is open. Refer to the *Troubleshooting* Section at the front of this manual for the cause of overheating.

5   If the engine has overheated, it might have leaking cylinder head gaskets, scuffed pistons and/or warped or cracked cylinder heads.

Replacement

Refer to illustrations 4.11a, 4.11b, 4.12 and 4.13

Warning: *On Sequoia models equipped with rear height control suspension, adjust the height control to the NORMAL mode, turn OFF the height control, then turn off the engine BEFORE raising the vehicle.*

6   Disconnect the negative cable from the battery (see Chapter 5).

7   Remove the engine splash shield (see Chapter 1) and drain the cooling system (see Chapter 1).

4.13 Install the thermostat

8   Remove the engine cover (see Chapter 1, **illustration 24.12a**).

9   On models with 2UZ-FE engines, remove the throttle body cover and air cleaner housing (see Chapter 4).

10   On 3UR-FE/3UR-FBE engines, disconnect and remove the air pump hoses (see Chapter 6).

Note: *On some models, it may be necessary to remove the water by-pass hose from the top of the housing for access to the air pump hose clamp.*

11   Detach the radiator hose, then remove the housing mounting nuts and remove the housing and thermostat from the engine **(see illustrations)**. Be prepared for some coolant to spill as the gasket seal is broken.

Note: *The thermostat and housing are an assembly and should not be disassembled.*

12   Thoroughly clean the sealing surfaces and install a new gasket onto the thermostat **(see illustration)**. Make sure it is evenly fitted all the way around.

13   Install the thermostat unit **(see illustration)** and tighten the housing fasteners to the torque listed in this Chapter's Specifications. Reinstall the remaining components in the reverse order of removal.

14   Refill the cooling system (see Chapter 1). Run the engine and check for leaks and proper operation.

5.10a Remove the shroud mounting bolt from right side . . .

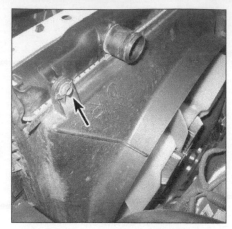

5.10b . . . then the left side

5.11 Unhook the fan shroud mounting tab (A) from the radiator (B) at the bottom of the shroud

5 Cooling fan and clutch - check, removal and installation

Check

Warning: *While checking the fan, make sure that the engine is NOT started. If it is, you could be severely injured. As a safeguard, remove the key from the ignition switch.*

Warning: *Before the fan clutch operation can be checked in Step 5, the engine must be warmed up to its normal operating temperature and then turned off. Even though the engine won't be running during this check, it's HOT! Make sure that you don't touch the engine itself during this check, or you could be burned.*

Warning: *Keep hands, tools and clothing away from the fan when the engine is running. To avoid injury or damage DO NOT operate the engine with a damaged fan. Do not attempt to repair fan blades - replace a damaged fan with a new one.*

Note: *The coolant overflow reservoir is molded into the fan shroud and is not removable from the shroud.*

1   Symptoms of fan clutch failure are continuous noisy operation, looseness, vibration and/or silicone fluid leaking from the clutch.

Cold engine checks

2   Rock the fan back and forth by hand to check for excessive bearing play.
3   With the engine cold, turn the blades by hand. The fan should turn freely.
4   Visually inspect for substantial fluid leakage from the fan clutch assembly, a deformed bi-metal spring or grease leakage from the cooling fan bearing. If any of these conditions exist, replace the fan clutch.

Hot engine check

5   Start the engine and allow it to warm up to its normal operating temperature. When the engine is fully warmed up, turn off the ignition switch. Turn the fan by hand. Some resistance should be felt. If the fan turns easily, replace the fan clutch.

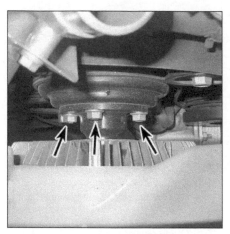

5.12a Unscrew the four nuts that retain the fan to the water pump (one hidden from view), detach the fan/clutch assembly . . .

Removal and installation

Refer to illustrations 5.10a, 5.10b, 5.11, 5.12a, 5.12b and 5.13

Warning: *On Sequoia models equipped with rear height control suspension, adjust the height control to the NORMAL mode, turn OFF the height control, then turn off the engine BEFORE raising the vehicle.*

6   Disconnect the negative cable from the battery (see Chapter 5).
7   Remove the engine splash shield (see Chapter 1) and drain the cooling system (see Chapter 1).
8   Remove the engine cover **(see Chapter 1, illustration 24.12a)** and disconnect the upper hose from the radiator.
9   Disconnect the overflow hose from the radiator.
10   Remove the fan shroud mounting bolts from the top of the shroud **(see illustrations)**.
11   Unhook the fan shroud mounting tabs from the radiator at the bottom of the shroud **(see illustration)**.
12   Remove the fan clutch-to-water pump mounting nuts and separate the fan from the water pump **(see illustration)**. Lift the fan and

5.12b . . . and remove the fan shroud and fan at the same time

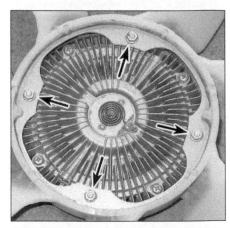

5.13 Remove these four nuts to separate the fan blades from the clutch (V8 model shown, V6 models similar)

shroud from the engine compartment **(see illustration)**.
13   To separate the fan blades from the viscous clutch hub, remove the mounting nuts **(see illustration)**.
14   Installation is the reverse of removal.

6.4 Use a pair of pliers to squeeze the hose clamp open, then slide the clamp back on the hose

6.8a Remove the right side deflector plastic retaining pins and deflector ...

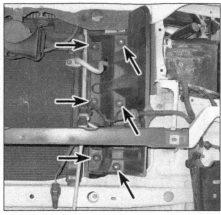

6.8b ... then remove the left side deflector plastic retaining pins and deflector

6.9 Location of the radiator mounting bolts

6.10 Carefully lift the radiator straight up and out of the vehicle

6 Radiator - removal and installation

Refer to illustrations 6.4, 6.8a, 6.8b, 6.9 and 6.10

Warning: *Do not start this procedure until the engine is completely cool.*

Warning: *On Sequoia models equipped with rear height control suspension, adjust the height control to the NORMAL mode, turn OFF the height control, then turn off the engine BEFORE raising the vehicle.*

1   Disconnect the negative battery cable (see Chapter 5).

2   Remove the engine splash shield (see Chapter 1) and drain the cooling system (see Chapter 1).

3   Remove the engine cover **(see Chapter 1, illustration 24.12a)** and, on 2014 models, remove the grille (see Chapter 11).

4   Detach the upper and lower radiator hoses from the radiator **(see illustration)**. If the hoses stick, twist them with a pair of large pliers to break the bond, but be careful not to damage the fittings on the radiator.

5   Disconnect the reservoir hose from the radiator filler neck.

6   Remove the cooling fan and fan shroud (see Section 5).

7   On Sequoia models, remove the front bumper and radiator grille (see Chapter 11). On Tundra models, remove the radiator grille (see Chapter 11).

8   Remove the radiator side deflector plastic retaining pins and the deflectors **(see illustrations)**.

Note: *Sequoia models are equipped with radiator side support seals that are removed the same way as the side deflectors.*

9   Remove the radiator mounting bolts **(see illustration)**.

10   Lift out the radiator **(see illustration)**. Be aware of dripping fluids and the sharp fins.

11   With the radiator removed, it can be inspected for leaks, damage and internal blockage. If in need of repairs, have a radiator shop or dealer service department perform the work as special techniques are required.

12   Bugs and dirt can be cleaned from the radiator with compressed air and a soft brush. Don't bend the cooling fins as this is done.

Warning: *Wear eye protection when using compressed air.*

13   Installation is the reverse of removal.

14   After installation, fill the cooling system with the proper mixture of antifreeze and water. Refer to Chapter 1 if necessary.

15   Start the engine and check for leaks. Allow the engine to reach normal operating temperature, indicated by both radiator hoses becoming hot. Recheck the coolant level and add more if required.

7 Water pump - replacement

Warning: *Wait until the engine is completely cool before beginning this procedure.*

Warning: *Do not allow antifreeze to come in contact with your skin or painted surfaces of the vehicle. Rinse off spills immediately with plenty of water. Antifreeze is highly toxic if ingested. Never leave antifreeze lying around in an open container or in puddles on the floor; children and pets are attracted by its sweet smell and may drink it. Check with local authorities on disposing of used antifreeze. Many communities have collection centers, which will see that antifreeze is disposed of safely. Never dump used antifreeze on the ground or into drains.*

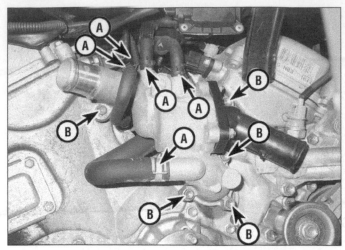

7.9 Water inlet assembly details on the 1GR-FE V6 engine

A   *Bypass hose clamp locations*
B   *Water inlet assembly mounting bolts*

7.10 Remove the number 1 (A) and number 2 (B) idler pulleys

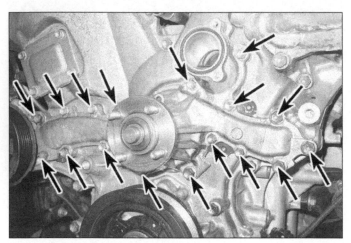

7.13 Location of the water pump mounting bolts on the V6 engine

7.14 Install a new gasket and clean the surface of the timing cover thoroughly

Warning: *On Sequoia models equipped with rear height control suspension, adjust the height control to the NORMAL mode, turn OFF the height control, then turn off the engine BEFORE raising the vehicle.*

1   Disconnect the cable from the negative terminal of the battery (see Chapter 5).
2   Drain the cooling system (see Chapter 1).
3   Remove the engine cover.
4   Remove the engine splash shield (see Chapter 1).
5   Remove the drivebelt(s) (see Chapter 1).
6   Remove the upper and lower radiator hoses (see Section 6).
7   Remove the fan shroud and cooling fan (see Section 5).

1GR-FE models

Refer to illustrations 7.9, 7.10, 7.13, 7.14, 7.15 and 7.16

8   Remove the air filter housing (see Chapter 4).

9   Remove the water inlet housing **(see illustration)**.
10   Remove the number 1 and number 2 idler pulley center bolts and remove the pulleys **(see illustration)**.
11   Remove the alternator (see Chapter 5).
12   Remove the drivebelt tensioner (see Chapter 1).
13   Remove the water pump mounting bolts **(see illustration)**.
Note: *Mark the location of the stud bolts for proper reassembly.*
14   Thoroughly clean the gasket mating surfaces and install a new gasket.
15   Install the water pump, tightening the bolts to the torque listed in this Chapter's Specifications **(see illustration)**.
16   Install a new O-ring on the water pump **(see illustration)**.
17   Install the water inlet assembly. Install new O-rings, then tighten the water inlet assembly mounting fasteners to the torque listed in this Chapter's Specifications.

18   The remainder of installation is the reverse of removal.
19   Refill the cooling system (see Chapter 1), run the engine and check for leaks.

2UZ-FE models

Refer to illustrations 7.21, 7.22 and 7.26

20   Remove the timing belt (see Chapter 2B). Remove the number 2 idler pulley (see Chapter 2B).
21   Disconnect the bypass hose from the water inlet assembly, remove the mounting bolts and separate the water inlet from the engine **(see illustration)**.
22   Remove the water pump mounting bolts **(see illustration)** and separate the pump from the engine block.
23   Remove the O-ring from the coolant bypass pipe directly behind the water pump. Replace the O-ring with a new one.
24   Install a new gasket and water pump. Tighten the water pump bolts to the torque

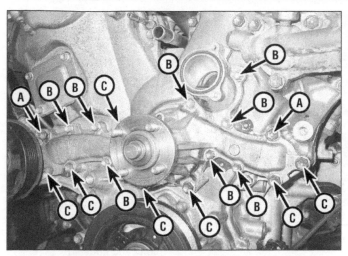

7.15 Water pump bolt identification - 2011 and later
1GR-FE engines only

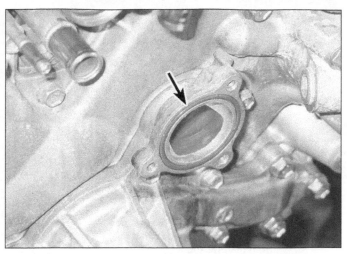

7.16 After installing the water pump, be sure to replace this
O-ring with a new one

7.21 Remove the water inlet assembly mounting bolts

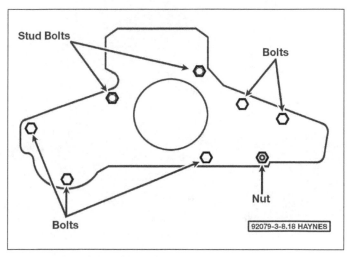

7.22 Location of the 5 bolts, the 2 stud bolts and the single nut on
the 2UZ-FE V8 water pump

listed in this Chapter's Specifications.

25   Thoroughly clean the gasket surface of the water inlet assembly.

26   Install a new O-ring, then apply a 3 mm bead of RTV sealant to the groove on the water inlet assembly **(see illustration).** Install the water pump within five minutes of sealant application. Tighten the water inlet assembly fasteners to the torque listed in this Chapter's Specifications.

27   The remainder of installation is the reverse of removal.

28   Refill the cooling system (see Chapter 1), run the engine and check for leaks.

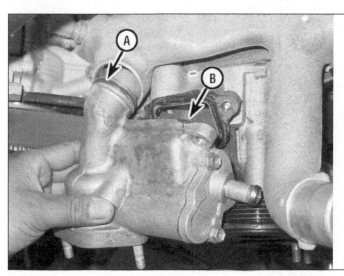

7.26 Install a new O-ring (A), then apply a 3 mm bead of RTV sealant to the groove on the backside of the water inlet assembly (B)

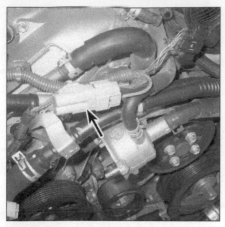

7.29 Disconnect the electrical harness from the bracket

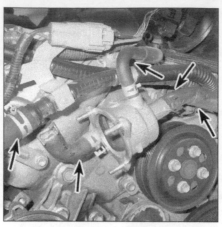

7.32a Disconnect the water inlet hoses

7.32b Remove the water inlet mounting bolts and housing

7.33 Hold the water pump pulley and remove the bolts

7.34 Location of the water pump mounting bolts - 1UR-FE, 3UR-FE and 3UR-FBE V8 models

7.35 Remove the water inlet O-ring

1UR-FE, 3UR-FE and 3UR-FBE models

Refer to illustrations 7.29, 7.32a, 7.32b, 7.33, 7.34, 7.35, 7.37 and 7.38

29   Disconnect the harness from the bracket **(see illustration)**.

30   If you're working on a 1UR-FE engine, remove the air conditioning compressor and position it aside (see Section 13). **Warning:** *Do not disconnect the refrigerant lines.*

31   Disconnect the oil cooler hoses from both ends of the tubes, then remove the cooler tube mounting bolts and tubes from the front of the engine (see Section 16).

32   Disconnect the bypass hoses from the water inlet assembly, remove the mounting bolts and separate the water inlet from the engine **(see illustrations)**.

33   Using a holding tool, remove the water pump pulley bolts **(see illustration)**.

34   Remove the water pump mounting bolts **(see illustration)** and separate the pump from the engine block.

35   Remove and discard the O-ring for the coolant bypass pipe from the water pump **(see illustration)**.

36   Thoroughly clean the gasket surface of the water pump and water inlet assembly.

37   Install a new gasket and water pump **(see illustration)**.

38   Tighten the water pump bolts to the torque listed in this Chapter's Specifications **(see illustration)**.

39   Install a new water inlet O-ring **(see illustration 7.35)**; if necessary to hold the O-ring in place, apply a 3 mm bead of RTV sealant to the groove, then install the O-ring.

40   Install the water inlet and tighten the water inlet assembly fasteners to the torque listed in this Chapter's Specifications.

41   The remainder of installation is the reverse of removal.

42   Refill the cooling system (see Chapter 1), run the engine and check for leaks.

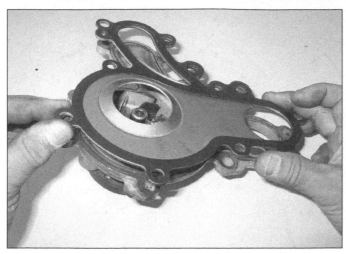

7.37 Install a new gasket and clean the surface of the timing chain cover thoroughly

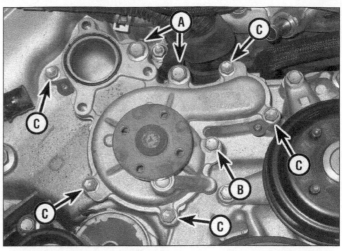

7.38 Water pump bolt identification - 1UR-FE, 3UR-FE and 3UR-FBE V8 models

8 Coolant temperature indicator system - check

Warning: *Wait until the engine is completely cool before beginning this procedure.*

1   The coolant temperature indicator system consists of a temperature gauge on the dash and a sensor mounted on the engine. On all models, an Engine Coolant Temperature (ECT) sensor (see Chapter 6), which is an information sensor for the Powertrain Control Module (PCM), provides a signal to the PCM which controls and actuates the temperature gauge.

2   If an overheating indication has occurred, first check the coolant level and mixture in the system (see Chapter 1). Also, refer to the *Troubleshooting* Section at the beginning of this Chapter before assuming that the temperature indicator is faulty.

3   Start the engine and warm it up for 10 minutes. If the temperature gauge has not moved from the C position, check the wiring harness connections going to the instrument cluster.

4   If there is a problem with the ECT sensor, it is very likely that the Malfunction Indicator Lamp will be illuminated and the sensor or circuit will need repair (see Chapter 6).

9 Air conditioning and heater blower motor - removal and installation

Warning: *The models covered by this manual are equipped with Supplemental Restraint systems (SRS), more commonly known as airbags. Always disable the airbag system before working in the vicinity of any airbag system components to avoid the possibility of accidental deployment of the airbag, which could cause personal injury (see Chapter 12).*
Note: *Some Sequoia models are equipped*

with rear heating and air conditioning. The rear A/C unit is mounted behind the rear quarter trim panel. The rear A/C unit houses the heater core (heater radiator) and the evaporator.
Note: *The fan cannot be separated from the blower motor; if the motor or fan is bad, they must be replaced as unit*

1   The blower unit is located in the cooling assembly under the dash. On Sequoia models with rear heater units, an additional blower motor is located behind the right rear quarter trim panel.

Front mounted blower motors

Refer to illustration 9.3

2   On Sequoia models, pull the lower trim panel from under the passenger's side instrument panel.

3   Disconnect the blower motor connector, then remove the three mounting screws **(see illustration)**.

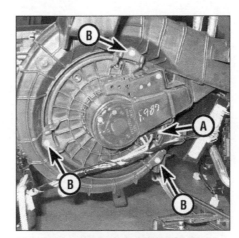

9.3 Disconnect the blower motor connector (A), then remove the three mounting screws (B) and remove the blower motor

4   Lower the blower assembly from the housing.
5   Installation is the reverse of removal.

Rear mounted blower motors (Sequoia models only)

6   Remove the rear seats then remove the rear seat belt outer anchors from the floor (see Chapter 11).
7   Remove the rear floor mat support plate.
8   Remove the rear door scuff plate (see Chapter 11).
9   Remove the trim from the right side window (see Chapter 11).
10   Remove the right, rear quarter trim panel.
11   Remove the blower motor connector, then remove the three mounting screws.
12   Remove the blower assembly from the housing.
13   Installation is the reverse of removal.

10 Heater core - removal and installation

Warning: *The models covered by this manual are equipped with Supplemental Restraint systems (SRS), more commonly known as airbags. Always disable the airbag system before working in the vicinity of any airbag system components to avoid the possibility of accidental deployment of the airbag, which could cause personal injury (see Chapter 12).*
Warning: *Do not allow antifreeze to come in contact with your skin or painted surfaces of the vehicle. Rinse off spills immediately with plenty of water. Antifreeze is highly toxic if ingested. Never leave antifreeze lying around in an open container or in puddles on the floor; children and pets are attracted by it's sweet smell and may drink it. Check with local authorities about disposing of used antifreeze. Many communities have collection centers which will see that antifreeze is dis-*

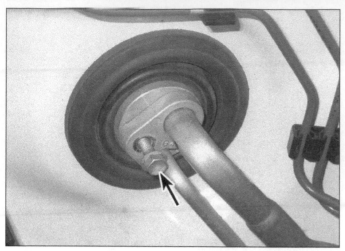

10.7a Remove the liquid tube and the suction tube mounting bolt

10.7b Rotate the mounting plate clockwise and disconnect the tubes from the cooling unit

10.8 With the help of an assistant, remove the instrument panel reinforcement brace and cooling unit as an assembly

10.10 Remove the defroster duct

posed of safely. Never dump used antifreeze on the ground or pour it into drains.

Warning: *The air conditioning system is under high pressure. Do not loosen any hose fittings or remove any components until the system has been discharged. Air conditioning refrigerant should be properly discharged into an EPA-approved recovery/recycling unit by a dealer service department or an automotive air conditioning repair facility. Always wear eye protection when disconnecting air conditioning system fittings.*

Note: *The factory recommends removal of the entire instrument panel to remove the heater core on front mounted systems. This involves disconnecting numerous electrical connectors and there is the potential for breakage of delicate plastic tabs on various components. This is a difficult procedure for the home mechanic. There are many hidden fasteners, difficult angles to work in and many electrical connectors to tag and dis-*

connect/connect. We recommend that this procedure be done only by an experienced do-it-yourselfer.

Note: *Some Sequoia models are equipped with rear heating and air conditioning. The rear A/C unit is mounted behind the rear quarter trim panel. The rear A/C unit houses the heater core (heater radiator) and the evaporator.*

1   Wait until the engine is completely cool before beginning this procedure.
2   Disconnect the negative cable from the battery.
3   Drain the cooling system (see Chapter 1).
4   Have the refrigerant discharged and recovered by an air conditioning technician.

Front-mounted heating system
Refer to illustrations 10.7a, 10.7b, 10.8, 10.10, 10.11a, 10.11b, 10.11c, 10.12, 10.13, 10.14 and 11.15

5   Working in the engine compartment, dis-

connect the heater hoses at the firewall. Push the rubber seal around the hoses toward the inside of the vehicle, releasing it from the sheetmetal. Plug the heater core tubes to prevent leakage when it is removed.
6   Remove the instrument panel (see Chapter 11).
7   Remove the liquid tube and the suction tube mounting bolt, then rotate the mounting plate and disconnect the tubes from the cooling unit **(see illustrations)**.
8   Remove the instrument panel reinforcement brace (see Chapter 11) with the cooling unit attached **(see illustration)**.
9   Remove the mounting screws and separate the cooling unit from the reinforcement brace.
10   Remove the defroster heater-to-register duct **(see illustration)**.
11   Remove the A/C amplifier, cover plate and damper servo from the cooler unit **(see illustrations)**.

10.11a Remove the A/C amplifier fasteners and amplifier

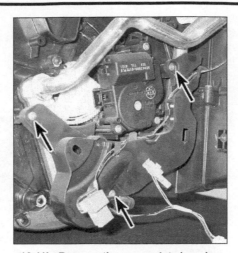

10.11b Remove the cover plate housing fasteners and cover

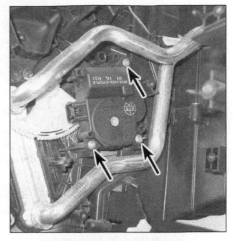

10.11c Remove the damper servo fasteners and servo

10.12 Remove the upper blower case from the unit

10.13 Remove the plastic retainer from the heater core tubes

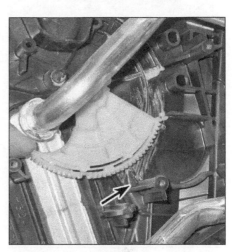

10.14 Location of the servo actuator gear

12   Remove the upper blower case fasteners and lift the case from the unit **(see illustration)**.
13   Remove the plastic retainer from the heater core tubes **(see illustration)**.
14   Slide the servo actuator gear up away from the heater core **(see illustration)**.
15   Pull the heater core from the case and maneuver it out of the unit **(see illustration)**. **Caution:** *Work slowly and carefully to avoid breaking any plastic components during removal.*
16   Installation is the reverse of removal. Refill the cooling system (see Chapter 1), reconnect the battery and run the engine. Check for leaks and proper system operation.

Rear-mounted heating system

17   Remove the right, rear quarter trim panel (see Chapter 11).
18   Remove the bolts from the air conditioning pressure lines and separate the liquid tube and suction tube from the A/C unit.
19   Remove the heater hoses from the coolant pipes.

20   Remove the clips and separate the air ducts from the vehicle.
21   Remove the mounting bolts and remove the A/C unit case from the rear of the vehicle.
22   Remove the blower motor (see Section 9).
23   Remove the power transistor and the thermistor.
24   Remove the screws and the bracket, then separate the coolant pipe from the rear A/C unit case.
25   Disconnect the coolant hoses and the coolant pipes from the outer case.
26   Remove the water valve mounting bolts and separate the water valve from the A/C unit case.
27   Remove the expansion valve.
28   Remove the evaporator.
29   Remove the air outlet servo motor.
30   Separate the rear A/C unit case halves and remove the heater core (heater radiator) from the A/C unit case.
31   Installation is the reverse of removal. Refill the cooling system (see Chapter 1), reconnect the battery and run the engine. Check for leaks and proper system operation.

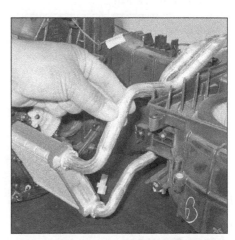

10.15 Maneuver the heater core and tube assembly out of the cooler unit

12.3 After the condenser is removed, remove the Allen plug and pull the drier from the tube on the condenser

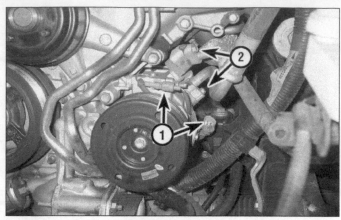

13.5 Typical air conditioning compressor electrical connector (1) and refrigerant line (2) locations

13.6 Remove the nuts, then unbolt the compressor mounting studs

11   Air conditioning and heater control assembly - removal and installation

Warning: *The models covered by this manual are equipped with Supplemental Restraint systems (SRS), more commonly known as airbags. Always disable the airbag system before working in the vicinity of any airbag system components to avoid the possibility of accidental deployment of the airbag, which could cause personal injury (see Chapter 12).*
Note: *Some Sequoia models are equipped with rear heating and air conditioning. The rear A/C unit is mounted behind the rear quarter trim panel. The rear A/C unit houses the heater core (heater radiator) and the evaporator.*
1   Disconnect the negative cable from the battery (see Chapter 5).

Front mounted control assembly

2   Remove the lower and center instrument panel trim and the center console (see Chapter 11).
3   Carefully pry the integration and control panel assembly (see Chapter 11) out from the instrument panel until the electrical

connectors can be disconnected.
4   Disconnect the electrical connectors and remove the control assembly.
Note: *On 2014 models, remove the control assembly–to-panel mounting screws and separate the air conditioning control assembly from the trim panel.*
5   Installation is the reverse of removal.
6   Run the engine and check for proper functioning of the heater (and air conditioning, if equipped).

Rear-mounted control assembly

7   Release the three clips on each side of the center console rear panel and pull the panel off the floor console.
8   Disconnect the electrical connectors and the two harness clamps.
9   Pull the control assembly out slightly and disconnect the connectors.
10   Remove the heater control assembly mounting screws then release the clips and remove the control assembly from the console end panel.
11   Installation is the reverse of removal.

12   Air conditioning receiver/drier - removal and installation

Refer to illustration 12.3
Warning: *The air conditioning system is under high pressure. Do not loosen any hose fittings or remove any components until the system has been discharged. Air conditioning refrigerant must be properly discharged into an EPA-approved recovery/recycling unit by a dealer service department or an automotive air conditioning repair facility. Always wear eye protection when disconnecting air conditioning system fittings.*
1   Have the refrigerant discharged by an air conditioning technician.
2   Disconnect the cable from the negative battery terminal (see Chapter 5).
3   Remove the condenser (see Section 14). Using an Allen wrench, detach the end plug

and remove the receiver/drier from the condenser **(see illustration)**.
4   Installation is the reverse of removal. Be sure to tighten the end plug to the torque listed in this Chapter's Specifications.
5   Have the system evacuated, charged and leak tested by the shop that discharged it. If the receiver was replaced, have them add the proper amount of refrigeration oil to the compressor. Use only compressor oil compatible with R-134a refrigerant.

13   Air conditioning compressor - removal and installation

Refer to illustrations 13.5 and 13.6
Warning: *The air conditioning system is under high pressure. Do not loosen any hose fittings or remove any components until the system has been discharged. Air conditioning refrigerant should be properly discharged into an EPA-approved recovery/recycling unit by a dealer service department or an automotive air conditioning repair facility. Always wear eye protection when disconnecting air conditioning system fittings.*
Warning: *On Sequoia models equipped with rear height control suspension, adjust the height control to the NORMAL mode, turn OFF the height control, then turn off the engine BEFORE raising the vehicle.*
1   Have the refrigerant discharged by an automotive air conditioning technician.
2   Remove the under-vehicle splash shield (see Chapter 1).
3   Raise the front of the vehicle and place it securely on jackstands. Remove the accessory drivebelt or serpentine drivebelt (see Chapter 1).
4   Remove the left front wheel, then the inner fender rubber seal plastic clips from the wheel well splash shield (see Chapter 11) and remove the seal.
5   Detach the electrical connector and disconnect the refrigerant lines **(see illustration)**.
6   Remove the mounting nuts, then unbolt the compressor mounting studs and remove the compressor **(see illustration)**.

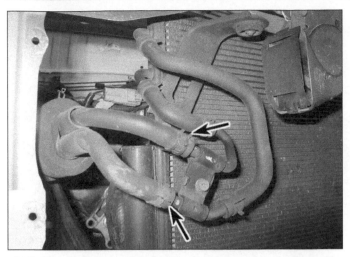

14.3 Disconnect the oil cooler hoses - on models equipped with an air cooled transmission oil cooler

14.6 A/C condenser details

1	Condenser mounting bolts	3	Pressure switch
2	Discharge line	4	Condenser line

7   If a new or rebuilt compressor is being installed, follow the directions supplied with the compressor regarding the proper amount of oil to add prior to installation.

8   Installation is the reverse of removal. Replace any O-rings with new ones specifically made for the type of refrigerant in your system and lubricate them with refrigerant oil, also designed specifically for your system (R-134a).

9   Have the system evacuated, recharged and leak tested by the shop that discharged it.

14 Air conditioning condenser - removal and installation

Refer to illustrations 14.3, 14.6 and 14.9

Warning: *The air conditioning system is under high pressure. Do not loosen any hose fittings or remove any components until the system has been discharged. Air conditioning refrigerant should be properly discharged into an EPA-approved recovery/recycling unit by a dealer service department or an automotive air conditioning repair facility. Always wear eye protection when disconnecting air conditioning system fittings.*

Note: *If the condenser is to be replaced with a new unit, add 1.4 to 1.7 fluid ounces of compressor oil to the system.*

1   Have the refrigerant discharged by an air conditioning technician.

2   Remove the front bumper cover and the radiator grille (see Chapter 11).

3   Disconnect the oil cooler hoses on models equipped with an air cooled transmission oil cooler **(see illustration).**

4   Remove the fan shroud and cooling fan (see Section 5) and the radiator (see Section 6).

14.9 Make sure the condenser rubber cushions fit in the mounting points

5   On Sequoia models, disconnect the condenser fan electrical connectors, then remove the mounting bolts and fan.

6   Disconnect the A/C pressure switch electrical connector then disconnect the line from the condenser and the discharge hose **(see illustration).**

7   Remove the condenser mounting bolts **(see illustration 14.6).**

8   On models without air cooled transmission oil coolers, pull the condenser forward and lift it out. On models with air cooled transmission oil coolers, push the condenser towards the engine and lift it out.

9   Install the condenser, with brackets (if removed) and bolts, making sure the rubber cushions fit properly on the mounting points **(see illustration).**

10   Reconnect the refrigerant lines, using new O-rings where needed. If a new condenser has been installed, add approximately 1.4 to 1.7 ounces (50 cc) of new refrigerant oil of the correct type (R-134a).

11   Reinstall the remaining parts in the

15.2 Disconnect the electrical connector from the pressure cycling switch

reverse order of removal.

12   Have the system evacuated, charged and leak tested by the shop that discharged it.

15 Air conditioning pressure cycling switch - removal and installation

Refer to illustration 15.2

Warning: *The air conditioning system is under high pressure. Do not loosen any hose fittings or remove any components until the system has been discharged. Air conditioning refrigerant should be properly discharged into an EPA-approved recovery/recycling unit by a dealer service department or an automotive air conditioning repair facility. Always wear eye protection when disconnecting air conditioning system fittings.*

1   Have the refrigerant discharged by an air conditioning technician.

2   Unplug the electrical connector from the pressure cycling switch **(see illustration).**

16.15 Disconnect the cooler hoses

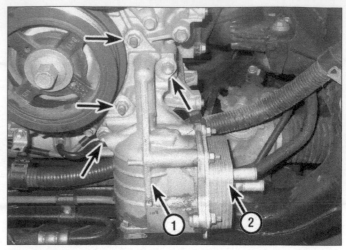

16.16a Remove the oil filter bracket fasteners; the oil filter bracket (1) can then be separated from the oil cooler (2)

16.16b Remove and discard the oil filter bracket O-rings

3   Unscrew the pressure cycling switch. Hold the fitting on the line with a wrench to prevent deforming the pressure line.
4   Lubricate the switch O-ring with clean refrigerant oil compatible with R134-a systems.
5   Screw the new switch onto the threads, then tighten it to the torque listed in this Chapter's Specifications.
6   Reconnect the electrical connector.
7   Have the system evacuated, charged and leak tested by the shop that discharged it.

16 Oil cooler - removal and installation

Warning: *Wait until the engine is completely cool before beginning this procedure.*

1   Drain the engine oil and remove the oil filter (see Chapter 1).
2   Drain the coolant into a container (see Chapter 1).

1GR-FE and 2UZ-FE engines

3   Disconnect the oil cooler hoses from the oil cooler. Position a drain pan directly below the cooler lines to catch any residual oil.
4   Remove the union bolt and separate the oil cooler from the oil filter adapter.
5   Installation is the reverse of removal.
6   Install a new O-ring and plate washer.
7   Tighten the union bolt to the torque listed in this Chapter's Specifications.
8   Install a new oil filter. Refill the engine with oil and the cooling system with the proper type and concentration of antifreeze (see Chapter 1). Check for coolant leakage and proper gauge function.

1UR-FE, 3UR-FE and 3UR-FBE engines

Refer to illustration 16.15, 16.16a and 16.16b

9   Disconnect the negative battery cable.
10   Remove the engine splash shield **(see Chapter 1, illustration 8.6a)**.
11   Remove the engine cover **(see Chapter 1, illustration 24.12a)**.
12   Remove the drivebelt (see Chapter 1).
13   Remove the cooling fan and shroud (see Section 5).
14   Remove the air conditioning compressor (see Section 13).
15   Disconnect the oil cooler hoses from the oil cooler **(see illustration)**. Position a drain pan directly below the cooler lines to catch any residual oil. Remove the tube mounting bracket bolts and move the tubes to the side.
16   Remove the oil filter bracket fasteners and remove the bracket **(see illustration)**. Remove and discard the O-rings **(see illustration)**.
17   Remove the oil cooler mounting nuts and remove the cooler from the oil filter bracket.
18   Remove the two oil cooler O-rings, then the spacer bolt, spacer and gasket.
19   Installation is the reverse of removal.
20   Install a new gasket and O-rings on the oil cooler spacer.
21   Install new O-rings on the oil filter bracket.
22   Torque the cooler nuts and bracket fasteners to the specifications listed in this Chapter.
23   Install a new oil filter element. Refill the engine with oil and the cooling system with the proper type and concentration of antifreeze (see Chapter 1). Check for coolant leakage and proper gauge function.

Chapter 4
Fuel and exhaust systems

Contents

Specifications

Fuel system

Fuel system pressure (at idle)
  1GR-FE
    2010 and earlier models .... 41 to 42 psi (281 to 287 kPa)
    2011 and later models .... 46.5 to 47.4 psi (321 to 327 kPa)
  1UR-FE .... 41 to 42 psi (281 to 287 kPa)
  2UZ-FE and 3UR-FE .... 38 to 44 psi (265 to 304 kPa)
  3UR-FBE
    Fuel pressure switching valve is OFF .... 39.2 to 47.9 psi (270 to 330 kPa)
    Fuel pressure switching valve is ON .... 53.7 to 62.4 psi (370 to 430 kPa)
Residual fuel system pressure (five minutes after engine turned off) ...... 21 psi (145 kPa) minimum
Injector resistance (approximate at 68-degrees F [20-degrees C]) .... 11.6 to 12.4 ohms

Torque specifications	**Ft-lbs** (unless otherwise indicated)	**Nm**

Note: *One foot-pound (ft-lb) of torque is equivalent to 12 inch-pounds (in-lbs) of torque. Torque values below approximately 15 ft-lbs are expressed in inch-pounds, since most foot-pound torque wrenches are not accurate at these smaller values.*

Pulsation damper pipe-to-fuel rail (2UZ-FE V8 engine)	29	39
Fuel pressure regulator bolts		
1GR-FE engines	80 in-lbs	9
2UZ-FE engines	66 in-lbs	7.5
1UR-FE, 3UR-FE and 3UR-FBE engines	84 in-lbs	9.5
Fuel rail		
1GR-FE engines (bolts)		
2010 and earlier models	132 in-lbs	15
2011 and later models	15	20
2UZ-FE engines (nuts)	15	20
1UR-FE, 3UR-FE and 3UR-FBE engines (bolts)	15	20
Fuel rail crossover pipe banjo bolts (2UZ-FE V8 engine)	29	39
Throttle body mounting bolts/nuts		
1GR-FE engines		
2010 and earlier models	96 in-lbs	11
2011 and later models	84 in-lbs	9.5
2UZ-FE engines	120 in-lbs	14
1UR-FE, 3UR-FE and 3UR-FBE engines	84 in-lbs	9.5
Fuel tank strap bolts	30	40

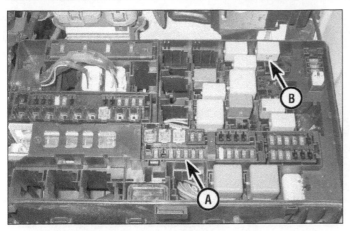

2.2 The fuel pump fuse (A) is located in the engine compartment fuse relay box; (B) is the fuel pump main relay

2.9 An automotive stethoscope is used to listen to the fuel injectors in operation

1   General information

Fuel system warnings

Gasoline is extremely flammable and repairing fuel system components can be dangerous. Consider your automotive repair knowledge and experience before attempting repairs which may be better suited for a professional mechanic.

- *Don't smoke or allow open flames or bare light bulbs near the work area*
- *Don't work in a garage with a gas-type appliance (water heater, clothes dryer)*
- *Use fuel-resistant gloves. If any fuel spills on your skin, wash it off immediately with soap and water*
- *Clean up spills immediately*
- *Do not store fuel-soaked rags where they could ignite*
- *Prior to disconnecting any fuel line, you must relieve the fuel pressure (see Section 3)*
- *Wear safety glasses*
- *Have a proper fire extinguisher on hand*

Fuel system

The fuel system consists of the fuel tank, electric fuel pump/fuel level sending unit (located in the fuel tank), fuel rail and fuel injectors. The fuel injection system is a multi-port system; multi-port fuel injection uses timed impulses to inject the fuel directly into the intake port of each cylinder. The Powertrain Control Module (PCM) controls the injectors. The PCM monitors various engine parameters and delivers the exact amount of fuel required into the intake ports.

Fuel is circulated from the fuel pump to the fuel rail through fuel lines running along the underside of the vehicle. Various sections of the fuel line are either rigid metal or nylon, or flexible fuel hose. The various sections of the fuel hose are connected either by quick-connect fittings or threaded metal fittings.

Exhaust system

The exhaust system consists of the exhaust manifold(s), catalytic converter(s), muffler(s), tailpipe and all connecting pipes, flanges and clamps. The catalytic converters are an emission control device added to the exhaust system to reduce pollutants.

2   Troubleshooting

Fuel pump

Refer to illustration 2.2

1   The fuel pump is located inside the fuel tank. Sit inside the vehicle with the windows closed, turn the ignition key to ON (not START) and listen for the sound of the fuel pump as it's briefly activated. You will only hear the sound for a second or two, but that sound tells you that the pump is working. Alternatively, have an assistant listen at the fuel filler cap.
2   If the pump does not come on, check the fuel pump fuse labeled "F/PMP" and the relay labeled "F/PMP" **(see illustration)**. If the fuse is okay, check the wiring back to the fuel pump. If the fuse and wiring are okay, the fuel pump module is probably defective. If the pump runs continuously with the ignition key in the ON position, the Powertrain Control Module (PCM) is probably defective. Have the PCM checked by a professional mechanic.

Fuel injection system

Refer to illustration 2.9

Note: *The following procedure is based on the assumption that the fuel pump is working and the fuel pressure is adequate (see Section 4).*

3   Check all electrical connectors that are related to the system. Check the ground wire connections for tightness.
4   Verify that the battery is fully charged (see Chapter 5).
5   Inspect the air filter element (see Chapter 1).

6   Check all fuses related to the fuel system (see Chapter 12).
7   Check the air induction system between the throttle body and the intake manifold for air leaks. Also inspect the condition of all vacuum hoses connected to the intake manifold and to the throttle body.
8   Remove the air intake duct from the throttle body and look for dirt, carbon, varnish, or other residue in the throttle body, particularly around the throttle plate. If it's dirty, clean it with carb cleaner, a toothbrush and a clean shop towel.
9   With the engine running, place an automotive stethoscope against each injector, one at a time, and listen for a clicking sound that indicates operation **(see illustration)**.
Warning: *Stay clear of the drivebelt and any rotating or hot components.*
10   If you can hear the injectors operating, but the engine is misfiring, the electrical circuits are functioning correctly, but the injectors might be dirty or clogged. Try a commercial injector cleaning product (available at auto parts stores). If cleaning the injectors doesn't help, replace the injectors.
11   If an injector is not operating (it makes no sound), disconnect the injector electrical connector and measure the resistance across the injector terminals with an ohmmeter. Compare this measurement to the other injectors. If the resistance of the non-operational injector is quite different from the other injectors, replace it.
12   If the injector is not operating, but the resistance reading is within the range of resistance of the other injectors, the PCM or the circuit between the PCM and the injector might be faulty.

3   Fuel pressure relief procedure

Warning: *Gasoline is extremely flammable. See **Fuel system warnings** in Section 1.*
1   Remove the fuel filler cap to relieve any pressure built-up in the fuel tank.
2   Remove the fuel pump relay from the

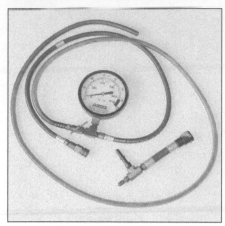

4.1a This fuel pressure testing kit contains all the necessary fittings and adapters, along with the fuel pressure gauge, to test most automotive systems

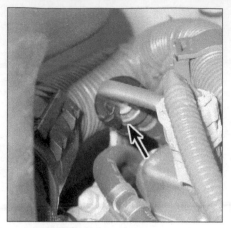

4.1b Attach the fuel pressure gauge between the two sides of the quick disconnect fitting; turn the ignition key ON and check the fuel pressure

6.1 A typical exhaust system hanger. Inspect regularly and replace at the first sign of damage or deterioration

underhood fuse/relay box **(see illustration 2.2)**.
3   Attempt to start the engine; it should immediately stall. Crank the engine several more times to ensure the fuel system has been completely relieved. Disconnect the cable from the negative terminal of the battery before working on the fuel system.
4   It's a good idea to cover any fuel connection to be disassembled with rags to absorb the residual fuel that may leak out. Properly dispose of the rags.

---

4   Fuel pressure - check

Refer to illustrations 4.1a and 4.1b
Warning: *Gasoline is extremely flammable. See* **Fuel system warnings** *in Section 1.*
Note: *The following procedure assumes that the fuel pump is receiving voltage and runs.*
1   Relieve the fuel system pressure (see Section 3). Disconnect the fuel line from the fuel rail, then use the proper adapter to connect the pressure gauge between the fuel line and fuel rail **(see illustrations)**.
2   Turn all the accessories Off and switch the ignition key On. The fuel pump should run for about two seconds; note the reading on the gauge. After the pump has stopped running, the pressure should hold steady. After five minutes it should not drop below the minimum listed in this Chapter's Specifications.
3   Start the engine and allow it to idle. Note the gauge reading as soon as the pressure stabilizes, and compare it with the pressure listed in this Chapter's Specifications.
4   If the fuel pressure is not within specifications, check the following:
   a) *Check for a restriction in the fuel system (kinked fuel line, plugged fuel pump inlet strainer or clogged fuel filter). If no restrictions are found, replace the fuel pump module (see Section 8).*
   b) *If the fuel pressure is higher than specified, replace the fuel pump module (see Section 8).*

5   After the reading is done, relieve the fuel pressure (see Section 3) and remove the fuel pressure gauge.

---

5   Fuel lines and fittings - general information and disconnection

Warning: *Gasoline is extremely flammable. See* **Fuel system warnings** *in Section 1.*
1   Relieve the fuel pressure before servicing fuel lines or fittings (see Section 3), then disconnect the cable from the negative battery terminal (see Chapter 5) before proceeding.
2   The fuel supply line connects the fuel pump in the fuel tank to the fuel rail on the engine. The Evaporative Emission (EVAP) system lines connect the fuel tank to the EVAP canister and connect the canister to the intake manifold.
3   Whenever you're working under the vehicle, be sure to inspect all fuel and evaporative emission lines for leaks, kinks, dents and other damage. Always replace a damaged fuel or EVAP line immediately.
Warning: *On Sequoia models equipped with rear height control suspension, adjust the height control to the NORMAL mode, turn OFF the height control, then turn off the engine BEFORE raising the vehicle.*
4   If you find signs of dirt in the lines during disassembly, disconnect all lines and blow them out with compressed air. Inspect the fuel strainer on the fuel pump pick-up unit for damage and deterioration.

Steel tubing
5   It is critical that the fuel lines be replaced with lines of equivalent type and specification.
6   Some steel fuel lines have threaded fittings. When loosening these fittings, hold the stationary fitting with a wrench while turning the tube nut.

Plastic tubing
7   When replacing fuel system plastic tubing, use only original equipment replacement

plastic tubing.
Caution: *When removing or installing plastic fuel line tubing, be careful not to bend or twist it too much, which can damage it. Also, plastic fuel tubing is NOT heat resistant, so keep it away from excessive heat.*

Flexible hoses
8   When replacing fuel system flexible hoses, use only original equipment replacements.
9   Don't route fuel hoses (or metal lines) within four inches of the exhaust system or within ten inches of the catalytic converter. Make sure that no rubber hoses are installed directly against the vehicle, particularly in places where there is any vibration. If allowed to touch some vibrating part of the vehicle, a hose can easily become chafed and it might start leaking. A good rule of thumb is to maintain a minimum of 1/4-inch clearance around a hose (or metal line) to prevent contact with the vehicle underbody.

---

6   Exhaust system servicing - general information

Refer to illustration 6.1
Warning: *Allow exhaust system components to cool before inspection or repair. Also, when working under the vehicle, make sure it is securely supported on jackstands.*
Warning: *On Sequoia models equipped with rear height control suspension, adjust the height control to the NORMAL mode, turn OFF the height control, then turn off the engine BEFORE raising the vehicle.*
1   The exhaust system consists of the exhaust manifolds, catalytic converter, muffler, tailpipe and all connecting pipes, flanges and clamps. The exhaust system is isolated from the vehicle body and from chassis components by a series of rubber hangers **(see illustration)**. Periodically inspect these hangers for cracks or other signs of deterioration, replacing them as necessary.

Disconnecting Fuel Line Fittings

Two-tab type fitting; depress both tabs with your fingers, then pull the fuel line and the fitting apart

On this type of fitting, depress the two buttons on opposite sides of the fitting, then pull it off the fuel line

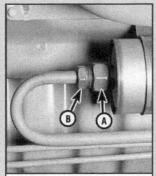

Threaded fuel line fitting; hold the stationary portion of the line or component (A) while loosening the tube nut (B) with a flare-nut wrench

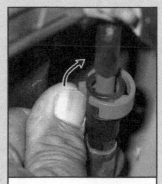

Plastic collar-type fitting; rotate the outer part of the fitting

Metal collar quick-connect fitting; pull the end of the retainer off the fuel line and disengage the other end from the female side of the fitting . . .

. . . insert a fuel line separator tool into the female side of the fitting, push it into the fitting and pull the fuel line off the pipe

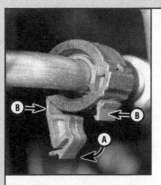

Some fittings are secured by lock tabs. Release the lock tab (A) and rotate it to the fully-opened position, squeeze the two smaller lock tabs (B) . . .

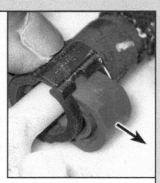

. . . then push the retainer out and pull the fuel line off the pipe

Spring-lock coupling; remove the safety cover, install a coupling release tool and close the tool around the coupling . . .

. . . push the tool into the fitting, then pull the two lines apart

Hairpin clip type fitting: push the legs of the retainer clip together, then push the clip down all the way until it stops and pull the fuel line off the pipe

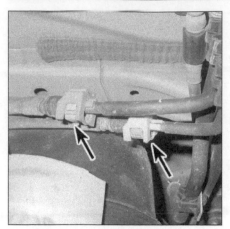

7.5 Disconnect the fuel supply and return hoses (for instructions regarding quick-connect fittings, see Section 5)

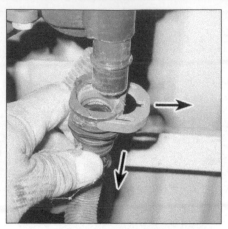

7.6 Pull the retainer clip back and disconnect the vent line from the EVAP

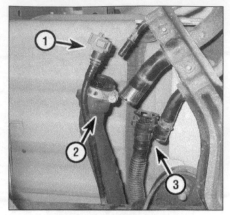

7.7 Before lowering the fuel tank, disconnect these hoses

1   Fuel tank breather tube (for instructions regarding quick-connect fittings, see Section 5)
2   Filler pipe hose
3   EVAP breather tube

2   Conduct regular inspections of the exhaust system to keep it safe and quiet. Look for any damaged or bent parts, open seams, holes, loose connections, excessive corrosion or other defects which could allow exhaust fumes to enter the vehicle. Do not repair deteriorated exhaust system components; replace them with new parts.

3   If the exhaust system components are extremely corroded, or rusted together, a cutting torch is the most convenient tool for removal. Consult a properly-equipped repair shop. If a cutting torch is not available, you can use a hacksaw, or if you have compressed air, there are special pneumatic cutting chisels that can also be used. Wear safety goggles to protect your eyes from metal chips and wear work gloves to protect your hands.

4   Here are some simple guidelines to follow when repairing the exhaust system:

a) *Work from the back to the front when removing exhaust system components.*
b) *Apply penetrating oil to the exhaust system component fasteners to make them easier to remove.*
c) *Use new gaskets, hangers and clamps.*
d) *Apply anti-seize compound to the*

threads of all exhaust system fasteners during reassembly.
e) *Be sure to allow sufficient clearance between newly installed parts and all points on the underbody to avoid overheating the floor pan and possibly damaging the interior carpet and insulation. Pay particularly close attention to the catalytic converter and heat shield.*

7   Fuel tank - removal and installation

Warning: *Gasoline is extremely flammable. See the* **Fuel system warnings** *in Section 1.*
Warning: *On Sequoia models equipped with rear height control suspension, adjust the height control to the NORMAL mode, turn OFF the height control, then turn off the engine BEFORE raising the vehicle.*

Removal

Refer to illustrations 7.5, 7.6, 7.7, 7.8a, 7.8b, 7.9a and 7.9b

1   Relieve the fuel system pressure (see Section 3) and remove the fuel tank cap.
2   Disconnect the cable from the negative terminal of the battery.

3   Raise the rear of the vehicle and place it securely on jackstands.
4   If equipped, remove the tank shield nut/bolts and lower the shield from the vehicle.
5   On 3UR-FBE engines, remove the fuel pressure switching valve mounting bolt, then on all models, disconnect the fuel supply and return hoses **(see illustration)**. If you're unfamiliar with quick-connect fittings, refer to Section 5.
6   Pull the retainer back and disconnect the vent line **(see illustration)**.
7   Disconnect and detach the fuel tank breather tube, the filler pipe hose and the breather tube **(see illustration)**.
8   Place a transmission jack or a floor jack under the fuel tank **(see illustration)**. If you're using a floor jack, put a piece of plywood between the jack head and the tank to protect the tank. Raise the jack just enough to take the weight of the tank off the fuel tank bands, then remove the bolts that secure the fuel tank bands **(see illustration)**.

7.8a Support the tank with a transmission jack (shown) or use a floor jack; if you use a floor jack, put a piece of plywood between the jack head and the fuel tank to protect the tank

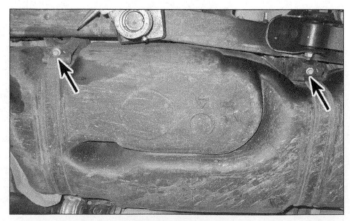

7.8b To detach the fuel tank, remove these tank band bolts

7.9a Detach the fuel pump/fuel level sending unit harness clip and remove the access cover . . .

7.9b . . . then disconnect the fuel pump/fuel level sending unit electrical connector

9   Carefully lower the tank just far enough to detach the fuel pump/fuel level sending unit wiring harness clip and to disconnect the fuel pump/fuel level sending unit electrical connector **(see illustrations) an**d the fuel pressure switching valve connector on 3UR-FBE engines.

10   Carefully lower the tank and remove it.

11   Installation is the reverse of removal.

8   Fuel pump/fuel level sending unit assembly – removal, component replacement and installation

Warning: *Gasoline is extremely flammable. See the* **Fuel system warnings** *in Section 1.*
Warning: *On Sequoia models equipped with rear height control suspension, adjust the height control to the NORMAL mode, turn OFF the height control, then turn off the engine BEFORE raising the vehicle.*

Removal
Refer to illustrations 8.3, 8.4a, 8.4b and 8.5
Note: *On 2009 and later 3UR-FBE and 2010*

and later 3UR-FE engines, if the fuel pump or fuel gauge sending unit is defective, the entire assembly must be replaced.

1   Relieve the system fuel pressure (see Section 3).

2   Remove the fuel tank (see Section 7).

3   Disconnect the electrical connector from the fuel pump/fuel level sending unit, then remove the two tube joint U-clips **(see illustration)**. Pull up on the fuel supply and return line fittings to disconnect them from the pump.

4   Use a pump retainer removal tool **(see illustration)** or large pair of channel lock pliers to loosen the threaded fuel pump retaining ring. As the retainer is being loosened, push the three locking tabs inwards to prevent them from locking against the tabs on the top **(see illustration)**.

Note: *The fuel pump/fuel level sending unit assembly is spring loaded; once the ring is removed, the assembly will spring up.*

5   Carefully withdraw the fuel pump/fuel level sending unit assembly from the fuel tank. Making sure you don't damage the fuel

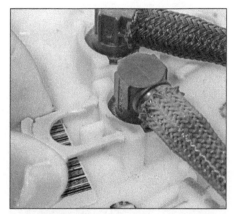

8.3 To disconnect the fuel supply and return line fittings, pull out the tube joint U-clips, then pull the tube fittings straight up

pump filter or bend the sending unit float arm, remove the O-ring seal **(see illustration)**.

8.4a To detach the fuel pump from the fuel tank, unscrew the retaining ring

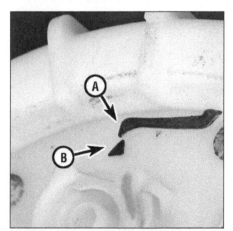

8.4b As the retainer is being loosened, push the locking tabs (A) inwards to prevent them from locking against the tabs (B) on the unit face (there are three tabs total)

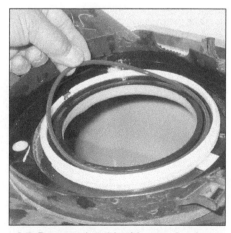

8.5 Remove the old rubber gasket from the fuel pump/fuel level sending unit flange

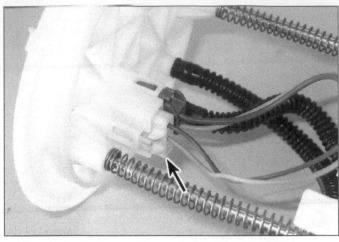

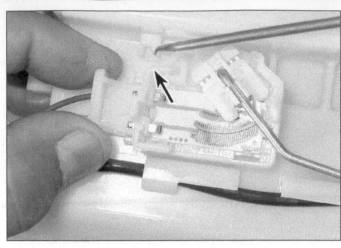

8.7 Disconnect the fuel gauge sending unit electrical connector

8.8 Release the lock tab and slide off the sender

6   If you want to replace either the fuel pump or the fuel level sending unit, you'll have to separate the two components.

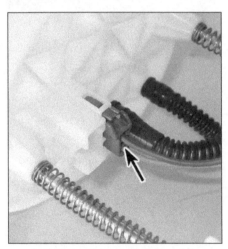

8.10 Disconnect the fuel pump electrical connector

Component replacement

Fuel level sending unit

Refer to illustrations 8.7 and 8.8

7   Disconnect the fuel gauge sending unit electrical connector from the underside of the fuel pump/fuel gauge sending unit mounting flange **(see illustration)**.

8   To detach the fuel gauge sending unit from the pump assembly, release the lock tab **(see illustration)** and slide off the sender.

Fuel pump

Refer to illustrations 8.10, 8.11, 8.13a, 8.13b, 8.14 and 8.16

9   Remove the fuel level sending unit (see Steps 7 and 8).

10   Disconnect the fuel pump electrical connector (the other connector) from the underside of the fuel pump/fuel gauge sending unit mounting flange **(see illustration)**.

11   Detach the suction filter hose from the fuel sub-tank **(see illustration)**.

12   Disengage the two plastic tubes from the tube guides on top of the fuel suction plate.

13   Using the tip of a screwdriver, disengage the three snap-claws **(see illustration)** that secure the fuel suction plate to the fuel sub-tank, then separate the fuel suction plate and fuel suction filter as a single assembly from the fuel sub-tank **(see illustration)**.

14   Using the screwdriver again, disengage the five snap-claws from the fuel suction filter and remove the suction filter from the fuel pump **(see illustration)**. If the suction filter is dirty, try cleaning it with solvent and an old toothbrush. If you're unable to clean the suction filter, replace it.

15   Disconnect the fuel pump harness electrical connector from the fuel pump.

16   Remove the O-ring from the fuel filter or fuel pump (it might be stuck to either component) and discard it **(see illustration)**.

Fuel pressure regulator (low pressure)

Refer to illustration 8.18

Note: *On some models, the pressure regulator and housing must be replaced as a unit. To remove this type, disengage the snap-claws and replace the unit.*

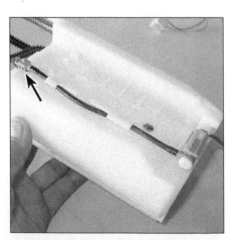

8.11 Detach the suction filter hose using a small screwdriver, then release the clip from the top of the unit

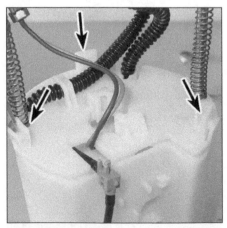

8.13a Disengage the three snap-claws . . .

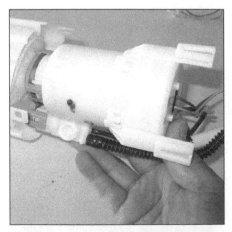

8.13b . . . then separate the fuel suction plate and fuel suction filter

8.14 Disengage the snap-claws and remove the suction filter from the fuel pump

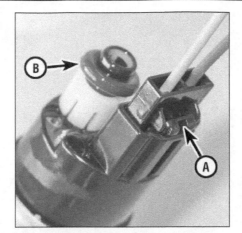

8.16 Disconnect the fuel pump harness electrical connector (A) and remove the O-ring (B) from the fuel pump

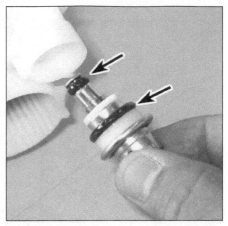

8.18 The low pressure regulator is a push fit in the fuel pump/sending unit assembly make sure the O-rings come out with it

17   Using the screwdriver again, disengage the five snap-claws from the fuel suction filter and remove the suction filter from the fuel pump **(see illustration 8.14)**.
18   Twist the regulator from the assembly **(see illustration)**.

Installation

Refer to illustrations 8.20 and 8.21

19   Push the regulator into the housing. Be sure to install new O-rings on the regulator.
20   Install the spacer and new O-ring to the pump **(see illustration)** and connect the electrical connector to the pump.
21   Coat the O-ring with gasoline and slide the fuel pump into the assembly, making sure to align the pump with the outlet **(see illustration)**.
22   Engage the five snap-claws of the fuel suction filter with the assembly.
23   The remainder of installation is the

reverse of removal.
Note: *Make sure the locking tab is seated and the electrical connector is installed on the fuel level sensor.*

9   Air filter housing - removal and installation

All models except 1GR-FE engines

Refer to illustrations 9.2, 9.3 and 9.4

1   Remove the engine cover **(see illustration 24.12a** in Chapter 1).
2   Disconnect the vent hose and the vacuum sensing hose, then loosen the clamps that secure the air intake duct to the air filter housing and throttle body **(see illustration)**. Pull the air intake duct away from the housing and throttle body.

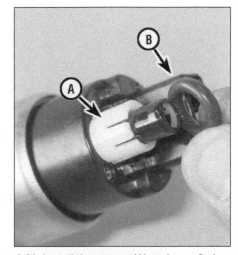

8.20 Install the spacer (A) and new O-ring (B) to the pump

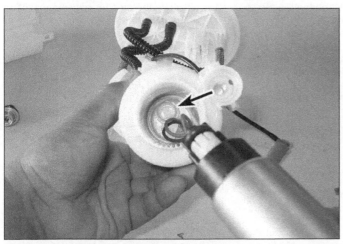

8.21 Slide the fuel pump into the assembly, making sure to align the pump with the outlet

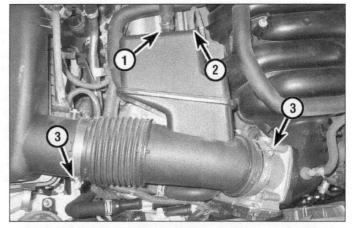

9.2 Air intake duct details (V8 engines):

1   *Vent hose*
2   *Vacuum sensing hose*
3   *Hose clamp screws*

9.3 Disconnect the MAF sensor electrical connector (A), then unclip the cover hold downs

9.4 To detach the air filter housing, remove these two bolts, lift up the housing and carefully work the air intake snorkel out from its grommet in the right fenderwell; do not try to detach the air intake snorkel from the housing

9.5 Loosen the hose clamp that secures the air intake duct to the air filter housing and disconnect the duct from the housing (1GR-FE V6 engine)

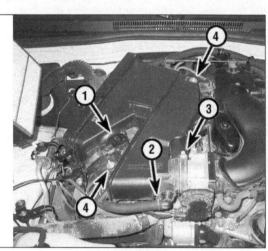

9.7 To detach the air filter housing from a 1GR-FE V6 engine:

1   *Disconnect the electrical connector from the MAF sensor*
2   *Loosen the spring clamp and disconnect the PCV fresh air intake hose*
3   *Loosen the screw on the hose clamp that secures the air filter housing to the throttle body*
4   *Remove the two air filter housing mounting bolts*

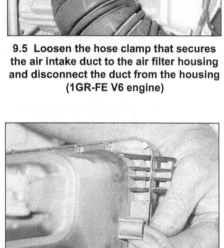

9.10 After removing the two filter housing mounting bolts, pull the housing forward slightly and disconnect this vacuum hose from the back of the housing

3   Disconnect the Mass Air Flow (MAF) sensor electrical connector and unclip the air filter cover **(see illustration)**.
4   Remove the two air filter housing mounting bolts, lift up on the air filter housing and carefully work the air intake snorkel free of its grommet in the right fenderwell **(see illustration)**.

1GR-FE V6 engine models

Refer to illustrations 9.5, 9.7 and 9.10

5   Loosen the hose clamp and disconnect the air intake duct from the air filter housing **(see illustration)**.
6   Remove the engine cover **(see illustration 24.12a** in Chapter 1).
7   Disconnect the electrical connector from the Mass Air Flow (MAF) sensor **(see illustration)**.
8   Loosen the spring-type clamp and disconnect the Positive Crankcase Ventilation (PCV) fresh air intake hose from the air filter housing **(see illustration 9.7)**.
9   Loosen the screw on the hose clamp that secures the intake manifold to the throttle

body **(see illustration 9.7)**.
10   Remove the air filter housing mounting bolts **(see illustration 9.7)**. Pull the air filter housing forward, disconnect the vacuum hose from the back of the housing **(see illustration)** and remove the air filter housing.

All models

11   Installation is the reverse of removal.

10   Throttle body - removal and installation

Warning: *Gasoline is extremely flammable. See the* **Fuel system warnings** *in Section 1.*
Warning: *Wait until the engine is completely cool before beginning this procedure.*
Caution: *Do not clean the Throttle Position (TP) sensor, the Idle Air Control (IAC) valve or the throttle control motor with solvent.*
Note: *All of these models are equipped with fully electronic throttle bodies; they have no accelerator cable.*

10.2 On 2UZ-FE engines, remove the throttle body cover fasteners and cover

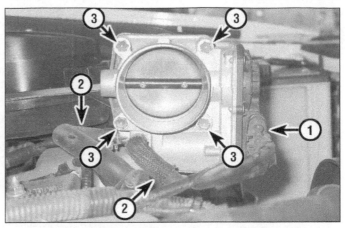

10.6 Throttle body details (1GR-FE and 2UZ-FE engines):

1   Throttle motor/TP sensor electrical connector
2   Water bypass hoses
3   Throttle body mounting bolts (V6, shown) or three bolts and a nut (V8, not shown)

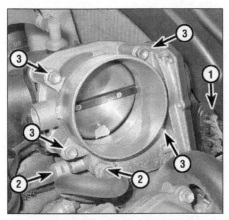

10.9 Throttle body details (1UR-FE, 3UR-FE and 3UR-FBE engines):

1   Throttle motor/TP sensor electrical connector
2   Water bypass hoses
3   Throttle body mounting bolts

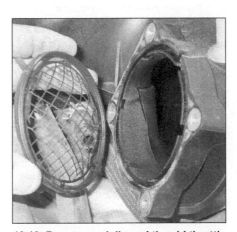

10.13 Remove and discard the old throttle body gasket; do NOT reuse this gasket or you will run the risk of an air leak

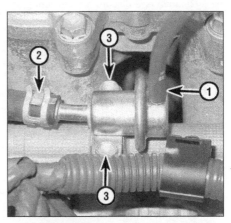

11.3 Fuel pressure regulator assembly details (2UZ-FE V8 models)

1   Vacuum hose
2   Fuel return hose clamp
3   Fuel pressure regulator mounting bolts

---

1   Remove the engine cover **(see illustration 24.12a** in Chapter 1**).**

1GR-FE (2010 and earlier) and 2UZ-FE engines

Refer to illustrations 10.2 and 10.6

2   On 2UZ-FE engines, remove the throttle body cover **(see illustration).**
3   Clamp off the coolant hoses to the throttle body.
4   Remove the air filter housing (see Section 9).
5   On 1GR-FE engines, remove the windshield wiper motor linkage (see Chapter 12) and cowl panel (see Chapter 11).
6   Disconnect the electrical connectors from the throttle body **(see illustration).**

1GR-FE (2011 and later), 1UR-FE, 3UR-FE and 3UR-FBE engines

Refer to illustration 10.9

7   Clamp off the coolant hoses to the throttle body.
8   Remove the air intake duct **(see illustration 9.2).**
9   Disconnect the electrical connectors from the throttle body **(see illustration).**

All models

Refer to illustration 10.13

10   Loosen the hose clamps and disconnect both water bypass hoses from the throttle body **(see illustration 10.6 or 10.9).**
11   Remove the throttle body mounting bolts **(see illustration 10.6 or 10.9).**
12   Remove the throttle body.
13   Remove and discard the old throttle body gasket **(see illustration).**
14   Installation is the reverse of removal.

Use a new gasket and tighten the throttle body mounting bolts and/or nut(s) to the torque listed in this Chapter's Specifications.
15   Check the coolant level, adding as necessary (see Chapter 1).

---

11   Fuel pressure regulator - removal and installation

Warning: *Gasoline is extremely flammable. See the* **Fuel system warnings** *in Section 1.*
1   Relieve the fuel system pressure (see Section 3).
2   Disconnect the negative battery cable.

2UZ-FE engines

Refer to illustrations 11.3 and 11.6

3   Disconnect the vacuum hose from the fuel pressure regulator **(see illustration).**

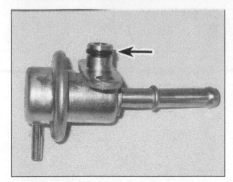

11.6 Remove and discard the old fuel pressure regulator O-ring (2UZ-FE V8 models)

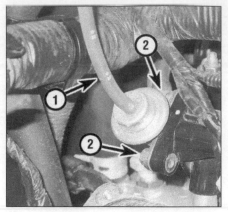

11.11 To detach the fuel pressure regulator from the fuel rail on a V6 model, disconnect the vacuum hose (1) and remove the two pressure regulator mounting bolts (2) (1GR-FE V6 models shown)

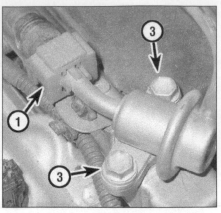

11.12 Fuel pressure regulator assembly details (1UR-FE, 3UR-FE and 3UR-FBE V8 models shown)

1   *Fuel return hose quick-connect fitting*
2   *Vacuum hose*
3   *Fuel pressure regulator mounting bolts*

4   Loosen the fuel return hose clamp and then disconnect the fuel return hose from the regulator.
5   Remove the two bolts that secure the fuel pressure regulator to the fuel rail and then remove the regulator.
6   Remove and discard the old fuel pressure regulator O-ring **(see illustration)**. Install a new one before reinstalling the regulator.
7   Installation is the reverse of removal. Tighten the fuel pressure regulator bolts to the torque listed in this Chapter's Specifications.

All other engines

Refer to illustrations 11.11 and 11.12

8   Remove the engine cover **(see illustration 24.12a in Chapter 1)**.
9   Remove the air filter housing (see Section 9).
10   Disconnect the fuel return hose. If you're unfamiliar with quick-connect fittings, refer to Section 5.
11   Disconnect the vacuum hose from the fuel pressure regulator **(see illustration)**.
12   Remove the fuel pressure regulator mounting bolts and remove the pressure regulator **(see illustration)**. Remove and discard the old pressure regulator O-ring.
13   Installation is the reverse of removal. Use a new O-ring and tighten the pressure regulator mounting bolts to the torque listed in this Chapter's Specifications.

12 Fuel rail and injectors - removal and installation

All models

Warning: *Gasoline is extremely flammable. See the* **Fuel system warnings** *in Section 1.*
1   Relieve the fuel system pressure (see Section 3).
2   Detach the cable from the negative terminal of the battery.
3   Remove the air intake duct (see Section 9).

1GR-FE models

4   Remove the upper intake manifold (see Chapter 2A).

5   Remove the windshield wiper motor linkage (see Chapter 12) and cowl panel (see Chapter 11).
6   Disconnect the quick-connect fittings for fuel pipe No. 1 (the fuel supply hose) and fuel pipe No. 2 (the fuel return hose). If you're unfamiliar with quick-connect fittings, refer to Section 5.
7   Disconnect the vacuum hose from the fuel pressure regulator.
8   Disconnect the six electrical connectors from the fuel injectors.
9   Remove the fuel rail mounting bolts and remove the fuel rail and injectors as a single assembly.

2UZ-FE models

Refer to illustrations 12.13, 12.18, 12.20a and 12.20b

10   Remove the throttle body cover and the air intake duct **(see illustrations 9.2 and 10.2)**.
11   Unscrew the pulsation damper, then remove and discard the old sealing washers.
12   Disconnect the PCV hose from the PCV valve (see Chapter 6).
13   Disconnect the electrical connector and the hoses from the Vacuum Switching Valve for the EVAP system (EVAP VSV), remove the EVAP VSV retaining bolt, then remove the accelerator cable bracket and the EVAP VSV from the intake manifold **(see illustration)**.
14   Remove the throttle body cover bracket bolt and the bracket.
15   Disconnect the fuel injector/ignition coil wiring harness clamps from the engine lifting hook and from the EVAP VSV bracket on the left side of the intake manifold, and from the two brackets bolted to the right fuel rail.
16   Disconnect the electrical connectors from the four ignition coils and from the four fuel injectors on each cylinder bank, then set the wiring harnesses aside.
17   Disconnect the vacuum hose from the fuel pressure regulator (see Section 11).

18   Remove the bolt that secures the fuel return pipe bracket to the left fuel rail **(see illustration)**.
19   Remove the left and right banjo bolts that connect the fuel rail crossover pipe to the front ends of the fuel rails, then remove the crossover pipe clamp bolt. Discard the old banjo bolt sealing washers.
20   Remove the two nuts that attach each fuel rail **(see illustrations)**.

1UR-FE, 3UR-FE and 3UR-FBE models

Refer to illustrations 12.22, 12.23 and 12.27

21   Remove the engine cover (see Chapter 1, **illustration 24.12a**).
22   Remove the fuel rail insulators from both sides of the engine **(see illustration)**.

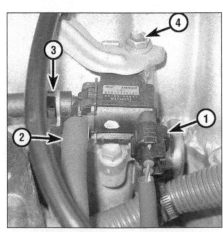

12.13 EVAP Vacuum Switching Valve details

1   *Electrical connector*
2   *Vacuum hoses*
3   *Retaining bolt*

12.18 Remove the bolt that attaches the fuel return pipe bracket to the left fuel rail

12.20a To detach the left fuel rail from the intake manifold, remove these two nuts

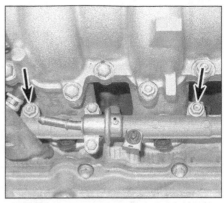

12.20b To detach the right fuel rail from the intake manifold, remove these two nuts

12.22 Remove the fuel rail insulator from each side of the engine

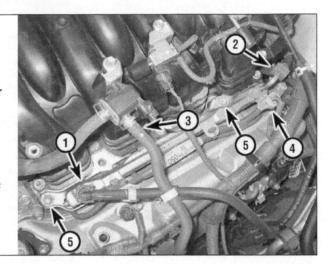

12.23 Fuel injector component details - driver's side shown, passenger's side similar

1   Fuel pipe quick-connect fitting
2   Fuel pipe quick-connect fitting
3   Vent hose
4   Fuel injector harness connector
5   Fuel injector rail mounting bolts

23   Disconnect the quick-connect fittings for fuel pipes **(see illustration)**. If you're unfamiliar with quick-connect fittings, refer to Section 5.
24   Disconnect the vent hose.
25   Disconnect the vacuum hose from the fuel pressure regulator **(see illustration 11.18)**.
26   Disconnect the fuel injector harness.

27   Remove the fuel rail mounting bolts and remove the fuel rail and injectors as a single assembly **(see illustration)**.

All models

Refer to illustrations 12.29, 12.30 and 12.31

28   Remove the two fuel rails and six or eight fuel injectors by pulling straight up on each fuel rail while wiggling the injectors free of their bores in the intake manifold.
29   On 1UR-FE, 3UR-FE and 3UR-FBE models, disconnect the electrical connectors from the fuel injectors **(see illustration)**.
30   Remove each injector from the fuel rail **(see illustration)**.

12.27 Remove the fuel rail and injector assembly from the engine

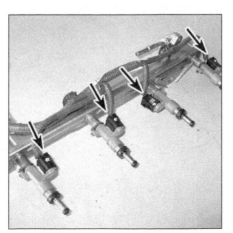

12.29 Disconnect the electrical connectors from each injector

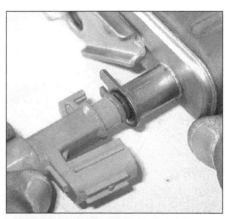

12.30 To work an injector out of its bore in the fuel rail, pull on it while wiggling it from side-to-side at the same time

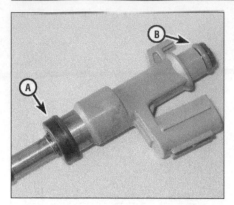

12.31 Remove the old grommet (A) and O-ring (B) from each injector

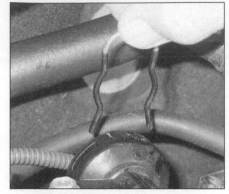

13.3a To remove the fuel pulsation damper from the fuel rail on V6 models, remove the retainer clip . . .

13.3b . . . then pull the pulsation damper out of the fuel rail

31 Remove the old O-ring and grommets from each injector **(see illustration)**.
32 Install a new O-ring and grommets on each injector. Coat the O-rings and grommets with a little clean engine oil to help them slide onto the injectors more easily.
33 Insert the injectors into the fuel rails, then push the injectors back into their bores in the intake manifold until they're fully seated. Coat the outer surfaces of the injector O-rings and grommets with a little clean engine oil to facilitate pushing the injectors back into the fuel rail and into their bores in the intake manifold. Install the fuel rail retaining fasteners and tighten them to the torque listed in this Chapter's Specifications.
34 Using new sealing washers, install the fuel rail crossover pipe and tighten the banjo bolts to the torque listed in this Chapter's Specifications.
35 Installation is otherwise the reverse of removal.

13 Fuel pressure pulsation damper (1GR-FE and 2UZ-FE models) - replacement

Warning: *Gasoline is extremely flammable. See the* **Fuel system warnings** *in Section 1.*

Note: *The fuel pressure pulsation damper is located at the rear end of the left fuel rail tube.*
1 Remove the engine cover (see Chapter 1).
2 Relieve the fuel system pressure (see Section 3).

1GR-FE models

Refer to illustrations 13.3a, 13.3b and 13.4

3 Remove the clip from the pulsation damper, then pull the pulsation damper out of the fuel rail **(see illustrations)**.
4 Remove and discard the old pulsation damper O-ring.
5 Installation is the reverse of removal. Use a new O-ring. Apply a light coat of oil to the new O-ring before installing it on the pulsation damper.

2UZ-FE models

Refer to illustrations 13.6a and 13.6b

6 Unscrew the pulsation damper, remove the fuel supply line banjo fitting, then remove and discard the old sealing washers **(see illustrations)**.
7 Installation is the reverse of removal. Use new sealing washers.

14 Fuel pressure switching valve (3UR-FBE engines) – removal and installation

Warning: *Gasoline is extremely flammable. See the Fuel system warnings in Section 1.*
Note: *The fuel pressure switching valve is located on top of the fuel tank.*
1 Remove the fuel pressure switching valve bracket mounting bolt and separate the bracket from the frame.
2 Remove the fuel tank (see Section 7).
Note: *Before the fuel tank can be completely lowered, the electrical connector to the fuel pressure switching valve must be disconnected (see Section 7, Step 5).*
3 Remove the fuel pressure switching valve protective cover mounting nut and cover.
4 Disconnect the three fuel line quick-connect fittings and separate the lines from the valve. If you're unfamiliar with quick-connect fittings, refer to Section 5.
5 Remove the fuel pressure switching valve.
6 Installation is the reverse of removal.

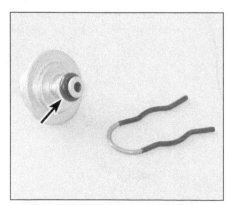

13.4 Remove and discard the old O-ring from the fuel pressure pulsation damper

13.6a Unscrew the pulsation damper . . .

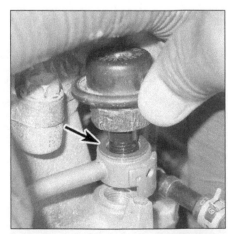

13.6b . . . and remove the fuel supply line banjo fitting. Discard the old sealing washers

Chapter 5
Engine electrical systems

Contents

Specifications

Charging system

Charging voltage	13.2 to 14.8 volts
Standard amperage	
No load	10 amps or less
With load	30 amps or more

Torque specifications

	Ft-lbs	Nm
Alternator mounting bolts		
2UZ-FE engines	29	39
1GR-FE, 1UR-FE, 3UR-FE and 3UR-FBE engines	32	43
Starter mounting bolts		
1GR-FE and 2UZ-FE engines	29	39
1UR-FE, 3UR-FE and 3UR-FBE engines	27	37

1   General information and precautions

General information

Ignition system

The electronic ignition system consists of the Crankshaft Position (CKP) sensor, the Camshaft Position (CMP) sensor, the Knock Sensor (KS), the Powertrain Control Module (PCM), the ignition switch, the battery, the individual ignition coils or a coil pack, and the spark plugs. For more information on the CKP, CMP and KS sensors, as well as the PCM, refer to Chapter 6.

Charging system

The charging system includes the alternator (with an integral voltage regulator), the Powertrain Control Module (PCM), the Body Control Module (BCM), a charge indicator light on the dash, the battery, a fuse or fusible link and the wiring connecting all of these components. The charging system supplies electrical power for the ignition system, the lights, the radio, etc. The alternator is driven by a drivebelt.

Starting system

The starting system consists of the battery, the ignition switch, the starter relay, the Powertrain Control Module (PCM), the Body Control Module (BCM), the Transmission Range (TR) switch, the starter motor and solenoid assembly, and the wiring connecting all of the components.

Precautions

Always observe the following precautions when working on the electrical system:

a) *Be extremely careful when servicing engine electrical components. They are easily damaged if checked, connected or handled improperly.*

b) *Never leave the ignition switched on for long periods of time when the engine is not running.*

c) *Never disconnect the battery cables while the engine is running.*

d) *Maintain correct polarity when connecting battery cables from another vehicle during jump starting - see the "Booster battery (jump) starting" Section at the front of this manual.*

e) *Always disconnect the cable from the negative battery terminal before working on the electrical system, but read the battery disconnection procedure first (see Section 3).*

It's also a good idea to review the safety-related information regarding the engine electrical systems located in the *Safety first!* Section at the front of this manual before beginning any operation included in this Chapter.

Electrical system components – V8 (3UR-FE) engine models

1 Battery	4 Ignition coil
2 Underhood fuse/relay box	5 Starter (bottom of the engine)
3 Alternator (bottom of the engine)	

2   Troubleshooting

Ignition system

1   If a malfunction occurs in the ignition system, do not immediately assume that any particular part is causing the problem. First, check the following items:

a) *Make sure that the cable clamps at the battery terminals are clean and tight.*
b) *Test the condition of the battery (see Steps 21 through 24). If it doesn't pass all the tests, replace it.*
c) *Check the ignition coil or coil pack connections.*
d) *Check any relevant fuses in the engine compartment fuse and relay box (see Chapter 12). If they're burned, determine the cause and repair the circuit.*

Check

Refer to illustration 2.3

Warning: *Because of the high voltage generated by the ignition system, use extreme care when performing a procedure involving ignition components.*

Note: *The ignition system components on these vehicles are difficult to diagnose. In the event of ignition system failure that you can't diagnose, have the vehicle tested at a dealer service department or other qualified auto repair facility.*

Note: *You'll need a spark tester for the following test. Spark testers are available at most auto supply stores.*

2   If the engine turns over but won't start, verify that there is sufficient ignition voltage to fire the spark plugs as follows.

3   On models with a coil-over-plug type ignition system, remove a coil and install the tester between the boot at the lower end of the coil and the spark plug **(see illustration)**. On models with spark plug wires, disconnect a spark plug wire from a spark plug and install the tester between the spark plug wire boot and the spark plug.

4   Crank the engine and note whether or not the tester flashes.

Caution: *Do NOT crank the engine or allow it to run for more than five seconds; running the*

engine for more than five seconds may set a *Diagnostic Trouble Code (DTC) for a cylinder misfire.*

Models with a coil-over-plug type ignition system

5   If the tester flashes during cranking, the coil is delivering sufficient voltage to the spark plug to fire it. Repeat this test for each cylinder to verify that the other coils are OK.

6   If the tester doesn't flash, remove a coil from another cylinder and swap it for the one being tested. If the tester now flashes, you know that the original coil is bad. If the tester still doesn't flash, the PCM or wiring harness is probably defective. Have the PCM checked out by a dealer service department or other qualified repair shop (testing the PCM is beyond the scope of the do-it-yourselfer because it requires expensive special tools).

7   If the tester flashes during cranking but a misfire code (related to the cylinder being tested) has been stored, the spark plug could be fouled or defective.

Models with spark plug wires

8   If the tester flashes during cranking, sufficient voltage is reaching the spark plug to fire it.

9   Repeat this test on the remaining cylinders.

10   Proceed on this basis until you have verified that there's a good spark from each spark plug wire. If there is, then you have verified that the coils in the coil pack are functioning correctly and that the spark plug wires are OK.

11   If there is no spark from a spark plug wire, then either the coil is bad, the plug wire is bad or a connection at one end of the plug wire is loose. Assuming that you're using new plug wires or known good wires, then the coil is probably defective. Also inspect the coil pack electrical connector. Make sure that it's clean, tight and in good condition.

12   If all the coils are firing correctly, but the engine misfires, then one or more of the plugs might be fouled. Remove and check the spark plugs or install new ones (see Chapter 1).

13   No further testing of the ignition system is possible without special tools. If the problem persists, have the ignition system tested

by a dealer service department or other qualified repair shop.

Charging system

14   If a malfunction occurs in the charging system, do not automatically assume the alternator is causing the problem. First check the following items:

a) *Check the drivebelt tension and condition, as described in Chapter 1. Replace it if it's worn or deteriorated.*
b) *Make sure the alternator mounting bolts are tight.*
c) *Inspect the alternator wiring harness and the connectors at the alternator and voltage regulator. They must be in good condition, tight and have no corrosion.*
d) *Check the fusible link (if equipped) or main fuse in the underhood fuse/relay box. If it is burned, determine the cause, repair the circuit and replace the link or fuse (the vehicle will not start and/or the accessories will not work if the fusible link or main fuse is blown).*
e) *Start the engine and check the alternator for abnormal noises (a shrieking or squealing sound indicates a bad bearing).*
f) *Check the battery. Make sure it's fully charged and in good condition (one bad cell in a battery can cause overcharging by the alternator).*
g) *Disconnect the battery cables (negative first, then positive). Inspect the battery posts and the cable clamps for corrosion. Clean them thoroughly if necessary (see Chapter 1). Reconnect the cables (positive first, negative last).*

Alternator - check

15   Use a voltmeter to check the battery voltage with the engine off. It should be at least 12.6 volts **(see illustration 2.21)**.

16   Start the engine and check the battery voltage again. It should now be approximately 13.5 to 15 volts.

17   If the voltage reading is more or less than the specified charging voltage, the voltage regulator is probably defective, which will require replacement of the alternator (the voltage regulator is not replaceable separately). Remove the alternator and have it bench tested (most auto parts stores will do this for you).

18   The charging system (battery) light on the instrument cluster lights up when the ignition key is turned to ON, but it should go out when the engine starts.

19   If the charging system light stays on after the engine has been started, there is a problem with the charging system. Before replacing the alternator, check the battery condition, alternator belt tension and electrical cable connections.

20   If replacing the alternator doesn't restore voltage to the specified range, have the charging system tested by a dealer service department or other qualified repair shop.

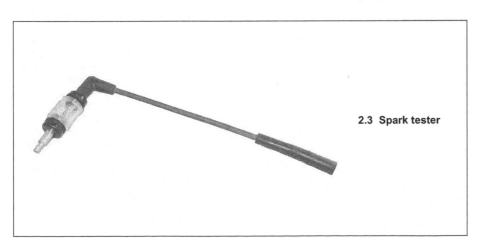

2.3 Spark tester

2.21 To test the open circuit voltage of the battery, touch the black probe of the voltmeter to the negative terminal and the red probe to the positive terminal of the battery; a fully charged battery should be at least 12.6 volts

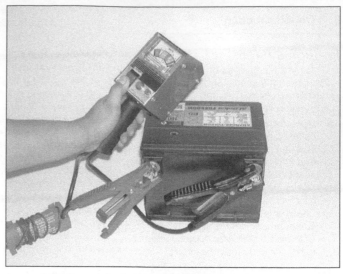

2.23 Connect a battery load tester to the battery and check the battery condition under load following the tool manufacturer's instructions

Battery - check

Refer to illustrations 2.21 and 2.23

21   Check the battery state of charge. Visually inspect the indicator eye on the top of the battery (if equipped with one); if the indicator eye is black in color, charge the battery as described in Chapter 1. Next perform an open circuit voltage test using a digital voltmeter.

Note: *The battery's surface charge must be removed before accurate voltage measurements can be made. Turn on the high beams for ten seconds, then turn them off and let the vehicle stand for two minutes.*

With the engine and all accessories Off, touch the negative probe of the voltmeter to the negative terminal of the battery and the positive probe to the positive terminal of the battery **(see illustration)**. The battery voltage should be 12.6 volts or slightly above. If the battery is less than the specified voltage, charge the battery before proceeding to the next test. Do not proceed with the battery load test unless the battery charge is correct.

22   Disconnect the negative battery cable, then the positive cable from the battery.

23   Perform a battery load test. An accurate check of the battery condition can only be performed with a load tester **(see illustration)**. This test evaluates the ability of the battery to operate the starter and other accessories during periods of high current draw. Connect the load tester to the battery terminals. Load test the battery according to the tool manufacturer's instructions. This tool increases the load demand (current draw) on the battery.

24   Maintain the load on the battery for 15 seconds and observe that the battery voltage does not drop below 9.6 volts. If the

battery condition is weak or defective, the tool will indicate this condition immediately. **Note:** *Cold temperatures will cause the minimum voltage reading to drop slightly. Follow the chart given in the manufacturer's instructions to compensate for cold climates. Minimum load voltage for freezing temperatures (32 degrees F) should be approximately 9.1 volts.*

Starting system

The starter rotates, but the engine doesn't

25   Remove the starter (see Section 8). Check the overrunning clutch and bench test the starter to make sure the drive mechanism extends fully for proper engagement with the flywheel ring gear. If it doesn't, replace the starter.

26   Check the flywheel ring gear for missing teeth and other damage. With the ignition turned off, rotate the flywheel so you can check the entire ring gear.

The starter is noisy

27   If the solenoid is making a chattering noise, first check the battery (see Steps 21 through 24). If the battery is okay, check the cables and connections.

28   If you hear a grinding, crashing metallic sound when you turn the key to Start, check for loose starter mounting bolts. If they're tight, remove the starter and inspect the teeth on the starter pinion gear and flywheel ring gear. Look for missing or damaged teeth.

29   If the starter sounds fine when you first turn the key to Start, but then stops rotating the engine and emits a zinging sound, the problem is probably a defective starter drive that's not staying engaged with the ring gear. Replace the starter.

The starter rotates slowly

30   Check the battery (see Steps 21 through 24).

31   If the battery is okay, verify all connections (at the battery, the starter solenoid and motor) are clean, corrosion-free and tight. Make sure the cables aren't frayed or damaged.

32   Check that the starter mounting bolts are tight so it grounds properly. Also check the pinion gear and flywheel ring gear for evidence of a mechanical bind (galling, deformed gear teeth or other damage).

The starter does not rotate at all

33   Check the battery (see Steps 21 through 24).

34   If the battery is okay, verify all connections (at the battery, the starter solenoid and motor) are clean, corrosion-free and tight. Make sure the cables aren't frayed or damaged.

35   Check all of the fuses in the underhood fuse/relay box.

36   Check that the starter mounting bolts are tight so it grounds properly.

37   Check for voltage at the starter solenoid "S" terminal when the ignition key is turned to the start position. If voltage is present, replace the starter/solenoid assembly. If no voltage is present, the problem could be the starter relay, the Transmission Range (TR) switch (see Chapter 6) or clutch start switch (see Chapter 8), or with an electrical connector somewhere in the circuit (see the wiring diagrams at the end of Chapter 12). Also, on many modern vehicles, the Powertrain Control Module (PCM) and the Body Control Module (BCM) control the voltage signal to the starter solenoid; on such vehicles a special scan tool is required for diagnosis.

3   Battery - disconnection and reconnection

Caution: *Always disconnect the cable from the negative battery terminal FIRST and hook it up LAST or the battery may be shorted by the tool being used to loosen the cable clamps.*

Some vehicle systems (radio, alarm system, power door locks, power windows etc.) require battery power all the time, either to enable their operation or to maintain control unit memory (Powertrain Control Module, automatic transmission control module, etc.), which would be lost if the battery were to be disconnected. So before you disconnect the battery, note the following points:

a) *Before connecting or disconnecting the cable from the negative battery terminal, make sure that you turn the ignition key and the lighting switch to their OFF positions. Failure to do so could damage semiconductor components.*

b) *On 2015 and later models equipped with a navigation system or rear seat entertainment system, after the ignition switch is turned off, the navigation receiver still records various types of memory and settings. Once the ignition switch is turned to the OFF position, be sure to wait at least 1 minute before disconnecting the cable from the negative battery terminal to prevent damage to the navigation system.*

c) *On a vehicle with power door locks, it is a wise precaution to remove the key from the ignition and to keep it with you, so that it does not get locked inside if the power door locks should engage accidentally when the battery is reconnected!*

d) *After the battery has been disconnected, then reconnected (or a new battery has been installed) on vehicles with an automatic transmission, the Transmission Control Module (TCM) will need some time to relearn its adaptive strategy. As a result, shifting might feel firmer than usual. This is a normal condition and will not adversely affect the operation or service life of the transmission. Eventually, the TCM will complete its adaptive learning process and the shift feel of the transmission will return to normal.*

e) *The engine management system's PCM has some learning capabilities that allow it to adapt or make corrections in response to minor variations in the fuel system in order to optimize drivability and idle characteristics. However, the PCM might lose some or all of this information when the battery is disconnected. The PCM must go through a relearning process before it can regain its former drivability and performance characteristics. Until it relearns this lost data, you might notice a difference in drivability, idle and/or shift "feel." To facilitate this relearning process, refer to "Enabling the PCM to relearn" below.*

f) *The power window system must be recalibrated before the windows will properly open and close To facilitate this calibration process, refer to "Power window calibration" below.*

The battery is located in the engine compartment. To disconnect the battery for service procedures requiring power to be cut from the vehicle, loosen the cable end bolt and disconnect the cable from the negative battery terminal. Isolate the cable end to prevent it from coming into accidental contact with the battery terminal.

Memory savers

Devices known as memory savers (typically, small 9-volt batteries) can be used to avoid some of the above problems. A memory saver is usually plugged into the cigarette lighter, and then you can disconnect the vehicle battery from the electrical system. The memory saver will deliver sufficient current to maintain security alarm codes and - maybe, but don't count on it! - PCM memory. It will also run unswitched (always on) circuits such as the clock and radio memory, while isolating the car battery in the event that a short circuit occurs while the vehicle is being serviced.

Warning: *If you're going to work around any airbag system components, disconnect the battery and do not use a memory saver. If you do, the airbag could accidentally deploy and cause personal injury.*

Caution: *Because memory savers deliver current to operate unswitched circuits when the battery is disconnected, make sure that the circuit that you're going to service is actually open before working on it!*

Enabling the PCM to relearn

After the battery has been reconnected, perform the following procedure in order to facilitate PCM relearning:

1   Start the engine and allow it to warm up to its normal operating temperature.
2   Drive the vehicle at part-throttle, under moderate acceleration and idle conditions, until normal performance returns.
3   Park the vehicle and apply the parking brake with the engine running.
4   Put the shift lever in DRIVE.
5   Allow the engine to idle for about two minutes or until the idle stabilizes. Make sure that the engine is at its normal operating temperature.

Power window calibration

After the battery has been reconnected, perform the following procedure on each window in order to calibrate the power windows:

1   Turn the ignition switch to the "RUN" position.
2   Using the individual window switches, press the window switch down position until the window is completely down and hold the switch down for an additional three seconds.
3   Using the same individual window switch, press the window switch up position until the window is completely up and hold the switch for an additional three seconds.
4   Check to see that the window go all the way up and down, then repeat this procedure for the remaining windows.

4   Battery - removal and installation

Refer to illustration 4.1

Note: *On 2015 and later models equipped with a navigation system or rear seat entertainment system, once the ignition switch is turned to the OFF position, wait at least 1 minute before disconnecting the cable from the negative battery terminal to prevent damage to the navigation system.*

1   Disconnect the cable from the negative battery terminal first, then disconnect the cable from the positive battery terminal **(see illustration)**.
2   Remove the battery hold-down clamp mounting nuts and clamp.
3   Lift out the battery. Be careful - it's heavy.

Note: *Battery straps and handlers are available at most auto parts stores for reasonable prices. They make it easier to remove and carry the battery.*

4   If you are replacing the battery, make sure you get one that's identical, with the same dimensions, amperage rating, cold cranking rating, etc. Also, be sure to remove the heat shield from the old battery and install it on the new battery.
5   Installation is the reverse of removal. Be sure to connect the positive cable first and the negative cable last.

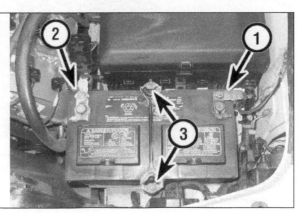

4.1 Battery details:

1   Negative battery cable
2   Positive battery cable
3   Battery hold-down clamp nuts

5   Battery cables - replacement

1   When removing the cables, always disconnect the cable from the negative battery terminal first and hook it up last, or you might accidentally short out the battery with the tool you're using to loosen the cable clamps. Even if you're only replacing the cable for the positive terminal, be sure to disconnect the negative cable from the battery first.

2   Disconnect the old cables from the battery, then trace each of them to their opposite ends and disconnect them. Be sure to note the routing of each cable before disconnecting it to ensure correct installation.

3   If you are replacing any of the old cables, take them with you when buying new cables. It is vitally important that you replace the cables with identical parts.

4   Clean the threads of the solenoid or ground connection with a wire brush to remove rust and corrosion. Apply a light coat of battery terminal corrosion inhibitor or petroleum jelly to the threads to prevent future corrosion.

5   Attach the cable to the solenoid or ground connection and tighten the mounting nut/bolt securely.

6   Before connecting a new cable to the battery, make sure that it reaches the battery post without having to be stretched.

7   Connect the cable to the positive battery terminal first, *and then* connect the ground cable to the negative battery terminal.

6   Ignition coils - replacement

1   Disconnect the negative cable from the battery.

V6 engines

2   Remove the engine cover mounting nuts, lift the cover up slightly and pull it forward to disengage the tabs at the rear of the cover. Remove the cover.

3   Disconnect and remove the air intake duct (see Chapter 4).

4   Remove the air filter housing (see Chapter 4).

5   On 2011 and later models, remove the emission control valve set (see Chapter 6).

V8 engines

2UZ-FE engines

6   Remove the two throttle body cover bolts and cover.

7   Disconnect and remove the air intake duct (see Chapter 4).

8   Remove the plastic clips from the rubber seal on the left front inner fenderwell splash shield (see Chapter 11), and remove the seal.

9   Disconnect the air switching valve hoses, then unbolts the tube and the air switching valve from the engine (see Chapter 6).

1UR-FE, 3UR-FE and 3UR-FBE engines

10   Remove the engine cover (see Chapter 1).

11   Disconnect and remove the air intake duct (see Chapter 4).

12   Remove the air filter housing (see Chapter 4).

All models

Refer to illustrations 6.13a and 6.13b

13   Each ignition coil/igniter assembly is secured by one bolt; unscrew the bolt, disconnect the electrical connector and pull the coil/igniter assembly out, using a twisting motion **(see illustrations)**.

14   Installation is the reverse of removal.

7   Alternator - removal and installation

Warning: *On Sequoia models equipped with rear height control suspension, adjust the height control to the NORMAL mode, turn OFF the height control, then turn off the engine BEFORE raising the vehicle.*

Removal

1   Disconnect the cable from the negative battery terminal (see Section 3).

V6 engines

2   Remove the battery (see Section 4).

3   Remove the engine cover (see Section 6, Step 2).

4   Remove the lower splash shield (see Chapter 1).

5   Working from under the front of the vehicle, remove the drivebelt (see Chapter 1).

6   On 2011 and later models, remove the idler pulley center bolt and pulley.

7   Remove the wiring harness clamp bolt then disconnect the electrical connectors to the alternator.

8   Remove the upper and lower alternator mounting bolts and lift the alternator from the engine.

Note: *On 2011 and later models, it may be necessary to remove the left side exhaust manifold heat shield to access the lower mounting bolt (see Chapter 2A).*

V8 engines

Refer to illustrations 7.16, 7.17a and 7.17b

9   Remove the engine cover (see Chapter 1).

10   Loosen the right-front wheel lug nuts. Raise the front of the vehicle and place it securely on jackstands. Remove the drivebelt (see Chapter 1).

11   Remove the air filter inlet duct (see Chapter 4).

12   On models with 1UR-FE, 3UR-FE or 3UR-FBE engines, remove the air filter housing (see Chapter 4).

13   On Sequoia models with 3UR-FE or 3UR-FBE engines, drain the cooling system, then remove the radiator hoses (see Chapter 1), and the cooling fan and shroud (see Chapter 3).

14   Remove the right front wheel. Remove the plastic clips from the rubber seal on the inner fender well splash shield (see Chapter 11) and remove the seal.

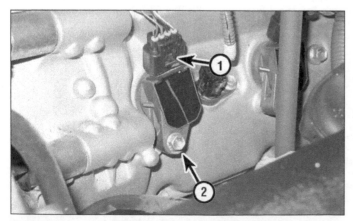

6.13a Ignition coil details

1   *Electrical connector retaining tab*

2   *Mounting bolt*

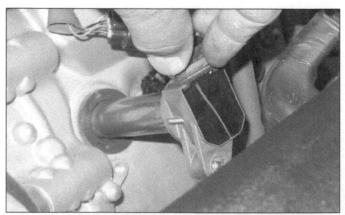

6.13b The individual ignition coil boot mounts directly over each spark plug

7.16 Before unbolting the alternator, disconnect the electrical connectors (3UR-FE model shown, others similar)

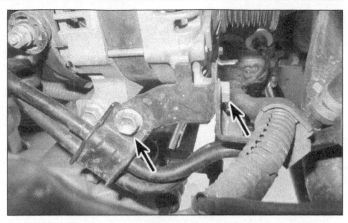

7.17a Remove the cooler tube bracket bolts and pull it out of the way

15　Remove the power steering pump (see Chapter 10).
16　Disconnect the electrical connectors from the alternator **(see illustration)**.
17　Remove the alternator mounting bolts and nut **(see illustrations)**.
18　Remove the alternator.

Installation

19　If you are replacing the alternator, take the old alternator with you when purchasing a replacement unit. Make sure that the new/rebuilt unit is identical to the old alternator. Look at the terminals - they should be the same in number, size and locations as the terminals on the old alternator. Finally, look at the identification markings - they will be stamped in the housing or printed on a tag or plaque affixed to the housing. Make sure that these numbers are the same on both alternators.
20　Many new/rebuilt alternators do not have a pulley installed, so you may have to switch the pulley from the old unit to the new/rebuilt one. When buying an alternator, find out the shop's policy regarding installation of pulleys - some shops will perform this service free of charge.
21　Installation is the reverse of removal. Tighten alternator mounting bolts to the torque listed in this Chapter's Specifications.

22　Check the charging voltage to verify that the alternator is operating correctly (see Section 2).

8　Starter motor - removal and installation

Warning: *On Sequoia models equipped with rear height control suspension, adjust the height control to the NORMAL mode, turn OFF the height control, then turn off the engine BEFORE raising the vehicle.*
1　Disconnect the cable from the negative battery terminal.

V6 models

Note: *The starter is located on the driver's side of the engine near the bottom of the engine.*
2　Raise the front of the vehicle and place it securely on jackstands.
3　Remove the engine under-cover.
4　Disconnect the electrical connectors from the starter motor/solenoid assembly.
5　Remove the starter motor mounting bolts and remove the starter motor.
6　Installation is the reverse of removal. Tighten the starter motor mounting bolts to

the torque listed in this Chapter Specification's Section.

V8 models

Note: *On 4.7L V8 (2UZ-FE) engines, the starter is located on top of the engine, between the cylinder heads, underneath the upper intake manifold.*

4.7L (2UZ-FE) engines

7　Remove the upper intake manifold (see Chapter 2B).

All other V8 engines

Refer to illustrations 8.12 and 8.14

8　Loosen the right front wheel lug nuts then raise the front of the vehicle and place it securely on jackstands. Remove the wheel and engine under-cover.
9　Remove the plastic clips from the rubber seal on the right front inner fender well splash shield (see Chapter 11), and remove the seal.
10　Remove the power steering pump (see Chapter 10).
11　Remove the alternator (see Section 7).
12　Remove the engine oil dipstick tube bracket bolt and remove the dipstick tube from the oil pan **(see illustration)**. Discard the O-ring from the end of the tube.

7.17b To remove the alternator, remove nut (A) and these bolts (3UR-FE model shown all other similar)

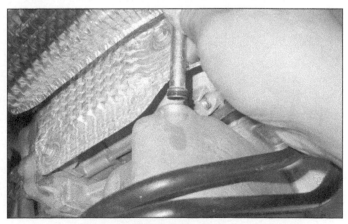

8.12 Remove the dipstick tube from the oil pan and discard the O-ring

13   Disconnect the right side exhaust pipe from the manifold and remove the right side exhaust manifold (see Chapter 2).

14   Remove the starter motor heat shield mounting bolts and shield **(see illustration)**.

All V8 models

Refer to illustrations 8.15a, 8.15b and 8.16

15   Disconnect the electrical connectors from the starter motor/solenoid assembly **(see illustrations)**.

16   Remove the starter motor mounting bolts and remove the starter motor **(see illustration)**.

17   Installation is the reverse of removal. Tighten the starter motor mounting bolts to the torque listed in this Chapter's Specifications.

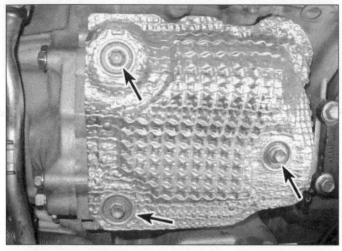

8.14 Remove the starter heat shield mounting bolts

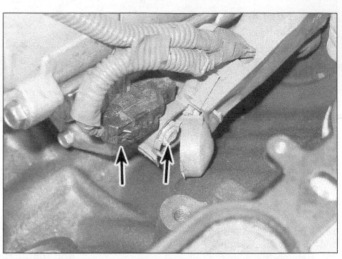

8.15a Before removing the starter assembly, disconnect the two electrical connectors - 4.7L (2UZ-FE) engine shown

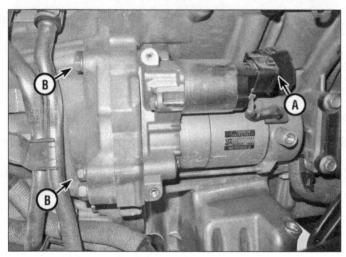

8.15b Disconnect and remove the electrical connector and battery wire (A); (B) indicates the starter motor mounting bolts (3UR-FE model shown)

8.16 To detach the starter motor on 4.7L (2UZ-FE) engines, remove the wiring protector bolt (lower arrow) and the two starter bolts (upper bolt indicates right starter bolt; left bolt not visible in this photo)

Chapter 6
Emissions and engine control systems

Contents

Specifications

Torque specifications*

	Ft-lbs (unless otherwise indicated)	Nm
Note: One foot-pound (ft-lb) of torque is equivalent to 12 inch-pounds (in-lbs) of torque. Torque values below approximately 15 foot-pounds are expressed in inch-pounds, because most foot-pound torque wrenches are not accurate at these smaller values.		
Engine Coolant Temperature (ECT) sensor	15	20
Knock sensor (2)	15	20
Oxygen sensor (2)	30	40

1   General information

To prevent pollution of the atmosphere from incompletely burned and evaporating gases, and to maintain good driveability and fuel economy, a number of emission control systems are incorporated. They include the:

Catalytic converter

A catalytic converter is an emission control device in the exhaust system that reduces certain pollutants in the exhaust gas stream. There are two types of converters: oxidation converters and reduction converters.

Oxidation converters contain a monolithic substrate (a ceramic honeycomb) coated with the semi-precious metals platinum and palladium. An oxidation catalyst reduces unburned hydrocarbons (HC) and carbon monoxide (CO) by adding oxygen to the exhaust stream as it passes through the substrate, which, in the presence of high temperature and the catalyst materials, converts the HC and CO to water vapor (H_2O) and carbon dioxide (CO_2).

Reduction converters contain a monolithic substrate coated with platinum and rhodium. A reduction catalyst reduces oxides of nitrogen (NOx) by removing oxygen, which in the presence of high temperature and the catalyst material produces nitrogen (N) and carbon dioxide (CO_2).

Catalytic converters that combine both types of catalysts in one assembly are known as "three-way catalysts" or TWCs. A TWC can reduce all three pollutants.

Evaporative Emissions Control (EVAP) system

The Evaporative Emissions Control (EVAP) system prevents fuel system vapors (which contain unburned hydrocarbons) from escaping into the atmosphere. On warm days, vapors trapped inside the fuel tank expand until the pressure reaches a certain threshold. Then the fuel vapors are routed from the fuel tank through the fuel vapor vent valve and the fuel vapor control valve to the EVAP canister, where they're stored temporarily until the next time the vehicle is operated. When the conditions are right (engine warmed up, vehicle up to speed, moderate or heavy load on the engine, etc.) the PCM opens the canister purge valve, which allows fuel vapors to be drawn from the canister into the intake manifold. Once in the intake manifold, the fuel vapors mix with incoming air before being drawn through the intake ports into the combustion chambers where they're burned up with the rest of the air/fuel mixture. The EVAP system is complex and virtually impossible to troubleshoot without the right tools and training.

Exhaust Gas Recirculation (EGR) system

Note: *This system is used on the 1UR-FE engine only.*

The EGR system reduces oxides of nitrogen by recirculating exhaust gases from the exhaust manifold, through the EGR valve and intake manifold, then back to the combustion chambers, where it mixes with the incoming air/fuel mixture before being consumed. These recirculated exhaust gases dilute the incoming air/fuel mixture, which cools the combustion chambers, thereby reducing NOx emissions.

The EGR system consists of the Powertrain Control Module (PCM), the EGR valve, the EGR valve position sensor and various other information sensors that the PCM uses to determine when to open the EGR valve. The degree to which the EGR valve is opened is referred to as "EGR valve lift." The PCM is programmed to produce the ideal EGR valve lift for varying operating conditions. The EGR valve position sensor, which is an integral part of the EGR valve, detects the amount of EGR valve lift and sends this information to the PCM. The PCM then compares it with the appropriate EGR valve lift for the operating conditions. The PCM increases current flow to the EGR valve to increase valve lift and reduces the current to reduce the amount of lift. If EGR flow is inappropriate to the operating conditions (idle, cold engine, etc.) the PCM simply cuts the current to the EGR valve and the valve closes.

Secondary Air Injection (AIR) system

Some models are equipped with a secondary air injection (AIR) system. The secondary air injection system is used to reduce tailpipe emissions on initial engine start-up. The system uses an electric motor/pump assembly, relay, vacuum valve/solenoid, air shut-off valve, check valves and tubing to inject fresh air directly into the exhaust manifolds. The fresh air (oxygen) reacts with the exhaust gas in the catalytic converter to reduce HC and CO levels. The air pump and solenoid are controlled by the PCM through the AIR relay. During initial start-up, the PCM energizes the AIR relay, the relay supplies battery voltage to the air pump and the vacuum valve/solenoid, engine vacuum is applied to the air shut-off valve which opens and allows air to flow through the tubing into the exhaust manifolds. The PCM will operate the air pump until closed loop operation is reached (approximately four minutes). During normal operation, the check valves prevent exhaust backflow into the system.

Powertrain Control Module (PCM)

The Powertrain Control Module (PCM) is the brain of the engine management system. It also controls a wide variety of other vehicle systems. In order to program the new PCM, the dealer needs the vehicle as well as the new PCM. If you're planning to replace the PCM with a new one, there is no point in trying to do so at home because you won't be able to program it yourself.

Positive Crankcase Ventilation (PCV) system

The Positive Crankcase Ventilation (PCV) system reduces hydrocarbon emissions by scavenging crankcase vapors, which are rich in unburned hydrocarbons. A PCV valve or orifice regulates the flow of gases into the intake manifold in proportion to the amount of intake vacuum available.

The PCV system generally consists of the fresh air inlet hose, the PCV valve or orifice and the crankcase ventilation hose (or PCV hose). The fresh air inlet hose connects the air intake duct to a pipe on the valve cover. The crankcase ventilation hose (or PCV hose) connects the PCV valve or orifice in the valve cover to the intake manifold.

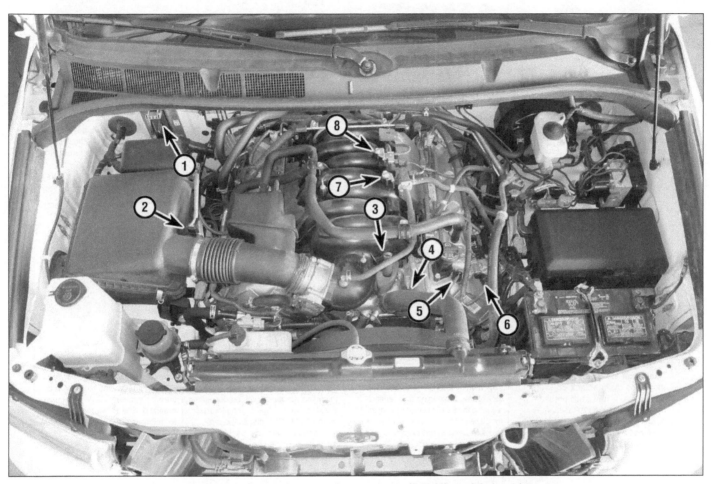

Emissions and engine control components (5.7L V8 model shown)

1 Powertrain Control Module (PCM) (located under cowl)
2 Mass Air Flow/Intake Air Temperature (MAF/IAT) sensor
3 Crankcase pressure regulator valve for PCV system
4 Engine Coolant Temperature (ECT) sensor
5 Camshaft Position (CMP) sensor
6 Camshaft timing oil control valve – 1 of 4
7 EVAP canister purge regulator valve
8 Acoustic Control Induction System Vacuum Switching Valve

Information Sensors

Accelerator Pedal Position (APP) sensor - as you press the accelerator pedal, the APP sensor alters its voltage signal to the PCM in proportion to the angle of the pedal, and the PCM commands a motor inside the throttle body to open or close the throttle plate accordingly

Camshaft Position (CMP) sensor - produces a signal that the PCM uses to identify the number 1 cylinder and to time the firing sequence of the fuel injectors

Crankshaft Position (CKP) sensor - produces a signal that the PCM uses to calculate engine speed and crankshaft position, which enables it to synchronize ignition timing with fuel injector timing, and to detect misfires

Engine Coolant Temperature (ECT) sensor - a thermistor (temperature-sensitive variable resistor) that sends a voltage signal to the PCM, which uses this data to determine the temperature of the engine coolant

Fuel tank pressure sensor - measures the fuel tank pressure and controls fuel tank pressure by signaling the EVAP system to purge the fuel tank vapors when the pressure becomes excessive

Intake Air Temperature (IAT) sensor - monitors the temperature of the air entering the engine and sends a signal to the PCM to determine injector pulse-width (the duration of each injector's on-time) and to adjust spark timing (to prevent spark knock)

Knock sensor - a piezoelectric crystal that oscillates in proportion to engine vibration which produces a voltage output that is monitored by the PCM. This retards the ignition timing when the oscillation exceeds a certain threshold

Manifold Absolute Pressure (MAP) sensor - monitors the pressure or vacuum inside the intake manifold. The PCM uses this data to determine engine load so that it can alter the ignition advance and fuel enrichment

Mass Air Flow (MAF) sensor - measures the amount of intake air drawn into the engine. It uses a hot-wire sensing element to measure the amount of air entering the engine

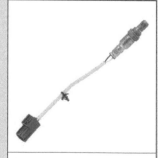

Oxygen sensors - generates a small variable voltage signal in proportion to the difference between the oxygen content in the exhaust stream and the oxygen content in the ambient air. The PCM uses this information to maintain the proper air/fuel ratio. A second oxygen sensor monitors the efficiency of the catalytic converter

Throttle Position (TP) sensor - a potentiometer that generates a voltage signal that varies in relation to the opening angle of the throttle plate inside the throttle body. Works with the PCM and other sensors to calculate injector pulse width (the duration of each injector's on-time)

Photos courtesy of Wells Manufacturing, except APP and MAF sensors.

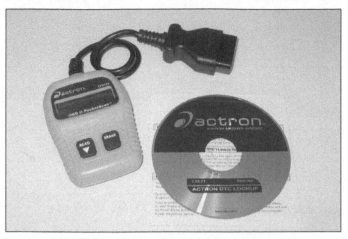

2.4a Simple code readers are an economical way to extract trouble codes when the CHECK ENGINE light comes on

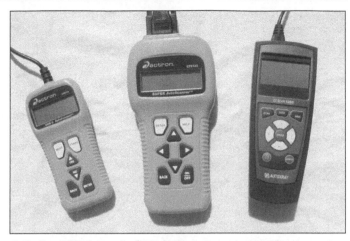

2.4b Hand-held scan tools like these can extract computer codes and also perform diagnostics

2 On Board Diagnosis (OBD) system

General description

1   All models are equipped with the second generation OBD-II system. This system consists of an on-board computer known as the Powertrain Control Module (PCM), and information sensors, which monitor various functions of the engine and send data to the PCM. This system incorporates a series of diagnostic monitors that detect and identify fuel injection and emissions control system faults and store the information in the computer memory. This system also tests sensors and output actuators, diagnoses drive cycles, freezes data and clears codes.

2   The PCM is the brain of the electronically controlled fuel and emissions system. It receives data from a number of sensors and other electronic components (switches, relays, etc.). Based on the information it receives, the PCM generates output signals to control various relays, solenoids (fuel injectors) and other actuators. The PCM is specifically calibrated to optimize the emissions, fuel economy and driveability of the vehicle.

3   It isn't a good idea to attempt diagnosis or replacement of the PCM or emission control components at home while the vehicle is under warranty. Because of a federally-mandated warranty which covers the emissions system components and because any owner-induced damage to the PCM, the sensors and/or the control devices may void this warranty, take the vehicle to a dealer service department if the PCM or a system component malfunctions.

Scan tool information

Refer to illustrations 2.4a and 2.4b

4   Because extracting the Diagnostic Trouble Codes (DTCs) from an engine management system is now the first step in troubleshooting many computer-controlled systems and components, a code reader, at the very

least, will be required **(see illustration)**. More powerful scan tools can also perform many of the diagnostics once associated with expensive factory scan tools **(see illustration)**. If you're planning to obtain a generic scan tool for your vehicle, make sure that it's compatible with OBD-II systems. If you don't plan to purchase a code reader or scan tool and don't have access to one, you can have the codes extracted by a dealer service department or an independent repair shop.
Note: *Some auto parts stores even provide this service.*

3 Obtaining and clearing Diagnostic Trouble Codes (DTCs)

All models covered by this manual are equipped with on-board diagnostics. When the PCM recognizes a malfunction in a monitored emission or engine control system, component or circuit, it turns on the Malfunction Indicator Light (MIL) on the dash. The PCM will continue to display the MIL until the problem is fixed and the Diagnostic Trouble Code (DTC) is cleared from the PCM's memory. You'll need a scan tool to access any DTCs stored in the PCM.

Before outputting any DTCs stored in the PCM, thoroughly inspect ALL electrical connectors and hoses. Make sure that all electrical connections are tight, clean and free of corrosion. And make sure that all hoses are correctly connected, fit tightly and are in good condition (no cracks or tears).

Accessing the DTCs

Refer to illustration 3.1

1   The Diagnostic Trouble Codes (DTCs) can only be accessed with a code reader or scan tool. Professional scan tools are expensive, but relatively inexpensive generic code readers or scan tools **(see illustrations 2.4a and 2.4b)** are available at most auto parts stores. Simply plug the connector of the scan tool into the diagnostic connector **(see illustration)**. Then follow the instructions included

with the scan tool to extract the DTCs.
2   Once you have output all of the stored DTCs, look them up on the accompanying DTC chart.
3   After troubleshooting the source of each DTC, make any necessary repairs or replace the defective component(s).

Clearing the DTCs

4   Clear the DTCs with the code reader or scan tool in accordance with the instructions provided by the tool's manufacturer.

Diagnostic Trouble Codes

5   The accompanying tables are a list of the Diagnostic Trouble Codes (DTCs) that can be accessed by a do-it-yourselfer working at home (there are many, many more DTCs available to professional mechanics with proprietary scan tools and software, but those codes cannot be accessed by a generic scan tool). If the problem persists after you have checked and repaired the connectors, wire harness and vacuum hoses (if applicable) for an emission-related system, component or circuit, have the vehicle checked by a dealer service department or other qualified repair shop.

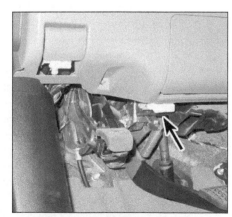

3.1 The Data Link Connector (DLC) is located under the lower left end of the instrument panel

Diagnostic Trouble Codes

Code	Code identification
P0010	Camshaft position A actuator circuit (Bank 1)
P0011	Camshaft position A, system performance problem or too much advance (Bank 1)
P0012	Camshaft position A, timing over-retarded (Bank 1)
P0013	Camshaft position "B" Actuator circuit / open (Bank 1), open or short in OCV for exhaust camshaft (for Bank 1) circuit
P0014	Camshaft position "B" - timing over-advanced or system performance (Bank 1), Valve Timing, oil control valve (OCV) for exhaust camshaft
P0015	Camshaft position "B" - timing over-retarded (Bank 1) - Valve Timing or OCV for exhaust camshaft
P0016	Crankshaft position/camshaft position correlation (Bank 1, sensor A)
P0017	Crankshaft position - Camshaft Position correlation (Bank 1, sensor B), valve timing or camshaft timing oil control valve assembly (Bank 1)
P0018	Crankshaft position/camshaft position correlation (Bank 2, sensor A)
P0019	Crankshaft position - Camshaft position correlation (Bank 2, sensor B), valve timing or camshaft timing oil control valve assembly (Bank 2)
P0020	Camshaft position A actuator circuit (Bank 2)
P0021	Camshaft position A, timing over-advanced/system performance problem (Bank 1)
P0022	Camshaft position A, timing over-retarded (Bank 1)
P0023	Camshaft position "B" actuator circuit / open (Bank 2), open or short
P0024	Camshaft position "B" - timing over-advanced or system performance
P0025	Camshaft position "B" - timing over-retarded (Bank 2) - valve timing
P0031	Oxygen sensor heater control circuit, low voltage (Bank 1, sensor 1)
P0032	Oxygen sensor heater control circuit, high voltage (Bank 1, sensor 1)
P0037	Oxygen sensor heater control circuit, low voltage (Bank 1, sensor 2)
P0038	Oxygen sensor heater control circuit, high voltage (Bank 1, sensor 2)
P0043	Oxygen sensor heater control circuit, low voltage (Bank 1, sensor 3)
P0044	Oxygen sensor heater control circuit, high voltage (Bank 1, sensor 3)
P0051	Oxygen sensor heater control circuit, low voltage (Bank 2, sensor 1)
P0052	Oxygen sensor heater control circuit, high voltage (Bank 2, sensor 1)
P0057	Oxygen sensor heater control circuit, low voltage (Bank 2, sensor 2)

Code	Code identification
P0058	Oxygen sensor heater control circuit, high voltage (Bank 2, sensor 2)
P006A	Manifold Absolute Pressure (MAP)/Mass Air Flow (MAF) correlation Bank 1, induction system leak or mass air flow meter problem
P0100	Mass Air Flow (MAF) sensor or circuit fault
P0101	Mass Air Flow (MAF) sensor range or performance problem
P0102	Mass Air Flow (MAF) sensor circuit, low input voltage
P0103	Mass Air Flow (MAF) sensor circuit, high input voltage
P0107	Manifold Absolute Pressure/Barometric Pressure circuit low input open or short in Manifold Absolute Pressure sensor circuit
P0108	Manifold Absolute Pressure/Barometric pressure circuit high input open or short in Manifold Absolute Pressure sensor circuit
P0110	Intake Air Temperature (IAT) sensor (in MAF sensor) or circuit fault
P0111	Intake Air Temperature (IAT) sensor gradient too high or Mass Air Flow (MAF) meter assembly
P011B	Engine Coolant Temperature/Intake Air Temperature(IAT), correlation fault
P0112	Intake Air Temperature (IAT) sensor circuit, low input voltage
P0113	Intake Air Temperature (IAT) sensor circuit, high input voltage
P0115	Engine Coolant Temperature (ECT) sensor or circuit fault
P0116	Engine Coolant Temperature (ECT) sensor range or performance problem
P0117	Engine Coolant Temperature (ECT) sensor, low input voltage
P0118	Engine Coolant Temperature (ECT) sensor, high input voltage
P0120	Throttle/Accelerator Pedal Position sensor A or circuit fault
P0121	Throttle/Accelerator Pedal Position sensor A, range/performance problem
P0122	Throttle/Accelerator Pedal Position sensor A circuit, low voltage input
P0123	Throttle/Accelerator Pedal Position sensor A circuit, high voltage input
P0125	Insufficient coolant temperature for closed-loop fuel control
P0128	Coolant temperature below thermostat regulating temperature
P0130	Pre-converter oxygen sensor circuit fault (Bank 1, sensor 1)
P0133	Pre-converter oxygen sensor circuit, slow response (Bank 1, sensor 1)
P0135	Pre-converter oxygen sensor heater or circuit fault (Bank 1, sensor 1)
P0136	Post-converter oxygen sensor circuit fault (Bank 1, sensor 2)

Diagnostic Trouble Codes (continued)

Code	Code identification
P0137	Post-converter oxygen sensor circuit, low voltage (Bank 1, sensor 2)
P0138	Post-converter oxygen sensor circuit, high voltage (Bank 1, sensor 2)
P0139	Oxygen sensor circuit slow response (Bank 1 sensor 2)
P013A	Oxygen sensor slow response - rich to lean (Bank 1 sensor 2)
P013C	Oxygen sensor slow response - rich to lean (Bank 2 sensor 2)
P014C	A/F sensor slow response, rich to lean (Bank 1, sensor 1)
P014D	A/F sensor slow response, lean to rich (Bank 1, sensor 1)
P014E	A/F sensor slow response, rich to lean (Bank 2, sensor 1)
P014F	A/F sensor slow response, lean to rich (Bank 2, sensor 1)
P0141	Post-converter oxygen sensor heater or circuit fault (Bank 1, sensor 2)
P0150	Pre-converter oxygen sensor circuit malfunction (Bank 2, sensor 1)
P0153	Pre-converter oxygen sensor circuit slow response (Bank 2, sensor 1)
P0155	Pre-converter oxygen sensor heater circuit malfunction (Bank 2, sensor 1)
P0156	Post-converter oxygen sensor circuit malfunction (Bank 2, sensor 2)
P0157	Post-converter oxygen sensor circuit, low voltage (Bank 2, sensor 2)
P0158	Post-converter oxygen sensor circuit, high voltage (Bank 2, sensor 2)
P0159	Oxygen sensor circuit slow response (Bank 2 sensor 2)
P015A	A/F sensor delayed response, rich to lean (Bank 1, sensor 1)
P015B	A/F sensor delayed response, lean to rich (Bank 1, sensor 1)
P015C	A/F sensor delayed response, rich to lean (Bank 2, sensor 1)
P015D	A/F sensor delayed response, lean to rich (Bank 2, sensor 1)
P0161	Post-converter oxygen sensor heater circuit problem (Bank 2, sensor 2)
P0171	Fuel injection system too lean (Bank 1)
P0172	Fuel injection system too rich (Bank 1)
P0174	Fuel injection system lean (Bank 2)
P0175	Fuel injection system rich (Bank 2)

Code	Code identification
P0220	Throttle/Accelerator Pedal Position sensor B circuit
P0222	Throttle/Accelerator Pedal Position sensor B circuit, low voltage input
P0223	Throttle/Accelerator Pedal Position sensor B circuit, high voltage input
P0230	Fuel pump primary circuit
P0300	Random/multiple cylinder misfire detected
P0301	Cylinder No. 1 misfire detected
P0302	Cylinder No. 2 misfire detected
P0303	Cylinder No. 3 misfire detected
P0304	Cylinder No. 4 misfire detected
P0305	Cylinder No. 5 misfire detected
P0306	Cylinder No. 6 misfire detected
P0307	Cylinder No. 7 misfire detected
P0308	Cylinder No. 8 misfire detected
P032C	Knock sensor 3 circuit low
P032D	Knock sensor 3 circuit high
P0325	Knock Sensor No. 1 or circuit fault
P0327	Knock Sensor No. 1 circuit, low voltage (Bank 1 or single sensor)
P0328	Knock Sensor No. 1 circuit, high voltage (Bank 1 or single sensor)
P0330	Knock sensor no. 2 or circuit fault
P0331	Knock Sensor No. 2 circuit, low voltage (Bank 2)
P0332	Knock Sensor No. 2 circuit, high voltage (Bank 2)
P0333	Knock Sensor No. 2 circuit, high input (Bank 2)
P0335	Crankshaft Position (CKP) sensor A circuit fault
P0337	Crankshaft Position (CKP) sensor "A" circuit low input
P0338	Crankshaft Position (CKP) sensor "A" circuit high input
P0339	Crankshaft Position (CKP) sensor A circuit intermittent
P0340	Camshaft Position (CMP) sensor A circuit (Bank 1 or single sensor)

Diagnostic Trouble Codes (continued)

Code	Code identification
P0342	Camshaft Position (CMP) sensor A circuit - low input
P0343	Camshaft Position (CMP) sensor A circuit - high input
P0345	Camshaft Position (CMP) sensor A circuit (Bank 2)
P0346	Camshaft Position (CMP) sensor A circuit range or performance problem
P0347	Camshaft Position (CMP) sensor A circuit - low input (bank 2)
P0348	Camshaft Position (CMP) sensor A circuit - range/performance problem (bank 2)
P0351	Ignition coil A primary/secondary circuit
P0352	Ignition coil B primary/secondary circuit
P0353	Ignition coil C primary/secondary circuit
P0354	Ignition coil D primary/secondary circuit
P0355	Ignition coil E primary/secondary circuit
P0356	Ignition coil F primary/secondary circuit
P0357	Ignition coil G primary/secondary circuit
P0358	Ignition coil H primary/secondary circuit
P033C	Knock sensor 4 circuit low input
P033D	Knock sensor 4 circuit high input
P0365	Camshaft position sensor "B" circuit (Bank 1)
P0367	Camshaft position sensor "B" circuit low input (Bank 1)
P0368	Camshaft position sensor "B" circuit high input (Bank 1)
P0390	Camshaft position sensor "B" circuit (Bank 2)
P0392	Camshaft position sensor "B" circuit low input (Bank 2)
P0393	Camshaft position sensor "B" circuit high input (Bank 2)
P0401	Exhaust gas recirculation flow insufficient detected - EGR valve assembly
P0403	Exhaust gas recirculation control circuit - open or short in EGR valve circuit
P0412	Air injection system air switching valve malfunction
P0415	Secondary air injection system switching valve "B" circuit - open in ASV drive circuit

Code	Code identification
P0416	Secondary air injection system switching valve "B" circuit open
P0417	Secondary air injection system switching valve "B" circuit shorted
P0418	Air injection system air pump malfunction
P0419	Secondary air injection system control "B" circuit, open in air pump drive circuit
P0420	Catalyst system efficiency below threshold (Bank 1)
P0430	Catalyst system efficiency below threshold (Bank 2)
P043E	Evaporative Emission Control (EVAP) system reference orifice clogged
P043F	Evaporative Emission Control (EVAP) system reference orifice high flow
P0440	Evaporative Emission Control (EVAP) system malfunction
P0441	Evaporative Emission Control (EVAP) system incorrect purge flow
P0443	Evaporative Emission Control (EVAP) system purge control valve circuit, open or short in purge VSV circuit
P0446	EVAP canister vent control valve circuit fault
P0450	EVAP system pressure sensor (fuel tank pressure sensor)
P0451	EVAP system pressure sensor range or performance problem
P0452	EVAP system pressure sensor, low voltage input
P0453	EVAP system pressure sensor, high voltage input
P0455	Evaporative Emission Control (EVAP) system, big leak detected
P0456	Evaporative Emission Control (EVAP) system, very small leak detected
P0500	Vehicle Speed Sensor (VSS) A malfunction
P050A	Cold start idle air control system performance problem
P050B	Cold start ignition timing performance problem
P0503	Vehicle Speed Sensor (VSS) A intermittent, erratic or and/or high voltage
P0504	Brake switch A/B correlation
P0505	Idle air control system
P0550	Power Steering Pressure (PPS) sensor circuit malfunction - open or short in power steering oil pressure sensor circuit
P0552	Power Steering Pressure (PPS) sensor circuit low input - open in power steering oil pressure sensor circuit
P0553	Power Steering Pressure (PPS) sensor circuit high input - short in power steering oil pressure sensor circuit

Diagnostic Trouble Codes (continued)

Code	Code identification
P0560	System voltage
P0604	Internal Control Module Random Access Memory (RAM) error
P0606	Powertrain Control Module (PCM) processor
P0607	Powertrain Control Module (PCM) performance problem
P0617	Starter relay circuit, high voltage
P0630	Vehicle Identification Number (VIN) not programmed or mismatched PCM
P0657	Actuator supply voltage, circuit open
P060A	Internal control module monitoring processor performance, ECM
P060B	Internal control module A/D processing performance
P060D	Internal control module Accelerator Pedal Position, performance
P060E	Internal control module, throttle position performance
P062F	Internal control module, EEPROM error
P0705	Transmission Range (TR) sensor circuit malfunction (PRNDL input)
P0710	Transmission fluid temperature sensor A circuit malfunction
P0711	Transmission fluid temperature sensor A performance problem
P0712	Transmission fluid temperature sensor A circuit, low input voltage
P0713	Transmission fluid temperature sensor A circuit, high input voltage
P0717	Input/turbine speed sensor A circuit, no signal
P0722	Output speed sensor circuit, no signal
P0724	Brake switch B circuit, high voltage
P0748	Pressure control solenoid A electrical problem (shift solenoid valve SL1)
P0751	Shift solenoid A performance problem (shift solenoid valve S1)
P0756	Shift solenoid B performance problem (shift solenoid valve S2)
P0771	Shift solenoid E performance problem (shift solenoid valve SR)
P0776	Pressure control solenoid B performance problem (shift solenoid valve SL2)
P0778	Pressure control solenoid B electrical (shift solenoid valve SL2)

Code	Code identification
P0781	1-2 shift (1-2 shift valve)
P0973	Shift solenoid A control circuit, low voltage (shift solenoid valve S1)
P0974	Shift solenoid A control circuit, high voltage (shift solenoid valve S1)
P0976	Shift solenoid B control circuit, low voltage (shift solenoid valve S2)
P0977	Shift solenoid B control circuit, high voltage (shift solenoid valve S2)
P0985	Shift solenoid E control circuit, low voltage (shift solenoid valve SR)
P0986	Shift solenoid E control circuit, high voltage (shift solenoid valve SR)

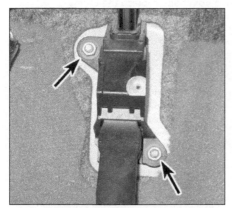

4.3 To detach the APP sensor, remove these two nuts

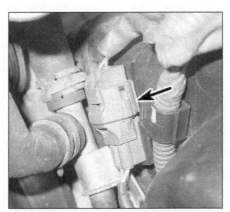

5.4 Disconnect the electrical connector for the CMP sensor (2UZ-FE engines)

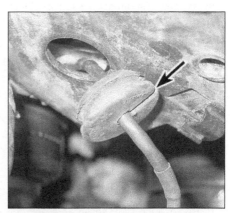

5.5 When removing the left No. 3 timing belt cover from the 2UZ-FE engines, pull the grommet for the CMP sensor electrical lead out of its hole in the cover, slide the lead out the slot in the grommet, then thread the lead through the hole in the cover

4 Accelerator Pedal Position (APP) sensor - replacement

Refer to illustration 4.3

Note: *The APP sensor is an integral component of the accelerator pedal assembly. To replace the APP sensor, you must replace the entire accelerator pedal assembly.*

1   Disconnect the cable from the negative battery terminal (see Chapter 5).
2   Working under the dash, disconnect the electrical connector from the top of the APP sensor.
3   Remove the two accelerator pedal/APP sensor mounting nuts **(see illustration)** and remove the accelerator pedal/APP sensor.
4   Installation is the reverse of removal.

5 Camshaft Position (CMP) sensor - replacement

V6 models

1   On 2010 and earlier V6 models, there are two CMP sensors, which are located on the inside walls of the cylinder heads, adjacent to the timing rotor installed on the front of each Variable Valve Timing-intelligent (VVT-i) con-

troller. Toyota refers to the CMP sensors on these models as VVT sensors, so we cover them in Section 20. On 2011 and later models, there are four CMP sensors identical to those used on the 1UR-FE, 3UR-FE and 3UR-FBE V8 engines (see Steps 9 through 11).

V8 models

2   Disconnect the cable from the negative battery terminal (see Chapter 5).

2UZ-FE engines

Refer to illustrations 5.4, 5.5 and 5.6

Note: *The CMP sensor is located on the front of the left cylinder head, adjacent to the intake camshaft timing belt sprocket.*

3   Remove the drivebelt (see Chapter 1).
4   Disconnect the CMP sensor electrical connector **(see illustration)**.
5   Remove the left No. 3 (the upper left) timing belt cover (see *Timing belt and sprockets - removal, inspection and installation* in Chapter 2B). When removing the cover, disengage the grommet for the CMP sensor electrical lead from its hole and slide the lead out the slot in the grommet, then pull the lead through the hole in the cover **(see illustration)**.
6   Remove the CMP sensor retaining bolt and stud **(see illustration)**.

7   Installation is the reverse of removal. Be sure to tighten the CMP sensor retaining bolt and stud securely.

5.6 To remove the CMP sensor on 2UZ-FE engines, remove the sensor retaining bolt and stud

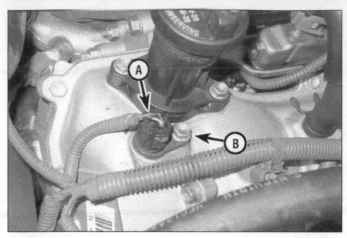

5.9 To remove the CMP sensor on 1UR-FE, 3UR-FE and 3UR-FBE V8 engines, disconnect the sensor electrical connector (A), then remove the sensor retaining bolt (B)

5.10 On 1UR-FE, 3UR-FE and 3UR-FBE V8 engines, lift the sensor from the valve cover and replace the O-ring

1GR-FE V6 (2010 and later), 1UR-FE, 3UR-FE and 3UR-FBE V8 engines

Refer to illustrations 5.9 and 5.10

Note: *The CMP sensor is located on the front of the left cylinder head, adjacent to the intake camshaft timing belt sprocket.*

8   Remove the engine cover (see Chapter 1, **illustration 24.12a**).
9   Disconnect the CMP sensor electrical connector **(see illustration)**.
10   Remove the CMP sensor retaining bolt and sensor from the driver's side valve cover **(see illustration)**.
11   Installation is the reverse of removal. Be sure to tighten the CMP sensor retaining bolt and stud securely.

6   Crankshaft Position (CKP) sensor - replacement

Refer to illustrations 6.6a and 6.6b

Note: *The CKP sensor is located at the front lower left corner of the engine.*

1   Disconnect the cable from the negative battery terminal (see Chapter 5).

2   Raise the vehicle and place it securely on jackstands.
Warning: *On Sequoia models equipped with rear height control suspension, adjust the height control to the NORMAL mode, turn the height control to OFF, then turn off the engine BEFORE raising the vehicle.*
3   Remove the engine under-cover (see Chapter 1, **illustration 8.6a and 8.6b**).
4   On V6 models, remove the alternator (see Chapter 5).
5   On V6 models, remove the bolt that secures the air conditioning suction line at the front of the engine, disconnect the electrical connector from the air conditioning compressor and unbolt the compressor (see Chapter 3).
Warning: *Do NOT disconnect the air conditioning hoses from the compressor. Move the compressor aside and support it with wire or rope.*
6   On V6 and 2UZ-FE V8 engines, disconnect the CKP sensor electrical connector **(see illustrations)**. On 1UR-FE, 3UR-FE and 3UR-FBE V8 engines, remove the sensor protector mounting bolts and protector from the driver's side lower rear corner of the engine then disconnect the sensor electrical connector.
7   Remove the CKP sensor retaining bolt.
8   Installation is the reverse of removal. Be

sure to tighten the CKP sensor retaining bolt securely.

7   Engine Coolant Temperature (ECT) sensor - replacement

Warning: *Wait until the engine has cooled completely before beginning this procedure.*
Caution: *Handle the Engine Coolant Temperature (ECT) sensor with care. Damage to the ECT sensor will affect the operation of the entire fuel injection system.*
1   Drain the engine coolant until the coolant level is below the sensor (see Chapter 1).

V6 models

Refer to illustration 7.3

Note: *The ECT sensor is located at the back of the engine, on the rear coolant crossover, which spans the valley between the two cylinder heads.*

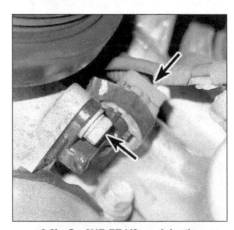

6.6b On 2UZ-FE V8 models, the Crankshaft Position (CKP) sensor is located on the left underside of the oil pump; to remove it, unplug the electrical connector and remove the sensor retaining bolt

6.6a On V6 models, the Crankshaft Position (CKP) sensor is located on the lower left side of the timing chain cover. To remove the CKP sensor, disconnect the electrical connector and remove the sensor mounting bolt

7.3 On V6 models, the ECT sensor is located on the rear coolant crossover between the cylinder heads

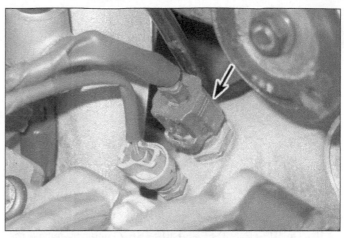

7.9 On 2UZ-FE V8 models, the Engine Coolant Temperature (ECT) sensor is located on the right end of the coolant crossover pipe under the throttle body

2   Remove the intake manifold (see Chapter 2A).
3   Disconnect the electrical connector from the ECT sensor **(see illustration)**.
4   Unscrew the ECT sensor from the coolant crossover with a deep socket.
5   Install a new sealing washer to the sensor.
6   Installation is otherwise the reverse of removal. Be sure to tighten the ECT sensor to the torque listed in this Chapter's Specifications.
7   Refill the cooling system (see Chapter 1).

V8 models
2UZ-FE engines

Refer to illustration 7.9

Note: *The ECT sensor is located on the front coolant crossover, which is located directly under the throttle body.*

8   Remove the throttle body cover (see Chapter 2B) and the air intake duct (see *Air filter housing - removal and installation* in Chapter 4).
9   Disconnect the electrical connector from the ECT sensor **(see illustration)**.
10   Unscrew and remove the ECT sensor from the front coolant crossover with a deep socket

1UR-FE, 3UR-FE and 3UR-FBE V8 engines

Refer to illustration 7.12

Note: *The ECT sensor is located on the front of the left cylinder head, just below the oil fill cap.*

11   Remove the engine cover (see Chapter 1, **illustration 24.12a**).
12   Disconnect the electrical connector from the ECT sensor **(see illustration)**.
13   Unscrew and remove the ECT sensor from the front coolant crossover with a deep socket

All models

14   Install a new sealing washer to the sensor.
15   Installation is otherwise the reverse of removal. Be sure to tighten the ECT sensor to the torque listed in this Chapter's Specifications.
16   Refill the cooling system (see Chapter 1).

8   Intake Air Temperature (IAT) sensor - replacement

The IAT sensor is an integral component of the Mass Air Flow (MAF) sensor. See Section 10.

9   Knock sensor - replacement

Warning: *Wait for the engine to cool completely before performing this procedure.*

V6 models

Note: *The knock sensors are located on top of the engine block, under the intake manifold.*
Note: *Accessing the knock sensors is extremely difficult because you have to remove the cylinder heads, so we recommend that you replace BOTH knock sensors even if only one of them is defective.*
1   Drain the engine coolant (see Chapter 1).
2   Remove the cylinder heads (see Chapter 2A).
3   Remove the heater coolant inlet hose and the coolant outlet pipe.
4   Disconnect the electrical connectors from the knock sensors.
5   Carefully note how the knock sensors are oriented, then remove the sensor retaining bolts and remove the sensors.

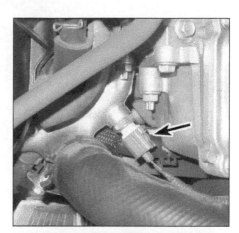

7.12 On 1UR-FE, 3UR-FE and 3UR-FBE V8 engines, the ECT sensor is located on the front of the left cylinder head, just below the oil fill cap

6   The orientation of the knock sensors is critical because they'll cause clearance problems during installation of the cylinder heads and intake manifold if installed incorrectly. So, when installing the knock sensors, make sure that each sensor is oriented in the same exact position that it was in before removal. When the knock sensors are correctly positioned, tighten the sensor retaining bolts to the torque listed in this Chapter's Specifications.
7   The remainder of installation is the reverse of removal.
8   Refill the engine cooling system (see Chapter 1).

V8 models

Refer to illustrations 9.10a, 9.10b and 9.11

Note: *The knock sensors are located on top of the engine block, under the intake manifold. On 2UZ-FE engines the air injection pump must be removed to access the two knock sensors. On 1UR-FE, 3UR-FE and 3UR-FBE V8 engines, there are four sensors.*

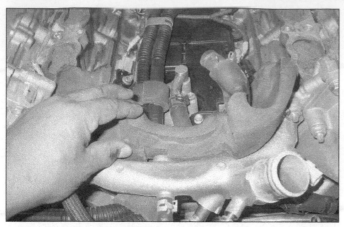

9.10a On 1UR-FE, 3UR-FE and 3UR-FBE V8 engines, remove the front sound insulator . . .

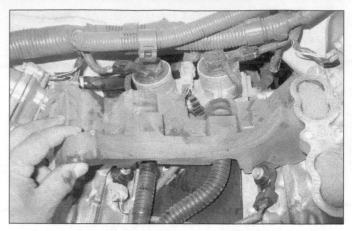

9.10b . . . . and the rear insulator

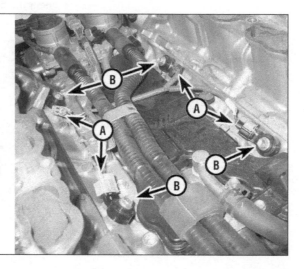

9.11 Locations of the knock sensor electrical connectors (A), and sensor retaining bolts (B). When installing the knock sensors, make sure that the centerline of each sensor is positioned within 15 degrees of the direction in which it was facing before removal (3UR-FE engine shown, all other models similar)

9   Remove the intake manifold (see Chapter 2B).
Note: On *2UZ-FE engines, it's not necessary to separate the upper and lower intake manifolds.*

10   On 1UR-FE, 3UR-FE and 3UR-FBE engines, remove the engine sound insulators **(see illustrations)**.
11   Disconnect the knock sensor electrical connector **(see illustration)**.

12   Remove the knock sensor retaining bolts and remove the sensors.
13   Installation is the reverse of removal. Be sure to tighten the knock sensor retaining bolts to the torque listed in this Chapter's Specifications.
14   Refill the engine cooling system (see Chapter 1).

10   Mass Air Flow/Intake Air Temperature (MAF/IAT) sensor - replacement

Refer to illustrations 10.2a and 10.2b
Note: *The MAF/IAT sensor is located on top of the air filter housing.*
1   Make sure the ignition key is in the Off position.
2   Disconnect the MAF/IAT sensor electrical connector **(see illustrations)**.
3   Remove the MAF/IAT sensor retaining screws and remove the sensor.
4   Installation is the reverse of removal.

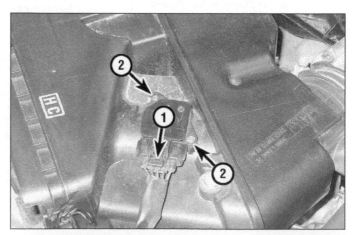

10.2a To remove the Mass Air Flow (MAF) sensor, unplug the electrical connector and remove the two MAF sensor retaining screws (V8 model shown, 2011 and later V6 models similar)

10.2b To remove the Mass Air Flow (MAF) sensor, depress this release tab (1), pull off the electrical connector, then remove the sensor retaining screws (2) (2010 and earlier V6 models)

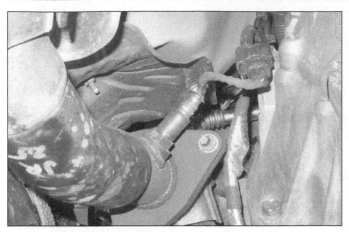

11.4 Typical driver's side (BANK 1), oxygen sensor locations (V8 models shown)

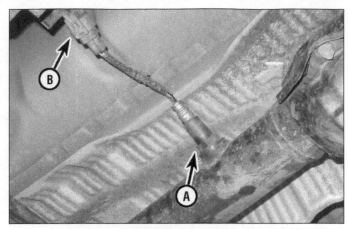

11.9 A typical downstream oxygen sensor (A); and sensor electrical connector (B)

11   Oxygen sensors - general information and replacement

General information

1   Use special care when servicing an oxygen sensor or air/fuel ratio sensor:

a) *Oxygen sensors and air/fuel ratio sensors have a permanently attached pigtail and electrical connector which cannot be removed from the sensor. Damage to or removal of the pigtail or electrical connector will ruin the sensor.*

b) *Grease, dirt and other contaminants should be kept away from the electrical connector and the louvered end of the sensor.*

c) *Do not use cleaning solvents of any kind on an oxygen sensor or air/fuel ratio sensor.*

d) *Do not drop or roughly handle an oxygen sensor or air/fuel ratio sensor.*

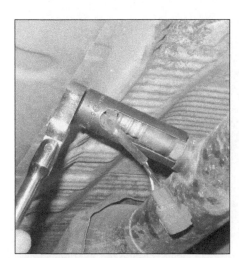

11.10 Use a special socket to unscrew the downstream oxygen sensor

e) *Be sure to install the silicone boot in the correct position to prevent the boot from melting and to allow the sensor to operate properly.*

Replacement

Note: *There are four oxygen sensors, one upstream sensor and one downstream sensor for each cylinder bank.*

Note: *Because it is installed in an exhaust manifold or exhaust pipe, which contract when cool, an oxygen sensor or air/fuel ratio sensor might be very difficult to loosen when the engine is cold. Rather than risk damage to the sensor, start and run the engine for a minute or two, then shut it off. Be careful not to burn yourself during the following procedure.*

2   Make sure the ignition key is in the OFF position.

3   Raise the vehicle and place it securely on jackstands.

Warning: *On Sequoia models equipped with rear height control suspension, adjust the height control to the NORMAL mode, turn the height control to OFF, then turn off the engine BEFORE raising the vehicle.*

Upstream oxygen sensors

Refer to illustration 11.4

Note: *The upstream sensors are located at the outlet ends of the exhaust manifolds, just above the integral upstream catalysts.*

4   Locate the upstream sensor **(see illustration)**, then trace the electrical lead to the sensor electrical connector and disconnect it.

5   Unscrew the upstream sensor with a wrench (there isn't room to use an oxygen sensor socket).

6   Apply anti-seize compound to the threads of the sensor to facilitate future removal.

Note: *Most new sensors already have anti-seize compound applied to the threads.*

7   Tighten the oxygen sensor to the torque listed in this Chapter's Specifications.

8   Installation is otherwise the reverse of removal.

Downstream oxygen sensor

Refer to illustrations 11.9 and 11.10

Note: *The downstream sensors are located in the front exhaust pipe assembly, ahead of the downstream catalysts.*

9   Locate the downstream oxygen sensor, then trace the electrical lead to the sensor electrical connector and disconnect it **(see illustration)**.

10   Unscrew the downstream sensor with an oxygen sensor socket or with a wrench **(see illustration)**.

11   Apply anti-seize compound to the threads of the sensor to facilitate future removal.

Note: *Most new sensors already have anti-seize compound applied to the threads.*

12   Tighten the oxygen sensor to the torque listed in this Chapter's Specifications.

13   Installation is otherwise the reverse of removal.

12   Transmission Range (TR) sensor - replacement

Even though the Society of Automotive Engineers (SAE) has recommended since 1996 that all manufacturers refer to the Park/ Neutral Position (PNP) switch as the Transmission Range (TR) sensor, Toyota continues to refer to the TR sensor as the PNP switch. To avoid confusion, we have therefore included the TR sensor/PNP switch in Chapter 7A.

13   Vehicle Speed Sensor (VSS) - replacement

Refer to illustration 13.2

1   On A750E and A750F transmissions, a single Vehicle Speed Sensor (VSS) is located on the right side of the transmission extension

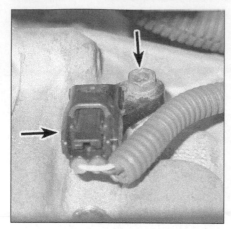

13.2 To remove VSS, simply unplug the electrical connector and remove the sensor retaining bolt

14.4a Depress the electrical harness release tab . . .

14.4b . . . then lift the connector up and off of the holder block

housing. On AB60E and AB60F transmissions, two Vehicle Speed Sensors (VSS) are used; they are both located on the right side of the transmission. One at the rear exten-

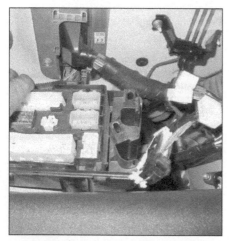

14.5 Remove the holder block retaining bolts and remove the block

sion housing and the other towards the front, above the oil cooler unit.

2   Disconnect the electrical connector from the VSS **(see illustration)**.

3   Remove the VSS retaining bolt **(see illustration 13.2)**, then remove the VSS from the transmission.

4   Remove and discard the old VSS O-ring.

5   Coat the new O-ring with clean ATF.

6   Installation is the reverse of removal.

14 Powertrain Control Module (PCM) - removal and installation

Refer to illustrations 14.4a, 14.4b, 14.5, 14.6a, 14.6b and 14.7

Warning: *The models covered by this manual are equipped with Supplemental Restraint systems (SRS), more commonly known as airbags. Always disable the airbag system before working in the vicinity of any airbag system components to avoid the possibility of accidental deployment of the airbag, which could cause personal injury (see Chapter 12).*

Caution: *To avoid electrostatic discharge damage to the PCM, handle the PCM only by its case. Do not touch the electrical terminals during removal and installation. If available, ground yourself to the vehicle with an anti-static ground strap, available at computer supply stores.*

Note: *The PCM is located in the engine compartment, in the cowl housing.*

1   Disconnect the cable from the negative battery terminal.

2   Remove the engine cover (see Chapter 1, **illustration 24.12a**).

3   Remove the air filter housing (see Chapter 4).

4   Disconnect the harness connectors from the holder block **(see illustrations)**.

5   Remove the three block fasteners and holder block **(see illustration)**.

6   Disconnect the lower, then the upper electrical connectors from the PCM **(see illustration)**. To release the PCM connectors, push the lever locks (at the end of each lever) in, while lifting the levers outwards and pull the PCM connectors out from the PCM **(see illustration)**.

14.6a Disconnect the lower connector (A) first, then upper connector (B) from the PCM

14.6b Push the lever locks in, while rotating the lever outwards, and pull the connectors out from the PCM

14.7 Remove the PCM retaining bolts and pull the PCM out of the cowl panel

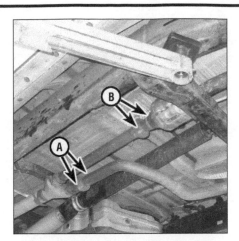

15.7a Rear pipe-to-catalytic converter nuts and bolts for the BANK 1 converters (A), and the nuts and bolts for the BANK 2 converters (B)

15.7b Remove the front pipe-to-exhaust manifold nuts and lower the converter assembly down

Caution: *The ignition switch must be turned OFF when pulling out or plugging in the electrical connectors to prevent damage to the PCM.*
7   Remove the PCM retaining bolts **(see illustration)** and remove the PCM.
Caution: *Avoid any static electricity damage to the computer by grounding yourself to the body before touching the PCM and using a special anti-static pad to store the PCM on once it is removed.*
8   Installation is the reverse of removal.

15   Catalytic converter

Note: *Because of the Federally mandated extended warranty which covers emissions-related components such as the catalytic converter, check with a dealer service department before replacing the converter at your own expense.*

General description

1   The catalytic converter is an emission control device added to the exhaust system to reduce pollutants from the exhaust gas stream. There are two types of converters: The oxidation catalyst reduces the levels of hydrocarbon (HC) and carbon monoxide (CO) by adding oxygen to the exhaust stream. The reduction catalyst lowers the levels of oxides of nitrogen (NOx) by removing oxygen from the exhaust gases. These two types of catalysts are combined into a three-way catalyst that reduces all three pollutants.

Check

2   The equipment for testing a catalytic converter is expensive. If you suspect that the converter on your vehicle is malfunctioning, take it to a dealer or authorized emissions inspection facility for diagnosis and repair.
3   Whenever the vehicle is raised for servicing underbody components, inspect the converter for leaks, corrosion, dents and other

damage. Inspect the welds/flange bolts that attach the front and rear ends of the converter to the exhaust system. If damage is discovered, the converter should be replaced.
4   Although catalytic converters don't break too often, they can become plugged. The easiest way to check for a restricted converter is to use a vacuum gauge to diagnose the effect of a blocked exhaust on intake vacuum.

a) *Connect a vacuum gauge to an intake manifold vacuum source (see Chapter 2C).*
b) *Warm the engine to operating temperature, place the transmission in Park and apply the parking brake.*
c) *Note and record the vacuum reading at idle.*
d) *Quickly open the throttle to near full throttle and release it shut. Note and record the vacuum reading.*
e) *Perform the test three more times, recording the reading after each test.*
f) *If the reading after the fourth test is more than one in-Hg lower than the reading recorded at idle, the exhaust system may be restricted (the catalytic converter could be plugged or an exhaust pipe or muffler could be restricted).*

Replacement

Refer to illustrations 15.5a and 15.5b

5   Raise the vehicle and place it securely on jackstands.
Warning: *On Sequoia models equipped with rear height control suspension, adjust the height control to the NORMAL mode, turn the height control to OFF, then turn off the engine BEFORE raising the vehicle.*
6   Spray the nuts on the exhaust flange studs with penetrant before removing them from the catalytic converter.
7   Remove the nuts and separate the catalytic converters from the exhaust system **(see illustrations)**.
8   Installation is the reverse of removal.

16   Evaporative emission control (EVAP) system - description and component replacement

General description

1   The fuel evaporative emission control (EVAP) system absorbs fuel vapors (unburned hydrocarbons) and, during engine operation, releases them into the intake manifold, from which they're drawn into the intake ports where they mix with the incoming air/fuel mixture.
2   When the engine is off, gasoline in the fuel tank, and residual fuel in the other components such as the intake manifold, warms up and evaporates, producing unburned hydrocarbon fuel vapors that would waste gas and pollute the air if released into the atmosphere. In an EVAP system, these vapors are routed from the fuel tank, throttle body and intake manifold through a system of hoses to the EVAP system's charcoal canister, where they're stored until the vehicle is operated again. The charcoal canister is able to store these vapors because it's filled with activated charcoal, a substance that can absorb many times its mass in hydrocarbon vapors.
3   The EVAP system consists of the charcoal canister, the canister closed valve, the EVAP system Vacuum Switching Valve (EVAP VSV) (more commonly referred to as the EVAP canister purge valve), the canister pump module (which consists of a vent valve, a leak detection pump and the canister pressure sensor), the air filter and the refueling valve. All components of the system except the purge valve are on or near the EVAP canister, which is located underneath the vehicle, behind the fuel tank. The EVAP canister purge valve is located in the engine compartment, on the intake manifold.
4   The vapor pressure sensor monitors the pressure inside the fuel tank. The vapor pressure sensor an integral component of the pump module, which is located at the EVAP

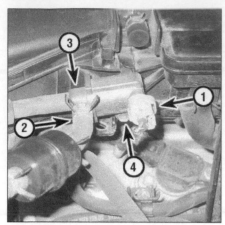

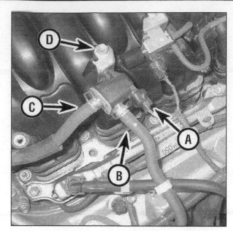

16.9a EVAP canister purge valve removal: disconnect the electrical connector (A), the canister purge hose (B) and the intake manifold hose (C), and the mounting bolt (D) (2010 and earlier V6 engines)

16.9b EVAP canister purge valve removal: disconnect the electrical connector (A), the canister purge hose (B) and the intake manifold hose (C), and remove the mounting bolt (D) (2011 and later V6 and all V8 engines)

16.14a Disconnect the electrical connector from the end of the canister - some models have a harness bracket mounted to the end of the EVAP canister

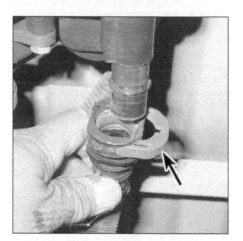

16.14b Disconnect the lines to the EVAP and the leak detection pump module

1   EVAP canister hose
2   Leak detection pump module electrical connector
3   Leak detection pump module hose

16.14c Pull the retaining clip out and disconnect the line

Component replacement

Warning: *Gasoline and gasoline vapors are extremely flammable, so take extra precautions when you work on any part of the fuel system. Don't smoke or allow open flames or bare light bulbs near the work area, and don't work in a garage where a gas-type appliance (such as a water heater or clothes dryer) is present. Since gasoline is carcinogenic, wear fuel-resistant gloves when there's a possibility of being exposed to fuel, and, if you spill any fuel on your skin, rinse it off immediately with soap and water. Mop up any spills immediately and do not store fuel-soaked rags where they could ignite. When you perform any kind of work on the fuel system, wear safety glasses and have a Class B type fire extinguisher on hand.*

EVAP canister purge valve

Refer to illustrations 16.9a and 16.9b

Note: *The EVAP canister purge valve is located on the left side of the intake manifold.*
8   Remove the engine cover (see Chapter 1, **illustration 24.12a**).
9   Disconnect the electrical connector from the purge valve **(see illustrations)**.
10   Disconnect the EVAP hoses from the purge valve **(see illustration 16.9a or 16.9b)**.
11   Remove the purge valve mounting bolt and remove the purge valve **(see illustration 16.9a or 16.9b)**.
12   Installation is the reverse of removal.

EVAP canister

Refer to illustrations 16.14a, 16.14b, 16.14c and 16.15

13   Raise the vehicle and place it securely on jackstands.
Warning: *On Sequoia models equipped with rear height control suspension, adjust the height control to the NORMAL mode, turn the height control to OFF, then turn off the engine BEFORE raising the vehicle.*
14   Disconnect the electrical connector(s) and hoses from the EVAP canister and the leak detection pump module **(see illustrations)**.

canister (on these models you cannot replace the vapor pressure sensor separately). When the vapor pressure sensor detects excessive pressure inside the fuel tank, the PCM commands the canister closed valve to open, which allows these vapors to migrate to the charcoal canister, where they're absorbed and stored. The canister closed valve also opens a hose that routes fresh (outside) air into the fuel tank as the vapors are routed to the canister to prevent the creation of a relative vacuum inside the tank as the vapors are vented to the canister. An air filter on this hose prevents dust and debris from entering the EVAP system and the fuel tank.
5   When the engine is running and the conditions are right, the PCM commands the purge valve to open, and intake vacuum pulls

the vapors stored in the charcoal canister out of the canister, through a purge line and into the manifold, where they're consumed by the engine. The PCM controls the volume of vapors drawn into the manifold in accordance with the driving conditions.
6   When the fuel tank is being filled, the refueling valve controls the flow rate of the vapors from the fuel tank to the canister.
7   The PCM monitors the EVAP system regularly for leaks by using the purge valve to induce a vacuum into the system and measuring how well the system can hold a vacuum. It does this by turning on the purge valve, which allows the intake manifold to pull vapor gases out of the canister and fuel tank. Normally, the canister closed valve would open the outside air hose to vent outside air into the system during normal purging. But when the system is being vacuum tested, the PCM directs the canister closed valve to remain closed, which produces the (relative) vacuum inside the system. If the PCM detects a leak, it outputs a Diagnostic Trouble Code (DTC).

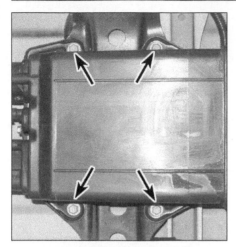

16.15 EVAP canister mounting bolt locations

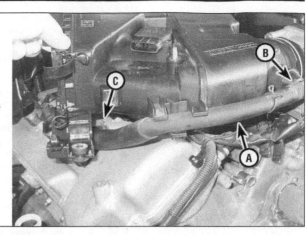

17.5 To remove the PCV fresh air inlet hose (A), slide back the spring-type clamps and disconnect the hose from the air filter housing (B) and from the pipe on the right valve cover (C)

15   Disconnect the wire from the top of the canister. Remove the EVAP canister mounting bolts **(see illustration)**, and remove the canister.

16   Installation is the reverse of removal.

EVAP canister pump module

17   Remove the EVAP canister and pump module lines **(see illustration 16.14b)**.

18   Disconnect the vacuum line and electrical connector to the pump.

19   Remove the pump mounting bolts and remove the pump from the canister.

20   Installation is the reverse of removal.

---

17   Positive Crankcase Ventilation (PCV) system - description, check and component replacement

Description

1   The Positive Crankcase Ventilation (PCV) system reduces hydrocarbon emissions by scavenging crankcase vapors. It does this by circulating fresh air from the air cleaner through the crankcase, where it mixes with blow-by gases, before being drawn through a PCV valve into the intake manifold.

2   The PCV system consists of the PCV valve and two hoses, one of which connects the air filter housing to the crankcase and the other (containing the PCV valve) which connects the crankcase to the intake manifold.

3   To maintain idle quality, the PCV valve restricts the flow when the intake manifold vacuum is high. If abnormal operating conditions (such as piston ring problems) arise, the system is designed to allow excessive amounts of blow-by gases to flow back through the crankcase vent tube into the air cleaner to be consumed by normal combustion.

Check

4   There is no scheduled inspection interval for the PCV valve or the PCV system hoses. But, over time the PCV system might become less efficient as an oil residue of sludge builds up inside the PCV valve and the hoses. One symptom of a clogged PCV system is leaking seals. When crankcase vapors can't escape, pressure builds inside the bottom end and eventually causes crankshaft seals to leak. Anytime that you're changing the oil, the air filter, the spark plugs, etc., it's a good idea to pull off the PCV hoses and inspect them and clean them out. If they're cracked, torn or deteriorated, replace them.

Component replacement

Refer to illustrations 17.5, 17.6 and 17.7

Note: *The photos accompanying this Section depict the PCV system on a V6 engine. The PCV system used on 2UZ-FE V8 engines is similar: the fresh air inlet hose connects a pipe on the air intake duct to a pipe on the right valve cover and the crankcase ventilation hose (PCV hose) connects the PCV valve on the left valve cover to a pipe between the throttle body and the intake manifold. The PCV valve on other V8*

engines is located underneath the front of the intake manifold (intake manifold removal is required to access the valve; see Chapter 2B).

5   To remove the PCV fresh air hose, slide back the spring-type clips and disconnect the ends of the hose from the air filter housing and the pipe on the right valve cover, respectively, then disengage the hose from the hose clips **(see illustration)**.

6   To remove the PCV hose, slide back the spring-type clamps and disconnect the hose from the PCV valve and the intake manifold **(see illustration)**.

7   To remove the PCV valve, remove the crankcase ventilation hose **(see illustration 17.6)**, then unscrew the PCV valve **(see illustration)** from the left valve cover.

8   Inspect the condition of the hoses. If they're clogged and/or dirty, blow them off with compressed air and wipe them off, then inspect them more carefully. If they're cracked, torn or deteriorated, replace them.

9   Inspect the condition of the PCV valve. If it's clogged, clean it with fresh solvent, then blow it out with compressed air. If you cannot remove the residue from the inside of the PCV valve, replace it.

10   Installation is the reverse of removal.

17.6 To remove the PCV hose (A), slide back the spring-type clamps and disconnect the hose from the PCV valve (B) and from the intake manifold (C)

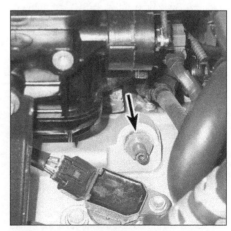

17.7 To remove the PCV valve, unscrew it with a deep socket or wrench

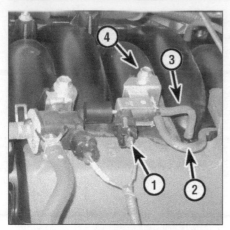

18.6 To remove the Acoustic Control Induction System Vacuum Switching Valve (ACIS VSV) from the intake manifold on a V8 model:

1   *Disconnect the electrical connector*
2   *Disconnect the vacuum hose that connects the VSV to the ACIS actuator*
3   *Disconnect the vacuum hose that directs manifold vacuum to its manifold vacuum source*
4   *Remove the VSV mounting bolt*

18   Acoustic Control Induction System (ACIS) - description and component replacement

Description

1   The PCM-controlled Acoustic Control Induction System (ACIS) varies the effective length of the intake manifold runners in response to engine speed and the angle of the throttle plate inside the throttle body. This capability increases efficiency and power at low and high speeds. Slightly different versions of the ACIS are used on V6 and V8 models.
2   The ACIS consists of a PCM-controlled Vacuum Switching Valve (VSV), an actuator and an intake air control valve. The VSV is a PCM controlled-device that controls the intake vacuum applied to the actuator. The actuator is a vacuum diaphragm that uses a pushrod and bellcrank to open and close the intake air control valve. The intake air control valve, which is an integral part of the intake manifold, opens and closes to alter the effective length of the intake manifold runners in two stages. On V6 models, there is one large intake air control valve for all six intake runners. On V8 models, there is one air control valve for each intake manifold runner.
3   At low-to-medium speeds, the PCM activates the VSV, sending vacuum to the actuator diaphragm. The actuator closes the intake air control valve, increasing the length of the intake manifold and improving intake efficiency.
4   At higher speeds, the PCM deactivates the VSV, cutting vacuum to the actuator diaphragm. The actuator opens the intake air control valve, decreasing the length of the intake manifold and improving engine power.

Component replacement

Refer to illustration 18.6

Note: *The actuator and the intake air control valve are integral components of the intake manifold. If either component fails, replace the intake manifold (see Chapter 2A or 2B). The VSV is the only component that you can replace at home.*

Note: *On V6 models, the VSV is located at the upper right rear corner of the intake manifold. On V8 models, the VSV is located at the left side of the intake manifold.*

5   Remove the engine cover (see Chapter 1, **illustration 24.12a**).
6   Disconnect the electrical connector from the VSV **(see illustration)**.
7   Clearly label both vacuum hoses, then disconnect both hoses from the VSV **(see illustration 18.6)**.
8   Remove the VSV mounting bolt and remove the VSV **(see illustration 18.6)**.
9   Installation is the reverse of removal.

19   Air injection system - description and component replacement

Description

1   Air injection helps the upstream catalytic converters reach their effective operating temperature more rapidly during warm-ups. The system consists of the PCM, an electronic air injection control driver, an electric air pump, an air pressure sensor, an electric air switching valve, two Vacuum Switching Valves (VSVs) and two vacuum-type air switching valves. These components are connected by an array of plastic and metal plumbing and wiring harnesses.
2   Using inputs from the Engine Coolant Temperature (ECT) sensor and Intake Air Temperature (IAT) sensor, the PCM determines whether the engine is cold or already warmed up. If the engine is already warmed up, the PCM doesn't turn on the air injection system. If the engine is cold or not sufficiently warmed up, the PCM calculates how much air is needed to warm up the catalysts. To do so, it uses inputs from the ECT, IAT and Mass Air Flow (MAF) sensors to calculate how long the air injection system should be turned on to bring the upstream catalysts up to their operating temperature.
3   When the PCM activates the air injection system, it turns on the two VSVs and it turns on the air injection control driver, which turns on the air injection pump and the electric air switching valve. The two VSVs send vacuum to the vacuum-type air switching valves, which allows the air pump to pump air through the electric air switching valve, through the two vacuum-type air switching valves, then through lines to the exhaust manifolds.
4   Using input from the air pressure sensor, the PCM monitors the air injection system and

uses this data to control the air injection control driver, which in turn controls the electric air switching valve.
5   When air is pumped into the exhaust manifolds, it helps to burn up any residual unburned fuel vapors, which are always present in the rich mixture used during warm-ups. Without this additional air, the engine exhaust would still warm up, but not as quickly, because the unburned fuel actually cools the exhaust temperature. But when the oxygen in the extra air combines with and promotes the burning of this unburned fuel, it heats up the exhaust gases, and the upstream catalysts, more quickly.

Component replacement

6   Because it is used only during cold-start warm-ups, the air injection system should be trouble-free for years. But if a component of the system fails, you can replace any of the critical components yourself. However, be aware that replacing some of these components - the air pressure sensor, the electric air switching valve and the air pump - is a little more difficult because you'll have to remove the intake manifold to access them.

Air injection control driver

Note: *The air injection control driver is located on the left side of the engine compartment, near the rear end of the left cylinder head.*
7   Disconnect the two electrical connectors from the air injection control driver.
8   Remove the two air injection control driver mounting bolts and remove the driver.
9   Installation is the reverse of removal.

Vacuum Switching Valves (VSVs)

Note: *The VSVs are located on the right side of the intake manifold.*
10   Disconnect the electrical connectors from the VSVs.
11   Disconnect the vacuum hoses from the VSVs.
12   Remove the VSV mounting bracket bolts and remove the VSV assembly.
13   Installation is the reverse of removal.

Vacuum-type air switching valves

Note: *The vacuum-type air switching valves are located on the back of the engine, where they're bolted to the rear coolant bypass housing. Toyota calls this component the rear water bypass joint.*
14   Remove the two nuts from each air tube flange and disconnect the two air tubes from the exhaust manifolds. Remove the two bolts from each air tube mounting flange and disconnect the air tubes from the switching valves. Remove both air tubes and remove and discard the old air tube mounting flange gaskets.
15   Remove the four air switching valve mounting bolts, pull back the switching valve assembly far enough to disconnect the two vacuum hoses, and remove the switching valve assembly.
16   Installation is the reverse of removal. Be sure to use new gaskets.

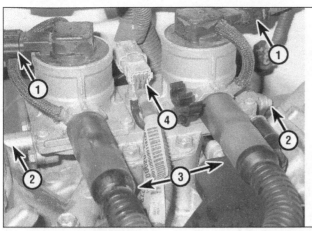

19.18 Air injection switching valve assembly details - 1UR-FE, 3UR-FE and 3UR-FBE engines

1 Air injection switching valve electrical connectors
2 Air injection tubes - to manifolds
3 Air injection tubes - to pump
4 Knock sensor harness

Air pressure sensor, electric air switching valve and air pump

Refer to illustration 19.18

17 Remove the intake manifold (see Chapter 2A or 2B).

18 On 1UR-FE, 3UR-FE and 3UR-FBE engines, disconnect the electrical connector the switching valve **(see illustration)**.

19 Remove the air tubes-to-switching valve mounting bolts.

20 Remove the switching valve-to-block bolts and remove the switching valve.

Air pressure sensor

21 Disconnect the electrical connector from the air pressure sensor

22 Remove the two pressure sensor mounting bolts and remove the pressure sensor from the air switching valve.

23 Installation is the reverse of removal.

Air switching valve

24 Disconnect the electrical connector from the air switching valve.

25 Clearly label the two hoses connected to the air switching valve, then disconnect them.

26 Remove the air switching valve mounting bolts and remove the air switching valve.

27 Installation is the reverse of removal.

Air pump only

Note: *On 1UR-FE, 3UR-FE and 3UR-FBE engines, the air pump is located under the passenger's side fender. To access the pump you must remove the front fender wheel splash shield (see Chapter 11).*

28 If you're replacing the air pump, remove the air pressure sensor and the air switching valve (see previous Steps).

29 Disconnect the electrical connector from the pump.

30 Disconnect the hoses from the pump.

31 Remove the three pump mounting bolts, bushings and spacers and remove the pump from its mounting bracket.

32 Installation is the reverse of removal.

Air pressure sensor/air switching valve/air pump assembly

33 If you're only removing the air pump to service something below it, you can remove the air pressure sensor, air switching valve and air pump as a single assembly.

34 Disconnect the electrical connectors from the air pressure sensor, the air switching valve and the air pump.

35 Disconnect all hoses from the air pressure sensor, the air switching valve and the air pump.

36 Remove the air pump assembly mounting bracket bolts and nuts and remove the air pressure sensor, air switching valve and air pump as a single assembly.

37 Installation is the reverse of removal.

20 Variable Valve Timing-intelligent (VVT-i) system - description and component replacement

Description

1 The Variable Valve Timing (VVT-i) system is used on all models. The VVT-i system varies intake and exhaust camshaft timing within an approximate range of 50 degrees (V6 models) or 44 degrees (V8 models) to produce valve timing that is optimized for the driving conditions. The VVT-i system achieves this by using engine oil pressure to advance or retard the controller on the front end the camshafts. **Note:** *2010 and earlier V6 models and all 2UZ-FE V8 engines only have controllers on the intake camshaft; 2011 and later V6 models and all 1UR-FE, 3UR-FE and 3UR-FBE V8 models have them on intake and exhaust camshafts.*

2 The VVT-i system consists of either two CMP/VVT-i sensors (2010 and earlier V6 models and all 2UZ-FE V8 engines), or four sensors (all 1URFE, 3UR-FE and 3UR-FBE models), the Powertrain Control Module (PCM), either two VVT-i oil control valves and two controllers (2010 and earlier V6 models and all 2UZ-FE V8 engines), or four timing oil control valves and four controllers (all 2011 and later 1GR-FE V6, 1UR-FE, 3UR-FE and 3UR-FBE engines). The VVT-i systems on V6 and V8 models are virtually identical.

3 Each controller consists of a housing, four vanes and a lock pin. Oil pressure directed into the housing from the advance or retard side of the intake cam causes the vane to rotate in relation to the intake camshaft, advancing or retarding the valve timing. The higher the oil pressure (or flow), the more the actuator assembly will rotate, thereby advancing or retarding the camshaft. When the engine is turned off, the intake cam is in its most retarded position to ensure easy starting. When the engine is first started and no oil pressure has yet been applied to the controller housing, the lock pin locks the housing and vane together to prevent it from making a knocking sound. When oil pressure enters the housing and is applied to the lock pin spring, the spring is compressed and the lock pin retracts, allowing the housing to rotate in relation to the four vanes.

4 When oil is applied to the advance side of the vane(s), the actuator advances the camshaft in a clockwise direction. When oil is applied to the retard side of the vanes, the actuator rotates the camshaft counterclockwise back to 0 degrees advance, which is the normal position of the actuator during engine operation under no-load or idle conditions. The PCM can also send a signal to the timing oil control valve to stop oil flow to both (advance and retard) passages to hold camshaft advance in its current position. Under light engine loads, the VVT system retards the camshaft timing to decrease valve overlap and stabilize engine output. Under medium engine loads, the VVT system advances the camshaft timing to increase valve overlap, thereby increasing fuel economy and decreasing exhaust emissions. Under heavy engine loads at low RPM, the VVT system advances the camshaft timing to help close the intake valve faster, which improves low to mid-range torque. Under heavy engine loads at high RPM, the VVT system retards the camshaft timing to slow the closing of the intake valve to improve engine horsepower.

5 The camshaft timing oil control valve is a PCM-controlled device that controls and directs the flow of oil to the advance or retard passages leading to the controller. There are two oil control valves, one for each intake camshaft. A spring-loaded spool valve inside the oil control valve directs oil pumped into the valve toward either the advance outlet port or the retard outlet port, depending on the engine speed, which determines oil pressure. When the engine is turned off, the spring is extended and the spool valve is in its most retarded state. Once the engine is started and oil pressure begins to rise with engine rpm, the spool is slowly pushed against its spring.

Component replacement

VVT sensors

Warning: *Wait until the engine is completely cool before beginning this procedure.*

Note: *On 2010 and earlier V6 models, there are two CMP/VVT sensors. The sensors are located on the inside walls of the cylinder heads, adjacent to the timing rotor installed on the front of each Variable Valve Timing-intelligent (VVT-i) controller. Toyota refers to these sensors as "CMP/VVT" sensors on earlier models, and simply as VVT sensors on later models. If you're buying a new sensor*

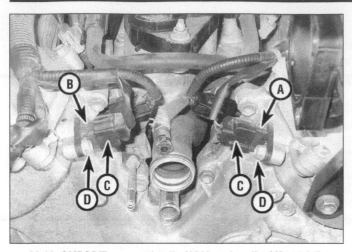

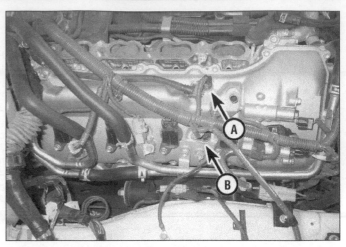

20.11 CMP/VVT sensor details (2010 and earlier V6 model):

A   Left CMP/VVT sensor   C   Electrical connector
B   Right CMP/VVT sensor   D   Mounting bolt

20.17a Location of the intake camshaft VVT sensor (A) and the exhaust camshaft VVT sensor (B) - right side shown, left side similar

at a Toyota parts department, use the Toyota terminology. On 2011 and later V6 models, Camshaft Position (CMP) sensors are used instead of the VVT sensor (see Section 5).
Note: There are two VVT sensors on 2UZ-FE V8 models. The sensors are located on the intake manifold sides of the valve covers. There are four VVT sensors used on all other V8 models.
6   Disconnect the cable from the negative battery terminal.
7   Remove the engine cover (see Chapter 1, **illustration 24.12a)** and on 2UZ-FE engines, remove the throttle body cover (see Chapter 4).
8   On V6 models, remove the air filter housing (see Chapter 4).
9   On V8 models, remove the air intake duct (see Chapter 4).

V6 models

Refer to illustration 20.11

10   If you're going to remove the CMP/VVT sensor from the left cylinder head, drain the engine coolant (see Chapter 1) and discon-

nect the two coolant bypass hoses from the throttle body.
Note: As an alternative to draining the coolant, the hoses can be clamped off using locking pliers.
11   Disconnect the electrical connector from the CMP/VVT sensor **(see illustration)**.
12   Remove the CMP/VVT sensor mounting bolt and remove the sensor (see illustration 20.11).
13   Installation is the reverse of removal.

2UZ-FE engines

14   Remove the intake manifold (see Chapter 2B).
15   Remove the CMP/VVT sensor mounting bolt and remove the sensor.
16   Installation is the reverse of removal.

1GR-FE V6, 1UR-FE, 3UR-FE and 3UR-FBE engines

Refer to illustrations 20.17a and 20.17b

17   Remove the VVT sensor mounting bolt, disconnect the electrical connector and remove the sensor **(see illustrations)**.
18   Installation is the reverse of removal.

Camshaft timing oil control valve

19   Remove the engine cover (see Chapter 1, **illustration 24.12a)**.
20   On 2UZ-FE engines, remove the throttle body cover.
21   Remove the air filter housing and the air intake duct (see Chapter 4).

2010 and earlier V6 models

Refer to illustration 20.23

22   Remove the wiper motor (see Chapter 12) and the cowl (see Chapter 11).
23   Disconnect the electrical connector from the camshaft timing oil control valve **(see illustration)**.
24   Remove the throttle body (see Chapter 4).
25   Remove the mounting bolt from the camshaft timing oil control valve, and remove the valve. Discard the O-ring.
26   Installation is the reverse of removal.

All other models

Refer to illustrations 20.27a and 20.27b

27   Disconnect the electrical connector to

20.17b Disconnect the VVT sensor connector (A), then remove the mounting bolt (B)

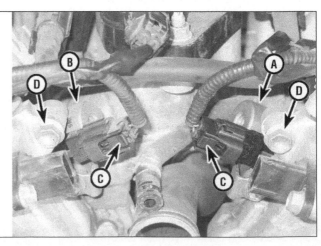

20.23 Camshaft timing oil control valve details (2010 and earlier V6 models):

A   Left camshaft timing oil control valve
B   Right camshaft timing oil control valve
C   Electrical connector
D   Mounting bolt

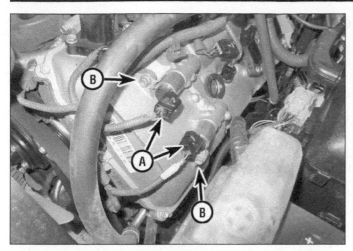

20.27a Disconnect the camshaft timing oil control valve electrical connector (A), then remove the mounting bolt

20.27b Slide the valve out of the valve cover and discard the O-ring - left side shown, right side identical

the camshaft timing oil control valve, remove the mounting bolt and the valve from the valve cover. Discard the O-ring **(see illustrations)**.
28   Installation is the reverse of removal.

21 Exhaust Gas Recirculation (EGR) system (1UR-FE engine) - component replacement

Warning: *The engine must be completely cool before beginning this procedure.*
1   Disconnect the cable from the negative terminal of the battery (see Chapter 5).
2   Remove the engine cover **(see Chapter 1, illustration 24.12a)**.
3   Drain the cooling system (see Chapter 1).

EGR valve
4   Disconnect the coolant hoses near the EGR valve that would interfere with removal.
5   Disconnect the electrical connector from the valve.
6   Remove the EGR valve mounting bolts and the four nuts from the EGR valve pipes.
7   Remove the EGR valve and retrieve the gaskets.
8   Remove the fasteners and detach the pipes from the EGR valve.
9   Clean all gasket mating surfaces thoroughly.
10   Installation is the reverse of removal, using new gaskets.
11   Refill the cooling system (see Chapter 1).

EGR valve cooler
12   Remove the intake manifold (see Chapter 2B).
13   Remove the coolant bypass pipe from the cylinder block valley.
14   Disconnect the hoses from the EGR cooler.
15   Unscrew the mounting fasteners and remove the EGR cooler assembly.
16   If necessary, remove the bolts and detach the pipe from the EGR valve cooler.
17   Installation is the reverse of removal, using new gaskets.
18   Refill the cooling system (see Chapter 1).

Notes

Chapter 7 Part A
Automatic transmission

Contents

Specifications

General

Transmission fluid type and capacity.. See Chapter 1

Torque specifications

Ft-lbs (unless otherwise indicated)   **Nm**

Note: *One foot-pound (ft-lb) of torque is equivalent to 12 inch-pounds (in-lbs) of torque. Torque values below approximately 15 ft-lbs are expressed in inch-pounds, because most foot-pound torque wrenches are not accurate at these smaller values.*

	Ft-lbs	Nm
Driveplate-to-torque converter bolts		
2014 and earlier models	35	47
2015 and later models	39	53
Transmission-to-engine bolts		
14 mm bolt head	52	71
17 mm bolt head	27	37
Valve body bolts	86 in-lbs	9.5
Crossmember-to-frame bolts	81	110
Transmission mount		
Mount-to-transmission	30	40
Mount-to-crossmember		
2010 and earlier models	144 in-lbs	16
2011 and later models	15	20

1   General information

All models are equipped with either a five-speed or six-speed automatic transmission. 2007 through 2009 vehicles are equipped with four different transmissions depending on the engine used: five-speed units designated as A750E (2WD) used with the 1GR-FE engines, or the A750F (4WD) used with 2UZ-FE engines; six-speed units designated as AB60E (2WD) used with 3UR-FE engines, or the AB60F (4WD) also used with 3UR-FE engines.

2010 and later vehicles are only equipped with six-speed units: designated as A760E (2WD) used with the 1GR-FE engines, the A760F (4WD) used with 2UZ-FE engines, the AB60E (2WD) used with 3UR-FE engines or the AB60F (4WD) also used with 3UR-FE engines.

All models are equipped with a lock-up torque converter, known as a Torque Converter Clutch (TCC). The clutch provides a direct connection between the engine and the drive wheels for improved efficiency and fuel economy.

Due to the complexity of the clutches and the hydraulic control system, and because of the special tools and expertise needed to overhaul an automatic transmission, diagnosis and repair of the transmission must be handled by a dealer service department or a transmission repair shop. The procedures in this Chapter are limited to general diagnosis, routine maintenance, adjustment and transmission removal and installation. However, even though the repair work must be done by a transmission specialist, you can save money by removing and installing the transmission yourself. You can also check and adjust the shift cable, replace the extension housing seal and check and replace the Park/Neutral position switch.

Caution: *If a vehicle with an automatic transmission is disabled, do NOT tow it at speeds greater than 30 mph or distances over 50 miles.*

2   Diagnosis - general

Note: *Automatic transmission malfunctions may be caused by five general conditions: poor engine performance, improper adjustments, hydraulic malfunctions, mechanical malfunctions or malfunctions in the computer or its signal network. Diagnosis of these problems should always begin with a check of the easily repaired items: fluid level and condition (see Chapter 1), shift linkage adjustment and throttle linkage adjustment. Next, perform a road test to determine if the problem has been corrected or if more diagnosis is necessary. If the problem persists after the preliminary tests and corrections are completed, additional diagnosis should be done by a dealer service department or transmission repair shop. Refer to the "Troubleshooting" Section at the front of this manual for information on symptoms of transmission problems.*

Preliminary checks

1   Drive the vehicle to warm the transmission to normal operating temperature.

2   Check the fluid level (see Chapter 1):

a) *If the fluid level is unusually low, add enough fluid to bring the level within the designated area of the dipstick, then check for external leaks (see below).*

b) *If the fluid level is abnormally high, drain off the excess, then check the drained fluid for contamination by coolant. The presence of engine coolant in the automatic transmission fluid indicates that a failure has occurred in the internal radiator walls that separate the coolant from the transmission fluid (see Chapter 3).*

c) *If the fluid is foaming, drain it and refill the transmission, then check for coolant in the fluid or a high fluid level.*

3   Check the engine idle speed.

Note: *If the engine is malfunctioning, do not proceed with the preliminary checks until it has been repaired and runs normally.*

4   Inspect the shift cable (see Section 4). Make sure it's properly adjusted and operates smoothly.

Fluid leak diagnosis

5   Most fluid leaks are easy to locate visually. Repair usually consists of replacing a seal or gasket. If a leak is difficult to find, the following procedure may help.

6   Identify the fluid. Make sure it's transmission fluid and not engine oil or brake fluid (automatic transmission fluid is a deep red color).

7   Try to pinpoint the source of the leak. Drive the vehicle several miles, then park it over a large sheet of cardboard. After a minute or two, you should be able to locate the leak by determining the source of the fluid dripping onto the cardboard.

8   Make a careful visual inspection of the suspected component and the area immediately around it. Pay particular attention to gasket mating surfaces. A mirror is often helpful for finding leaks in areas that are hard to see.

9   If the leak still cannot be found, clean the suspected area thoroughly with a degreaser or solvent, then dry it.

10   Drive the vehicle for several miles at normal operating temperature and varying speeds. After driving the vehicle, visually inspect the suspected component again.

11   Once the leak has been located, the cause must be determined before it can be properly repaired. If a gasket is replaced but the sealing flange is bent, the new gasket will not stop the leak. The bent flange must be straightened.

12   Before attempting to repair a leak, check to make sure the following conditions are corrected or they may cause another leak.

Note: *Some of the following conditions cannot be fixed without highly specialized tools and expertise. Such problems must be referred to a transmission repair shop or a dealer service department.*

Gasket leaks

13   Check the pan periodically. Make sure the bolts are tight, no bolts are missing, the gasket is in good condition and the pan is flat (dents in the pan may indicate damage to the valve body inside).

14   If the pan gasket is leaking, the fluid level or the fluid pressure may be too high, the vent may be plugged, the pan bolts may be too tight, the pan sealing flange may be warped, the sealing surface of the transmission housing may be damaged, the gasket may be damaged or the transmission casting may be cracked or porous. If sealant instead of gasket material has been used to form a seal between the pan and the transmission housing, it may be the wrong sealant.

Seal leaks

15   If a transmission seal is leaking, the fluid level or pressure may be too high, the vent may be plugged, the seal bore may be damaged, the seal itself may be damaged or improperly installed, the surface of the shaft protruding through the seal may be damaged or a loose bearing may be causing excessive shaft movement.

16   Make sure the dipstick tube seal is in good condition and the tube is properly seated. Periodically check the area around the speedometer gear or sensor for leakage. If transmission fluid is evident, check the O-ring for damage.

Case leaks

17   If the case itself appears to be leaking, the casting is porous and will have to be repaired or replaced.

18   Make sure the oil cooler hose fittings are tight and in good condition.

Fluid comes out vent pipe or fill tube

19   If this condition occurs, the transmission is overfilled, there is coolant in the fluid, the case is porous, the dipstick is incorrect, the vent is plugged or the drain back holes are plugged.

3   Output shaft oil seal (2WD) - replacement

Refer to illustrations 3.4 and 3.5

1   Oil leaks frequently occur due to wear of the extension housing oil seal. Replacement of this seal is relatively easy, since the repair can usually be performed without removing the transmission from the vehicle.

2   The extension housing oil seal is located at the extreme rear of the transmission, where the driveshaft is attached. Raise the vehicle and support it securely on jackstands. If the seal is leaking, transmission lubricant will be built up on the front of the driveshaft and may be dripping from the rear of the transmission.

3   Remove the driveshaft (see Chapter 8).

4   Using a puller or seal removal tool, carefully remove the oil seal out of the rear of the transmission **(see illustration)**. Do not damage the splines on the transmission output shaft.

5   Using a large section of pipe or a very large deep socket as a drift, install the new oil seal **(see illustration)**. Drive it into the bore squarely and make sure it's completely seated.

6   Lubricate the splines of the transmission output shaft and the outside of the driveshaft slip yoke with light-weight grease, then install the driveshaft. Be careful not to damage the lip of the new seal.

7   Check the lubricant level in the transmission, adding as necessary (see Chapter 1).

4   Shift cable - removal, installation and adjustment

Warning: *These models are equipped with airbags. Always disable the airbag system before working in the vicinity of any airbag system component to avoid the possibility of accidental deployment of the airbag(s), which could cause personal injury (see Chapter 12).*

1   Disconnect the negative battery cable from the battery (see Chapter 5).

2   Remove the cowl trim panel and the kick panel from the driver's side (see Chapter 11).

3   Remove the upper and lower steering column covers and the knee bolster trim panel (see Chapter 11).

Upper cable end
Column mount

Refer to illustration 4.7

4   Remove the steering wheel (see Chapter 10).

5   Remove the front seat and center console (see Chapter 11).

6   Pull the carpeting back from the firewall and remove the two cable housing mounting retainers and access plate bolts from the interior.

7   Place the shift lever into the "N" (neutral) position, then detach the upper end of the shift cable from the steering column shift lever **(see illustration)**.

8   Disengage the cable from the cable bracket.

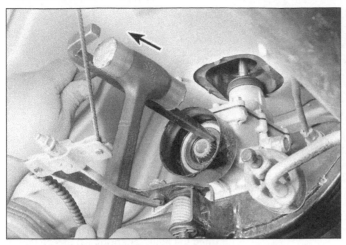

3.4 Using a seal removal tool to remove the seal

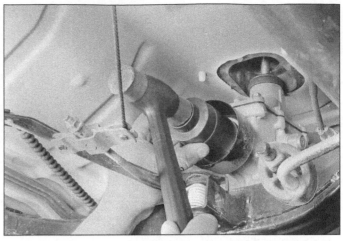

3.5 To install the new seal, tap it into place with a large socket and hammer

Floor mount

Refer to illustration 4.11

9   Remove the center console and shifter console (see Chapter 11).

10   Pull the carpeting back from the firewall and remove the two cable housing mounting retainers and access plate bolts from the interior.

11   Place the shift lever into the "N" (neutral) position, then detach the upper end of the shift cable from the shift lever **(see illustration)**.

12   Remove the cable retaining clip from the center housing.

Lower cable end

Refer to illustration 4.14

13   Raise the vehicle and place it securely on jackstands.

Warning: *On Sequoia models equipped with rear height control suspension, adjust the height control to the NORMAL mode,* turn OFF the height control, then turn off the engine BEFORE raising the vehicle.

14   Remove the nut that attaches the shift cable to the manual lever and disconnect the shift cable from the manual lever **(see illustration)**.

15   Remove the retainer clip and disengage the cable from the cable bracket. Remove the cable housing bracket bolt from under the vehicle.

16   Pull the cable through the floor and remove it.

17   Installation for both ends of the cable is the reverse of removal. Make sure the grommet is properly seated. Check and adjust the cable.

Adjustment

Refer to illustrations 4.19, 4.20 and 4.21

18   Place the shift lever into the "N" neutral position.

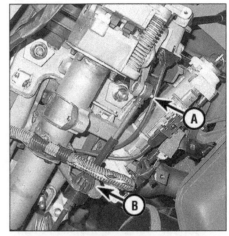

4.7 Pop the upper end of the shift cable from the ball socket on shifter lever (A), then squeeze the tabs on the cable casing and detach the cable from the column (B)

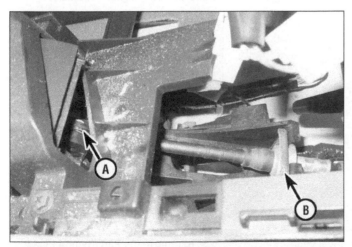

4.11 Pop the upper end of the shift cable from the ball socket on shifter arm (A), then remove the clip on the cable casing and lift the cable from the center housing (B)

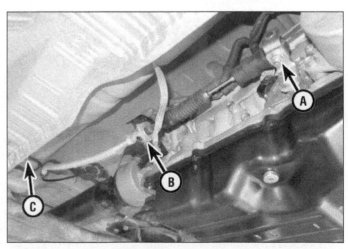

4.14 Remove the nut that attaches the shift cable (A) to the shifter lever, the retainer clip (B) and disengage the cable from the cable bracket and the bolt (C) that mount the cable housing to the body

4.19 Pull the boot back to expose the adjuster

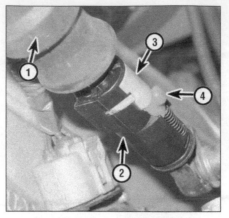

4.20 Shift cable adjuster details

1   *Boot*
2   *Shift cable adjuster case*
3   *Lock piece*
4   *Slider*

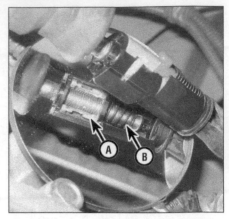

4.21 Press the lock piece into place (A) once the adjuster spring (B) has centered the cable in the housing

19   Pull the cable adjuster case boot back **(see illustration)**.
20   Push and hold the slider away from the lock piece, then press the lock piece out from the back side to release the lock **(see illustration)**.
21   Allow the adjuster case's spring to apply enough tension to center the cable, press the lock piece into the case, and release the slider to set the lock **(see illustration)**.
22   Shift the shift lever from the "N" (neutral) position through all of the gear ranges.
23   The lever should shift smoothly and the position indicator should line up with the shift lever position indicator. If the indicator and shift lever position do not match, the shift cable must be adjusted again.

5   Park/Neutral Position (PNP) switch/Transmission Range (TR) sensor - removal, installation and adjustment

Removal and installation

Refer to illustration 5.3, 5.4 and 5.5
1   The Park/Neutral Position (PNP) switch

prevents the engine from starting in any gear other than Park or Neutral. If the engine starts in any position other than Park or Neutral, it's either out of adjustment or defective.
2   Raise the vehicle and place it securely on jackstands.
Warning: *On Sequoia models equipped with rear height control suspension, adjust the height control to the NORMAL mode, turn OFF the height control, then turn off the engine BEFORE raising the vehicle.*
3   Remove the transmission control cable mounting nut from the shaft lever and remove the cable **(see illustration)**.
4   Unplug the electrical connector from the PNP switch, then remove the shaft lever nut, washer and lever **(see illustration)**.
5   Use a screwdriver to bend back the locking tabs and remove the locknut **(see illustration)**.
6   Remove the PNP switch retaining bolt **(see illustration 5.5)** and remove the switch.
7   Installation is the reverse of removal. Check and, if necessary, adjust the switch.

Adjustment

Refer to illustration 5.10
8   Raise the vehicle and place it securely on jackstands.
Warning: *On Sequoia models equipped with rear height control suspension, adjust the height control to the NORMAL mode, turn OFF the height control, then turn off the engine BEFORE raising the vehicle.*
9   Loosen the switch retaining bolt **(see illustration 5.5)** and place the switch in the N (neutral) position.
10   Align the groove and neutral basic line **(see illustration)**.
11   Holding the switch in this position, tighten the switch retaining bolt.

6   Transmission oil cooler - removal and installation

Warning: *Wait until the engine is completely cool before beginning this procedure.*

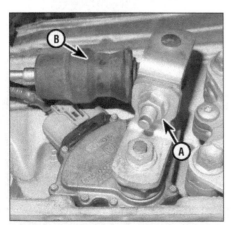

5.3 Remove the control cable mounting nut (A) from the shaft lever and remove the cable (B)

5.4 Unplug the electrical connector (A), then remove the shaft lever nut (B), washer and lever

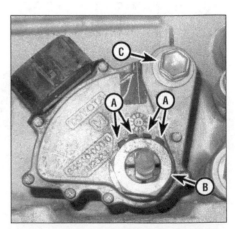

5.5 To remove the Park/Neutral Position switch, pry back the lock washer tabs (A) and remove the nut (B), then remove the retaining bolt (C)

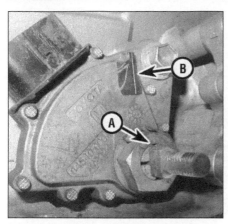

5.10 Align the groove (A) with the neutral basic line (B)

Models without a trailer towing system

Refer to illustration 6.5

1   Disconnect the negative battery cable from the battery.

2   Raise the vehicle and place it securely on jackstands.

Warning: *On Sequoia models equipped with rear height control suspension, adjust the height control to the NORMAL mode, turn OFF the height control, then turn off the engine BEFORE raising the vehicle.*

3   Drain the cooling system (see Chapter 1).

4   If equipped, remove the heat shield fastener and heat shield.

5   Place a drain pan under the oil cooler and disconnect the water-bypass hoses from the cooler **(see illustration)**, disconnect the pipes and allow the remaining coolant to drain.

6   Disconnect the two transmission oil cooler hoses **(see illustration 6.5)**.

7   Unscrew the three mounting bolts and remove the cooler **(see illustration 6.5)**.

8   Discard and replace the two oil cooler-to-oil cooler spacer O-rings.

9   Installation is the reverse of removal. Tighten all cooler clamps and lines securely.

10   Refill the cooling system (see Chapter 1).

11   Check the automatic transmission fluid level and add some, if necessary (see Chapter 1).

Models with a trailer towing system

12   Remove the front bumper cover and grille (see Chapter 12). Disconnect the hoses to the cooler from the engine side.

13   Unscrew the three mounting bolts and remove the cooler.

14   Installation is the reverse of removal. Tighten all cooler fasteners and cooler lines securely.

15   Check the automatic transmission fluid level and add some, if necessary (see Chapter 1).

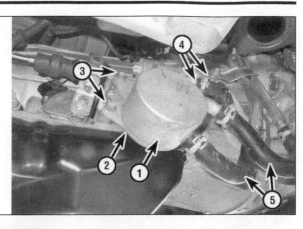

6.5 Typical transmission cooler details

1   *Transmission oil cooler*
2   *Spacer*
3   *Oil cooler mounting bolts (two of three shown)*
4   *Water-bypass hoses*
5   *Transmission oil cooler hoses*

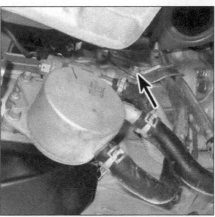

7.5 Remove the transmission cooler lines bracket bolt

7   Automatic transmission - removal and installation

Warning: *Wait until the engine is completely cool before beginning this procedure.*

Removal

Refer to illustrations 7.5, 7.13 and 7.18

1   Disconnect the negative cable from the battery.

2   Raise the vehicle and place it securely on jackstands.

Warning: *On Sequoia models equipped with rear height control suspension, adjust the height control to the NORMAL mode, turn OFF the height control, then turn off the engine BEFORE raising the vehicle.*

3   Remove any engine/transmission undercovers that are in the way.

4   Drain the engine coolant (see Chapter 1).

5   Remove the transmission cooler line bracket bolts, detach all brackets and clamps and disconnect the two cooler lines from the transmission **(see illustration)**.

6   Disconnect the driveshaft(s) (see Chapter 8).

7   Remove the front and center exhaust pipes (see Chapter 4).

8   Disconnect the shift cable from the transmission (see Section 4).

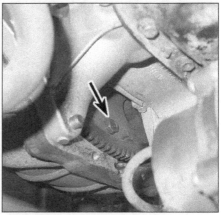

7.13 Remove the six driveplate-to-torque converter bolts by turning the crankshaft (in a clockwise direction only, viewed from the front) for access to each bolt – 4WD model shown

9   Drain the transmission fluid (see Chapter 1), then reinstall the pan.

10   Remove the starter and disconnect the wiring harness from the transmission (see Chapter 5).

11   On 4WD models, remove the torque converter access cover.

12   Mark the torque converter and the driveplate with a scribe or chalk so they can be installed in the same position.

13   Remove the six driveplate-to-torque converter bolts **(see illustration)**. Turn the crankshaft (in a clockwise direction only, viewed from the front) for access to each bolt.

Note: *On 2WD models, the access for torque converter bolts is through the starter opening.*

14   Remove the ground strap mounting bolt and strap from the transmission.

15   Support the transmission with a jack - preferably a jack made for this purpose. Safety chains will help steady the transmission on the jack.

16   Support the engine with a jack. Use a block of wood under the oil pan to spread the load.

17   Remove the bolts securing the transmission mount to the frame crossmember or, on 4WD models, the transfer case mount to the crossmember.

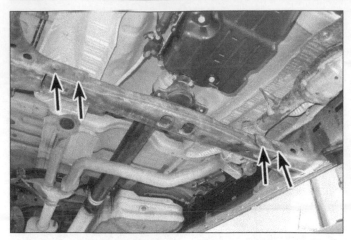

7.18 Remove the four transmission mount-to-crossmember bolts

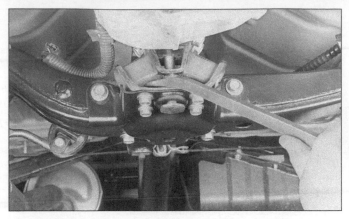

8.1 To check the transmission mount, insert a prybar or large screwdriver between the mount rubber and the mount bracket, then try to pry the transmission up off its mount; if the transmission moves significantly, replace the mount

18   Raise the transmission enough to allow removal of the crossmember, then remove the nuts/bolts securing the crossmember to the frame side rails. Remove the crossmember and the mount **(see illustration)**.

19   Remove the transmission-to-engine bolts.

20   Slowly lower the jack until you can remove the upper bolts securing the transmission to the engine. Several long extensions may have to be used to reach the upper bolts.

21   Unplug the electrical connectors from the transmission, then detach all wire harnesses from the transmission and set them aside.

22   Move the transmission to the rear to disengage it from the engine block dowel pins and make sure the torque converter is detached from the driveplate. Secure the torque converter to the transmission so it won't fall out during removal.

23   On 4WD models, once the transmission in out of the vehicle, remove the transfer case mounting bolts and separate the transfer case from the transmission.

Installation

24   Prior to installation, make sure the torque converter hub is securely engaged in the front pump of the transmission. This can be confirmed by pushing in on the converter and turning it; if it isn't seated completely it will drop into place as this is done.

25   On 4WD models, mount the transfer case to the transmission and tighten the bolts (see Chapter 7B).

26   With the transmission secured to the jack, raise it into position. Keep it level so the torque converter doesn't slide out.

27   Turn the torque converter until the marks on the converter and driveplate are aligned.

28   Move the transmission forward carefully until the dowel pins engage with the holes in the bellhousing.

29   Install the transmission-to-engine bolts. Tighten them to the torque listed in this Chapter's Specifications.

30   Install the driveplate-to-torque converter

bolts and tighten them to the torque listed in this Chapter's Specifications.
Note: *Install all of the bolts before tightening any of them.*

31   Install the crossmember and tighten the bolts and nuts securely.

32   Lower the transmission extension housing until the mount is seated on, and aligned with, the crossmember, then tighten the fasteners securely.

33   The remainder of installation is the reverse of removal.

34   Remove all jacks supporting the transmission and engine and lower the vehicle.

35   Fill the transmission with the specified fluid (Chapter 1), run the engine and check for fluid leaks.

---

8   Transmission mount - check and replacement

Check

Refer to illustration 8.1

1   Insert a large screwdriver or prybar into the space between the transmission and the crossmember and try to pry the transmission up slightly **(see illustration)**.

2   The transmission should not move away from the insulator much. If there is any separation of the rubber, the mount is worn out.

Replacement

Refer to illustration 8.3

3   Support the transmission with a jack and remove the transmission mount-to-crossmember retaining bolts **(see illustration)**.

4   Raise the transmission slightly with the jack and remove the mount-to-transmission fasteners.
Note: *On 4WD models, the mount is attached to the transfer case.*

5   Installation is the reverse of the removal procedure. Tighten the fasteners to the torque listed in this Chapter's Specifications.

9   Automatic transmission overhaul - general information

In the event of a fault occurring, it will be necessary to establish whether the fault is electrical, mechanical or hydraulic in nature, before repair work can be contemplated. Diagnosis requires detailed knowledge of the transmission's operation and construction, as well as access to specialized test equipment, and so is deemed to be beyond the scope of this manual. It is therefore essential that problems with the automatic transmission are referred to a dealer service department or other qualified repair facility for assessment.

Note that a faulty transmission should not be removed before the vehicle has been assessed by a knowledgeable technician equipped with the proper tools, as troubleshooting must be performed with the transmission installed in the vehicle.

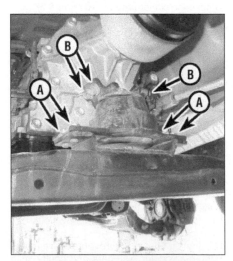

8.3 Location of the transmission mount-to-crossmember retaining bolts (A) and mount-to-transmission bolts (B) – (2WD model shown, 4WD model similar)

Chapter 7 Part B
Transfer case

Contents

Specifications

General

Transfer case - lubricant type .... See Chapter 1

Torque specifications

	Ft-lbs (unless otherwise indicated)	Nm

Note: *One foot-pound (ft-lb) of torque is equivalent to 12 inch-pounds (in-lbs) of torque. Torque values below approximately 15 foot-pounds are expressed in inch-pounds, because most foot-pound torque wrenches are not accurate at these smaller values.*

	Ft-lbs	Nm
Companion flange nut (front or rear)		
Sequoia models		
2016 and earlier models	94	127
2017 and later models	82	112
Tundra models	94	127
Transfer case-to-transmission bolts	30	40

1 General information

Four-wheel drive (4WD) models are equipped with a transfer case mounted on the rear of the transmission. Drive is transmitted from the engine, through the transmission and the transfer case to the front and rear axles by driveshafts.

The transfer case shift actuator shifts the transfer case into 4WD "LOW" and "HI" gear. The shift actuator cannot be removed from the transfer case without disassembling the transfer case.

We don't recommend trying to rebuild a transfer case at home. It's difficult to overhaul without special tools, and rebuilt units are available for less than it would cost to rebuild your own. However, there are a number of components that you can check, adjust and/or replace - and those are the items covered in this Chapter.

2 Oil seals - replacement

Refer to illustrations 2.3a, 2.3b, 2.4, 2.5 and 2.7

Note 1: *This procedure applies to both the front and rear seals. Although the accompanying photos show the transfer case out of the vehicle, it's not necessary to remove the transfer case to replace a seal.*

Note 2: *On WF1AM units (used on all 2014 and later Tundra 4WD models) the front companion flange oil seal can only be replaced by disassembling the transfer case (the front companion flange is held in place by an internal snap-ring). We recommend this repair be done by a transmission repair shop.*

1 Raise the vehicle and support it securely on jackstands.

Warning: *On Sequoia models equipped*

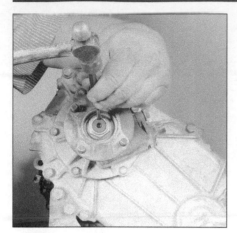

2.3a Unstake the companion flange retaining nut . . .

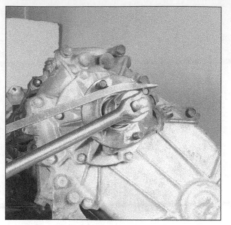

2.3b . . . then break the nut loose while holding the flange as shown

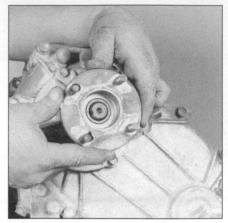

2.4 Pull the companion flange off the output shaft - it may be necessary to use a small puller

with rear height control suspension, adjust the height control to the NORMAL mode, turn OFF the height control, then turn off the engine BEFORE raising the vehicle.

2   If you're replacing the front seal, remove the front driveshaft; if you're replacing the rear seal, remove the rear driveshaft (see Chapter 8).

3   Unstake and remove the companion flange retaining nut **(see illustrations)** then remove and discard the O-ring.

4   Remove the companion flange **(see illustration)**. If the flange is difficult to remove from the output shaft, use a puller.

5   Pry out the seal with a screwdriver or a seal removal tool **(see illustration)**. Don't damage the seal bore.

6   Lubricate the new seal lip with multi-purpose grease.

7   Drive the seal into place with a large socket **(see illustration)**. The outside diameter of the socket should be slightly smaller than the outside diameter of the seal.

8   Install the companion flange, place the O-ring over the shaft and install the flange nut.

9   Tighten the companion flange nut to the torque listed in this Chapter's Specifications.

10   The remainder of installation is the reverse of removal.

---

3   4WD Touch Shift Selector System - description and control switch replacement

Description

1   The 4WD Shift Selector System allows the driver to switch the settings of the front differential and transfer case by selecting 2WD, High 4WD or Low 4WD with a 2-4 selector switch located on the dashboard. The system consists of the Electronic Control Unit (ECU), ADD Actuator (see Chapter 8), dash-mounted control switch, the actuator assembly, the 4WD detection switch, the Neutral position switch, the limit switch, and the 4WD indicator light (located on the instrument cluster, see Chapter 12).

Control switch

2   Remove the instrument cluster trim panel (see Chapter 11).

3   Push the three plastic claws inward and remove the switch from the panel.

4   Installation is the reverse of removal.

---

4   Transfer case - removal and installation

Removal

Warning: *On Sequoia models equipped with rear height control suspension, adjust the height control to the NORMAL mode, turn OFF the height control, then turn off the engine BEFORE raising the vehicle.*

1   Remove the transmission and transfer case as an assembly (see Chapter 7A).

2   Drain the transfer case lubricant (see Chapter 1).

3   Stand the transmission up on the bell housing end.

4   Remove the eight bolts securing the transfer case to the transfer case adapter

5   Lift the transfer case up until the transfer case is clear of the transmission output shaft splines. Keep the transfer case level as this is

2.5 Pry out the seal with a screwdriver or a seal removal tool

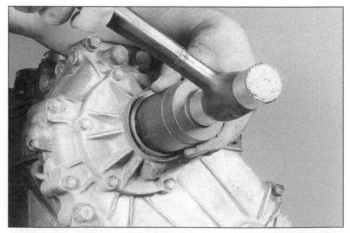

2.7 Drive the seal into place with a seal installer tool or a large socket

done. Once the input shaft is clear, it can be rotated and removed.

6 Installation is the reverse of removal. Tighten the transfer case-to-transfer transmission bolts to the torque listed in this Chapter's Specifications.

7 Install the transmission/transfer case assembly (see Chapter 7A).

8 Fill the transfer case with the specified fluid (Chapter 1), drive the vehicle and check for fluid leaks.

5 Transfer case overhaul - general information

1 Overhauling a transfer case is a difficult job for the do-it-yourselfer. It involves the disassembly and reassembly of many small parts. Numerous clearances must be precisely measured and, if necessary, changed with select-fit spacers and snap-rings. As a result, if transfer case problems arise, it can be removed and installed by a competent do-it-yourselfer, but overhaul should be left to a transmission repair shop. Rebuilt transfer cases may be available - check with your dealer parts department and auto parts stores. At any rate, the time and money involved in an overhaul is almost sure to exceed the cost of a rebuilt unit.

2 Nevertheless, it's not impossible for an inexperienced mechanic to rebuild a transfer case if the special tools are available and the job is done in a deliberate step-by-step manner so nothing is overlooked.

3 The tools necessary for an overhaul include internal and external snap-ring pliers, a bearing puller, a slide hammer, a set of pin punches, a dial indicator and possibly a hydraulic press. In addition, a large, sturdy workbench and a vise or transfer case stand will be required.

4 During disassembly of the transfer case, make careful notes of how each piece comes off, where it fits in relation to other pieces and what holds it in place. Noting how they are installed when you remove the parts will make it much easier to get the transfer case back together.

5 Before taking the transfer case apart for repair, it will help if you have some idea what area of the transfer case is malfunctioning. Certain problems can be closely tied to specific areas in the transfer case, which can make component examination and replacement easier. Refer to the *Troubleshooting* Section at the front of this manual for information regarding possible sources of trouble.

Notes

Chapter 8
Driveline

Contents

Specifications

General

Front driveaxle length
  Tundra .... 24.657 inches (62.63 cm)
  Sequoia .... 24.64 inches (62.59 cm)
Rear driveaxle length (Sequoia)
  LH side .... 31.67 inches (80.44 cm)
  RH side .... 33.16 inches (84.23 cm)

Torque specifications

Ft-lbs (unless otherwise indicated)   **Nm**

Note: *One foot-pound (ft-lb) of torque is equivalent to 12 inch-pounds (in-lbs) of torque. Torque values below approximately 15 ft-lbs are expressed in inch-pounds, because most foot-pound torque wrenches are not accurate at these smaller values.*

Driveshaft

Flange bolts/nuts
  Front (4WD) 59   80
  Rear
    2WD 52   70
    4WD 52   70
Center support bearing bolts 30   40
Center support bearing yoke or flange nut 94   128

Torque specifications (continued)   **Ft-lbs** (unless otherwise indicated)   **Nm**

Note: *One foot-pound (ft-lb) of torque is equivalent to 12 inch-pounds (in-lbs) of torque. Torque values below approximately 15 ft-lbs are expressed in inch-pounds, because most foot-pound torque wrenches are not accurate at these smaller values.*

Front driveaxle (4WD models)

Driveaxle hub nut	249	338

Rear driveaxle (Sequoia models)

Driveaxle hub nut	251	340

Front differential

Differential carrier		
2012 and earlier models	89	120
2013 and later models	83	113
ADD actuator	16	21

Rear differential (Sequoia models)

Front mounting bolts	89	120
Rear mounting bolts	148	200

Rear axle

Pinion bearing preload (1GR-FE, 2UZ-FE and 1UR-FE engines)		
Old bearing	8 to 12 in-lbs	0.9 to 1
New bearing	10 to 14 in-lbs	1 to 1.5
Brake backing plate fasteners		
Tundra	44	60
Sequoia	64	86
Pinion nut torque (1GR-FE, 2UZ-FE and 1UR-FE engines)	325	441

1   General information

The Sections in this Chapter deal with the components from the rear of the engine to the rear wheels (except for the transmission and transfer case, which are dealt with in Chapter 7) and forward to the front wheels on four-wheel drive (4WD) models. In this Chapter, the components are grouped into two categories: driveshaft(s) and axle(s). Separate Sections within this Chapter cover checks and repair procedures for components in each of these two groups.

Since nearly all these procedures involve working under the vehicle, make sure it's safely supported on sturdy jackstands or a hoist where the vehicle can be safely raised and lowered.

2   Driveshaft(s) and universal joints - general information

1   A driveshaft is a tube, or a pair of tubes, that transmits power between the transmission (or transfer case on 4WD models) and the differential. Universal joints are located at either end of the driveshaft; a third U-joint is employed just behind the center on two-piece driveshafts. The driveshaft is attached to the rear differential by a companion flange; on 4WD models, the front driveshaft is attached to the front differential the same way.

2   Driveshafts on 2WD models employ a splined yoke, known as a slip yoke or sleeve yoke, at the front, which slips into the extension housing of the transmission. This arrangement allows the driveshaft to slide back-and-forth within the transmission during vehicle operation. An oil seal prevents leakage of fluid at this point and keeps dirt from entering the transmission. If leakage is evident at the front of the driveshaft, replace the oil seal (see Chapter 7A).

3   On 4WD models, each driveshaft is attached to the transfer case by a companion flange. Once a front or rear driveshaft has been removed, either companion flange can be removed from the transfer case to replace the companion seal(s) (each companion flange uses two seals: one seal between the companion flange and the transfer case, and another, smaller, seal inside the companion flange itself). Refer to Chapter 7B for the transfer case seal replacement procedure.

4   On two-piece driveshafts, center bearings support the driveline. The center bearing is a ball-type bearing mounted in a rubber cushion attached to a frame crossmember. The bearing is pre-lubricated and sealed at the factory. On two-piece driveshafts, a sleeve yoke is employed in the rear driveshaft section.

5   The driveshaft assembly requires periodic lubrication. See Chapter 1 for the lubrication procedure and maintenance interval.

6   Since the driveshaft is a balanced unit, it's important that no undercoating, mud, etc. be allowed to stay on it. When the vehicle is raised for service it's a good idea to clean the driveshaft and inspect it for any obvious damage. Also, make sure the small weights used to originally balance the driveshaft are in place and securely attached. Whenever the driveshaft is removed it must be reinstalled in the same relative position to preserve the balance.

7   Problems with the driveshaft are usually indicated by a noise or vibration while driving the vehicle. A road test should verify if the problem is the driveshaft or another vehicle component. Refer to the *Troubleshooting* Section at the front of this manual. If you suspect trouble, inspect the driveline (see Section 3).

3   Driveline inspection

1   Raise the rear of the vehicle and support it securely on jackstands. Block the front wheels to keep the vehicle from rolling off the stands.

Warning: *On Sequoia models equipped with rear height control suspension, adjust the height control to the NORMAL mode, turn OFF the height control, then turn off the engine BEFORE raising the vehicle.*

2   Crawl under the vehicle and visually inspect the driveshaft. Look for any dents or cracks in the tubing. If any are found, the driveshaft must be replaced.

3   Check for oil leakage at the front and rear of the driveshaft. Leakage where the driveshaft enters the transmission or transfer case indicates a defective transmission/transfer case seal (see Chapter 7). Leakage where the driveshaft joins the differential indicates a defective pinion seal (see Section 9).

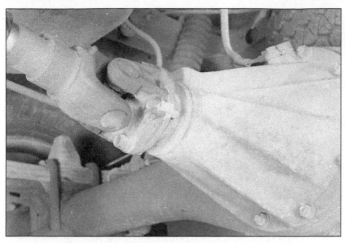

4.2 Mark the relationship of the driveshaft U-joint to the pinion flange

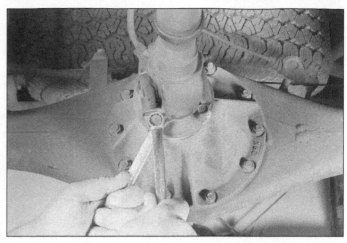

4.3 Using a backup wrench to hold each bolt, break loose all four nuts securing the flange yoke to the differential

4   While under the vehicle, have an assistant rotate a rear wheel so the driveshaft will rotate. As it does, make sure the universal joints are operating properly without binding, noise or looseness. Listen for any noise from the center bearing (if equipped), indicating it's worn or damaged. Also check the rubber portion of the center bearing for cracking or separation, which will necessitate replacement.

5   The universal joint can also be checked with the driveshaft motionless, by gripping your hands on either side of the joint and attempting to twist the joint. Any movement at all in the joint is a sign of considerable wear. Lifting up on the shaft will also indicate movement in the universal joints.

6   Finally, check the driveshaft mounting bolts at the ends to make sure they're tight.

7   On 4WD models, the above driveshaft checks should be repeated on all driveshafts. In addition, check for grease leakage around the sleeve yoke, indicating failure of the yoke seal.

8   Check for leakage where the driveshafts connect to the transfer case and front differential. Leakage indicates worn oil seals.

9   At the same time, check for looseness in the joints of the front driveaxles. Also check for grease or oil leakage from around the driveaxles by inspecting the rubber boots and both ends of each axle. Oil leakage at the differential junction indicates a defective side oil seal. Leakage at the wheel side indicates a defective front hub seal, while leakage at the boots means a damaged rubber boot. For servicing of these components, see the appropriate Sections.

---

4   Driveshaft - removal and installation

1   Raise the vehicle and support it securely on jackstands. Place the transmission in Neutral with the parking brake off. Block the front wheels to prevent the vehicle from rolling.

Warning: *On Sequoia models equipped with rear height control suspension, adjust the height control to the NORMAL mode, turn OFF the height control, then turn off the engine BEFORE raising the vehicle.*

Removal

Refer to illustrations 4.2, 4.3 and 4.4

2   Using a scribe, a hammer and punch, or paint, make marks on the driveshaft and the differential flange in line with each other **(see illustration)**. This is to make sure the driveshaft is reinstalled in the same position to preserve the balance.

3   Remove the bolts securing the flange yoke to the rear differential **(see illustration)**. Turn the driveshaft (or wheels) as necessary to bring the bolts into the most accessible position.

4   On vehicles with a two-piece driveshaft, remove the center bearing protector and remove the bolts, nuts and washers from the center support bearing bracket **(see illustration)**.

5   Lower the rear of the driveshaft. Slide the front of the driveshaft out of the transmis-

sion or transfer case or, if equipped with a companion flange, separate the flange at the transfer case.

6   On 2WD models, wrap a plastic bag over the transmission extension housing and hold it in place with a rubber band. This will prevent loss of fluid and protect against contamination while the driveshaft is out.

Installation

7   Remove the plastic bag from the transmission or transfer case and wipe the area clean. Inspect the oil seal carefully. Slide the front of the driveshaft into the transmission (2WD models) or bolt the U-joint flange yoke to the companion flange, installing the fasteners finger-tight (4WD models).

8   Raise the center bearing (if equipped) into place and screw the retaining bolts in a few turns. Raise the rear of the driveshaft into position, checking to be sure the marks are in alignment. If not, turn the rear wheels to match the pinion flange and the driveshaft.

9   Tighten all nuts to the torque listed in this Chapter's Specifications. Remove the jackstands and lower the vehicle.

4.4 Mark the relationship of the driveshaft to the yoke, remove the two mounting bolts, and detach the center bearing (two-piece driveshafts)

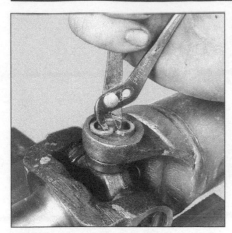

5.3 Outer type snap-rings can be removed with a small pair of pliers

5.4 To remove the U-joint from the driveshaft, use a vise as a press; the small socket will push the cross and bearing cup into the large socket

5.5 Grip the bearing cup with locking pliers and remove it from the yoke

5   Universal joints - replacement

Refer to illustration 5.3, 5.4 and 5.5

Warning: *On Sequoia models equipped with rear height control suspension, adjust the height control to the NORMAL mode, turn OFF the height control, then turn off the engine BEFORE raising the vehicle.*

Note: *A press or large vise will be required for this procedure. It may be a good idea to take the driveshaft to a repair or machine shop where the U-joints can be replaced for you, usually at a reasonable charge.*

1   Remove the driveshaft (see Section 4).
2   Place the driveshaft on a bench equipped with a vise.
3   Mark the shaft and yoke for proper reassembly, then remove the snap-rings from the U-joint **(see illustration)**.

Note: *The driveshafts covered in this manual are equipped with U-joints that use either outer snap-rings or inner snap-rings to retain the bearing cap. Outer snap-rings can be removed using a pair of pliers while inner snap-rings require a punch and a hammer to dislodge the snap-ring from the groove on the bearing cap.*

4   Place a piece of pipe or a large socket with the same inside diameter over one of the bearing cups. Position a socket which is of slightly smaller diameter than the cup on the opposite bearing cup **(see illustration)** and use the vise to force the cup out (inside the pipe or large socket), stopping just before it comes completely out of the yoke.
5   Use the vise or large pliers to work the cup the rest of the way out **(see illustration)**.
6   Transfer the sockets to the other side and press the opposite bearing cup out in the same manner.
7   After the bearing cups have been removed, lift the U-joint from the yoke and thoroughly clean all dirt and debris from the yokes on both ends of the driveshaft. Remove any metal burrs from the yoke bores.

8   Pack the new U-joint bearing cups with grease, this will allow the needle bearings to be held in place while your installing the bearing cups. Ordinarily, specific instructions for lubrication will be included with the U-joint servicing kit and should be followed carefully.
9   Position the U-joint body in the yoke and partially install one bearing cup in the yoke. If the U-joint is equipped with a grease fitting, be sure it points in the same direction as the grease fitting on the opposite end of the driveshaft.
10   Start the U-joint body into the bearing cup and partially install the other cup. Align the U-joint body between the bearing cups and press the bearing cups into position, being careful not to damage the dust seals.
11   Install the snap-rings. If difficulty is encountered in seating the snap-rings, strike the driveshaft yoke sharply with a hammer. This will spring the yoke ears slightly and allow the snap-rings to seat in the groove. This should also be done to center the U-joint after assembly.

Note: *If you still have difficulty seating the snap-rings, one of the small needle bearings may have become stuck between the bearing cap and the end of the spider. Disassemble and inspect the joint.*

12   Install the driveshaft (see Section 4).
13   If the U-joint is equipped with a grease fitting, lubricate it (see Chapter 1).
14   Remove the jackstands and lower the vehicle.

6   Center bearing - removal and installation

1   Raise the vehicle and support it securely on jackstands.

Warning: *On Sequoia models equipped with rear height control suspension, adjust the height control to the NORMAL mode, turn OFF the height control, then turn off the engine BEFORE raising the vehicle.*

2   Remove the driveshaft (see Section 4).
3   Clamp the driveshaft securely into a bench vise.
4   Separate the intermediate (front) part of the driveshaft from the rear part by marking the relationship of the center U-joint flange yoke and disassembling the U-joint (see Section 5).
5   Unstake the center yoke retaining nut and remove it.
6   Mark the relationship of the intermediate shaft to the yoke.
7   Remove the yoke from the intermediate shaft.
8   Remove the center bearing from the intermediate shaft.

Note: *This usually requires a puller or a punch and a hammer to separate the center bearing from the intermediate shaft.*

9   Holding the center bearing assembly in one hand, turn the bearing with the other hand and verify that it operates freely and smoothly. If it's stiff or noisy, replace it.
10   Installation is the reverse of removal. Tighten the yoke retaining nut to the torque listed in this Chapter's Specifications, then stake it.

7   Axles - description and check

Description

1   The rear axle assembly is a hypoid, semi-floating type (the centerline of the pinion gear is below the centerline of the ring gear). When the vehicle goes around a corner, the differential allows the outer rear tire to turn more quickly than the inner tire. The axleshafts are splined to the differential side gears, so when the vehicle goes around a corner, the inner tire, which turns more slowly than the outer tire, turns its side gear more slowly than the outer tire turns its side gear. The differential pinion gears roll around the slower side gear, driving the outer side gear - and tire - more quickly. The differential is housed within a

8.6 To detach the axleshaft from the rear axle housing, remove the four backing plate nuts

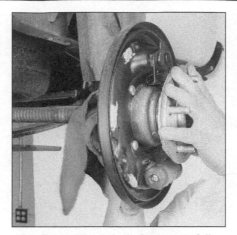

8.7 Extract the axleshaft very carefully from the axle housing, especially if you don't want to replace the axleshaft seal

8.8 Remove this O-ring from the rear axle housing; discard the old O-ring and install a new one before installing the axleshaft

casting with a pressed steel cover, known as the carrier. The steel axle tubes are pressed into and welded to the carrier.

2   A locking limited-slip rear axle is used on some models. This differential allows for normal operation until one wheel loses traction. A limited-slip unit is similar in design to a conventional differential, except for the addition of a pair of clutch cones which slow the rotation of the differential case when one wheel is on a firm surface and the other on a slippery one. The difference in wheel rotational speed produced by this condition applies additional force to the pinion gears and through the cone, which is splined to the axleshafts, equalizes the rotation speed of the axleshaft driving the wheel with traction.

3   On Sequoia models, a fully independent rear axle assembly is used. This consists of a differential and a pair of driveaxles. Each driveaxle has an inner and outer constant velocity (CV) joint.

Check

4   Often, a suspected axle problem lies elsewhere. Do a thorough check of other possible causes before assuming the axle is the problem.

5   The following noises are those commonly associated with axle diagnosis procedures:

a) *Road noise is often mistaken for mechanical faults. Driving the vehicle on different surfaces will show whether the road surface is the cause of the noise. Road noise will remain the same if the vehicle is under power or coasting.*

b) *Tire noise is sometimes mistaken for mechanical problems. Tires which are worn or low on pressure are particularly susceptible to emitting vibrations and noises. Tire noise will remain about the same during varying driving situations, where axle noise will change during coasting, acceleration, etc.*

c) *Engine and transmission noise can be deceiving because it will travel along the driveline. To isolate engine and transmission noises, make a note of the engine speed at which the noise is most pronounced. Stop the vehicle and place the transmission in Neutral and run the engine to the same speed. If the noise is the same, the axle is not at fault.*

6   Because of the special tools needed, overhauling the differential isn't cost effective for a do-it-yourselfer. The procedures included in this Chapter describe axleshaft removal and installation, axleshaft oil seal replacement, axleshaft bearing replacement and removal of the entire unit for repair or replacement. Any further work should be left to a dealer service department or other qualified repair shop.

Note: *If the rear axle must be replaced, refer to the identification code and manufacturer's code stamped on the front side of the right rear axle tube. This number contains information on the rear axle ratio, differential type, manufacturer and build date information, all of which are necessary to ensure that you get the right axle.*

8   Axleshaft, bearing and oil seals (rear, Tundra models) - removal and installation

Refer to illustrations 8.6, 8.7, 8.8 and 8.9

1   Release the parking brake. Raise the rear of the vehicle, support it securely on jackstands and block the front wheels. Remove the wheel and, if equipped, the brake drum.

2   If the vehicle is equipped with rear disc brakes, remove the caliper and disc (see Chapter 9).

3   If the vehicle is equipped with ABS, remove the ABS sensor (see Chapter 9).

4   Remove the brake assembly (see Chapter 9).

8.9 Use a seal removal tool or a big screwdriver to pry out the old axleshaft seal; use a seal installer or a big socket to install the new seal

5   Disconnect the parking brake cable and the hydraulic brake line to the wheel cylinder (see Chapter 9).

6   Remove the four backing plate mounting nuts **(see illustration)**.

7   Pull the axleshaft out of the rear axle housing **(see illustration)**.

8   Remove the O-ring from the rear axle housing **(see illustration)**.

9   Remove the axleshaft inner oil seal from the axle housing with a seal removal tool or a big screwdriver **(see illustration)**.

10   Further disassembly of the axleshaft assembly requires special tools and a hydraulic press. If the axleshaft, bearing or outer oil seal needs to be replaced, take the axleshaft assembly to an automotive machine shop.

11   Drive a new axleshaft inner seal into the end of the axle tube with a seal installer or a big socket. Coat the lip of the seal with clean oil or multi-purpose grease.

12   Install a new axle housing O-ring. Apply a light coat of oil to the new O-ring.

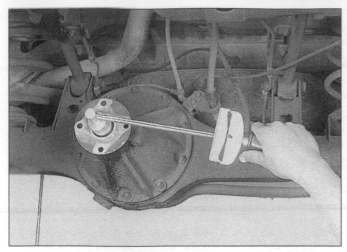

9.3 Use an inch-pound torque wrench to check the torque necessary to rotate the pinion shaft

9.4 Mark the relative positions of the pinion, nut and flange before removing the nut

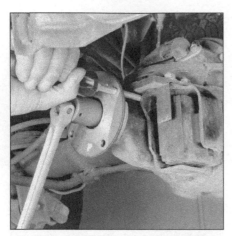

9.6 If you don't have a flange holding tool, lock the flange by jamming a large screwdriver through a bolt hole in the flange and wedge it underneath a bracket as shown, or under a reinforcement rib on the differential carrier

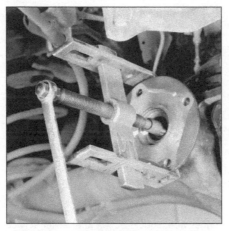

9.8 If you can't pull off the pinion flange by hand, remove it with a small puller

9.9 Pry out the old pinion seal with a seal removal tool or a big screwdriver or tap it out with a small punch

13   Make sure the axleshaft is clean and there are no burrs or metal splinters on it. Deburr any surface irregularities so the axleshaft doesn't damage the seal during installation. Lightly coat the axleshaft with clean oil, then insert it into the axle housing. Make sure the splined inner end of the axleshaft doesn't damage the lip of the new axleshaft seal.

14   Installation is the reverse of removal. Tighten the four backing plate mounting nuts to the torque listed in this Chapter's Specifications.

9   Differential pinion seal - replacement

Refer to illustration 9.3

Warning: *On Sequoia models equipped with rear height control suspension, adjust the height control to the NORMAL mode,* turn OFF the height control, then turn off the engine BEFORE raising the vehicle.

Note: *This procedure applies to the rear pinion seal on all vehicles as well as the front pinion seal on 4WD models.*

1   Loosen the wheel lug nuts, raise the vehicle and support it securely on jackstands. Block the wheels at the opposite end to keep the vehicle from rolling off the stands. Remove the wheels (this will allow you to obtain a more accurate pinion shaft preload reading).

2   Disconnect the driveshaft from the differential (see Section 4) and fasten it out of the way.

3   Use an inch-pound torque wrench to check the torque required to rotate the pinion **(see illustration)**. Record it for use later.

3UR-FE and 3UR-FBE engines

Refer to illustrations 9.4, 9.6, 9.8, 9.9 and 9.10

4   Scribe or punch alignment marks on the pinion shaft, nut and flange **(see illustration)**.

5   Count the number of threads visible between the end of the nut and the end of the pinion shaft and jot it down for later use.

6   A special flange holding tool is the best way to keep the companion flange from moving while the pinion nut is loosened. If you're unable to obtain a flange holding tool, immobilize the flange by inserting a big screwdriver through one of the U-joint bolt holes in the flange and wedge it against a bracket **(see illustration)** or reinforcement rib on the differential carrier.

7   Remove the pinion nut.

8   Withdraw the companion flange. It may be necessary to use a puller to draw it out **(see illustration)**. Do NOT attempt to pry behind the flange or hammer on the flange or the end of the pinion shaft.

9   Pry out the old seal **(see illustration)** and discard it.

10   Lubricate the lips of the new seal with high-temperature grease and tap it evenly into position with a seal installation tool or a large socket. Make sure it enters the housing

squarely and is tapped in to its full depth **(see illustration)**.

11 Align the mating marks made before disassembly and install the companion flange. If necessary, tighten the pinion nut to draw the flange into place. Do not try to hammer the flange into position.

12 Apply non-hardening sealant to the ends of the splines visible in the center of the flange so oil will be sealed in.

13 Install the washer (if equipped) and pinion nut. Tighten the nut carefully, until the original number of threads are exposed and the marks made in Step 4 are aligned.

14 Measure the torque required to rotate the pinion and tighten the nut in small increments until it matches the figure recorded in Step 3. In order to compensate for the drag of the new oil seal, the nut should be tightened more until the rotational torque of the pinion slightly exceeds what was recorded earlier, but not by more than 5 in-lbs.

15 Connect the driveshaft, install the wheels and lower the vehicle. Tighten the lug nuts to the torque listed in the Chapter 1 Specifications.

1GR-FE, 2UZ-FE and 1UR-FE engines

16 Scribe or punch alignment marks on the pinion shaft, nut and flange **(see illustration 9.4)**.

17 Using a hammer and chisel, unstake the pinion nut.

18 A special flange holding tool is the best way to keep the companion flange from moving while the pinion nut is loosened. If you're unable to obtain a flange holding tool, immobilize the flange by inserting a big screwdriver through one of the U-joint bolt holes in the flange and wedge it against a bracket **(see illustration 9.6)** or reinforcement rib on the differential carrier.

19 Remove the pinion nut.

20 Withdraw the companion flange. It may be necessary to use a puller to draw it out **(see illustration 9.8)**. Do NOT attempt to pry behind the flange or hammer on the flange, or the end of the pinion shaft.

21 Pry out the old seal and discard it.

22 Remove the oil slinger then use a two jaw puller mounted to the outer edge of the front pinion shaft bearing inner race and draw the bearing out.

23 Reverse the two jaw puller and draw out the outer race from the housing.

24 Using needle-nose pliers remove the drive pinion bearing spacer or crush sleeve and discard it.

25 Install a new drive pinion bearing spacer on to the pinion and slide it in as far as it will go.

Note: *If either the front pinion bearing or race was damaged while being removed from the housing, they must be replaced as a pair.*

26 Slide the front bearing outer race onto the pinion shaft and use a pipe or deep socket to install the race.

27 Slide the front inner race and bearing

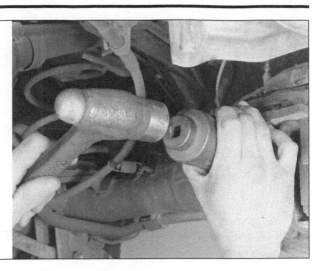

9.10 Lubricate the lips of the new pinion seal and seat it squarely in the bore, then drive it into the carrier with a seal driver or a large socket

onto the pinion shaft and use a pipe or deep socket to install into the outer race.

28 Install the oil slinger on to the pinion shaft.

29 Lubricate the lips of the new seal with high-temperature grease and tap it evenly into position with a seal installation tool or a large socket. Make sure it enters the housing squarely and is tapped in to its full depth **(see illustration 9.10)**.

30 Align the mating marks made before disassembly and install the companion flange. If necessary, tighten the pinion nut to draw the flange into place. Do not try to hammer the flange into position.

31 Apply non-hardening sealant to the ends of the splines visible in the center of the flange so oil will be sealed in.

32 Install the washer (if equipped) and a NEW pinion nut.

33 Tighten the nut carefully, to the torque listed in this Chapters Specifications.

34 Measure the preload or torque required to rotate the pinion and tighten the nut in small increments until it matches the figure recorded in Step 3. In order to compensate for the drag of the new oil seal, the nut should be tightened more until the rotational torque of the pinion slightly exceeds what was recorded earlier, but not by more than 5 in-lbs. Stake the collar of the nut.

Note: *If a new bearing has been installed, see the torque listed in this Chapters Specifications for the correct preload.*

35 Connect the driveshaft, install the wheels and lower the vehicle. Tighten the lug nuts to the torque listed in the Chapter 1 Specifications.

10 Axle (rear) - removal and installation

Warning: *On Sequoia models equipped with rear height control suspension, adjust the height control to the NORMAL mode, turn OFF the height control, then turn off the engine BEFORE raising the vehicle.*

1 Loosen the rear wheel lug nuts, raise the rear of the vehicle and support it securely on

jackstands placed under the frame (not under the axle). Block the front wheels to keep the vehicle from rolling off the stands. Remove the rear wheels.

2 Position a jack under the rear axle differential housing.

3 Disconnect the driveshaft from the differential (see Section 4). Fasten the driveshaft out of the way with a piece of wire from the underbody.

Tundra models

4 Disconnect the load sensing proportioning and bypass valve (LSP & BV) height sensing spring from the axle (see Chapter 9).

5 Detach all brake hoses and/or lines from the axle housing, then plug them to prevent fluid leakage.

6 Remove the rear brake assemblies (see Chapter 9).

7 Disconnect the parking brake cables from the brake assemblies and detach the cables from the rear axle housing (see Chapter 9).

8 Detach the vent hose from the axle housing.

9 Disconnect the shock absorbers from the spring seats or from the axle brackets (see Chapter 10).

10 Disconnect the leaf spring U-bolts and remove the spacers, bumpers and spring seats (see Chapter 10).

Sequoia models

11 Remove the rear driveaxles (see Section 11).

12 Remove the rear differential mounting fasteners (two at the rear and four at the front).

All models

13 Lower the jack under the differential, then remove the rear axle assembly/differential from under the vehicle.

14 Installation is the reverse of removal. Tighten all suspension fasteners to the torque listed in the Chapter 10 Specifications. On Sequoia models, tighten the four front mounting bolts first (rearmost bolts first, then the for-

11.2 Using a hammer and chisel, tap the grease cap from the hub

ward bolts), followed by the two rear mounting bolts (lower bolt first, then the upper bolt) to the torque listed in this Chapter's Specifications.

15   On Tundra models, bleed the brakes (see Chapter 9).

11   Driveaxle - removal and installation

Warning: *On Sequoia models equipped with rear height control suspension, adjust the height control to the NORMAL mode, turn OFF the height control, then turn off the engine BEFORE raising the vehicle.*

Front

Refer to illustrations 11.2, 11.3a and 11.3b

1   Loosen the front wheel lug nuts, raise the front of the vehicle and support it securely on jackstands. Block the rear wheels to keep the vehicle from rolling off the stands. Remove the wheels. Drain the differential (see Chapter 1).

2   Remove the grease cap **(see illustration).**

3   Remove the cotter pin and nut lock, place a prybar or large screwdriver between the wheel studs to hold the driveaxle and break the driveaxle/hub nut loose with a large breaker bar, or have an assistant apply the brakes. Remove the nut.

4   Disconnect the lower control arm from the balljoint (see Chapter 10). If you're removing the left driveaxle, remove the left side shock absorber (see Chapter 10).

5   Knock the driveaxle loose from the steering knuckle with a *brass* drift and hammer. Do NOT use a steel punch or strike the end of the driveaxle with a steel hammer; a steel punch or hammer will damage the threads or the splines on the end of the driveaxle.

6   Swing the steering knuckle outward and pull the driveaxle assembly out of the steering knuckle, then detach the driveaxle from the differential. If you're removing the right driveaxle, tap the inner CV joint out of the differential with a hammer and a brass drift; if you're removing the left driveaxle, a slide hammer with a special hooked adapter (available at most auto parts stores) will be needed to pull the inner CV joint from the differential.

7   Installation is the reverse of removal. Tighten the driveaxle/hub nut to the torque listed in this Chapter's Specifications, then install the nut lock and a new cotter pin. Tighten the lug nuts to the torque listed in the Chapter 1 Specifications. Tighten all suspension fasteners to the torque listed in the Chapter 10 Specifications.

Rear (Sequoia models)

8   Loosen the rear wheel lug nuts, raise the rear of the vehicle and support it securely on jackstands. Block the front wheels to keep the vehicle from rolling off the stands. Remove the wheels. Drain the differential (see Chapter 1).

9   Remove the grease cap.

10   Remove the cotter pin and nut lock, place a prybar or large screwdriver between the wheel studs to hold the driveaxle and break the driveaxle/hub nut loose with a large breaker bar, or have an assistant apply the brakes. Remove the nut.

11   Knock the driveaxle loose from the knuckle with a *brass* drift and hammer. Do NOT use a steel punch or strike the end of the driveaxle with a steel hammer; a steel punch or hammer will damage the threads or the splines on the end of the driveaxle.

12   Disconnect and remove the rear knuckle assembly (see Chapter 10).

13   Swing the knuckle outward and pull the driveaxle assembly out of the steering knuckle, then detach the driveaxle from the differential. Attach a slide hammer with a special hooked adapter (available at most auto parts stores) will be needed to pull the inner CV joint from the differential.

14   Installation is the reverse of removal. Tighten the driveaxle/hub nut to the torque listed in this Chapter's Specifications, then install the nut lock and a new cotter pin. Tighten the lug nuts to the torque listed in the Chapter 1 Specifications. Tighten all suspension fasteners to the torque listed in the Chapter 10 Specifications.

11.3a Remove the cotter pin and the nut lock

11.3b Place a prybar between two of the wheel studs, then loosen the driveaxle/hub nut

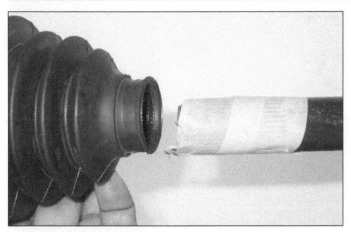

12.6 Wrap the splined area of the axleshaft with tape to prevent damage to the boots when removing or installing them

12.10 Equalize the pressure inside the boot by inserting a small, dull screwdriver between the boot and the outer race

12   Driveaxle boot - replacement

Warning: *On Sequoia models equipped with rear height control suspension, adjust the height control to the NORMAL mode, turn OFF the height control, then turn off the engine BEFORE raising the vehicle.*
1   Remove the driveaxle (see Section 11).

Disassembly

Refer to illustration 12.6
2   Mount the driveaxle in a vise with wood lined jaws (to prevent damage to the axleshaft). Check the CV joint for excessive play in the radial direction, which indicates worn parts. Check for smooth operation throughout the full range of motion for each CV joint. If a boot is torn, disassemble the joint, clean the components and inspect for damage due to loss of lubrication and possible contamination by foreign matter.
3   Using a pair of pliers, squeeze the retaining tabs of the large inboard joint boot clamp and slide it away from the CV joint. Using a pair of side cutters cut the boot clamps. Old and worn boots can be cut off.
4   Mark the inboard joint to the shaft to ensure that they are reassembled properly.
5   Using a pair of snap-ring pliers, expand the snap-ring that connects the outboard joint shaft to the inboard joint and separate the shafts.
6   If you haven't already cut them off, remove both boots. Wrap the splines on the inner end the axleshaft with electrical or duct tape to protect the boots from the sharp edges of the splines **(see illustration).**

Check

7   Thoroughly clean all components, including the outer CV joint assembly, with solvent until the old CV joint grease is completely removed. Inspect all visible bearing surfaces for cracks, pitting, scoring and other signs of wear. If the inner CV joint is worn, you can buy a new inner CV joint and install it on the old axleshaft; if the outer CV joint is worn, you'll have to purchase a new outer CV joint and axleshaft (they're sold pre-assembled).

Reassembly

Refer to illustration 12.10
8   Slide the clamps and boot(s) onto the axleshaft. Pack the outboard and inboard joint assemblies and boots with grease. Align the matchmarks and place the inboard joint on the shaft, then using a pair of snap-ring pliers, expand the snap-ring and seat the joint on the shaft.
9   Slide the boot into place, making sure both ends seat in their grooves. Adjust the length of the driveaxle to the dimension in this Chapter's Specifications.
10   Equalize the pressure in the boot, then tighten and secure the boot clamps **(see illustration).**
11   Install the driveaxle assembly (see Section 11).

13   Automatic Disconnecting Differential (ADD) (4WD models) - description, removal and installation

Description

1   The Automatic Disconnecting Differential (ADD) connects the power flow through the right axleshaft when 4WD mode is selected, and disconnects the power flow when 2WD mode is selected. Although ADD-equipped vehicles make selecting 2WD or 4WD more convenient (there are no locking hubs to deal with), they also increase wear on the CV joints and dust boots, as well as some of the axle and differential components, which rotate all the time, even in 2WD. If shifting into or out of 4WD becomes a problem, have the ADD system checked out by a dealer service department or other qualified repair shop that specializes in 4WD vehicles.

Removal and installation
Differential carrier
2   Loosen the wheel lug nuts, raise the vehicle and support it securely on jackstands. Remove the wheels.
Warning: *On Sequoia models equipped with rear height control suspension, adjust the height control to the NORMAL mode, turn OFF the height control, then turn off the engine BEFORE raising the vehicle.*
3   Remove the driveaxles (see Section 11).
4   Remove the engine under cover (see Chapter 1).
5   Drain the lubricant from the differential (see Chapter 1).
6   Disconnect the driveshaft from the front differential (see Section 4) and support the front end of the driveshaft with a piece of wire.
7   Disconnect the breather hose, detach the fasteners that retain the vacuum tubing bracket to the differential, unplug the actuator electrical connector and detach the actuator vacuum hoses. Remove the tube and wire harness assembly from the differential.
8   Support the differential with a transmission jack or a floor jack.
9   Remove the differential rear mounting bolts (three on the left rear, two on the right rear) and the two differential front mounting bolts, and lower the differential.
10   Installation is the reverse of removal. Refill the differential with the proper lubricant (see Chapter 1) and tighten the lug nuts to the torque listed in the Chapter 1 Specifications.

ADD actuator
11   Disconnect the actuator hose and electrical connector.
12   Remove the four retaining bolts.
13   Remove the actuator.
14   Installation is the reverse of removal. Before installing the actuator, remove the old RTV sealant from the mating surfaces of the differential and the actuator and apply a thin bead of new RTV sealant to those surfaces. Tighten the actuator bolts to the torque listed in this Chapter's Specifications.

Notes

Chapter 9 Brakes

Contents

Specifications

General
Brake fluid type	DOT 3 brake fluid

Disc brakes
Brake pad minimum lining thickness	See Chapter 1
Disc lateral runout limit	
Front	0.002 inch (0.051 mm)
Rear	0.008 inch (0.203 mm)
Disc minimum (discard) thickness	Refer to the dimension cast into the disc

Parking brake
Brake shoe minimum lining thickness	See Chapter 1
Maximum drum diameter	Cast into outside of the disc/drum
Parking brake pedal travel	6 to 9 clicks

Torque specifications   Ft-lbs (unless otherwise indicated)   Nm

Note: *One foot-pound (ft-lb) of torque is equivalent to 12 inch-pounds (in-lbs) of torque. Torque values below approximately 15 ft-lbs are expressed in inch-pounds, since most foot-pound torque wrenches are not accurate at these smaller values.*

	Ft-lbs (unless otherwise indicated)	Nm
ABS wheel speed sensor bolt		
Front	96 in-lbs	11
Rear	73 in-lbs	8
Brake line-to-rear caliper banjo bolt	22	30
Caliper mounting bolts		
Front		
2015 and earlier models	73	99
2016 and later models*	133	180
Rear	65	88
Caliper mounting bracket bolts (rear)		
2015 and earlier models	70	95
2016 and later models*		
Sequoia	118	160
Tundra	123	167
Master cylinder mounting nuts	18	25
Power brake booster-to-firewall nuts	132 in-lbs	15
Wheel lug nuts	See Chapter 1	

*Use new bolts

1   General information

The vehicles covered by this manual are equipped with hydraulically operated front and rear brake systems. The front and rear brakes are disc type. Both the front and rear brakes are self-adjusting. The disc brakes automatically compensate for pad wear.

Hydraulic system

The hydraulic system consists of two separate circuits. The master cylinder has separate reservoirs for the two circuits, and, in the event of a leak or failure in one hydraulic circuit, the other circuit will remain operative.

Power brake booster

The power brake booster, utilizing engine manifold vacuum and atmospheric pressure to provide assistance to the hydraulically operated brakes, is mounted on the firewall in the engine compartment.

Parking brake

The parking brake operates the rear brakes only, through cable actuation. It's activated by either a foot pedal mounted on the driver's side with a release handle or an electronic parking brake that has a switch in the center console. The switch takes the place of the manual foot pedal and release handle and activates a control module that operates the cable.

Service

After completing any operation involving disassembly of any part of the brake system, always test drive the vehicle to check for proper braking performance before resuming normal driving. When testing the brakes, perform the tests on a clean, dry, flat surface. Conditions other than these can lead to inaccurate test results.

Test the brakes at various speeds with both light and heavy pedal pressure. The vehicle should stop evenly without pulling to one side or the other.

Tires, vehicle load and wheel alignment are factors which also affect braking performance.

Precautions

There are some general cautions and warnings involving the brake system on this vehicle:

a) *Use only brake fluid conforming to DOT 3 specifications.*

b) *The brake pads contain fibers which are hazardous to your health if inhaled. Whenever you work on brake system components, clean all parts with brake system cleaner. Do not allow the fine dust to become airborne. Also, wear an approved filtering mask.*

c) *Safety should be paramount whenever any servicing of the brake components is performed. Do not use parts or fasteners which are not in perfect condition, and be sure that all clearances and torque specifications are adhered to. If you are at all unsure about a certain procedure, seek professional advice. Upon completion of any brake system work, test the brakes carefully in a controlled area before putting the vehicle into normal service. If a problem is suspected in the brake system, don't drive the vehicle until it's fixed.*

2 Troubleshooting

PROBABLE CAUSE	CORRECTIVE ACTION

No brakes - pedal travels to floor

1 Low fluid level 2 Air in system	1 and 2 Low fluid level and air in the system are symptoms of another problem - a leak somewhere in the hydraulic system. Locate and repair the leak
3 Defective seals in master cylinder	3 Replace master cylinder
4 Fluid overheated and vaporized due to heavy braking	4 Bleed hydraulic system (temporary fix). Replace brake fluid (proper fix)

Brake pedal slowly travels to floor under braking or at a stop

1 Defective seals in master cylinder	1 Replace master cylinder
2 Leak in a hose, line, caliper or wheel cylinder	2 Locate and repair leak
3 Air in hydraulic system	3 Bleed the system, inspect system for a leak

Brake pedal feels spongy when depressed

1 Air in hydraulic system	1 Bleed the system, inspect system for a leak
2 Master cylinder or power booster loose	2 Tighten fasteners
3 Brake fluid overheated (beginning to boil)	3 Bleed the system (temporary fix). Replace the brake fluid (proper fix)
4 Deteriorated brake hoses (ballooning under pressure)	4 Inspect hoses, replace as necessary (it's a good idea to replace all of them if one hose shows signs of deterioration)

Brake pedal feels hard when depressed and/or excessive effort required to stop vehicle

1 Power booster faulty	1 Replace booster
2 Engine not producing sufficient vacuum, or hose to booster clogged, collapsed or cracked	2 Check vacuum to booster with a vacuum gauge. Replace hose if cracked or clogged, repair engine if vacuum is extremely low
3 Brake linings contaminated by grease or brake fluid	3 Locate and repair source of contamination, replace brake pads or shoes
4 Brake linings glazed	4 Replace brake pads or shoes, check discs and drums for glazing, service as necessary
5 Caliper piston(s) or wheel cylinder(s) binding or frozen	5 Replace calipers or wheel cylinders
6 Brakes wet	6 Apply pedal to boil-off water (this should only be a momentary problem)
7 Kinked, clogged or internally split brake hose or line	7 Inspect lines and hoses, replace as necessary

Excessive brake pedal travel (but will pump up)

1 Drum brakes out of adjustment	1 Adjust brakes
2 Air in hydraulic system	2 Bleed system, inspect system for a leak

Excessive brake pedal travel (but will not pump up)

1 Master cylinder pushrod misadjusted	1 Adjust pushrod
2 Master cylinder seals defective	2 Replace master cylinder
3 Brake linings worn out	3 Inspect brakes, replace pads and/or shoes
4 Hydraulic system leak	4 Locate and repair leak

Troubleshooting (continued)

PROBABLE CAUSE	CORRECTIVE ACTION

Brake pedal doesn't return

PROBABLE CAUSE	CORRECTIVE ACTION
1 Brake pedal binding	1 Inspect pivot bushing and pushrod, repair or lubricate
2 Defective master cylinder	2 Replace master cylinder

Brake pedal pulsates during brake application

PROBABLE CAUSE	CORRECTIVE ACTION
1 Brake drums out-of-round	1 Have drums machined by an automotive machine shop
2 Excessive brake disc runout or disc surfaces out-of-parallel	2 Have discs machined by an automotive machine shop
3 Loose or worn wheel bearings	3 Adjust or replace wheel bearings
4 Loose lug nuts	4 Tighten lug nuts

Brakes slow to release

PROBABLE CAUSE	CORRECTIVE ACTION
1 Malfunctioning power booster	1 Replace booster
2 Pedal linkage binding	2 Inspect pedal pivot bushing and pushrod, repair/lubricate
3 Malfunctioning proportioning valve	3 Replace proportioning valve
4 Sticking caliper or wheel cylinder	4 Repair or replace calipers or wheel cylinders
5 Kinked or internally split brake hose	5 Locate and replace faulty brake hose

Brakes grab (one or more wheels)

PROBABLE CAUSE	CORRECTIVE ACTION
1 Grease or brake fluid on brake lining	1 Locate and repair cause of contamination, replace lining
2 Brake lining glazed	2 Replace lining, deglaze disc or drum

Vehicle pulls to one side during braking

PROBABLE CAUSE	CORRECTIVE ACTION
1 Grease or brake fluid on brake lining	1 Locate and repair cause of contamination, replace lining
2 Brake lining glazed	2 Deglaze or replace lining, deglaze disc or drum
3 Restricted brake line or hose	3 Repair line or replace hose
4 Tire pressures incorrect	4 Adjust tire pressures
5 Caliper or wheel cylinder sticking	5 Repair or replace calipers or wheel cylinders
6 Wheels out of alignment	6 Have wheels aligned
7 Weak suspension spring	7 Replace springs
8 Weak or broken shock absorber	8 Replace shock absorbers

Brakes drag (indicated by sluggish engine performance or wheels being very hot after driving)

PROBABLE CAUSE	CORRECTIVE ACTION
1 Brake pedal pushrod incorrectly adjusted	1 Adjust pushrod
2 Master cylinder pushrod (between booster and master cylinder) incorrectly adjusted	2 Adjust pushrod
3 Obstructed compensating port in master cylinder	3 Replace master cylinder
4 Master cylinder piston seized in bore	4 Replace master cylinder
5 Contaminated fluid causing swollen seals throughout system	5 Flush system, replace all hydraulic components
6 Clogged brake lines or internally split brake hose(s)	6 Flush hydraulic system, replace defective hose(s)

PROBABLE CAUSE

CORRECTIVE ACTION

Brakes drag (indicated by sluggish engine performance or wheels being very hot after driving) (continued)

PROBABLE CAUSE	CORRECTIVE ACTION
7 Sticking caliper(s) or wheel cylinder(s)	7 Replace calipers or wheel cylinders
8 Parking brake not releasing	8 Inspect parking brake linkage and parking brake mechanism, repair as required
9 Improper shoe-to-drum clearance	9 Adjust brake shoes
10 Faulty proportioning valve	10 Replace proportioning valve

Brakes fade (due to excessive heat)

PROBABLE CAUSE	CORRECTIVE ACTION
1 Brake linings excessively worn or glazed	1 Deglaze or replace brake pads and/or shoes
2 Excessive use of brakes	2 Downshift into a lower gear, maintain a constant slower speed (going down hills)
3 Vehicle overloaded	3 Reduce load
4 Brake drums or discs worn too thin	4 Measure drum diameter and disc thickness, replace drums or discs as required
5 Contaminated brake fluid	5 Flush system, replace fluid
6 Brakes drag	6 Repair cause of dragging brakes
7 Driver resting left foot on brake pedal	7 Don't ride the brakes

Brakes noisy (high-pitched squeal)

PROBABLE CAUSE	CORRECTIVE ACTION
1 Glazed lining	1 Deglaze or replace lining
2 Contaminated lining (brake fluid, grease, etc.)	2 Repair source of contamination, replace linings
3 Weak or broken brake shoe hold-down or return spring	3 Replace springs
4 Rivets securing lining to shoe or backing plate loose	4 Replace shoes or pads
5 Excessive dust buildup on brake linings	5 Wash brakes off with brake system cleaner
6 Brake drums worn too thin	6 Measure diameter of drums, replace if necessary
7 Wear indicator on disc brake pads contacting disc	7 Replace brake pads
8 Anti-squeal shims missing or installed improperly	8 Install shims correctly

Note: *Other remedies for quieting squealing brakes include the application of an anti-squeal compound to the backing plates of the brake pads, and lightly chamfering the edges of the brake pads with a file. The latter method should only be performed with the brake pads thoroughly wetted with brake system cleaner, so as not to allow any brake dust to become airborne.*

Brakes noisy (scraping sound)

PROBABLE CAUSE	CORRECTIVE ACTION
1 Brake pads or shoes worn out; rivets, backing plate or brake shoe metal contacting disc or drum	1 Replace linings, have discs and/or drums machined (or replace)

Brakes chatter

PROBABLE CAUSE	CORRECTIVE ACTION
1 Worn brake lining	1 Inspect brakes, replace shoes or pads as necessary
2 Glazed or scored discs or drums	2 Deglaze discs or drums with sandpaper (if glazing is severe, machining will be required)
3 Drums or discs heat checked	3 Check discs and/or drums for hard spots, heat checking, etc. Have discs/drums machined or replace them
4 Disc runout or drum out-of-round excessive	4 Measure disc runout and/or drum out-of-round, have discs or drums machined or replace them
5 Loose or worn wheel bearings	5 Adjust or replace wheel bearings
6 Loose or bent brake backing plate (drum brakes)	6 Tighten or replace backing plate

Troubleshooting (continued)

PROBABLE CAUSE	CORRECTIVE ACTION

Brakes chatter (continued)

7 Grooves worn in discs or drums	7 Have discs or drums machined, if within limits (if not, replace them)
8 Brake linings contaminated (brake fluid, grease, etc.)	8 Locate and repair source of contamination, replace pads or shoes
9 Excessive dust buildup on linings	9 Wash brakes with brake system cleaner
10 Surface finish on discs or drums too rough after machining (especially on vehicles with sliding calipers)	10 Have discs or drums properly machined
11 Brake pads or shoes glazed	11 Deglaze or replace brake pads or shoes

Brake pads or shoes click

1 Shoe support pads on brake backing plate grooved or excessively worn	1 Replace brake backing plate
2 Brake pads loose in caliper	2 Loose pad retainers or anti-rattle clips
3 Also see items listed under Brakes chatter	

Brakes make groaning noise at end of stop

1 Brake pads and/or shoes worn out	1 Replace pads and/or shoes
2 Brake linings contaminated (brake fluid, grease, etc.)	2 Locate and repair cause of contamination, replace brake pads or shoes
3 Brake linings glazed	3 Deglaze or replace brake pads or shoes
4 Excessive dust buildup on linings	4 Wash brakes with brake system cleaner
5 Scored or heat-checked discs or drums	5 Inspect discs/drums, have machined if within limits (if not, replace discs or drums)
6 Broken or missing brake shoe attaching hardware	6 Inspect drum brakes, replace missing hardware

Rear brakes lock up under light brake application

1 Tire pressures too high	1 Adjust tire pressures
2 Tires excessively worn	2 Replace tires
3 Defective proportioning valve	3 Replace proportioning valve

Brake warning light on instrument panel comes on (or stays on)

1 Low fluid level in master cylinder reservoir (reservoirs with fluid level sensor)	1 Add fluid, inspect system for leak, check the thickness of the brake pads and shoes
2 Failure in one half of the hydraulic system	2 Inspect hydraulic system for a leak
3 Piston in pressure differential warning valve not centered	3 Center piston by bleeding one circuit or the other (close bleeder valve as soon as the light goes out)
4 Defective pressure differential valve or warning switch	4 Replace valve or switch
5 Air in the hydraulic system	5 Bleed the system, check for leaks
6 Brake pads worn out (vehicles with electric wear sensors - small probes that fit into the brake pads and ground out on the disc when the pads get thin)	6 Replace brake pads (and sensors)

PROBABLE CAUSE	CORRECTIVE ACTION

Brakes do not self adjust

Disc brakes

1 Defective caliper piston seals	1 Replace calipers. Also, possible contaminated fluid causing soft or swollen seals (flush system and fill with new fluid if in doubt)
2 Corroded caliper piston(s)	2 Same as above

Rapid brake lining wear

1 Driver resting left foot on brake pedal	1 Don't ride the brakes
2 Surface finish on discs or drums too rough	2 Have discs or drums properly machined
3 Also see Brakes drag	

---

3 Anti-lock Brake System (ABS) - general information

Description

1   The Anti-lock Brake System (ABS) is designed to maintain vehicle maneuverability, directional stability and optimum deceleration under severe braking conditions on most road surfaces. It does so by monitoring the rotational speed of the wheels and controlling the brake line pressure during braking. This prevents the wheels from locking up.

Electronic control unit (ECU)

2   The electronic control unit (ECU) for the Anti-lock Brake System is an integral part of the ABS actuator in the engine compartment.

3   The ECU monitors the rotation of each wheel with four wheel speed sensors, processes this information and avoids wheel lockup by controlling the hydraulic line pressure accordingly. Here's how it works: When the brakes are applied too firmly during a panic stop, hydraulic line pressure inside the brake lines builds to such a high level that it locks up the wheels, causing the vehicle to skid out of control. On a vehicle equipped with ABS, the ECU prevents the hydraulic pressure from reaching this dangerously high level by monitoring the rotational speed of the wheels. When a wheel begins to slow down (lock up) in relation to the other wheels, the ECU energizes the solenoid (inside the actuator) controlling the hydraulic brake fluid circuit to that wheel. The energized solenoid opens the circuit, allowing some of the brake fluid into a reservoir, thereby lowering the pressure and preventing the wheel from locking up. As soon as the rotation speed of the wheel equals that of the other wheels, the solenoid closes and pressure begins to build again. This cycle of opening and closing the circuit occurs many times a second at each wheel. The ECU can regulate the pressure to a single wheel, or to two, three or all four wheels, simultaneously.

4   The ECU also monitors the ABS system for malfunctions. If the ECU detects a problem, the ABS warning light on the instrument cluster lights up and a diagnostic code is stored which, when retrieved by a service technician, will indicate the problem area or component. When the engine is started, the ABS warning light glows for about three seconds (indicating that the ECU is monitoring the system for faults), then goes out; if the ABS light remains on, there's a problem in the ABS system. Take the vehicle to a dealer service department or an authorized service facility.

Actuator

5   The actuator assembly, which is mounted inside the engine compartment, houses the solenoids which regulate brake fluid pressure in response to signals from the ECU.

Speed sensors

6   Each wheel has its own speed sensor. The speed sensor is a small variable reluctance sensor (pick-up coil) which sends a variable voltage signal (actually, an alternating current sine wave output) to the ECU. The ECU converts this analog signal into a digital code which it compares to its map (program), then either ignores it (if the wheel is rotating at the same speed as the other wheels) or executes a command to open a solenoid for the circuit to that wheel (if the wheel is starting to slow down in relation to the other wheels).

Brake light switch

7   The brake light switch signals the control unit when the driver steps on the brake pedal. Without this signal the anti-lock system won't activate.

Diagnosis and repair

8   If the ABS warning light on the instrument cluster comes on and stays on, make sure the parking brake is released and there's no problem with the brake hydraulic system. If neither of these is the cause, the anti-lock system is probably malfunctioning. Although special test procedures are necessary to properly diagnose the system, the home mechanic can perform a few preliminary checks before taking the vehicle to a dealer service department.

a) *Make sure the brakes, calipers and wheel cylinders are in good condition.*

3.11 Unplug the electrical connector from the wheel speed sensor

b) *Inspect the electrical connectors at the ECU. Make sure they're clean and tight.*
c) *Check the fuses.*
d) *Follow the wiring harness to the speed sensor(s) and brake light switch, and make sure all connections are clean and tight and the wiring isn't damaged.*

9   If the above preliminary checks don't rectify the problem, the vehicle should be diagnosed by a dealer service department or other qualified repair shop.

Wheel speed sensor - removal and installation

Warning: *On Sequoia models equipped with rear height control suspension, adjust the height control to the NORMAL mode, turn OFF the height control, then turn off the engine BEFORE raising the vehicle.*

Front wheel speed sensor

Refer to illustration 3.11

10   Loosen the wheel lug nuts, raise the front of the vehicle and support it securely on jackstands. Block the wheels at the opposite end.

11   Unplug the electrical connector from the sensor **(see illustration)**.

3.15 Unplug the electrical connector (A) from the wheel speed sensor, then remove the sensor mounting nut (B) and detach the sensor from the axle housing

4.2 Wash the disc and caliper with brake system cleaner to remove the brake dust; DO NOT blow off the brake dust with compressed air

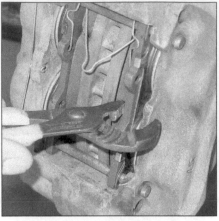

4.4a Squeeze the pads towards the caliper frame to free-up the pads and depress the pistons into their bores.

Note: *Replace one pad at a time to prevent the pistons on the other side of the caliper from coming out*

4.4b Detach the pin retaining clip

4.4c Pull out the pad retaining pins

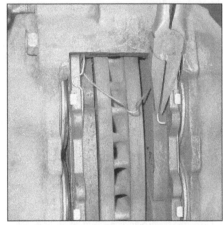

4.4d Remove the anti-rattle spring

12   Remove the sensor mounting bolt and detach the sensor from the steering knuckle.
13   Installation is the reverse of removal. Tighten the sensor mounting bolt to the torque listed in this Chapter's Specifications. Tighten the wheel lug nuts to the torque listed in the Chapter 1 Specifications.

Rear wheel speed sensor

Refer to illustration 3.15

14   Loosen the wheel lug nuts, raise the rear of the vehicle and support it securely on jackstands. Block the wheels at the opposite end.
15   Unplug the electrical connector from the wheel speed sensor **(see illustration)**.
16   Remove the sensor mounting nut and detach the sensor from the axle housing.
17   Installation is the reverse of the removal procedure. Tighten the sensor mounting bolt to the torque listed in this Chapter's Specifications. Tighten the wheel lug nuts to the torque listed in the Chapter 1 Specifications.

4   Disc brake pads - replacement

Refer to illustration 4.2

Warning: *Disc brake pads must be replaced on both wheels at the same time - never replace the pads on only one wheel. Also, brake system dust is hazardous to your health. DO NOT blow it out with compressed air and DO NOT inhale it. An approved filtering mask should be worn when working on the brakes. DO NOT use gasoline or solvents to remove the dust. Use brake system cleaner only!*
Warning: *On Sequoia models equipped with rear height control suspension, adjust the height control to the NORMAL mode, turn OFF the height control, then turn off the engine BEFORE raising the vehicle.*

1   Loosen the wheel lug nuts, raise the vehicle and support it securely on jackstands. Block the wheels at the opposite end. Remove the wheels.
2   Remove about two-thirds of the fluid

from the master cylinder reservoir and discard it; as the pistons are pushed in to make room for the new pads, the fluid will be forced back into the reservoir. Position a drain pan under the brake assembly and clean the caliper and surrounding area with brake system cleaner **(see illustration)**.
3   While the pads are removed, inspect the caliper for brake fluid leaks and ruptures in the piston boot(s). Replace the caliper if it's damaged or leaking (see Section 5). Also inspect the brake disc carefully (see Section 6). If machining is necessary, follow the information in that Section to remove the disc.

Front brake pads

Refer to illustrations 4.4a through 4.4f

4   To replace the front brake pads, follow the accompanying photos, beginning with **illustration 4.4a**. Be sure to stay in order and read the caption under each illustration. Work on one brake assembly at a time so you'll have something to refer to if you get in trouble.

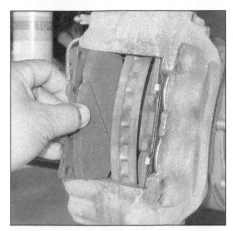

4.4e Remove the inner brake pad and push the inner pistons back into their bores to provide room for the new pad

4.4f Apply anti-squeal compound to the back of the new pads, then stick the anti-squeal shims to them. Install the new inner pad, then replace the outer pad the same way. Reinstall the pad retaining pins, anti-rattle spring and the pin retaining clip, then proceed to Step 7

4.5 To make room for the new pads, use a C-clamp to depress the piston into its bore before removing the caliper (typical)

4.6a Always wash the brakes with brake cleaner before disassembling anything

4.6b Using an opened wrench, loosen the lower bolt . . .

4.6c . . . then remove the mounting bolt or "guide pin"

Rear brake pads

Refer to illustrations 4.5, 4.6a through 4.6j

5   Using a C-clamp, depress the caliper piston into its bore to make room for the new, thicker pads **(see illustration)**.

6   To replace the rear brake pads, follow the accompanying photos, beginning with **illustration 4.6a**. Be sure to stay in order and read the caption under each illustration. Work on one brake assembly at a time so you'll have something to refer to if you get in trouble.

Front or rear brake pads

7   Install the brake pads on the opposite wheel, then install the wheels and lower the vehicle. Tighten the lug nuts to the torque listed in the Chapter 1 Specifications.

8   Add brake fluid to the reservoir until it's full (see Chapter 1). Pump the brakes several times to seat the pads against the disc, then check the fluid level again.

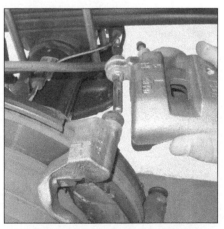

4.6d Rotate the caliper up until the upper mounting bolt can be slid out of the bracket

4.6e Remove the inner pad and anti-squeal shim

4.6f Remove the outer pad and anti-squeal shim

4.6g Remove the upper and lower support plates; inspect them for damage and replace if necessary (if they're weak or distorted, they should be replaced)

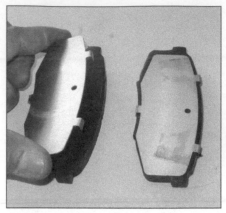

4.6h Install the anti-squeal shims to the brake pads

4.6i Install the pads; make sure the ears on the pads are properly engaged with the pad support plates as shown. Install the new inner pad, then the outer pad the same way, and install the caliper over them

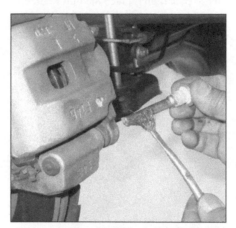

4.6j Before installing the caliper mounting bolts, clean and check them for corrosion and damage. If significantly corroded or damaged, replace them. Lubricate the sliding surfaces of the bolts with high-temperature brake grease, install and tighten them to the torque listed in this Chapter's Specifications, then proceed to Step 7

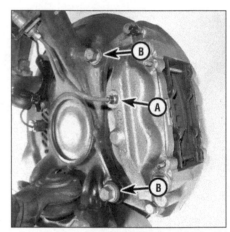

5.3 To remove a front caliper, unscrew the hydraulic line fitting (A) with a flare-nut wrench, then remove the caliper mounting bolts (B)

9   Check the operation of the brakes before driving the vehicle in traffic. Try to avoid heavy brake applications until the brakes have been applied lightly several times to seat the pads.

5   Disc brake caliper - removal and installation

Warning: *On Sequoia models equipped with rear height control suspension, adjust the height control to the NORMAL mode, turn OFF the height control, then turn off the engine BEFORE raising the vehicle.*
Note: *If caliper replacement is indicated (usually because of fluid leaks, a stuck piston or broken bleeder screw) explore your options. New and factory rebuilt calipers are available on an exchange basis.*
1   Loosen the front wheel lug nuts, raise the front of the vehicle and support it securely on jackstands. Apply the parking brake. Remove the wheels. Position a drain pan under the

brake assembly and clean the caliper and surrounding area with brake system cleaner **(see illustration 4.2)**.
2   Remove about two-thirds of the fluid from the master cylinder reservoir and discard it; as the pistons are pushed in for clearance to allow the pads to be removed, the fluid will be forced back into the reservoir. Position a drain pan under the brake assembly and clean the caliper and surrounding area with brake system cleaner.

Front caliper

Refer to illustration 5.3
3   Unscrew the tube nut fitting and detach the brake line from the caliper, then remove the caliper mounting bolts **(see illustration)**.
Note: *Use a flare-nut wrench, if available, to prevent rounding-off the corners of the fitting.*
If the caliper can't be pulled off, the brake

pads are hanging up on the ridge around the circumference of the disc; remove the brake pads (see Section 4).

Rear caliper

Refer to illustrations 5.5a and 5.5b
4   Push the piston back into the bore with a C-clamp to provide room for the new brake pads **(see illustration 4.5)**. As the piston is depressed to the bottom of the caliper bore, the fluid in the master cylinder will rise. Make sure it doesn't overflow. If necessary, siphon off some of the fluid.
5   Remove the brake hose-to-caliper union bolt **(see illustration)**.
Note: *If you're removing the caliper for access to other components, leave the hose connected.*
Discard the sealing washer set on each side of the banjo fitting; use a new set when you reattach the brake hose to the caliper. Plug the banjo fitting with a piece of rubber hose **(see illustration)**.
6   Remove the caliper mounting bolts and detach the caliper from the mounting bracket.

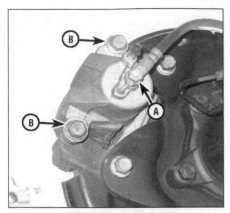

5.5a Remove the brake hose-to-caliper union bolt (A), then remove the caliper mounting bolts (B)

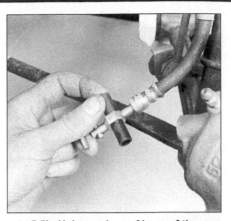

5.5b Using a piece of hose of the appropriate size, plug the banjo fitting to prevent brake fluid from dripping out of the hose and to prevent contaminants from entering the brake system

6.2 Hang the caliper with a piece of wire - don't let it hang by the hose or brake line!

6.3 The brake pads on this vehicle were obviously neglected - they wore out completely and cut deep grooves into the disc; if the disc is worn this severely, replace it

6.4a Measure the brake disc runout with a dial indicator; if the reading exceeds the maximum allowable runout limit, the disc must be resurfaced or replaced

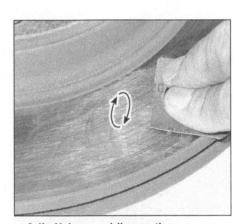

6.4b Using a swirling motion, remove the glaze from the disc surface with sandpaper or emery cloth

Front or rear caliper

7   If you're planning to install the same caliper, clean the caliper with brake system cleaner. DO NOT use kerosene or petroleum-based solvents. Carefully inspect the caliper for leaks and damage. Do NOT install a caliper that is leaking or damaged.

8   Install the caliper and tighten the caliper mounting bolts to the torque listed in this Chapter's Specifications.

9   Connect the brake hose or line to the caliper; use new sealing washers if you're installing a rear caliper and tighten the union bolt to the torque listed in this Chapter's Specifications.

10   Bleed the brake system (see Section 9). If you capped or plugged the line or hose and not much fluid was lost, you'll probably only have to bleed the circuit to the caliper that was removed. If the brake fluid hose was not disconnected (caliper removed for access to other components), the brakes will not require bleeding.

11   Install the wheel and lug nuts, lower the vehicle and tighten the lug nuts to the torque listed in the Chapter 1 Specifications. Check brake operation carefully before placing the vehicle into service.

6   Brake disc - inspection, removal and installation

Warning: *On Sequoia models equipped with rear height control suspension, adjust the height control to the NORMAL mode, turn OFF the height control, then turn off the engine BEFORE raising the vehicle.*

Inspection

Refer to illustrations 6.2, 6.3, 6.4a, 6.4b, 6.5a, 6.5b, 6.6, 6.7a, 6.7b and 6.7c

1   Loosen the wheel lug nuts, raise the front of the vehicle and support it securely on jackstands. Apply the parking brake. Remove the wheel. Reinstall the lug nuts to hold the disc in place against the hub (washers may have to be used to allow the nuts to apply pressure to the disc).

2   Remove the brake caliper and hang it with a length of wire (don't disconnect the brake line or hose from the caliper) **(see illustration)**. If you're removing a front caliper, unclip the brake line from the bracket on the steering knuckle.

Caution: *Don't let the caliper hang by the brake hose or line.*

If you're checking a rear disc, remove the brake pads (see Section 3).

3   Visually inspect the disc surface for score marks and other damage **(see illustration)**. Light scratches and shallow grooves are normal after use and won't affect brake operation. Deep grooves require disc removal and refinishing by an automotive machine shop. Check both sides of the disc.

4   To check disc runout, place a dial indicator at a point about 1/2-inch from the outer edge of the disc **(see illustration)**. Set the indicator to zero and turn the disc. The indicator reading should not exceed the allowable runout listed in this Chapter's Specifications. If it does, the disc should be refinished by an automotive machine shop.

Note: *To produce a smooth finish and ensure a perfectly smooth surface - thereby eliminating brake pedal pulsation or any other undesirable symptoms - the discs should be resurfaced regardless of the dial indicator reading. If you elect not to have the discs resurfaced, deglaze them with sandpaper or emery cloth* **(see illustration)**.

6.5a Measure the brake disc thickness at several points with a micrometer

6.5b The minimum allowable thickness dimension is cast into the back side of the disc

6.6 To remove the rear caliper brackets, remove the mounting bolts and remove the bracket

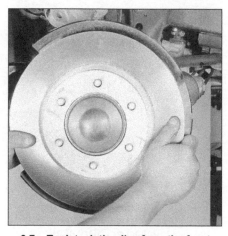

6.7a To detach the disc from the front hub, simply pull it off

6.7b Insert a screwdriver into the access hole and loosen the star wheel adjuster and remove the disc

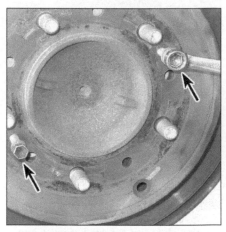

6.7c Thread two bolts into the hub face and slowly, evenly tighten the bolts until the disc can be pulled off by hand

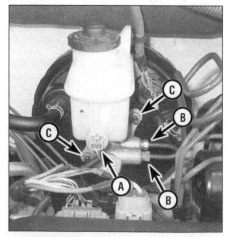

7.2 Master cylinder mounting details

A Fluid level sensor electrical connector
B Brake line fittings
C Mounting nuts

5   The disc must not be machined to a thickness less than the specified minimum thickness, which is cast into the disc. Measure disc thickness with a micrometer **(see illustrations)**.

Removal and installation

Note: *If you're planning on reusing the brake disc, mark the relationship between the brake disc and hub so it can be installed it its original location.*

6   Remove the brake caliper, if not already done (see Section 5). If you're removing a rear disc, also remove the caliper mounting bracket **(see illustration)**.

7   Remove the lug nuts that were installed in Step 1 and pull off the disc **(see illustration)**. On rear disc models, if the disc won't come off, remove the rubber plug on the hub face and insert a screwdriver into the hole and push the star wheel adjuster downwards **(see illustration)** until the drum is loosened then remove the disc. In the event disc cannot be removed once the adjuster wheel has been loosened,

insert two bolts into the hub section of the disc then slowly and evenly tighten them until the drum can be removed **(see illustration)**.

8   Installation is the reverse of removal. Tighten all brake fasteners to the torque values listed in this Chapter's Specifications. Tighten the wheel lug nuts to the torque listed in the Chapter 1 Specifications.

7   Master cylinder - removal and installation

Removal

Refer to illustration 7.2

1   Place rags under the brake line fittings and prepare caps or plastic bags to cover the ends of the lines once they're disconnected. **Caution:** *Brake fluid will damage paint. Cover all painted surfaces and avoid spilling fluid during this procedure.*

2   Unplug the brake fluid level warning switch electrical connector **(see illustration)**.

7.8 The best way to bleed air from the master cylinder before installing it on the vehicle is with a pair of bleeder tubes that direct brake fluid into the reservoir during bleeding

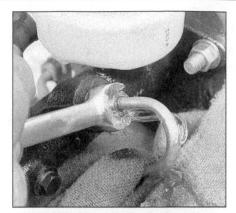

7.17 Have an assistant depress the brake pedal and hold it down, then loosen the fitting nut, allowing the air and fluid to escape; repeat this procedure on both fittings until the fluid is clear of air bubbles

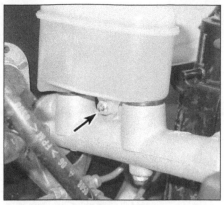

7.20 The master cylinder fluid reservoir is secured by a single TORX screw

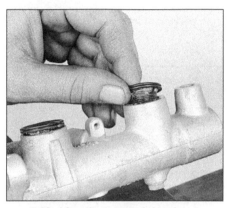

7.22 After the reservoir has been removed, pull the grommets from the master cylinder body; if they're hard, cracked or damaged, or have been leaking, replace them

3   Loosen the tube nuts at the ends of the brake lines where they enter the master cylinder. To prevent rounding off of the flats on these nuts, a flare-nut wrench, which wraps around the nut, should be used. Pull the brake lines away from the master cylinder slightly and plug the ends to prevent contamination. Be careful not to kink the hydraulic lines.
4   Remove the master cylinder mounting nuts **(see illustration 7.2)**. Remove the master cylinder from the vehicle.
5   Remove the reservoir cap and discard any fluid remaining in the reservoir.
6   Check the O-ring on the end of the master cylinder, replacing it if it is cracked or hardened.

Installation

Refer to illustrations 7.8 and 7.17

Warning: *If you're replacing the master cylinder assembly or master cylinder pressure sensor(s), the zero point calibration of the steering angle, master cylinder pressure, yaw rate and deceleration sensors for the Vehicle Skid Control (VSC) system must be performed. This will require taking the vehicle to a dealer service department or other qualified repair shop equipped with the special tool necessary to carry out this procedure. The brake hydraulic system will work as a conventional braking system, but the ABS, Traction Control and VSC system may not function properly.*

7   Bench bleed the master cylinder before installing it. Because it will be necessary to apply pressure to the master cylinder piston and, at the same time, control flow from the brake line outlets, it is recommended that the mater cylinder be mounted in a vise, with the jaws of the vise clamping on the mounting flange.
8   Attach a pair of bleeder tubes (available at most auto parts stores) to the outlet ports of the master cylinder **(see illustration)**.
9   Fill the reservoir with brake fluid of the recommended type (see Chapter 1).
10   Slowly push the pistons into the master cylinder (a large Phillips screwdriver can be

used for this) - air will be expelled from the pressure chambers and into the reservoir. Because the tubes are submerged in fluid, air can't be drawn back into the master cylinder when you release the pistons.
11   Repeat the procedure until no more air bubbles are present.
12   Remove the bleed tubes, one at a time, and install plugs in the open ports to prevent fluid leakage and air from entering. Install the reservoir cap.
13   Install the master cylinder over the studs on the power brake booster and tighten the nuts only finger-tight at this time. Don't forget to use a new gasket.
14   Thread the brake line fittings into the master cylinder. Since the master cylinder is still a bit loose, it can be moved slightly so the fittings thread in easily. Don't strip the threads as the fittings are tightened.
15   Tighten the mounting nuts to the torque listed in this Chapter's Specifications. Tighten the brake line fittings securely.
16   Plug in the electrical connector to the fluid level warning switch.
17   Fill the master cylinder reservoir with fluid, then bleed the master cylinder and the brake system (see Section 9). To bleed the master cylinder on the vehicle, have an assistant depress the brake pedal and hold it down. Loosen the fitting to allow air and fluid to escape **(see illustration)**. Tighten the fitting, then allow your assistant to return the pedal to its rest position. Repeat this procedure on both fittings until the fluid is free of air bubbles. Check the operation of the brake system carefully before driving the vehicle.

Reservoir/grommet replacement

Refer to illustrations 7.20 and 7.22

Note: *The brake fluid reservoir can be replaced separately from the master cylinder body if it becomes damaged. If there is leakage between the reservoir and the master cylinder body, the grommets on the reservoir can be replaced.*

18   Remove as much fluid as possible from the reservoir with a suction gun, large syringe or a poultry baster.
Warning: *If a poultry baster is used, never again use it for the preparation of food.*
19   Place rags under the master cylinder to absorb any fluid that may spill out once the reservoir is detached from the master cylinder.
Caution: *Brake fluid will damage paint. Cover all body parts and be careful not to spill fluid during this procedure.*
20   Remove the TORX screw that retains the reservoir to the master cylinder **(see illustration)**.
21   Pull the reservoir out of the master cylinder body.
22   Pull the grommets out of the master cylinder **(see illustration)**.
23   Lubricate the new grommets with clean brake fluid, then press them into place.
24   Push the reservoir into the grommets and secure it with the screw.
25   Refill the reservoir with the recommended brake fluid (see Chapter 1) and check for leaks.
26   Bleed the master cylinder **(see illustration 7.17)**, followed by the remainder of the system (see Section 9).

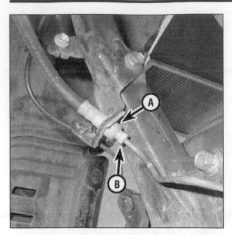

8.3 Hold the hose fitting (A) with a wrench to prevent twisting the line, then loosen the tube nut (B) with a flare-nut wrench to prevent rounding off the corners of the nut

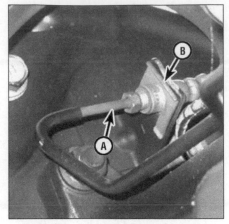

8.4 Once the tube nut (A) has been completely loosened, pull off the clip (B) with a pair of pliers, or pry it off with a screwdriver

9.8 When bleeding the brakes, a hose is connected to the bleed screw at the caliper or wheel cylinder and then submerged in brake fluid - air will be seen as bubbles in the tube and container (all air must be expelled before moving to the next wheel)

8 Brake hoses and lines - check and replacement

Warning: *On Sequoia models equipped with rear height control suspension, adjust the height control to the NORMAL mode, turn OFF the height control, then turn off the engine BEFORE raising the vehicle.*

1   About every six months, with the vehicle raised and placed securely on jackstands, the flexible hoses which connect the steel brake lines with the front and rear brake assemblies should be inspected for cracks, chafing of the outer cover, leaks, blisters and other damage. These are important and vulnerable parts of the brake system and inspection should be complete. A light and mirror will be needed for a thorough check. If a hose exhibits any of the above defects, replace it with a new one.

Flexible hoses

Refer to illustrations 8.3 and 8.4

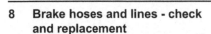

2   Clean all dirt away from the ends of the hose.
3   Disconnect the brake line from the hose fitting **(see illustration)**. Be careful not to bend the frame bracket or line. If necessary, soak the connections with penetrating oil.
4   Remove the U-clip from the female fitting at the bracket **(see illustration)** and remove the hose from the bracket.
5   If you're removing a rear brake hose, disconnect the hose from the caliper, discarding the copper washer unit from both sides of the fitting. Using a new copper washer unit, slide the unit over the end of the hose then attach the new brake hose to the caliper.
6   Pass the female fitting through the frame or frame bracket. With the least amount of twist in the hose, install the fitting in this position.
Note: *The weight of the vehicle must be on the suspension, so the vehicle should not be raised while positioning the hose.*

7   Install the U-clip in the female fitting at the frame bracket.
8   Attach the brake line to the hose fitting using a back-up wrench on the fitting. Tighten the tube nut securely.
9   Carefully check to make sure the suspension or steering components don't make contact with the hose. Have an assistant push down on the vehicle and also turn the steering wheel lock-to-lock during inspection.
10   Bleed the brake system (see Section 9).

Metal brake lines

11   When replacing brake lines, be sure to use the correct parts. Don't use copper tubing for any brake system components. Purchase steel brake lines from a dealer parts department or auto parts store.
12   Prefabricated brake lines, with the ends already flared and fittings installed, are available at auto parts stores and dealer service departments. If necessary, carefully bend the line to the proper shape. A tube bender is recommended for this.
Caution: *Don't crimp or damage the line.*
13   When installing the new line make sure it's well supported in the brackets and has plenty of clearance between moving or hot components.
14   After installation, check the master cylinder fluid level and add fluid as necessary. Bleed the brake system as outlined (see Section 9) and test the brakes carefully before placing the vehicle into normal operation.

9 Brake hydraulic system - bleeding

Refer to illustration 9.8

Warning: *Wear eye protection when bleeding the brake system. If the fluid comes in contact with your eyes, immediately rinse them with water and seek medical attention.*

Note: *Bleeding the brake system is necessary to remove any air that's trapped in the system when it's opened during removal and installation of a hose, line, caliper, wheel cylinder or master cylinder.*

1   It will probably be necessary to bleed the system at all four brakes if air has entered the system due to low fluid level, or if the brake lines have been disconnected at the master cylinder.
2   If a brake line was disconnected only at a wheel, then only that caliper or wheel cylinder must be bled.
3   If a brake line is disconnected at a fitting located between the master cylinder and any of the brakes, that part of the system served by the disconnected line must be bled.
4   Remove any residual vacuum from the brake power booster by applying the brake several times with the engine off.
5   Remove the master cylinder reservoir cap and fill the reservoir with brake fluid. Reinstall the cap.
Note: *Check the fluid level often during the bleeding operation and add fluid as necessary to prevent the fluid level from falling low enough to allow air bubbles into the master cylinder.*
6   Have an assistant on hand, as well as a supply of new brake fluid, an empty clear plastic container, a length of clear tubing to fit over the bleeder valve and a wrench to open and close the bleeder valve.
7   Beginning at the right rear wheel, loosen the bleeder screw slightly, then tighten it to a point where it's snug but can still be loosened quickly and easily.
8   Place one end of the tubing over the bleeder screw fitting and submerge the other end in brake fluid in the container **(see illustration)**.
9   Have the assistant slowly push down on the brake pedal, then hold the pedal firmly depressed.

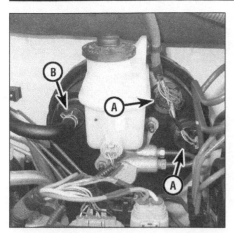

10.6 Disconnect the electrical connectors to the booster (A), then disconnect the vacuum hose (B)

10.10 To disconnect the power brake booster pushrod from the brake pedal, remove the retaining clip and slide out the clevis pin

10.11a To detach the booster from the firewall, remove nuts on the right side . . .

10 While the pedal is held depressed, open the bleeder screw just enough to allow a flow of fluid to leave the valve. Watch for air bubbles to exit the submerged end of the tube. When the fluid flow slows after a couple of seconds, tighten the screw and have your assistant release the pedal slowly.

11 Repeat Steps 9 and 10 until no more air is seen leaving the tube, then tighten the bleeder screw and proceed to the left rear wheel, the right front wheel and the left front wheel, in that order, and perform the same procedure. Check the fluid in the master cylinder reservoir frequently.

12 Never use old brake fluid. It contains moisture which can boil, rendering the brakes useless.

13 Refill the master cylinder with fluid at the end of the operation.

14 Check the operation of the brakes. The pedal should feel solid when depressed, with no sponginess. If necessary, repeat the entire process.

Warning: *Do not operate the vehicle if you are in doubt about the effectiveness of the brake system.*

---

10 Power brake booster - check, removal and installation

Operating check

1 Depress the brake pedal several times with the engine off and make sure there's no change in the pedal reserve distance.

2 Depress the pedal and start the engine. If the pedal goes down slightly, operation is normal.

Airtightness check

3 Start the engine and turn it off after one or two minutes. Depress the brake pedal slowly several times. If the pedal depresses less each time, the booster is airtight.

4 Depress the brake pedal while the engine is running, then stop the engine with the pedal

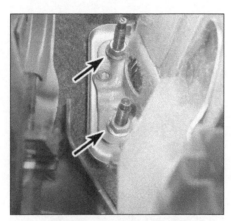

10.11b . . . and the left side

depressed. If there's no change in the pedal reserve travel after holding the pedal for 30 seconds, the booster is airtight.

Removal

Refer to illustrations 10.6, 10.10, 10.11a and 10.11b

5 Power brake booster units shouldn't be disassembled. They require special tools not normally found in most automotive repair stations or shops. Because of its critical relationship to brake performance, the booster should be replaced with a new or rebuilt one.

6 Disconnect the electrical connectors from the booster **(see illustration)**.

7 Disconnect the vacuum hose leading from the engine to the booster **(see illustration 10.6)**. Be careful not to damage the hose when removing it from the booster fitting.

8 Remove the brake master cylinder (see Section 7).

9 Remove the left side lower finish panel from the instrument panel (see Chapter 11).

10 Locate the pushrod clevis connecting the booster to the brake pedal **(see illustration)**. Remove the clevis pin retaining clip with pliers and pull out the pin.

11 Remove the four nuts holding the brake

booster to the firewall **(see illustration)**; you may need a light to see them. Slide the booster straight out from the firewall until the studs clear the holes.

Installation

12 Installation procedures are basically the reverse of removal. Make sure the clevis pin retaining clip is fully seated and the booster mounting nuts are tightened to the torque listed in this Chapter's Specifications.

13 After the final installation of the master cylinder and brake lines, the system must be bled (see Section 9) and the brake pedal height and freeplay must be checked (see Chapter 1).

14 Carefully test the operation of the brakes before placing the vehicle in normal operation.

---

11 Parking brake shoes - replacement

Refer to illustrations 11.4a through 11.4q, and 11.5

Warning: *Parking brake shoes must be replaced on both wheels at the same time - never replace the shoes on only one wheel. Also, the dust created by the brake system is harmful to your health. Never blow it out with compressed air and don't inhale any of it. An approved filtering mask should be worn when working on the brakes. Do not, under any circumstances, use petroleum-based solvents to clean brake parts. Use brake system cleaner only!*

Warning: *On Sequoia models equipped with rear height control suspension, adjust the height control to the NORMAL mode, turn OFF the height control, then turn off the engine BEFORE raising the vehicle.*

1 Loosen the wheel lug nuts, raise the rear of the vehicle and support it securely on jackstands. Block the front wheels to keep the vehicle from rolling.

2 Release the parking brake.

3 Remove the wheel then remove the disc **(see illustrations 6.6 and 6.7b)**.

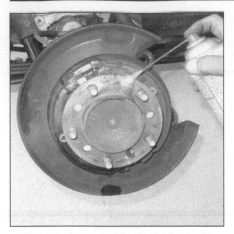

11.4a Wash the parking brake assembly with brake system cleaner to remove the brake dust; DO NOT blow off the brake dust with compressed air

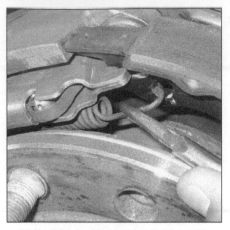

11.4b With locking pliers, remove one end of the parking brake shoe return spring, then remove the spring from the other shoe

11.4c Remove the parking brake shoe adjuster spring

11.4d Remove the parking brake shoe adjuster from the ends of the parking brake shoes

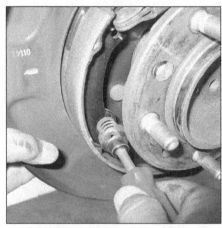

11.4e Using pliers or a hold-down tool, rotate the hold-down spring retainer and free the anchor pin, then slowly let the hold down spring up and remove them from both the shoes

11.4f Disconnect the parking brake shoe actuating lever from the shoes

11.4g Remove both of the shoes

11.4h Slide the actuating lever outwards to expose the end of the parking brake cable

Note: *All four parking brake shoes must be replaced at the same time, but to avoid mixing up parts, work on only one brake assembly at a time.*

4   Follow the accompanying illustrations for the parking brake shoe replacement procedure **(see illustrations 11.4a through 11.4q)**. Be sure to stay in order and read the caption under each illustration.

5   Before reinstalling the rear disc, the drum portion of the disc should be checked for cracks, score marks, deep scratches and hard spots, which will appear as small discolored areas. If the hard spots cannot be removed with fine emery cloth or if any of the other conditions listed above exist, the disc must be taken to an automotive machine shop to have it resurfaced.

Note: *Professionals recommend resurfacing the drum portion each time a brake job is done. Resurfacing will eliminate the possibility of out-of-round drums. If the drums are worn*

11.4i Slide the cable end over the actuator tips and off

11.4j Reconnect the parking brake cable to the actuator and slide the arm over the cable end

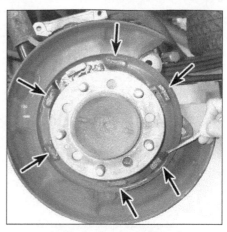

11.4k Apply a light film of high-temperature grease to the backing plate where the shoes ride against it

11.4l Install both brake shoes and the retaining pin . . .

11.4m . . . then place the hold-down spring and retainer onto the shoes, and lock the retainer onto the end of the retaining pin

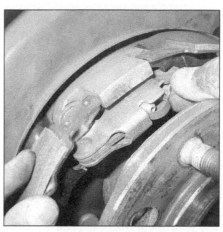

11.4n Place the end of the shoes into the actuator

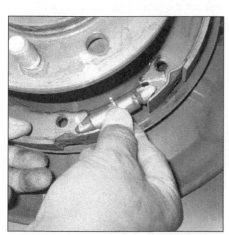

11.4o Lube the threads of the star wheel adjuster with high temperature grease, then install it between the shoes

11.4p Hook the adjuster spring to both shoes

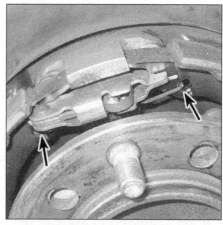

11.4q Hook the shoe return spring to both shoes

11.5 Cast mark locations are on the outside of the drum

12.3 Location of the parking brake cable adjuster nuts

13.1 The brake light switch is located at the top of the brake pedal assembly; to remove it, pull the boot back and unplug the electrical connector, then rotate the switch counterclockwise and remove it from the bracket

so much that they can't be resurfaced without exceeding the maximum allowable diameter (cast into the drum), then new drums will be required **(see illustration)**. At the very least, if you elect not to have the drums resurfaced, remove the glaze from the surface with emery cloth using a swirling motion.

6   Install the brake disc/drum on the axle flange, then turn the star wheel adjuster with a screwdriver until the shoes lock against the drum **(see illustration 6.7b)**. Then back the star wheel adjuster off eight full clicks. Apply the parking brake a couple of times, then turn the rotor by hand it should rotate smoothly and install the rubber plug. If it doesn't repeat this step again until it does.

7   Install the caliper bracket, rear brake pads and caliper.

8   Install the wheel and lug nuts and check to see if the wheel rotates freely. Apply the parking brake. The wheel should not be able to turn. If it does, apply the parking brake three times and check it again.

Note: *The parking brake is adjusted automatically by applying the park brake at least three times.*

9   Lower the vehicle and tighten the lug nuts to the torque listed in the Chapter 1 Specifications.

10   Check the operation of the brakes carefully before driving the vehicle. After driving the vehicle, check the operation of the parking brake and if necessary, apply the parking brake to adjust the brakes until satisfactory action is obtained.

---

12   Parking brake - adjustment

Refer to illustration 12.3

Warning: *On Sequoia models equipped with rear height control suspension, adjust the height control to the NORMAL mode, turn OFF the height control, then turn off the*

engine BEFORE raising the vehicle.

1   The parking brake is operated by a pedal under the left end of the instrument panel on all models. All parking brake systems are self-adjusting; automatic adjusters compensate for brake shoe wear. However, additional adjustment may be needed in the event of cable stretch, which occurs as the vehicle ages. The pedal should apply the parking brake within the number of clicks listed in this Chapter's Specifications. If the handle or pedal applies the parking brake system in less than the specified number of clicks, the cable is too tight; if it travels more than the specified number of clicks, the cable is too loose.

2   Release the parking brake, raise the rear of the vehicle until the wheels are off the ground and support it securely on jackstands. Block the front wheels to prevent the vehicle from rolling.

3   Cable tension is adjusted at the pedal **(see illustration)**. Remove the left-side under-dash panel, the left kick panel and the heater duct. Using one wrench on the adjuster nut, use another wrench to loosen the locknut, then tighten or loosen the adjuster nut until the proper pedal travel is obtained.

Note: *If the threads appear rusty, apply penetrating oil before attempting adjustment.*

4   Tighten the locknut securely after adjustment. Make sure the rear wheels turn freely when the parking brake is released.

5   Remove the jackstands and lower the vehicle.

---

13   Brake light switch - check and replacement

Check

Refer to illustration 13.1

1   The brake light switch **(see illustration)** is located on the brake pedal bracket. You'll

need to remove the trim panel beneath the steering column to get to the switch and connector (see Chapter 11).

2   With the brake pedal in the fully released position, the switch opens the brake light circuit. When the brake pedal is depressed, the switch closes the circuit and sends current to the brake lights.

3   If the brake lights are inoperative, check the fuse and the bulbs (see Chapter 12).

4   If the fuse and bulbs are okay, verify that voltage is available at the switch.

5   If there's no voltage to the switch, search for an open circuit condition between the fuse block and the switch. If there is voltage to the switch, close the switch (depress the brake pedal) and verify that there's voltage on the other side of the switch.

6   If there's no voltage on the other side of the switch, replace the switch. If there is voltage but the brake lights still don't work, look for an open circuit condition between the switch and the brake lights.

Note: *There is always the possibility that all of the brake light bulbs are burned out, but this is not very likely.*

Replacement

7   Remove the trim panel below the steering column (see Chapter 11).

8   Unplug the electrical connector from the switch **(see illustration 13.1)**.

9   Rotate the switch counterclockwise.

10   Remove the switch from its bracket.

11   Installation is the reverse of removal.

12   Adjust the switch (see Chapter 1).

Chapter 10
Suspension and steering systems

Contents

Specifications

General

Power steering fluid type	See Chapter 1
Balljoint stud turning torque	
Stabilizer bar links	4.4 to 30 in-lbs
Upper control arm balljoint	9 to 39 in-lbs
Lower control arm balljoint	27 to 89 in-lbs

Torque specifications	Ft-lbs (unless otherwise indicated)	Nm
Front suspension		
Shock absorber-to-lower control arm bolt/nut............................	144	195
Shock absorber upper mounting nuts		
2013 and earlier models............................	21	28
2014 and later models............................	33	45
Shock absorber-to-axle housing............................	66	89
Upper control arm-to-frame bolt/nut............................	173	235
Lower control arm-to-frame bolt............................	207	280
Stabilizer bar-to-link nut............................	111	150
Stabilizer bar link-to-lower control arm nut............................	89	120
Stabilizer bar bracket-to-frame nut/bolt............................	51	69
Upper balljoint-to-steering knuckle nut............................	81	109
Lower balljoint-to-steering knuckle		
Bolts............................	221	300
Nut............................	123	167
Hub and wheel bearing assembly-to-knuckle bolts		
Front............................	73	99
Rear	64	86
Rear suspension		
Tundra		
Shock absorber-to-frame nut	21	28
Shock absorber-to-axle bolt/nut............................	66	90
Leaf spring		
To shackle bolt/nut............................	77	104
Shackle-to-frame bolt/nut............................	77	104
To frame bracket (front) bolt/nut............................	77	104
U-bolt-to-spring seat nuts	74	100
Sequoia		
Rear stabilizer bar		
Stabilizer bracket-to-body............................	20	27
Link-to-suspension arm	72	97
Control arms............................	137	186
Steering		
Power steering pressure line-to-pump union bolt............................	38	51
Power steering pump mounting bracket............................	21	28
Steering wheel nut............................	37	50
Steering gear............................	89	121
Steering column mounting nuts		
Sequoia............................	16	21
Tundra............................	15	20
Intermediate shaft-to-steering pinch bolts............................	26	35
Tie-rod end-to-steering knuckle nut............................	51	69

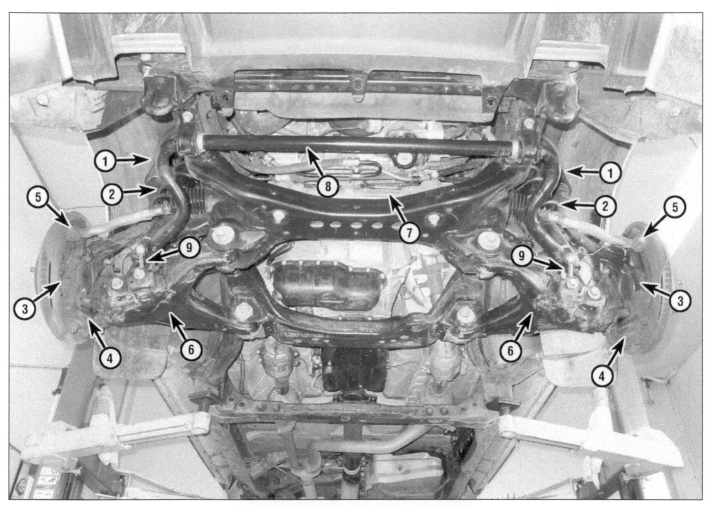

1.1 Front suspension and steering components

1	Upper control arm	4	Lower balljoint	7	Steering gear
2	Shock absorber/coil spring assembly	5	Tie-rod end	8	Stabilizer bar
3	Steering knuckle	6	Lower control arm	9	Stabilizer bar link

1   General information

Front suspension

Refer to illustration 1.1

The front suspension system is fully independent **(see illustration)**. The steering knuckles are connected to the upper and lower control arms by balljoints. The control arms are bolted to the frame. The shock absorbers and coil springs are integral assemblies; the upper ends are bolted to brackets on the frame and the lower ends are bolted to the lower control arms. All models use a front stabilizer bar to reduce vehicle roll during cornering.

Sequoia models are equipped with an air suspension system that uses a front absorber control actuator mounted to the top of the front shock absorbers. The system includes rear air cylinders, height sensors, and an air compressor that uses data from the ECU and suspension control ECU to control ride height and suspension control.

1.2 Rear suspension components - Tundra

1	Leaf spring shackle	3	Leaf spring	5	Spring hanger
2	Shock absorber	4	Spring seat	6	Rear axle housing

Rear suspension

Refer to illustration 1.2

The rear suspension on Tundra models consists of a pair of multi-leaf springs and two shock absorbers **(see illustration)**. The rear axle assembly is attached to the leaf springs by U-bolts. The front ends of the springs are attached to the frame at the front hangers, through rubber bushings. The rear ends of the springs are attached to the frame by shackles which allow the springs to alter their length when the vehicle is in operation.

The rear suspension on Sequoia models consists of a pair of coil springs, two shock absorbers, six suspension arms (four lower, two upper), two rear suspension air bags, two height sensors and an air compressor. A sta-bilizer bar, bolted to the axle and connected to the frame by a pair of links, reduces vehicle roll during cornering.

Warning: *On Sequoia models equipped with rear height control suspension, adjust the height control to the NORMAL mode, turn OFF the height control, then turn off the engine BEFORE raising the vehicle.*

2.2 To detach the lower end of the shock absorber/coil spring assembly from the lower control arm, remove this nut and bolt

2.3 To detach the upper end of the shock absorber/coil spring assembly from its mounting bracket, remove these three nuts (NOT the nut in the middle, which is the damper rod nut; it must never be removed unless the spring is compressed with a spring compressor)

2.7 Install the spring compressor(s) in accordance with the tool manufacturer's instructions; compress the spring until you can wiggle it before removing the damper rod nut

Steering

All models are equipped with power-assisted rack-and-pinion steering systems. The steering gear is bolted to the front crossmember and is connected to the steering knuckles by a pair of tie-rods.

Frequently, when working on the suspension or steering system components, you may come across fasteners which seem impossible to loosen. These fasteners on the underside of the vehicle are continually subjected to water, road grime, mud, etc., and can become rusted or frozen, making them extremely difficult to remove. In order to unscrew these stubborn fasteners without damaging them (or other components), use lots of penetrating oil and allow it to soak in for a while. Using a wire brush to clean exposed threads will also ease removal of the nut or bolt and prevent damage to the threads. Sometimes a sharp blow with a hammer and punch is effective in breaking the bond between a nut and bolt threads, but care must be taken to prevent the punch from slipping off the fastener and ruining the threads. Heating the stuck fastener and surrounding area with a torch sometimes helps too, but isn't recommended because of the obvious dangers associated with fire. Long breaker bars and extension, or cheater, pipes will increase leverage, but never use an extension pipe on a ratchet - the ratcheting mechanism could be damaged. Sometimes, turning the nut or bolt in the tightening (clockwise) direction first will help to break it loose. Fasteners that require drastic measures to unscrew should always be replaced with new ones.

Since most of the procedures that are dealt with in this Chapter involve jacking up the vehicle and working underneath it, a good pair of jackstands will be needed. A hydraulic floor jack is the preferred type of jack to lift the vehicle, and it can also be used to support certain components during various operations.

Warning: *Never, under any circumstances, rely on a jack to support the vehicle while*

working on it. Also, whenever any of the suspension or steering fasteners are loosened or removed they must be inspected and, if necessary, replaced with new ones of the same part number or of original equipment quality and design. Torque specifications must be followed for proper reassembly and component retention. Never attempt to heat or straighten suspension or steering components. Instead, replace bent or damaged parts with new ones.

---

2   Shock absorber/coil spring (front) - removal, component replacement and installation

Refer to illustrations 2.2, 2.3, 2.7 and 2.8

Warning: *Before undertaking the following procedure, be aware that disassembling the shock absorber/coil spring assemblies is a potentially dangerous job. Careless or unsafe work can cause serious injury. Use only a high-quality spring compressor and follow the spring compressor manufacturer's instructions. After removing the compressed spring, set it aside in a safe, isolated place.*

Warning: *On Sequoia models equipped with rear height control suspension, adjust the height control to the NORMAL mode, turn OFF the height control, then turn off the engine BEFORE raising the vehicle.*

Note: *If the shock absorber/coil spring assemblies must be replaced, you can save time by simply installing new complete shock/coil assemblies. Or, you can replace just the shocks or just the springs. But, to do so, you will have to disassemble the shock absorber/coil spring assemblies. Therefore, before deciding which way you want to go, find out the cost of each option. You may find that the cost for two assembled complete shock/coil*

assemblies is only slightly higher than the cost for two new shock absorbers or coil springs.

1   Loosen the front wheel lug nuts. Raise the vehicle and support it securely on jackstands. Remove the front wheels.

2   Remove the nut and bolt attaching the lower end of the shock absorber to the lower control arm **(see illustration)**. On 4WD models, push the bolt through the shock absorber and bracket as far as you can (but be careful not to nick the driveaxle), then pry the lower control arm down and remove the bolt.

3   On models equipped with front absorber control actuators, remove the two the actuator fasteners, actuator(s) and the bracket(s). Remove the nuts that attach the upper end of the shock to the frame bracket **(see illustration)**.

4   Remove the shock absorber/coil spring assembly.

5   Inspect the shock absorber for leaking fluid, dents, cracks and other damage. Inspect the coil spring for chips and cracks which could cause premature failure. Inspect the spring seats for hardness and general deterioration. If either the shock or the spring is worn or damaged, replace it. If you're installing new complete units, proceed to Step 14; if you're going to install new shocks or coil springs, proceed to the next Step.

6   Secure the shock/coil assembly in a bench vise. If you're planning to reuse the old shock absorbers, line the jaws of the vise with wood or shop rags to protect the shock bodies. Don't tighten the jaws any more than necessary; over-tightening the vise may crush the shock body.

7   Install a spring compressor in accordance with the tool manufacturer's instructions **(see illustration)**. (You can buy a spring compressor at most auto parts stores or rent one from most equipment yards on a daily basis.) Compress the spring far enough to relieve all pressure from the spring seat; when you can wiggle the spring, it's compressed enough to disassemble the shock/spring assembly.

2.8 Remove the damper shaft nut and the compressed spring assembly

3.2 Remove the stabilizer bar link-to-bar nut (A) then the link-to-control arm bolt (B) from each side

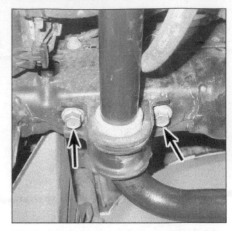

3.3 To separate the stabilizer bar from the frame, remove the bushing clamp bolts from both clamps

Warning: *Don't compress the spring any more than necessary.*

8   Hold the damper rod with a wrench and remove the damper rod nut **(see illustration)**.
9   Remove the upper retainer, upper cushion and suspension support.
10   Remove the compressed spring assembly and set it safely aside.
11   Remove the lower retainers and cushion.
12   Inspect the cushions for cracks and tears and general deterioration. Replace them if they're damaged. Make sure the retainers are in good condition too. If they're distorted or otherwise damaged, replace them.
13   Reassembly is the reverse of disassembly. Make sure the lower end of the coil spring is correctly seated in the low spot in the lower spring seat. Tighten the damper rod nut to the torque listed in this Chapter's Specifications before releasing tension on the spring.
14   Installation is the reverse of removal. Tighten the upper and lower fasteners to the torque listed in this Chapter's Specifications.

15   Tighten the lug nuts to the torque listed in the Chapter 1 Specifications.

3   Stabilizer bar and bushings (front) - removal and installation

Refer to illustrations 3.2 and 3.3
Warning: *On Sequoia models equipped with rear height control suspension, adjust the height control to the NORMAL mode, turn OFF the height control, then turn off the engine BEFORE raising the vehicle.*
1   Raise the vehicle and support it securely on jackstands.
2   Remove the nuts from the stabilizer bar links and detach the links from the bar. Also remove the bolts attaching the links to the lower control arms, then remove the links **(see illustration)**.
3   Remove the stabilizer bar bushing bracket nuts and/or bolts **(see illustration)**.
4   Remove the stabilizer bar.
5   Remove the rubber bushings from the

stabilizer bar.
6   Inspect the rubber bushings for cracks, tears and deterioration. If they're worn or damaged, replace them.
7   Check the balljoint on the lower end of each link for looseness or other signs of excessive wear. You can check the rotational torque required to turn the balljoint by threading the nut onto the ballstud and turning it with an inch-pound torque wrench and comparing your reading to the value listed in this Chapter's Specifications.
8   When you install the rubber bushings on the stabilizer bar, position them so the slits face toward the front of the vehicle.
9   Installation is otherwise the reverse of removal. Tighten all fasteners to the torque listed in this Chapter's Specifications.

4   Upper control arm (front) - removal and installation

Refer to illustrations 4.4a, 4.4b and 4.5
Warning: *On Sequoia models equipped with rear height control suspension, adjust the height control to the NORMAL mode, turn OFF the height control, then turn off the engine BEFORE raising the vehicle.*
1   Loosen the wheel lug nuts, raise the front of the vehicle and support it securely on jackstands. Apply the parking brake. Remove the wheel.
2   If equipped with ABS, disconnect the speed sensor wiring harness from the upper control arm and steering knuckle.
3   Remove the inner fender apron seal.
4   Remove the cotter pin, then loosen the upper balljoint nut a few turns. Using a two-jaw puller or a picklefork-type balljoint separator, separate the balljoint from the upper control arm **(see illustrations)**.
Caution: *Don't allow the steering knuckle to fall outward, as the brake hose may be damaged. It's a good idea to wire the steering knuckle to the coil spring so this doesn't happen.*

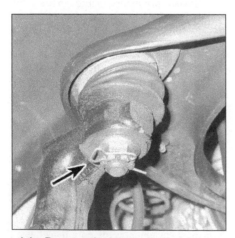

4.4a Remove the cotter pin, then loosen the upper balljoint nut a few turns

4.4b If you don't have a two-jaw puller, a picklefork-type balljoint separator can be used, but keep in mind that this tool will probably damage the balljoint seal

4.5 The upper control arm is attached to the frame with a single, long pivot bolt; remove the nut (A) and slide the bolt (B) out towards the rear of the vehicle

5.6 Remove the cotter pin, back off - but don't remove - the ballstud nut, then use a two-jaw puller to separate the ballstud from the lower control arm

5   Remove the nut, washer and pivot bolt and detach the upper control arm from the frame **(see illustration)**. Remove the arm.

6   Inspect the bushings for wear and deterioration. If they're cracked or damaged, take the arm to an automotive machine shop and have new bushings installed.

7   Installation is the reverse of removal. Tighten all suspension fasteners to the torque listed in this Chapter's Specifications, and use a new cotter pin on the upper control arm balljoint nut. If necessary, tighten the balljoint nut a little more to align the hole in the ballstud with the slots in the nut - don't loosen the nut to achieve this alignment.

Note: *The pivot bolt/nut should be tightened with the vehicle at normal ride height. This can be done after the vehicle has been lowered to the ground (on vehicles with adequate clearance), or can be simulated by raising the lower control arm with a floor jack.*

8   Tighten the lug nuts to the torque listed in the Chapter 1 Specifications.

9   It's a good idea to have the wheel alignment checked and, if necessary, adjusted.

5   Lower control arm - removal and installation

Refer to illustrations 5.6 and 5.7

Warning: *On Sequoia models equipped with rear height control suspension, adjust the height control to the NORMAL mode, turn OFF the height control, then turn off the engine BEFORE raising the vehicle.*

1   Loosen the wheel lug nuts, raise the front of the vehicle and support it securely on jackstands. Apply the parking brake. Remove the wheel.

2   Remove the under-vehicle splash shield (see Chapter 1, **illustrations 8.6a and 8.6b**).

3   Unbolt the steering gear assembly from the frame (see Section 21).

4   Unbolt the shock absorber/coil spring assembly from the lower control arm (see Section 2).

5   Remove the bolts and detach the stabilizer bar links from the lower control arms (see Section 3).

6   Remove the cotter pin, then loosen the

lower balljoint nut a few turns. Using a two-jaw puller or a picklefork-type balljoint separator, separate the balljoint from the lower control arm **(see illustration)**.

7   Make alignment marks on the front and rear adjusting cams **(see illustration)**. Hold the adjusting cam plate nuts with a wrench and unscrew the pivot bolts. Remove the bolts and detach the control arm from the frame. **Note:** *When removing the rear pivot bolts, you'll have to reposition the steering gear up and back to make room for the bolts to slide out* **(see illustration 21.8)**.

8   Inspect the bushings for wear and deterioration. If they're cracked or damaged, take the arm to an automotive machine shop and have new bushings installed.

9   Installation is the reverse of removal. Make sure that the alignment marks you made prior to disassembly are lined up. Tighten all suspension fasteners to the torque listed in this Chapter's Specifications, and use a new cotter pin on the lower control arm balljoint nut. If necessary, tighten the balljoint nut a little more to align the hole in the ballstud with the slots in the nut - don't loosen the nut to achieve this alignment.

Note: *The pivot bolts should be tightened with the vehicle at normal ride height. This can be done after the vehicle has been lowered to the ground (on vehicles with adequate clearance), or can be simulated by raising the lower control arm with a floor jack.*

10   Tighten the lug nuts to the torque listed in the Chapter 1 Specifications.

11   Have the wheel alignment checked and, if necessary, adjusted.

6   Steering knuckle - removal and installation

Warning: *On Sequoia models equipped with rear height control suspension, adjust the height control to the NORMAL mode,*

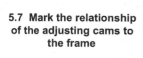

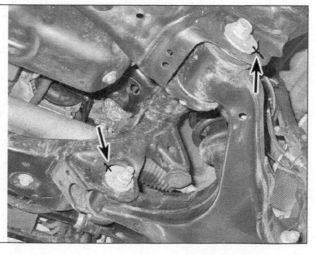

5.7 Mark the relationship of the adjusting cams to the frame

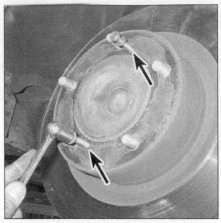

7.3 If the brake disc will not slide off of the bearing hub, install two bolts and tighten them equally until the disc is free of the hub

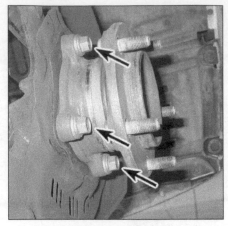

7.4 Front hub and wheel bearing assembly mounting bolt locations – three of four bolts shown

turn OFF the height control, then turn off the engine BEFORE raising the vehicle.

1   Loosen the wheel lug nuts. Raise the front of the vehicle and support it securely on jackstands. Apply the parking brake. Remove the wheel.

2   Detach all brake hose brackets from the steering knuckle. On vehicles with ABS, detach the wheel speed sensor from the steering knuckle.

3   Remove the brake caliper and brake disc (see Chapter 9). Hang the caliper with a length of wire - don't let it hang by the brake hose.

4   Disconnect the tie-rod end from the steering knuckle (see Section 19).

5   On 4WD models, remove the driveaxle/hub nut (see Chapter 8).

6   Disconnect the upper and lower control arms from the steering knuckle (see Sections 4 and 5), then remove the steering knuckle. On 4WD models, guide the driveaxle out of the hub, being careful to not overextend the inner CV joint. Support the driveaxle with a length of wire - don't let it hang by the inner CV joint.

7   Installation is the reverse of removal. Tighten all suspension fasteners to the torque values listed in this Chapter's Specifications.

8   Tighten the lug nuts to the torque listed in the Chapter 1 Specifications.

7   Hub and wheel bearing (front) - removal, bearing replacement and installation

Warning: *On Sequoia models equipped with rear height control suspension, adjust the height control to the NORMAL mode, turn OFF the height control, then turn off the engine BEFORE raising the vehicle.*

Removal

Refer to illustrations 7.3 and 7.4
Warning: *On 4WD models, the manufacturer recommends replacing the driveaxle/hub nuts*

with new ones whenever they are removed.

1   Remove the dust cover or grease cap and, on 4WD models, loosen the driveaxle/hub nut (see Chapter 8).

2   Loosen the wheel bolts, raise the vehicle and support it securely on jackstands. Remove the wheel.

3   Remove the brake caliper, the caliper mounting bracket and the brake disc from the knuckle (see Chapter 9).
Caution: *Be sure to support the brake caliper with a length of wire or rope.*
Note: *It may be necessary to press the brake disc from the hub assembly using two bolts* **(see illustration)**.

4   Remove the hub/bearing assembly mounting bolts **(see illustration)** from the front of the steering knuckle.

5   Remove the hub/bearing assembly from the steering knuckle and replace the O-ring on the hub.

Bearing replacement

6   Due to the special tools and expertise required to press the hub from the bearing, this job should be left to a professional mechanic. Take the hub and bearing assembly to an automotive machine shop or other qualified repair facility for service.

Installation

7   Make sure that the mounting surfaces inside the steering knuckle and on the driveaxle splines (4WD models) are smooth and free of burrs and nicks prior to installing the hub/bearing assembly.

8   On 2WD models, install the hub/bearing assembly to the steering knuckle and tighten the bolts to the torque listed in this Chapter's Specifications.

9   On 4WD models, lubricate the driveaxle splines with multi-purpose grease, then place the hub/bearing assembly onto the steering knuckle, tightening the bolts to the torque listed in this Chapter's Specifications. Install a new driveaxle/hub nut, but don't attempt to tighten it yet.

10   Install the brake disc, the caliper mounting bracket and the caliper; tighten the fasteners to the torque values listed in the Chapter 9 Specifications.

11   Install the wheel, remove the jackstands and lower the vehicle.

12   Tighten the driveaxle/hub nut to the torque listed in the Chapter 8 Specifications.
Note: *Have an assistant apply the brakes while tightening the driveaxle/hub nut.*

13   Tighten the lug nuts to the torque listed in the Chapter 1 Specifications.

8   Balljoints - check and replacement

Warning: *On Sequoia models equipped with rear height control suspension, adjust the height control to the NORMAL mode, turn OFF the height control, then turn off the engine BEFORE raising the vehicle.*

Check

1   Inspect the upper and lower balljoints for looseness whenever the vehicle is raised for any reason. You can check the balljoints with the suspension assembled as follows.

2   Raise the front of the vehicle and support it securely on jackstands.

3   Wipe the balljoints clean and inspect the seals for cuts and tears. If a balljoint seal is damaged, replace the balljoint.

Lower balljoint

4   With the vehicle raised and supported on jackstands and the suspension hanging free, attach a dial indicator to the lower control arm, with the plunger of the dial indicator touching the steering knuckle. Pull up and push down on the steering knuckle with a force of 65 pounds each way and check the dial indicator. If there is more than 0.020-inch of play, replace the balljoint.

Upper balljoint

5   With the vehicle raised and supported on jackstands, and a floor jack supporting the lower control arm (slightly raised), pry up and down on the upper control arm while feeling for play in the balljoint. If there is significant play, replace the balljoint.

All balljoints

6   Balljoints should also be checked whenever they're separated from the steering knuckle or the upper or lower control arm. See if you can turn the ballstud in its socket with your fingers. If the balljoint is loose, or if the ballstud can be turned, replace the balljoint.

7   Toyota also specifies a more accurate version of this bench test. Flip the balljoint stud back and forth five times, then install the nut. Using an inch-pound torque wrench, measure the turning torque as follows: turn the nut continuously at a rate of one turn every two to four seconds. On the fifth turn, read the indicated torque and compare it to the acceptable torque

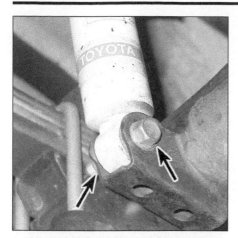

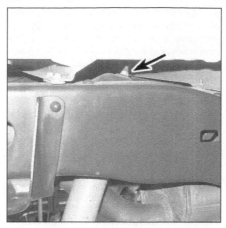

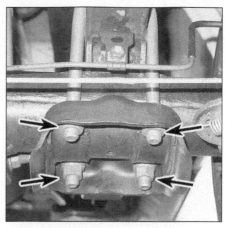

9.4 Shock absorber lower mounting bolt and nut (Tundra shown; on Sequoia models, the bolt threads into a boss on the rear axle or the lower control arm)

9.5 To detach the upper end of the shock absorber from the frame, remove this nut

10.2 To detach the spring from the axle, remove the four U-bolt nuts and washers

range listed in this Chapter's Specifications. If the indicated turning torque is not within the specified range, replace the balljoint.

Replacement

8   The balljoints are not serviceable from the control arm and must be replaced as an assembly.

9   Rear shock absorber - removal and installation

Refer to illustrations 9.4 and 9.5

Warning: *On Sequoia models equipped with rear height control suspension, adjust the height control to the NORMAL mode, turn OFF the height control, then turn off the engine BEFORE raising the vehicle.*

1   Loosen the rear wheel lug nuts. Raise the rear of the vehicle and support it securely

on jackstands. Block the front wheels so the vehicle doesn't roll off the stands. Remove the rear wheels.
2   Support the rear axle or control arm with a floor jack, placed near the shock absorber to be changed.
3   On Sequoia models, mark the height control sensor link-to-bracket, remove the nut and disconnect the sensor link from the bracket.
4   Remove the shock absorber lower mounting bolt **(see illustration)**.
Note: *On Sequoia models, it may be necessary to disconnect the rear exhaust system to access the mounting bolts.*
5   Remove the shock absorber upper mounting nut **(see illustration)**. Remove the shock absorber. Note the arrangement of the retainers and cushions.
6   Installation is the reverse of removal. Tighten the mounting fasteners to the torque listed in this Chapter's Specifications. Tighten

the lug nuts to the torque listed in the Chapter 1 Specifications.

10   Leaf spring (Tundra models) - removal and installation

Refer to illustrations 10.2, 10.4 and 10.5

1   Loosen the rear wheel lug nuts. Raise the rear of the vehicle and support it securely on jackstands. Block the front wheels to keep the vehicle from rolling off the stands. Remove the wheel.
2   Support the axle with a floor jack and raise it just enough to take the weight and relieve the tension on the leaf springs. Remove the four U-bolt nuts and washers **(see illustration)**.
3   Remove the spring seat and U-bolts. Note the location and position of any caster adjusting wedges.
4   Remove the nut from the hanger pin bolt, then remove the bolt **(see illustration)**.
5   Remove the nuts and washers from the shackle bolts **(see illustration)**. Remove the bolts and detach the shackle from the frame, then remove the spring assembly from the vehicle.
6   If the bushings at the ends of the spring are worn or deteriorated, an automotive machine shop can press the old ones out and press new ones in.
7   Installation is the reverse of removal. Gradually tighten the U-bolt nuts in a criss-cross pattern to the torque listed in this Chapter's Specifications. The hanger pin and shackle pin bolts/nuts should also be tightened to the torque listed in this Chapter's Specifications, but this should be done with the vehicle resting at normal ride height.
8   Tighten the lug nuts to the torque listed in the Chapter 1 Specifications.

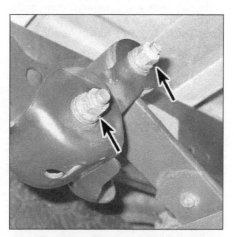

10.4 To detach the front end of the spring from the frame, remove the nut from the hanger pin bolt, then remove the bolt

10.5 To detach the rear end of the spring, remove the nuts from the shackle bolts, pull the bolts out, then detach the spring and shackle from the vehicle

11 Coil springs (rear, Sequoia models) - removal and installation

Warning: *On Sequoia models equipped with rear height control suspension, adjust the height control to the NORMAL mode, turn OFF the height control, then turn off the engine BEFORE raising the vehicle.*

1   Loosen the rear wheel lug nuts. Raise the rear of the vehicle and support it securely on jackstands. Block the front wheels to keep the vehicle from rolling off the stands. Remove the wheels.

2   Remove the shock absorbers (see Section 9).

3   Loosen, but don't remove all three of the upper control arm mounting bolts (see Section 14).

4   Place a block of wood on a floor jack then place the jack under the No. 2 lower control arm on the side that's being removed. Slowly raise the arm to the point just before the vehicle is starting to be lifted.

5   Install a spring compressor onto the spring in accordance with the tool manufacturer's instructions. (You can buy a spring compressor at most auto parts stores or rent one from most equipment yards on a daily basis.) Compress the spring far enough to relieve all pressure from the spring seat; when you can wiggle the spring.

Warning: *Don't compress the spring any more than necessary.*

6   Slowly lower the jack and arm until the spring and spring compressor can be removed through the opening.

7   Installation is the reverse of removal. Make sure that the lower end of each coil spring is seated properly on the spring seat. Tighten all suspension fasteners to the torque listed in this Chapter's Specifications.

Note: *Before tightening the upper control arm bolts, raise the suspension with a floor jack to simulate normal ride height.*

8   Tighten the lug nuts to the torque listed in the Chapter 1 Specifications.

12 Stabilizer bar (rear, Sequoia models) - removal and installation

Warning: *On Sequoia models equipped with rear height control suspension, adjust the height control to the NORMAL mode, turn OFF the height control, then turn off the engine BEFORE raising the vehicle.*

1   Raise the rear of the vehicle and support it securely on jackstands. Block the front wheels to keep the vehicle from rolling off the stands.

2   Remove the stabilizer bar link nuts and remove both links.

3   Detach the bushing bracket bolts.

4   Remove the stabilizer bar.

5   Inspect the rubber bushings for cracks, tears and deterioration. If they're worn or damaged, replace them.

6   Check the balljoint on the lower end of each link for looseness or other signs of excessive wear. You can check the rotational torque required to turn the balljoint by threading the nut onto the ballstud and turning it with an inch-pound torque wrench and comparing your reading to the value listed in this Chapter's Specifications.

7   When you install the rubber bushings on the stabilizer bar, position them so the slits face toward the front of the vehicle.

8   Installation is otherwise the reverse of removal. Tighten all fasteners to the torque listed in this Chapter's Specifications.

13 Lower suspension arms (rear, Sequoia models) - removal and installation

Warning: *Remove and install one control arm at a time to prevent the axle from shifting.*
Warning: *On Sequoia models equipped with rear height control suspension, adjust the height control to the NORMAL mode, turn OFF the height control, then turn off the engine BEFORE raising the vehicle.*
Caution: *Always support the lower control arm before removing the lower shock mounting bolts, because the lower control arm will drop about 3 inches.*
Note: *Inspect the bushings for wear and deterioration. If they're cracked or damaged, take the arm to an automotive machine shop and have new bushings installed.*

1   Loosen the rear wheel lug nuts, raise the rear of the vehicle and support it securely on jackstands. Block the front wheels to keep the vehicle from rolling off the stands. Remove the wheel.

Lower control arm (No. 1)

2   Remove the rear shock absorber (see Section 9).

3   Make alignment marks on the front and outer adjusting cams. Hold the adjusting cam plate nuts with a wrench and unscrew the pivot bolts and rear bolt. Remove all the bolts and detach the lower control arm from the frame.

4   Installation is the reverse of removal. Make sure that the alignment marks you made prior to disassembly are lined up. Tighten all suspension fasteners to the torque listed in this Chapter's Specifications.

Note: *The pivot bolts should be tightened with the vehicle at normal ride height. This can be done after the vehicle has been lowered to the ground (on vehicles with adequate clearance), or can be simulated by raising the lower control arm with a floor jack.*

Lower control arm (No. 2)

5   Remove the rear coil spring (see Section 11) or air bag (see Section 16).

6   If equipped remove the rear air bags assembly (see Section 16).

7   Remove the cotter pin, then loosen the

lower balljoint nut a few turns. Using a two-jaw puller or a picklefork-type balljoint separator, separate the balljoint from the lower control arm.

8   Make alignment marks on the adjusting cams. Hold the adjusting cam plate nut with a wrench and unscrew the bolt. Remove all the bolt and detach the lower control arm from the frame.

9   Installation is the reverse of removal. Make sure that the alignment marks you made prior to disassembly are lined up. Tighten all suspension fasteners to the torque listed in this Chapter's Specifications.

Note: *The pivot bolts should be tightened with the vehicle at normal ride height. This can be done after the vehicle has been lowered to the ground (on vehicles with adequate clearance), or can be simulated by raising the lower control arm with a floor jack.*

14 Upper control arm (rear, Sequoia models) - removal and installation

Warning: *Remove and install one control arm at a time to prevent the axle from shifting.*
Warning: *On Sequoia models equipped with rear height control suspension, adjust the height control to the NORMAL mode, turn OFF the height control, then turn off the engine BEFORE raising the vehicle.*
Caution: *Always support the lower control arm before removing the lower shock mounting bolts, because the lower control arm will drop about 3 inches.*
Note: *Inspect the bushings for wear and deterioration. If they're cracked or damaged, take the arm to an automotive machine shop and have new bushings installed.*

1   Loosen the rear wheel lug nuts, raise the rear of the vehicle and support it securely on jackstands. Block the front wheels to keep the vehicle from rolling off the stands. Remove the wheel.

2   Remove the fuel tank (see Chapter 4).

3   Mark the height control sensor link-to-bracket then remove the nut and disconnect the sensor link from the bracket.

4   Remove the rear brake disc (see Chapter 9).

5   Remove the rear brake hose three mounting bracket bolts and brackets then move the hose out of the way.

Note: *DO not disconnect the hose or the brake system will need to be bleed.*

6   Disconnect the rear wheel speed sensor (see Chapter 9), then remove the harness fasteners and move the electrical harness out of the way.

7   Remove the lower arm mounting bolts and nuts and maneuver the arm out of the vehicle.

8   Installation is the reverse of removal. Make sure that the alignment marks you made prior to disassembly are lined up. Tighten all suspension fasteners to the torque listed in this Chapter's Specifications.

Note: *The pivot bolts should be tightened with the vehicle at normal ride height. This can be done after the vehicle has been lowered to the ground (on vehicles with adequate clearance), or can be simulated by raising the lower control arm with a floor jack.*

15 Hub and bearing assembly (rear, Sequoia models) - removal and installation

Warning: *On Sequoia models equipped with rear height control suspension, adjust the height control to the NORMAL mode, turn OFF the height control, then turn off the engine BEFORE raising the vehicle.*

Removal

Warning: *The manufacturer recommends replacing driveaxle/hub nuts with new ones whenever they are removed.*

1   Loosen the wheel bolts, raise the vehicle and support it securely on jackstands. Remove the wheel. Block the front wheels to keep the vehicle from rolling off the stands.
2   Remove the rear coil spring (see Section 11).
3   Remove the rear suspension air bags (see Section 16).
4   Remove the brake caliper, the caliper mounting bracket and the brake disc (see Chapter 9).
Caution: *Be sure to support the brake caliper with a length of wire or rope.*
Note: *It may be necessary to press the brake disc from the hub assembly using two bolts **(see illustration 7.3)**.*
5   Remove the rear wheel speed sensor and harness (see Chapter 9).
6   Remove the grease cap using a screwdriver and hammer to pry it out of the hub.
7   Remove the cotter pin and adjusting lock cap, then remove the driveaxle/hub nut (see Chapter 8).
8   Remove the No.2 lower control arm (see Section 13).
9   Disconnect the upper control arm from the knuckle and slide the knuckle off of the driveaxle.
10   Remove the hub/bearing assembly mounting bolts from the back of the knuckle.
11   Remove the hub/bearing assembly from the steering knuckle and replace the O-ring on the hub.

Bearing replacement

12   Due to the special tools and expertise required to press the hub from the bearing, this job should be left to a professional mechanic. Take the hub and bearing assembly to an automotive machine shop or other qualified repair facility for service.

Installation

13   Make sure that the mounting surfaces inside the steering knuckle and on the driveaxle splines are smooth and free of burrs and nicks prior to installing the hub/bearing assembly.

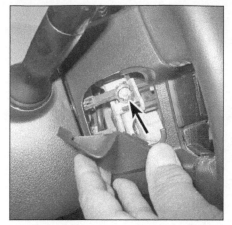

17.3a Pry off the covers on the left side . . .

14   Lubricate the driveaxle splines with multi-purpose grease, then place the hub/bearing assembly onto the knuckle and tighten the bolts to the torque listed in this Chapter's Specifications. Install the knuckle, making sure the driveaxle slides through the bearing. Install a new driveaxle/hub nut, but don't attempt to tighten it yet.
15   Installation is the reverse of removal. Tighten all suspension fasteners to the torque listed in this Chapter's Specifications.
Note: *The pivot bolts should be tightened with the vehicle at normal ride height. This can be done after the vehicle has been lowered to the ground (on vehicles with adequate clearance), or can be simulated by raising the lower control arm with a floor jack.*
16   Install the wheels, remove the jackstands and lower the vehicle.
17   Tighten the driveaxle/hub bolt to the torque listed in the Chapter 8 Specifications.
Note: *Have an assistant apply the brakes while tightening the driveaxle/hub bolt.*
18   Tighten the lug nuts to the torque listed in the Chapter 1 Specifications.

16 Suspension air bags (rear, Sequoia models) - removal and installation

Warning: *On Sequoia models equipped with rear height control suspension, adjust the height control to the NORMAL mode, turn OFF the height control, then turn off the engine BEFORE raising the vehicle.*
1   Raise the rear of the vehicle and support it securely on jackstands. Block the front wheels to keep the vehicle from rolling off the stands.
2   Place a block of wood on a floor jack, then place the jack under the No. 2 lower control arm on the side that's being removed. Slowly raise the arm to the point the airbag is compressed.
3   Remove the airbag retaining bolt from the bottom of the control arm.
4   Remove the clip from the upper side of the airbag.

17.3b . . . then the right side of the steering wheel and, using a Torx bit of the correct size, loosen the Torx screws that retain the airbag module; back off the screws until the grooves in their circumference catch on the screw cases (when you feel the screws catch, don't force them beyond that point)

5   Disconnect the height control "air" tube from the air bag using special tool # 09730-00010, or equivalent.
Note: *Always replace the tube O-rings.*
6   Carefully lower the jack and remove the air bag.
7   Installation is the reverse of removal. Tighten both nuts to the torque listed in this Chapter's Specifications (the vehicle should be at normal ride height when the bolts are tightened).

17 Steering wheel - removal and installation

Refer to illustrations 17.3a, 17.3b, 17.4, 17.5, 17.6, 17.7 and 17.9

Warning: *The models covered by this manual are equipped with airbags. The airbag is armed and can deploy (inflate) anytime the battery is connected. To prevent accidental deployment (and possible injury), turn the ignition key to LOCK and disconnect the negative battery cable whenever working near airbag components. After the battery is disconnected, wait at least two minutes before beginning work (the system has a back-up capacitor that must fully discharge). For more information see Chapter 12.*
1   Park the vehicle with the front wheels pointed straight ahead and the steering wheel centered. Disconnect the cable from the negative terminal of the battery (see Chapter 5).
2   Disable the airbag system (see Chapter 12).
3   Pry out the two small covers, one on each side of the steering wheel, and loosen the two Torx screws that retain the airbag module **(see illustrations)**. Back out the screws until the grooves along the screw circumferences catch on the screw cases.

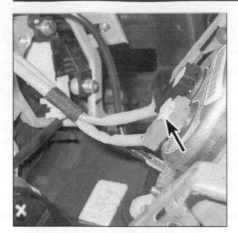

17.4 To unplug the airbag connector, lift the locking tab up, then disconnect the connector from the airbag

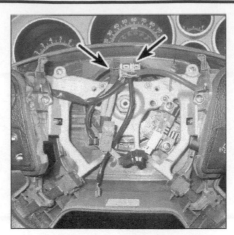

17.5 Unplug any other electrical connectors that would interfere with steering wheel removal

17.6 Remove the steering wheel nut, then mark the relationship of the steering wheel to the steering shaft

4   Lift the airbag module off the steering wheel and disconnect the airbag connector **(see illustration)**.
Warning: *When handling the airbag module, hold it with the trim side facing away from you. Set the airbag module down in a safe location with the trim side facing up.*
5   Unplug any other electrical connectors, such as the one for the horn **(see illustration)**.
6   Remove the steering wheel retaining nut and mark the position of the steering wheel to the shaft, if marks don't already exist or don't line up **(see illustration)**.
7   Use a puller to detach the steering wheel from the shaft **(see illustration)**. Don't hammer on the shaft to dislodge the wheel.
Warning: *While the steering wheel is removed, do NOT turn the steering shaft. If the steering shaft is turned, the spiral cable will be un-centered and the harness will break, rendering the airbag inoperative. If the spiral cable is accidentally un-centered, it must be*

centered before installing the steering wheel.
8   Remove the steering column covers (see Chapter 11).
9   Make sure the spiral cable is centered, as follows: Verify that the front wheels are pointing straight ahead. Turn the spiral cable housing counterclockwise by hand until it becomes hard to turn. Turn the spiral cable clockwise about 2-1/2 turns and align the marks **(see illustration)**. If it's necessary to remove the spiral cable from the combination switch, disengage the retaining clips, follow the wiring harness down the steering column and disconnect the electrical connector, then remove the spiral cable unit.
10   To install the wheel, align the mark on the steering wheel hub with the mark on the shaft and slide the wheel onto the shaft. Install the nut and tighten it to the torque listed in this Chapter's Specifications.
11   Installation is otherwise the reverse of removal.

18   Steering column - removal and installation

Warning: *The models covered by this manual are equipped with airbags. The airbag is armed and can deploy (inflate) anytime the battery is connected. To prevent accidental deployment (and possible injury), turn the ignition key to LOCK and disconnect the negative battery cable whenever working near airbag components. After the battery is disconnected, wait at least two minutes before beginning work (the system has a back-up capacitor that must fully discharge). For more information see Chapter 12.*

Removal

Refer to illustrations 18.6, 18.7 and 18.8
1   Park the vehicle with the wheels pointing straight ahead. Disconnect the cable from the negative terminal of the battery.

17.7 Use a steering wheel puller to remove the steering wheel

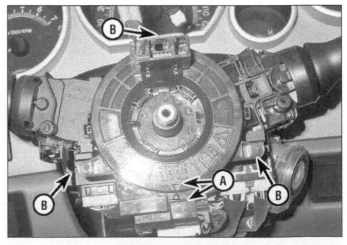

17.9 Spiral cable details

A   *Alignment marks*   B   *Retaining clips*

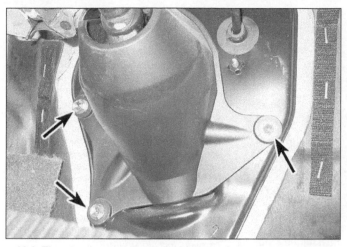

18.6 The steering column hole cover is retained by three bolts

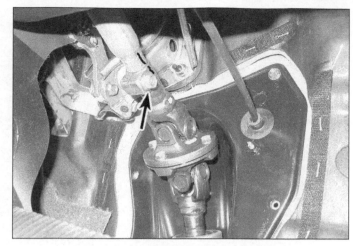

18.7 Mark the U-joint to the steering shaft, then
remove the pinch bolt

2   Remove the steering wheel (see Section 17), then turn the ignition key to the LOCK position to prevent the steering shaft from turning.
Caution: *If this is not done, the airbag spiral cable could be damaged.*
3   Remove the lower finish panel under the column (see Chapter 11) and the heater duct behind it.
4   Remove the steering column covers (see Chapter 11).
5   Disconnect the electrical connectors for the steering column harness. Disconnect the shift cable (see Chapter 7A).
6   Remove the column hole cover at the firewall **(see illustration)**.
7   Mark the relationship of the intermediate shaft U-joint to the steering shaft, then remove the pinch bolt **(see illustration)**.
8   Remove the steering column mounting fasteners **(see illustration)**, lower the column and pull it to the rear, making sure noth-

ing is still connected. Separate the intermediate shaft from the steering shaft and remove the column.

Installation
9   Guide the steering column into position, then align the marks made in Step 7 and connect the intermediate shaft.
10   Install and tighten the column mounting fasteners to the torque listed in this Chapter's Specifications.
11   Install the pinch bolt, tightening it to the torque listed in this Chapter's Specifications.
12   The remainder of installation is the reverse of removal.
Warning: *If you're installing the steering column on a Sequoia model, the zero point calibration of the steering angle, master cylinder pressure, yaw rate and deceleration sensors for the Vehicle Skid Control (VSC) system must be performed. This will require taking the vehicle to a dealer service depart-*

ment or other qualified repair shop equipped with the special tool necessary to carry out this procedure.

19   Tie-rod ends - removal and installation

Refer to illustrations 19.2a, 19.2b, 19.3a and 19.3b
Warning: *On Sequoia models equipped with rear height control suspension, adjust the height control to the NORMAL mode, turn OFF the height control, then turn off the engine BEFORE raising the vehicle.*
1   Loosen the wheel lug nuts, raise the vehicle and place it securely on jackstands. Remove the wheel.
2   Loosen the tie-rod end locknut and mark the position of the tie-rod end on the threaded portion of the tie-rod **(see illustrations)**.

18.8 Steering column mounting fasteners

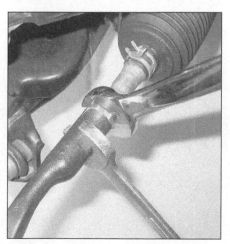

19.2a Loosen the tie-rod end locknut . . .

19.2b . . . and mark the position of the tie-
rod end on the threaded portion
of the tie-rod

19.3a Remove the cotter pin . . .

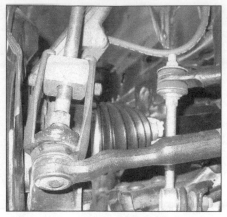

19.3b . . . loosen (don't remove) the castle nut from the tie-rod end balljoint stud, then install a puller and separate the tie-rod end from the steering knuckle

3   Remove the cotter pin and loosen the castle nut from the tie-rod end balljoint stud, then install a puller and separate the tie-rod end from the steering knuckle **(see illustrations)**. Remove the nut and detach the tie-rod end from the steering knuckle arm.
4   Unscrew the old tie-rod end and install the new one. Make sure the new tie-rod end is aligned with the mark you made on the threads of the tie-rod.
5   Installation is the reverse of removal. Tighten the tie-rod end balljoint nut to the torque listed in this Chapter's Specifications. Tighten the locknut securely.
6   Tighten the lug nuts to the torque listed in the Chapter 1 Specifications.

20 Steering gear boots - replacement

Refer to illustration 20.4
Warning: *On Sequoia models equipped with rear height control suspension, adjust the height control to the NORMAL mode, turn OFF the height control, then turn off the*

engine BEFORE raising the vehicle.
1   If a steering gear boot is torn, dirt and moisture can damage the steering gear. Replace it.
2   Loosen the wheel lug nuts, raise the vehicle and place it securely on jackstands. Remove the front wheels.
3   Remove the tie-rod end and locknut (see Section 19).
4   Remove the boot clamps **(see illustration)** and slide the boot off the tie-rod.
5   Installation is the reverse of removal. Use new clamps on the boot. Tighten the lug nuts to the torque listed in the Chapter 1 Specifications.

21 Steering gear - removal and installation

Refer to illustrations 21.5, 21.7 and 21.9
Warning: *The models covered by this manual are equipped with airbags. The airbag is armed and can deploy (inflate) anytime the*

battery is connected. To prevent accidental deployment (and possible injury), turn the ignition key to LOCK and disconnect the negative battery cable whenever working near airbag components. After the battery is disconnected, wait at least two minutes before beginning work (the system has a back-up capacitor that must fully discharge). For more information, see Chapter 12.
Warning: *If you're installing the steering gear on a Sequoia model and some Tundra models, the zero point calibration of the steering angle, master cylinder pressure, yaw rate and deceleration sensors for the Vehicle Skid Control (VSC) system must be performed. This will require taking the vehicle to a dealer service department or other qualified repair shop equipped with the special tool necessary to carry out this procedure.*
Warning: *Make sure the steering column shaft is not turned while the steering gear is removed or you could damage the airbag system clockspring. To prevent the shaft from turning, turn the ignition key to the lock position before beginning work, and run the seat belt through the steering wheel and clip it into its latch.*
Warning: *On Sequoia models equipped with rear height control suspension, adjust the height control to the NORMAL mode, turn OFF the height control, then turn off the engine BEFORE raising the vehicle.*
Note: *On 2008 through 2015 Sequoia models and 2010 through 2017 Tundra models, the engine must be removed before the steering gear can be removed from the vehicle (see Chapter 2C).*
1   Park the vehicle with the front wheels pointing straight ahead.
2   Loosen the wheel lug nuts. Raise the front of the vehicle and support it securely on jackstands. Apply the parking brake. Remove the wheels. Remove the under-vehicle splash shield **(see illustrations 8.6a and 8.6b** in Chapter 1).
3   Disconnect the tie-rod ends from the steering knuckles (see Section 19).
4   Remove the stabilizer bar (see Section 3).

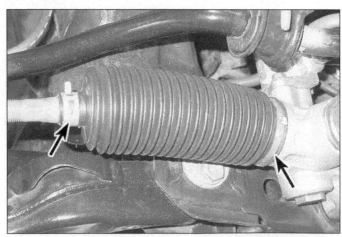

20.4 The outer clamp on the steering gear boot can be removed with a pair of pliers - the inner clamp must be cut off

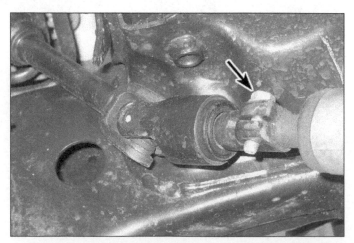

21.5 Mark the relationship of the intermediate shaft U-joint coupler to the steering gear and intermediate shaft, then remove the coupler pinch bolts and detach it from the steering gear

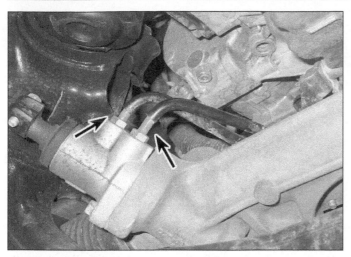

21.7 Steering gear pressure and return lines

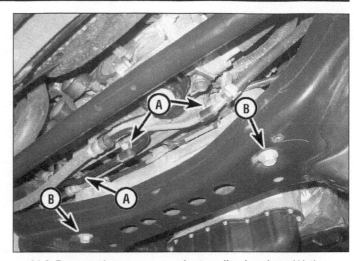

21.8 Remove the pressure and return line brackets (A) then remove the steering gear mounting fasteners (B)

5   Mark the relationship of the U-joint to the intermediate shaft and the steering gear **(see illustration)**, remove the U-joint pinch bolts, then slide the U-joint up and off the steering gear input shaft.
6   Remove the oil filter bracket on 1UR-FE, 3UR-FE and 3UR-FBE engines (see Chapter 2B).
7   Disconnect the pressure and return lines from the steering gear **(see illustration)**.
8   On 2016 and later models, remove the front differential assembly (see Chapter 8).
9   Remove the steering gear mounting fasteners (see illustration).
10   Move the steering gear to the right until the left side of the steering gear can be dropped down, then slide the steering gear down and out of the vehicle.
11   Installation is the reverse of removal. Align the matchmarks made in Step 5 and tighten all suspension and steering gear fasteners to the

torque listed in this Chapter's Specifications. Tighten the wheel lug nuts to the torque listed in the Chapter 1 Specifications.
12   Bleed the power steering system (see Section 23).

---

22   Power steering pump - removal and installation

Refer to illustrations 22.7, 22.8 and 22.9
Warning: *On Sequoia models equipped with rear height control suspension, adjust the height control to the NORMAL mode, turn OFF the height control, then turn off the engine BEFORE raising the vehicle.*
1   Disconnect the cable from the negative terminal of the battery.
2   Remove the engine cover (see Chapter 1).

3   Raise the front of the vehicle and support it securely on jackstands. Remove the accessory drivebelt or serpentine drivebelt (see Chapter 1).
4   Remove the right front wheel. Remove the plastic clips from the the rubber seal on the inner fenderwell splash shield (see Chapter 11), and remove the seal.
5   Remove the drivebelt (see Chapter 1).
6   Disconnect the electrical connector from the power steering pressure switch.
7   Position a drain pan under the power steering pump. Disconnect the pressure and return hoses from the backside of the pump **(see illustration)**. Plug the hoses to prevent contaminants from entering. Discard the sealing washers - new ones should be used during installation.
8   Rotate the pump pulley to access the two mounting bolts through the pulley **(see illustration)**.

22.7 Power steering pump mounting details

1   *Power steering pressure switch electrical connector*
2   *Return hose*
3   *Pressure hose*

22.8 Rotate the pump pulley until the bolts are exposed, then remove the bolts from the front

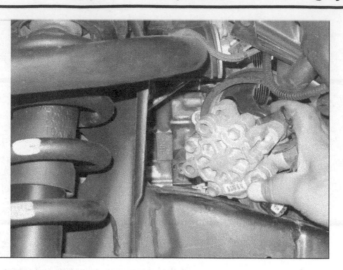

22.9 Remove the power steering pump through the right side wheel well

9   Remove the pump mounting fasteners and lift the pump out through the wheelwell **(see illustration)**, taking care not to spill fluid on the painted surfaces.

10   Installation is the reverse of removal. Use new sealing washers on either side of the pressure line fitting. Tighten the pressure line union bolt and the pump mounting bolts and nuts to the torque listed in this Chapter's Specifications.

11   Fill the power steering reservoir with the recommended fluid (see Chapter 1) and bleed the system (see Section 23).

23   Power steering system - bleeding

1   Following any operation in which the power steering fluid lines have been disconnected, the power steering system must be bled to remove all air and obtain proper steering performance.

2   With the front wheels in the straight ahead position, check the power steering fluid level and, if low, add fluid until it reaches the Cold mark on the dipstick.

3   Start the engine and allow it to run at fast idle. Recheck the fluid level and add more if necessary to reach the Cold mark on the dipstick.

4   Bleed the system by turning the wheels from side-to-side, without hitting the stops. This will work the air out of the system. Keep the reservoir full of fluid as this is done.

5   When the air is worked out of the system, return the wheels to the straight ahead position and leave the vehicle running for several more minutes before shutting it off.

6   Road test the vehicle to be sure the steering system is functioning normally and noise free.

7   Recheck the fluid level to be sure it's up to the Hot mark on the dipstick while the engine is at normal operating temperature. Add fluid if necessary (see Chapter 1).

24   Wheels and tires - general information

Refer to illustration 24.1

All vehicles covered by this manual are equipped with metric-size fiberglass or steel belted radial tires **(see illustration)**. Use of other size or type of tires may affect the ride and handling of the vehicle. Don't mix different types of tires, such as radials and bias belted, on the same vehicle as handling may be seriously affected. It's rec-

ommended that tires be replaced in pairs on the same axle, but if only one tire is being replaced, be sure it's the same size, structure and tread design as the other.

Because tire pressure has a substantial effect on handling and wear, the pressure on all tires should be checked at least once a month or before any extended trips (see Chapter 1).

Wheels must be replaced if they're bent, dented, leak air, have elongated bolt holes, are heavily rusted, out of vertical symmetry or if the lug nuts won't stay tight. Wheel repairs that use welding or peening are not recommended.

Tire and wheel balance is important to the overall handling, braking and performance of the vehicle. Unbalanced wheels can adversely affect handling and ride characteristics as well as tire life. Whenever a tire is installed on a wheel, the tire and wheel should be balanced by a shop with the proper equipment.

25   Front end alignment - general information

Refer to illustration 25.1

A front end alignment refers to the adjustments made to the front wheels so they're in proper angular relationship to the suspension and the ground **(see illustra-**

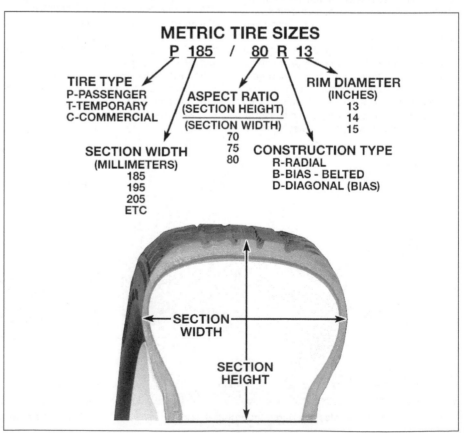

24.1 Metric tire size code

tion). Front wheels that are out of proper alignment not only affect steering control, but also increase tire wear.

Getting the proper front wheel alignment is a very exacting process, one in which complicated and expensive machines are necessary to perform the job properly. Because of this, you should have a technician with the proper equipment perform these tasks. We will, however, use this space to give you a basic idea of what is involved with front end alignment so you can better understand the process and deal intelligently with the shop that does the work.

Toe-in is the turning in of the front wheels. The purpose of a toe specification is to ensure parallel rolling of the front wheels. In a vehicle with zero toe-in, the distance between the front edges of the wheels will be the same as the distance between the rear edges of the wheels. The actual amount of toe-in is normally only a fraction of an inch. Toe-in adjustment is controlled by the tie-rod length. Incorrect toe-in will cause the tires to wear improperly by making them scrub against the road surface.

Camber is the tilting of the front wheels from vertical when viewed from the front of the vehicle. When the wheels tilt out at the top, the camber is said to be positive (+). When the wheels tilt in at the top the camber is negative (-). The amount of tilt is measured in degrees from vertical and this measurement is called the camber angle. This angle affects the amount of tire tread which contacts the road and compensates for changes in the suspension geometry when the vehicle is cornering or traveling over an undulating surface. Camber is adjusted by rotating cam-shaped adjusters at the front and rear of the lower control arm.

Caster is the tilting of the top of the front steering axis from the vertical. A tilt toward the rear is positive caster and a tilt toward the front is negative caster. Caster is also adjusted by rotating cam-shaped adjusters at the front and rear of the lower control arm.

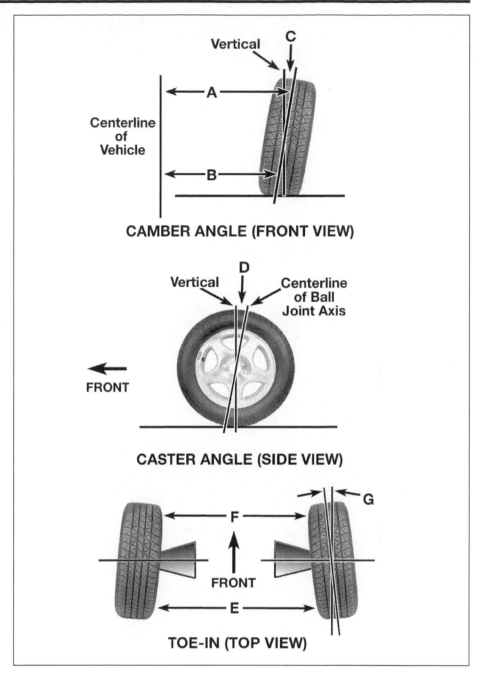

25.1 Front end alignment details

A minus B = C (degrees camber)
D = degrees caster

E minus F = toe-in (measured in inches)
G = toe-in (expressed in degrees)

Notes

Chapter 11   Body

Contents

1   General information

Warning: *The models covered by this manual are equipped with Supplemental Restraint Systems (SRS), more commonly known as airbags. Always disable the airbag system before working in the vicinity of any airbag system components to avoid the possibility of accidental deployment of the airbags, which could cause personal injury (see Chapter 12).*

Certain body components are particularly vulnerable to accident damage and can be unbolted and repaired or replaced. Among these parts are the hood, doors, tailgate, liftgate, bumpers and front fenders.

Only general body maintenance practices and body panel repair procedures within the scope of the do-it-yourselfer are included in this Chapter.

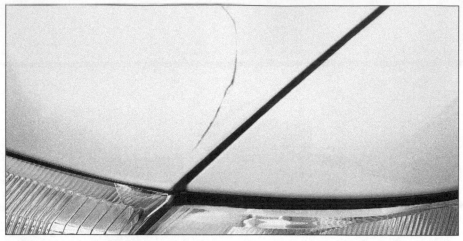

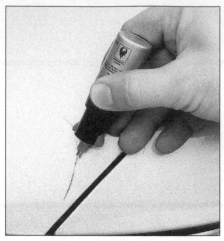

Make sure the damaged area is perfectly clean and rust free. If the touch-up kit has a wire brush, use it to clean the scratch or chip. Or use fine steel wool wrapped around the end of a pencil. Clean the scratched or chipped surface only, not the good paint surrounding it. Rinse the area with water and allow it to dry thoroughly

Thoroughly mix the paint, then apply a small amount with the touch-up kit brush or a very fine artist's brush. Brush in one direction as you fill the scratch area. Do not build up the paint higher than the surrounding paint

2   Repair minor paint scratches

No matter how hard you try to keep your vehicle looking like new, it will inevitably be scratched, chipped or dented at some point. If the metal is actually dented, seek the advice of a professional. But you can fix minor scratches and chips yourself. Buy a touch-up paint kit from a dealer service department or an auto parts store. To ensure that you get the right color, you'll need to have the specific make, model and year of your vehicle and, ideally, the paint code, which is located on a special metal plate under the hood or in the door jamb.

3   Body repair - minor damage

Plastic body panels

The following repair procedures are for minor scratches and gouges. Repair of more serious damage should be left to a dealer service department or qualified auto body shop. Below is a list of the equipment and materials necessary to perform the following repair procedures on plastic body panels.

Wax, grease and silicone removing solvent
Cloth-backed body tape
Sanding discs
Drill motor with three-inch disc holder
Hand sanding block
Rubber squeegees
Sandpaper
Non-porous mixing palette
Wood paddle or putty knife
Curved-tooth body file
Flexible parts repair material

Flexible panels (bumper trim)
1   Remove the damaged panel, if necessary or desirable. In most cases, repairs can be car-

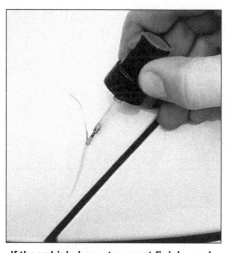

If the vehicle has a two-coat finish, apply the clear coat after the color coat has dried

ried out with the panel installed.
2   Clean the area(s) to be repaired with a wax, grease and silicone removing solvent applied with a water-dampened cloth.
3   If the damage is structural, that is, if it extends through the panel, clean the backside of the panel area to be repaired as well. Wipe dry.
4   Sand the rear surface about 1-1/2 inches beyond the break.
5   Cut two pieces of fiberglass cloth large enough to overlap the break by about 1-1/2 inches. Cut only to the required length.
6   Mix the adhesive from the repair kit according to the instructions included with the kit, and apply a layer of the mixture approximately 1/8-inch thick on the backside of the panel. Overlap the break by at least 1-1/2 inches.
7   Apply one piece of fiberglass cloth to the adhesive and cover the cloth with additional adhesive. Apply a second piece of fiberglass

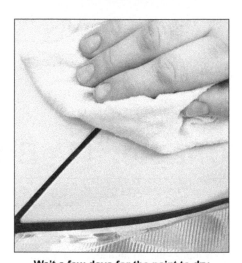

Wait a few days for the paint to dry thoroughly, then rub out the repainted area with a polishing compound to blend the new paint with the surrounding area. When you're happy with your work, wash and polish the area

cloth to the adhesive and immediately cover the cloth with additional adhesive in sufficient quantity to fill the weave.
8   Allow the repair to cure for 20 to 30 minutes at 60-degrees to 80-degrees F.
9   If necessary, trim the excess repair material at the edge.
10   Remove all of the paint film over and around the area(s) to be repaired. The repair material should not overlap the painted surface.
11   With a drill motor and a sanding disc (or a rotary file), cut a "V" along the break line approximately 1/2-inch wide. Remove all dust and loose particles from the repair area.
12   Mix and apply the repair material. Apply a light coat first over the damaged area; then continue applying material until it reaches a level

slightly higher than the surrounding finish.

13   Cure the mixture for 20 to 30 minutes at 60-degrees to 80-degrees F.

14   Roughly establish the contour of the area being repaired with a body file. If low areas or pits remain, mix and apply additional adhesive.

15   Block sand the damaged area with sandpaper to establish the actual contour of the surrounding surface.

16   If desired, the repaired area can be temporarily protected with several light coats of primer. Because of the special paints and techniques required for flexible body panels, it is recommended that the vehicle be taken to a paint shop for completion of the body repair.

Steel body panels

See photo sequence

Repair of dents

17   When repairing dents, the first job is to pull the dent out until the affected area is as close as possible to its original shape. There is no point in trying to restore the original shape completely as the metal in the damaged area will have stretched on impact and cannot be restored to its original contours. It is better to bring the level of the dent up to a point that is about 1/8-inch below the level of the surrounding metal. In cases where the dent is very shallow, it is not worth trying to pull it out at all.

18   If the backside of the dent is accessible, it can be hammered out gently from behind using a soft-face hammer. While doing this, hold a block of wood firmly against the opposite side of the metal to absorb the hammer blows and prevent the metal from being stretched.

19   If the dent is in a section of the body which has double layers, or some other factor makes it inaccessible from behind, a different technique is required. Drill several small holes through the metal inside the damaged area, particularly in the deeper sections. Screw long, self-tapping screws into the holes just enough for them to get a good grip in the metal. Now pulling on the protruding heads of the screws with locking pliers can pull out the dent.

20   The next stage of repair is the removal of paint from the damaged area and from an inch or so of the surrounding metal. This is easily done with a wire brush or sanding disk in a drill motor, although it can be done just as effectively by hand with sandpaper. To complete the preparation for filling, score the surface of the bare metal with a screwdriver or the tang of a file or drill small holes in the affected area. This will provide a good grip for the filler material. To complete the repair, see the Section on filling and painting.

Repair of rust holes or gashes

21   Remove all paint from the affected area and from an inch or so of the surrounding metal using a sanding disk or wire brush mounted in a drill motor. If these are not available, a few sheets of sandpaper will do the job just as effectively.

22   With the paint removed, you will be able to determine the severity of the corrosion and decide whether to replace the whole panel, if possible, or repair the affected area. New body panels are not as expensive as most people think and it is often quicker to install a new panel than to repair large areas of rust.

23   Remove all trim pieces from the affected area except those which will act as a guide to the original shape of the damaged body, such as headlight shells, etc. Using metal snips or a hacksaw blade, remove all loose metal and any other metal that is badly affected by rust. Hammer the edges of the hole in to create a slight depression for the filler material.

24   Wire-brush the affected area to remove the powdery rust from the surface of the metal. If the back of the rusted area is accessible, treat it with rust inhibiting paint.

25   Before filling is done, block the hole in some way. This can be done with sheet metal riveted or screwed into place, or by stuffing the hole with wire mesh.

26   Once the hole is blocked off, the affected area can be filled and painted. See the following subsection on filling and painting.

Filling and painting

27   Many types of body fillers are available, but generally speaking, body repair kits which contain filler paste and a tube of resin hardener are best for this type of repair work. A wide, flexible plastic or nylon applicator will be necessary for imparting a smooth and contoured finish to the surface of the filler material. Mix up a small amount of filler on a clean piece of wood or cardboard (use the hardener sparingly). Follow the manufacturer's instructions on the package, otherwise the filler will set incorrectly.

28   Using the applicator, apply the filler paste to the prepared area. Draw the applicator across the surface of the filler to achieve the desired contour and to level the filler surface. As soon as a contour that approximates the original one is achieved, stop working the paste. If you continue, the paste will begin to stick to the applicator. Continue to add thin layers of paste at 20-minute intervals until the level of the filler is just above the surrounding metal.

29   Once the filler has hardened, the excess can be removed with a body file. From then on, progressively finer grades of sandpaper should be used, starting with a 180-grit paper and finishing with 600-grit wet-or-dry paper. Always wrap the sandpaper around a flat rubber or wooden block, otherwise the surface of the filler will not be completely flat. During the sanding of the filler surface, the wet-or-dry paper should be periodically rinsed in water. This will ensure that a very smooth finish is produced in the final stage.

30   At this point, the repair area should be surrounded by a ring of bare metal, which in turn should be encircled by the finely feathered edge of good paint. Rinse the repair area with clean water until all of the dust produced by the sanding operation is gone.

31   Spray the entire area with a light coat of primer. This will reveal any imperfections in the surface of the filler. Repair the imperfections with fresh filler paste or glaze filler and once more smooth the surface with sandpaper. Repeat this spray-and-repair procedure until you are satisfied that the surface of the filler and the feathered edge of the paint are perfect. Rinse the area with clean water and allow it to dry completely.

32   The repair area is now ready for painting. Spray painting must be carried out in a warm, dry, windless and dust free atmosphere. These conditions can be created if you have access to a large indoor work area, but if you are forced to work in the open, you will have to pick the day very carefully. If you are working indoors, dousing the floor in the work area with water will help settle the dust that would otherwise be in the air. If the repair area is confined to one body panel, mask off the surrounding panels. This will help minimize the effects of a slight mismatch in paint color. Trim pieces such as chrome strips, door handles, etc., will also need to be masked off or removed. Use masking tape and several thickness of newspaper for the masking operations.

33   Before spraying, shake the paint can thoroughly, then spray a test area until the spray painting technique is mastered. Cover the repair area with a thick coat of primer. The thickness should be built up using several thin layers of primer rather than one thick one. Using 600-grit wet-or-dry sandpaper, rub down the surface of the primer until it is very smooth. While doing this, the work area should be thoroughly rinsed with water and the wet-or-dry sandpaper periodically rinsed as well. Allow the primer to dry before spraying additional coats.

34   Spray on the top coat, again building up the thickness by using several thin layers of paint. Begin spraying in the center of the repair area and then, using a circular motion, work out until the whole repair area and about two inches of the surrounding original paint is covered. Remove all masking material 10 to 15 minutes after spraying on the final coat of paint. Allow the new paint at least two weeks to harden, then use a very fine rubbing compound to blend the edges of the new paint into the existing paint. Finally, apply a coat of wax

---

4   Body repair - major damage

1   Major damage must be repaired by an auto body shop specifically equipped to perform body and frame repairs. These shops have the specialized equipment required to do the job properly.

2   If the damage is extensive, the frame must be checked for proper alignment or the vehicle's handling characteristics may be adversely affected and other components may wear at an accelerated rate.

3   Due to the fact that all of the major body components (hood, fenders, etc.) are separate and replaceable units, any seriously damaged components should be replaced rather than repaired. Sometimes the components can be found in a wrecking yard that specializes in used vehicle components, often at considerable savings over the cost of new parts.

These photos illustrate a method of repairing simple dents. They are intended to supplement *Body repair - minor damage* in this Chapter and should not be used as the sole instructions for body repair on these vehicles.

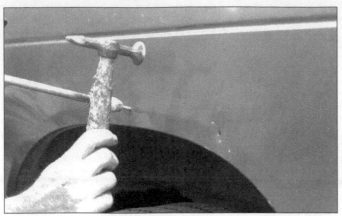

1 If you can't access the backside of the body panel to hammer out the dent, pull it out with a slide-hammer-type dent puller. Tap with a hammer near the edge of the dent to help 'pop' the metal back to its original shape, about 1/8-inch below the surface of the surrounding metal

2 Using coarse-grit sandpaper, remove the paint down to the bare metal. Clean the repair area with wax/silicone remover.

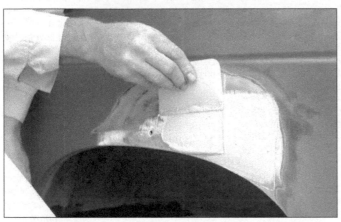

3 Following label instructions, mix up a batch of plastic filler and hardener, then quickly press it into the metal with a plastic applicator. Work the filler until it matches the original contour and is slightly above the surrounding metal

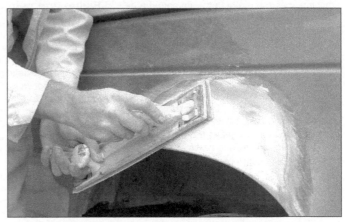

4 Let the filler harden until you can just dent it with your fingernail. File, then sand the filler down until it's smooth and even. Work down to finer grits of sandpaper - always using a board or block - ending up with 360 or 400 grit

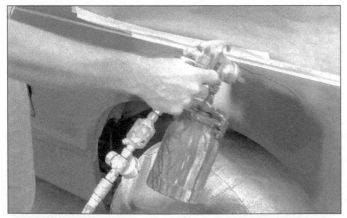

5 When the area is smooth to the touch, clean the area and mask around it. Apply several layers of primer to the area. A professional-type spray gun is being used here, but aerosol spray primer works fine

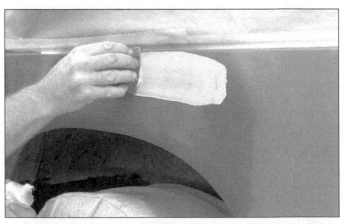

6 Fill imperfections or scratches with glazing compound. Sand with 360 or 400-grit and re-spray. Finish sand the primer with 600 grit, clean thoroughly, then apply the finish coat. Don't attempt to rub out or wax the repair area until the paint has dried completely (at least two weeks)

5 Upholstery, carpets and vinyl trim - maintenance

Upholstery and carpets

1   Every three months remove the floormats and clean the interior of the vehicle (more frequently if necessary). Use a stiff whiskbroom to brush the carpeting and loosen dirt and dust, then vacuum the upholstery and carpets thoroughly, especially along seams and crevices.

2   Dirt and stains can be removed from carpeting with basic household or automotive carpet shampoos available in spray cans. Follow the directions and vacuum again, then use a stiff brush to bring back the "nap" of the carpet.

3   Most interiors have cloth or vinyl upholstery, either of which can be cleaned and maintained with a number of material-specific cleaners or shampoos available in auto supply stores. Follow the directions on the product for usage, and always spot-test any upholstery cleaner on an inconspicuous area (bottom edge of a backseat cushion) to ensure that it doesn't cause a color shift in the material.

4   After cleaning, vinyl upholstery should be treated with a protectant.
Note: *Make sure the protectant container indicates the product can be used on seats - some products may make a seat too slippery.*
Caution: *Do not use protectant on vinyl-covered steering wheels.*

5   Leather upholstery requires special care. It should be cleaned regularly with saddle-soap or leather cleaner. Never use alcohol, gasoline, nail polish remover or thinner to clean leather upholstery.

6   After cleaning, regularly treat leather upholstery with a leather conditioner, rubbed in with a soft cotton cloth. Never use car wax on leather upholstery.

7   In areas where the interior of the vehicle is subject to bright sunlight, cover leather seating areas of the seats with a sheet if the vehicle is to be left out for any length of time.

Vinyl trim

8   Don't clean vinyl trim with detergents, caustic soap or petroleum-based cleaners. Plain soap and water works just fine, with a soft brush to clean dirt that may be ingrained. Wash the vinyl as frequently as the rest of the vehicle.

9   After cleaning, application of a high-quality rubber and vinyl protectant will help prevent oxidation and cracks. The protectant can also be applied to weather-stripping, vacuum lines and rubber hoses, which often fail as a result of chemical degradation, and to the tires.

6 Fastener and trim removal

Refer to illustrations 6.3 and 6.4

1   There is a variety of plastic fasteners used to hold trim panels, splash shields and other parts in place in addition to typical screws, nuts and bolts. Once you are familiar with them, they can usually be removed without too much difficulty.

2   The proper tools and approach can prevent added time and expense to a project by minimizing the number of broken fasteners and/or parts.

3   The following illustration shows various types of fasteners that are typically used on most vehicles and how to remove and install them **(see illustration)**. Replacement fasteners are commonly found at most auto parts stores, if necessary.

Fasteners

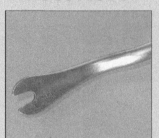

This tool is designed to remove special fasteners. A small pry tool used for removing nails will also work well in place of this tool

A Phillips head screwdriver can be used to release the center portion, but light pressure must be used because the plastic is easily damaged. Once the center is up, the fastener can easily be pried from its hole

Here is a view with the center portion fully released. Install the fastener as shown, then press the center in to set it

This fastener is used for exterior panels and shields. The center portion must be pried up to release the fastener. Install the fastener with the center up, then press the center in to set it

This type of fastener is used commonly for interior panels. Use a small blunt tool to press the small pin at the center in to release it . . .

. . . the pin will stay with the fastener in the released position

Reset the fastener for installation by moving the pin out. Install the fastener, then press the pin flush with the fastener to set it

This fastener is used for exterior and interior panels. It has no moving parts. Simply pry the fastener from its hole like the claw of a hammer removes a nail. Without a tool that can get under the top of the fastener, it can be very difficult to remove

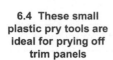

6.4 These small plastic pry tools are ideal for prying off trim panels

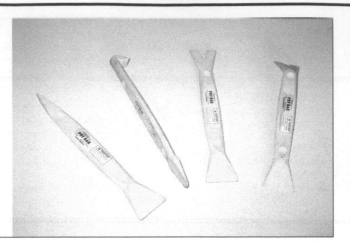

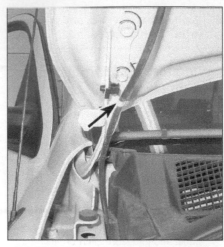

7.2 Detach the windshield washer fluid line from the connection

4   Trim panels are typically made of plastic and their flexibility can help during removal. The key to their removal is to use a tool to pry the panel near its retainers to release it without damaging surrounding areas or breaking-off any retainers. The retainers will usually snap out of their designated slot or hole after force is applied to them. Stiff plastic tools designed for prying on trim panels are available at most auto

parts stores **(see illustration)**. Tools that are tapered and wrapped in protective tape, such as a screwdriver or small pry tool, are also very effective when used with care.

---

7   Hood - removal, installation and adjustment

Note: *The hood is somewhat awkward to remove and install - at least two people should perform this procedure.*

Removal and installation

Refer to illustrations 7.2, 7.3, 7.4 and 7.5

1   Open the hood and place rags or covers over the windshield and fenders to protect them during the removal procedure.
2   Disconnect the windshield washer fluid lines **(see illustration)**.
3   Mark the relationship of the hood to the hinges **(see illustration)**.
4   Use a screwdriver to detach the hood support struts **(see illustration)**.
5   Have an assistant support one side of the hood, then take turns removing the hinge-to-hood bolts **(see illustration)**. Remove the hood.
6   Installation is the reverse of removal.

Adjustment

Refer to illustration 7.8

7   If necessary after installation, the entire hood latch assembly can be adjusted up-and-down as well as from side-to-side on the upper radiator support so the hood closes securely and is flush with the fenders. To do this, scribe a line around the hood latch mounting bolts to provide a reference point. Then loosen the bolts and reposition the latch as necessary.
Note: *The factory bolts are centering type that will not allow adjustment. To adjust the hood in relation to the hinges, these bolts must be replaced with standard bolts with flat washers and lock washers.*
Following adjustment, retighten the mounting bolts.
8   Finally, adjust the hood bumpers, so when closed, it is flush with the fenders **(see illustration)**.
9   The hood latch, as well as the hinges, should be periodically lubricated with white lithium-based grease to prevent sticking and wear.

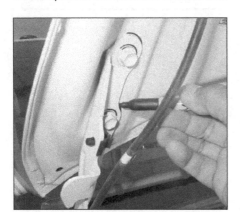

7.3 Before removing the hood hinge bolts, mark the relationship of the hood to the hood hinges

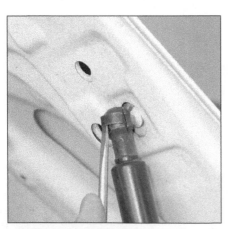

7.4 Use a screwdriver to pry the clip off and detach the hood support strut

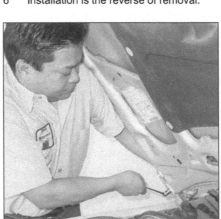

7.5 Support the hood with your shoulder while removing the hood bolts

7.8 Screw the hood bumpers in or out to adjust the hood flush with the fenders – left side shown, right side identical

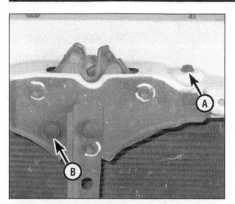

8.1a Remove the cover retaining screw (A), remove the plastic cap (B) and loosen the nut

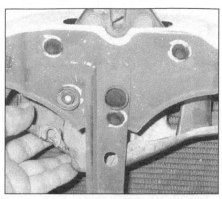

8.1b Remove the hood latch cover from under the latch

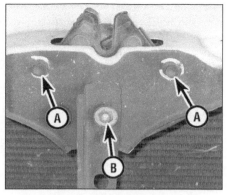

8.1c Remove the retaining bolts (A), remove the plastic cap then the nut (B) and detach the latch

8 Hood latch and release cable - removal and installation

Warning: *The models covered by this manual are equipped with Supplemental Restraint systems (SRS), more commonly known as airbags. Always disable the airbag system before working in the vicinity of any airbag system component to avoid the possibility of accidental deployment of the airbag(s), which could cause personal injury (see Chapter 12).*

Latch

Refer to illustrations 8.1a, 8.1b, 8.1c and 8.2

1   Remove the latch cover for access **(see illustrations)**, scribe a line around the latch to aid alignment when installing, then remove the retaining bolts securing the hood latch to the radiator support **(see illustration)**. Remove the latch.
2   Disconnect the hood release cable by disengaging the cable from the latch assembly **(see illustration)**.
3   Installation is the reverse of removal.
Note: *Adjust the latch so the hood engages securely when closed and the hood bumpers are slightly compressed.*

Cable

Refer to illustration 8.5

4   Working in the passenger compartment, remove the driver's side kick panel.
5   Lift the hood release handle lever upward, then remove the screws and disengage the cable from the hood release lever handle **(see illustration)**.
6   Attach a piece of thin wire or string to the end of the cable.
7   Working in the engine compartment, disconnect the hood release cable from the latch assembly (see Steps 1 and 2). Unclip all the cable retaining clips on the radiator support and the inner fenderwell.
8   Pull the cable forward into the engine compartment until you can see the wire or string. Remove the wire or string from the old cable and fasten it to the new cable.

9   With the new cable attached to the wire or string, pull the wire or string back through the firewall until the new cable reaches the inside handle.
10   Working in the passenger compartment, reinstall the new cable into the hood release lever, making sure the cable housing fits snugly into the notch in the handle bracket.
Note: *Pull on the cable with your fingers from the passenger compartment until the cable stop seats correctly in the grommet on the firewall.*
11   The remainder of installation is the reverse of removal.

9 Radiator grille - removal and installation

Refer to illustration 9.1

Warning: *The models covered by this manual are equipped with Supplemental Restraint systems (SRS), more commonly known as airbags. Always disable the airbag system before working in the vicinity of any airbag system component to avoid the possibility of accidental deployment of the airbag(s), which could cause personal injury (see Chapter 12).*

2013 and earlier models

1   Open the hood and remove the hood-to-grille retaining fasteners **(see illustration)**.

8.5 Lift the release lever for access and remove the lever retaining screws

2   Detach the radiator grille from the hood.
3   Installation is the reverse of removal.

2014 and later models

4   Open the hood, then remove the grille-to-radiator support mounting bolts and clips.
5   Use a trim tool to carefully pry the grille up, then outwards, to release the plastic fasteners.
6   Detach the grille from the radiator support and front bumper.
7   Installation is the reverse of removal.

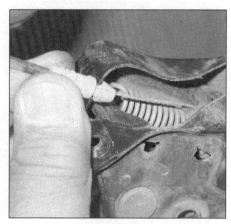

8.2 Disengage the cable from the latch assembly

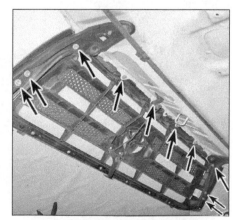

9.1 Typical grille fastener locations

10 Bumpers - removal and installation

Warning: *The models covered by this manual are equipped with Supplemental Restraint systems (SRS), more commonly known as airbags. Always disable the air-bag system before working in the vicinity of any airbag system component to avoid the possibility of accidental deployment of the airbag(s), which could cause personal injury (see Chapter 12).*

1   Disconnect the negative cable from the battery.

Front

2013 and earlier models

Refer to illustrations 10.4a, 10.4b, 10.5a, 10.5b, 10.6a, 10.6b and 10.7

2   Remove the fasteners for the front section of the inner fender splash shield and remove the shield **(see illustration 14.3b)**.

3   Disconnect the fog light electrical connectors (see Chapter 12), if equipped.

4   Remove the two headlight housing trim clips **(see illustration)**, then carefully pry the trim panel out, disengaging the two

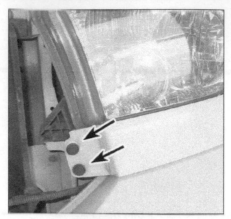

10.4a   Remove the two headlight housing trim clips

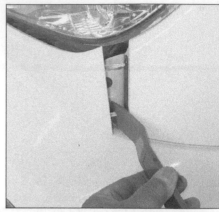

10.4b   Carefully pry the trim panel out to disengage the two remaining clips

remaining clips on the back of the panel **(see illustration)**.

Steel bumper models

5   Remove the bumper cover fasteners **(see illustrations)** and remove the cover.
Note: *The bumper cover must be unclipped where it contacts the front fenders.*

6   Remove the bumper upper and lower fasteners **(see illustrations)**.
Note: *Some models may have more lower fasteners located in the middle of the bumper.*

7   Remove the bumper from the vehicle **(see illustration)**.

8   Installation is the reverse of removal. Tighten all fasteners securely.

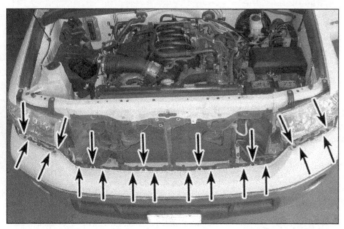

10.5a   Remove the front bumper cover fasteners . . .

10.5b   . . . and the two corner fasteners

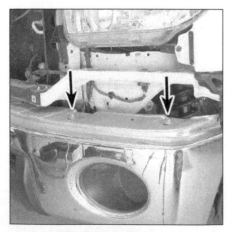

10.6a   Remove the bumper upper fasteners . . .

10.6b   . . . and lower fasteners from each side

10.7   Carefully remove the bumper from the vehicle

10.22 Remove the bolts from the frame rails if you're planning to remove the Tundra bumper, brackets and tow hitch as a unit

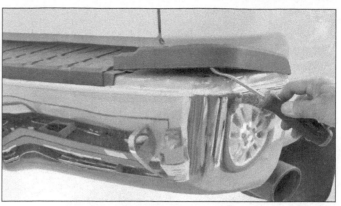

10.23 If you're planning to remove only the Tundra bumper, carefully pry the rear bumper pads out

Resin bumper cover models

9   Remove the push pin clips from along the top edge of the cover.
10   Remove the mounting screws along the bottom edge and ends of the bumper cover.
11   Remove the one piece cover from the vehicle.
12   Installation is the reverse of removal. Tighten all fasteners securely.

2014 and later models

13   Remove the radiator grille (see Section 9).
14   Remove the headlight housing front end trim panel(s) mounting bolts from each end of the panel(s) then carefully pry out the panel(s), disengaging the remaining plastic fasteners.
15   Remove the mounting screws from the ends of the bumper, located inside the inner fenderwell along the edge.
16   Remove the fasteners along the bottom edge of the bumper.
Note: *There are three different types of fasteners used to secure the bumper along the bottom edge.*
17   Working under the top edge of the bumper, remove the mounting bolts just below the headlight housings.
18   Disconnect the electrical connectors to the fog lights and parking assist sensors.

19   Remove the bumper from the vehicle.
20   Installation is the reverse of removal.

Rear

Tundra

21   Remove the license plate lights from the bumper (see Chapter 12) and disconnect any electrical connectors to the bumper.

Steel bumper models

Refer to illustrations 10.21, 10.23, 10.24a and 10.24b

22   If you're removing the bumper with the brackets and tow hitch, unbolt the bumper brackets from the frame **(see illustration)**.
23   If you're removing just the bumper, pry the rear bumper pads up, disengaging the clips and remove the pads **(see illustration)**.
24   Remove the bumper fasteners and remove the bumper **(see illustrations)** from the brackets.
25   Installation is the reverse of removal. Tighten all fasteners securely.

Resin bumper cover models

26   Remove the license plate from the bumper.
27   Remove the fasteners around the perimeter of the bumper cover and pull the cover off the vehicle.

28   Installation is the reverse of removal. Tighten all fasteners securely.

Sequoia

29   Remove the rear quarter panel mud guard fasteners and the guards.
30   Open the liftgate then remove the six fasteners on the top of the bumper.
31   Remove the six fasteners from underneath along the perimeter of the bumper cover and disconnect any electrical connectors to the bumper from below.
32   Remove the bumper cover from the vehicle.
33   Installation is the reverse of removal. Tighten all fasteners securely.

---

11   Door trim panel - removal and installation

Warning: *The models covered by this manual are equipped with Supplemental Restraint systems (SRS), more commonly known as airbags. Always disable the airbag system before working in the vicinity of any airbag system component to avoid the possibility of accidental deployment of the airbag(s), which could cause personal injury (see Chapter 12).*

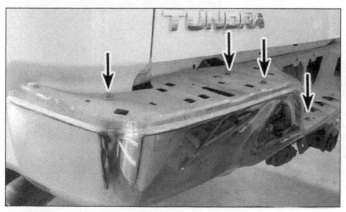

10.24a Remove the bumper upper mounting fasteners . . .

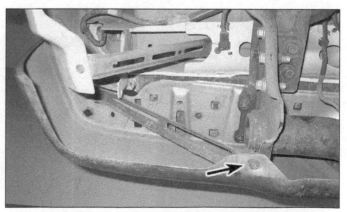

10.24b . . . then remove the lower fasteners

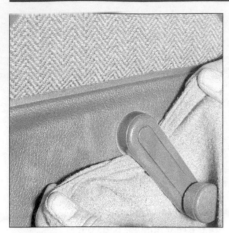

11.2 To remove the manual window regulator handle, work a clean shop rag between the handle and the door as shown, then pull up on the rag to pop off the snap-ring

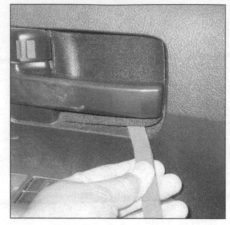

11.3a Pry up the cover . . .

11.3b . . . then remove the door handle screw

Removal

Refer to illustrations 11.2, 11.3a, 11.4a, 11.4b, 11.5, 11.6, 11.7, 11.8, 11.9a and 11.9b

1   On models equipped with power windows, disconnect the negative cable at the battery.

2   To remove the window regulator handle on models without power windows, work a shop rag between the handle and the door trim panel, then pull up on both ends of the rag and pop loose the snap-ring that secures the handle to the regulator shaft **(see illustration)**.

3   Remove the screw from the inside door handle trim cover, detach the cover from the door **(see illustrations)**.

4   To remove the switch control panel on power window equipped vehicles, carefully pry it upward to detach the clips at each end, then pull it out and unplug the electrical connector **(see illustrations)**.

5   Remove the pull handle screw **(see illustration)**.

11.4a On models with power windows, use a screwdriver to pry out the switch panel from the door trim panel . . .

6   Carefully pry out the window sail panel **(see illustration)** and disconnect the electrical connector to the speaker, if equipped. Use a small screwdriver to carefully pry out the clip.

7   Pry out the door light from the bottom

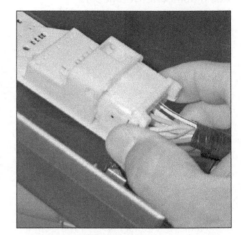

11.4b . . . then lift the panel out and disconnect the electrical connector

and disconnect the electrical connector **(see illustration)**.

8   To detach the door trim panel, carefully pry around the edge to detach the retaining clips **(see illustration)**, then pull it away from

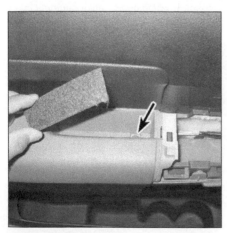

11.5 Remove the trim from inside the armrest and remove the screw

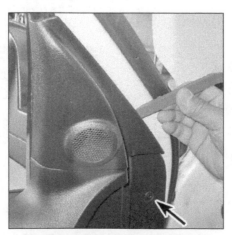

11.6 Pry out the sail panel, disconnect the electrical connector and remove the retaining clip just below the panel

11.7 Pry out and disconnect the door panel light

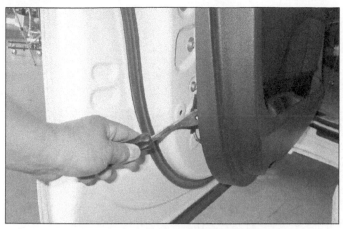

11.8 Insert a putty knife or trim panel tool between the door and the trim panel, then carefully pry the clips out

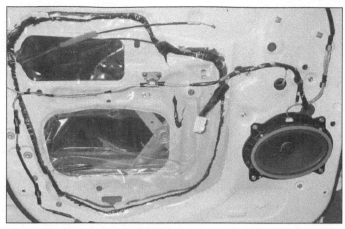

11.9a Before removing the watershield, remove any components that could damage it

the door, disconnect the electrical harness from the panel and detach the cables to the door handle (see Section 21).

9   For access to the inner door, carefully peel back the plastic water shield **(see illustrations)**.

Installation

10   Before installing the door trim panel, make sure the plastic watershield is correctly installed **(see illustration 11.9a)**. If the shield is torn, patch it with tape. If the edge fails to adhere to the door anywhere, seal it with a little silicone sealant. Install any grommets, clips or other fasteners which may have fallen out of the trim panel or the door when you removed the panel.

11   Place the door trim panel in position and push all the clips into their respective grommets until they're fully seated. Install the trim panel retaining screws and snap the plastic covers into place.

12   Install the inside handle trim cover and screw.

13   Install the pull handle screws.

14   Install the regulator handle or the switch control panel. Plug in all electrical connectors.

12 Mirrors - removal and installation

Outside mirror

Refer to illustration 12.2

1   Remove the door trim panel (see Section 11).

2   Remove the mirror-to-door retaining bolts/nuts and unplug the electrical connector **(see illustration)**.

3   Remove the mirror assembly.

4   Installation is the reverse of removal.

Outside mirror glass

Refer to illustrations 12.5, 12.6 and 12.7

5   Push the upper part of the mirror inwards at the top and hold it. Using a plastic prying tool, pry the base of the glass out from the bottom until the four clips are disengaged **(see illustration)**.

6   Remove the mirror glass and disconnect the heater electrical connectors **(see illustration)**, if equipped.

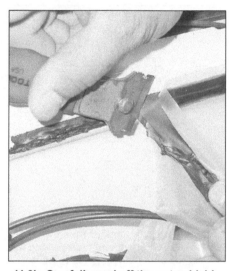

11.9b Carefully peel off the watershield - if the shield is torn during removal, repair it with tape or replace it. Make sure the shield is sealed all the way around the edge

12.2 Disconnect the electrical connector (A), then remove the mirror retaining nuts

12.5 Pry or pull the base of the glass out from the bottom until the four clips are disengaged

12.6 Disconnect the heater electrical connectors (A) and center the clips (B) on the base

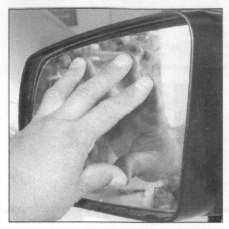

12.7 Center the clips on the mirror base and press the glass evenly inwards

12.9 Remove the Torx screw and slide the mirror off the base

13.3 Remove the bolt from the door stay

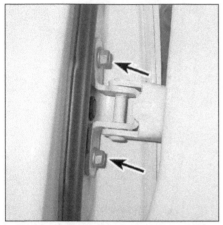

13.4a To detach the door from the body, scribe or draw alignment marks along the edges of the hinges, then remove the upper . . .

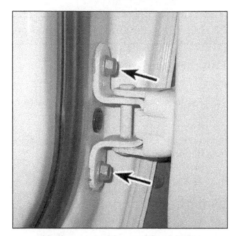

13.4b . . . and lower hinge bolts

7   To install the glass, connect the electrical connectors the center the clips on the mirror base and press the glass inwards **(see illustration)** until the clips snap in place.

Inside mirror

Refer to illustration 12.9

8   Disconnect the sensor cover clips, remove the cover(s), and disconnect electrical connector, if equipped.
9   To remove/install the mirror, remove/install the Torx screw at the base on the windshield **(see illustration)**.
10   If the mount plate itself has come off the windshield, adhesive kits are available at auto parts stores to re-secure it. Follow the instructions included with the kit.

13   Door - removal and installation

Refer to illustrations 13.3, 13.4a, 13.4b and 13.7
Warning: *The models covered by this manual are equipped with Supplemental Restraint*
systems (SRS), more commonly known as airbags. Always disable the airbag system before working in the vicinity of any airbag system component to avoid the possibility of accidental deployment of the airbag(s), which could cause personal injury (see Chapter 12).
1   Disconnect the negative cable from the battery.
2   Remove the door trim panel (see Section 11). On doors with power components, unplug all electrical connections (it's a good idea to label all connections to aid the reassembly process) and remove the electrical harness from the door. On models equipped with side airbags, the side impact sensor is mounted to the door. It can be identified by the bright yellow connector; take care not to hit or damage the sensor.
3   Open the door all the way and support it on jacks covered with cloth or pads to prevent damaging the paint, then remove the door stay bolt **(see illustration)**.
4   Mark the relationship of the door hinges to the body by scribing or drawing a line around the hinges. With an assistant supporting the door, remove the door hinge bolts, then carefully lift off the door **(see illustrations)**.

5   When installing the door, make sure the hinges are aligned with the outlines you made, then tighten the hinge bolts securely.
6   Installation is otherwise the reverse of removal.
7   If the door does not close properly after installation, slightly loosen the door latch striker bolts and carefully tap the striker up, down or sideways as necessary to provide positive engagement with the latch mechanism **(see illustration)**.

14   Front fender - removal and installation

Refer to illustrations 14.3a, 14.3b, 14.3c, 14.3d, 14.4a, 14.4b, 14.5a, 14.5b, 14.5c and 14.5d
Warning: *The models covered by this manual are equipped with Supplemental Restraint systems (SRS), more commonly known as airbags. Always disable the airbag system before working in the vicinity of any airbag system component to avoid the possibility of*

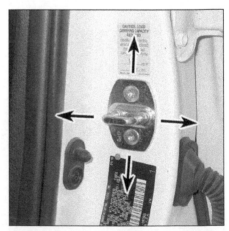

13.7 To adjust the door lock striker, slightly loosen the screws and tap the striker up, down or sideways as necessary until the door lock latch engages it correctly

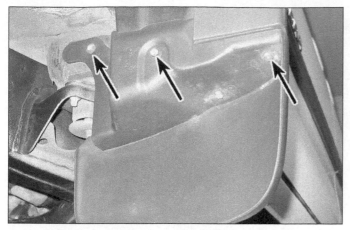

14.3a Remove the bolts retaining the mud flaps

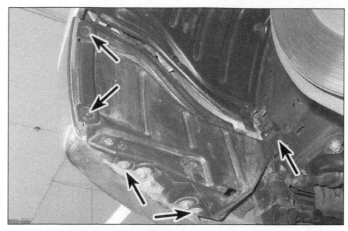

14.3b Fasteners locations for the front section of the inner fender splash shield

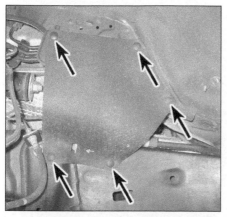

14.3c Fastener locations for the inner fender rubber seal

14.3d After removing the plastic clips, detach the wheel housing splash shields from the fender – the amount of clips vary by year and trim options

accidental deployment of the airbag(s), which could cause personal injury (see Chapter 12). **Warning:** *On Sequoia models equipped with rear height control suspension, adjust the height control to the NORMAL mode, turn OFF the height control, then turn off the engine BEFORE raising the vehicle.*

1   Loosen the front wheel lug nuts. Raise the vehicle, support it securely on jackstands and remove the front wheel.
2   Remove the front bumper (see Section 10).
3   Detach the mud flaps, fender flares (if equipped), inner fender rubber seal and wheel housing splash shields from the fender **(see**

illustrations). The splash shields are attached between the frame and the fender with plastic screws and clips. Remove the plastic Phillips head screw in the center, then pry the clips out.
4   Pry up the retaining pins and lift the weather strip off, then remove the retaining bolts from the top edge of the fender **(see illustration)**.

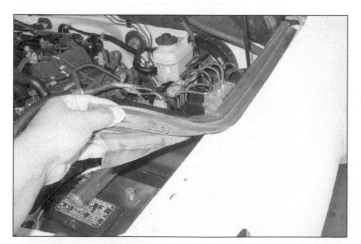

14.4a Pry the retaining clips up as the weather strip is removed

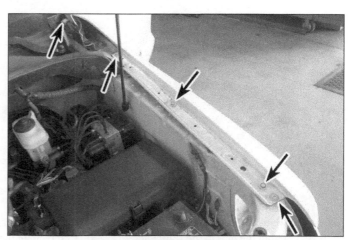

14.4b To detach the upper edge of the fender, remove the bolts shown

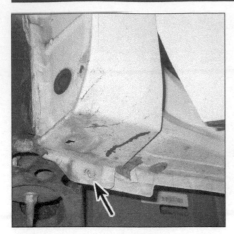

14.5a Remove the bolt securing the lower rear edge of the fender

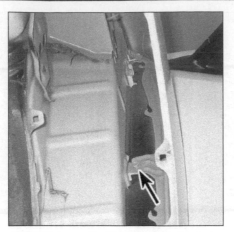

14.5b Remove the bolt retaining the inner rear edge of the fender

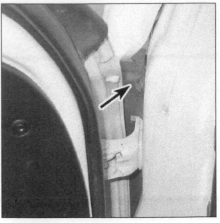

14.5c Open the door and remove the bolt retaining the rear upper edge of the fender

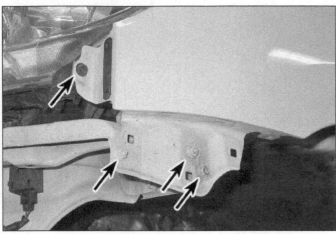

14.5d Remove the fasteners from the front of the fender

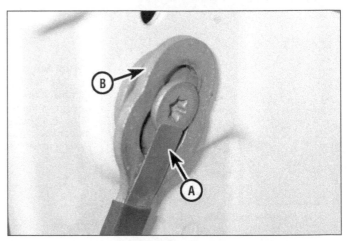

15.1a To remove the entire tailgate assembly from the vehicle, lift the lock tab (A) over the bolt and slide the cable ends (B) off, then raise the liftgate . . .

5 Remove the remaining bolts and detach the fender from the vehicle (**see illustrations**).
6 Remove the fender. It is a good idea to have an assistant support the fender while it's

being moved away from the vehicle to prevent damage to the surrounding body panels.
7 Installation is the reverse of removal. Tighten all fasteners securely.

15 Tailgate and latch (Tundra models) - removal and installation

Refer to illustrations 15.1a, 15.1b, 15.2, 15.3a, 15.3b, 15.4, 15.5a and 15.5b

1 To remove the tailgate assembly, disengage the two tailgate cables, raise the tailgate until the slot in the right side hinge lines up with the flat on the hinge pin, and lift the tailgate out (**see illustrations**).
2 To replace the tailgate latch strikers, mark their relationship to the body, then remove the striker bolts (**see illustration**).
3 To replace the tailgate handle or the latches, remove the plastic liner (if equipped), then remove the metal access panel (**see illustrations**).
4 Disengage the latch control rods (**see illustration**) from the tailgate handle or latch.
5 Unbolt the handle from the tailgate or detach the latches from the tailgate (**see illustrations**) and pull them out.
Note: *On models equipped with a back up*

15.1b . . . until the slot in the right side hinge lines up with the flat on the hinge pin, and lift the tailgate out

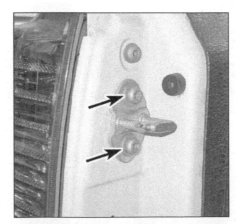

15.2 To remove the tailgate striker, mark its location and remove the bolts

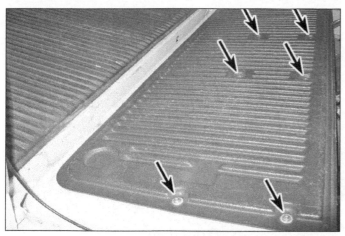

15.3a Remove the screws and detach the plastic tailgate liner, if equipped (six of eight screws shown)

15.3b Remove the metal access panel (if there is no liner, remove the panel fasteners)

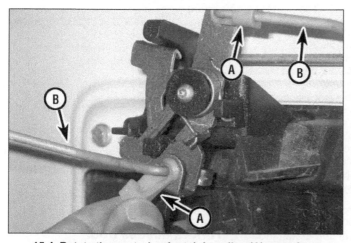

15.4 Rotate the control rod retaining clips (A) away from the rod (B) and pull the rod out of the clip

15.5a Unscrew the two screws and pull out the handle

camera, disconnect the camera connector and remove the mounting bolts and camera.

6   To replace the tailgate rods, rotate the rod retaining clip away from the rod, then pull the rod out of the clip.

7   Disengage the latch control rods from the tailgate handle, unbolt the handle from the tailgate or detach the latches from the tailgate, and pull the rods out.

8   To replace the tailgate cables, remove the Torx screws securing each end of the cable.

9   Installation is the reverse of removal. Tighten all fasteners securely.

16   Liftgate and spoiler (Sequoia models) - removal, installation and adjustment

Removal and installation

Refer to illustration 16.4

Note: *The liftgate is heavy and somewhat awkward to remove and install - at least two people should perform this procedure.*

1   Disconnect the negative cable from the battery.

2   Open the liftgate and support it securely. Remove the liftgate struts or power door rod **(see illustration 7.4).**

3   Disconnect the wire harness between the liftgate and the body.

4   Mark the relationship of the hinges to the liftgate, then unbolt them **(see illustration).**

5   Installation is the reverse of removal. If necessary, adjust the liftgate (see Step 6).

15.5b To remove a latch, unscrew the two screws and pull out the latch (the control rod must be disconnected from the back of the latch mechanism)

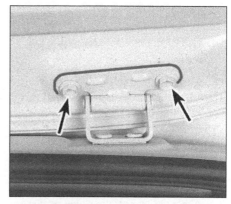

16.4 Remove the liftgate hinge bolts

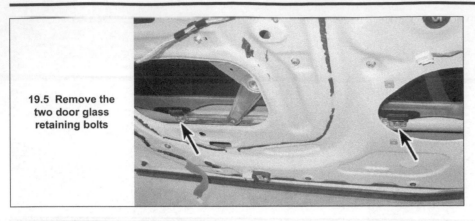

19.5 Remove the two door glass retaining bolts

Adjustment

6   If the liftgate requires adjustment to close properly, there are four adjustments available. To move the liftgate forward, rearward or vertically in relation to the body, loosen the nuts and bolts that attach the hinges to the body and lightly tap the hinges with a small plastic hammer in the direction you want to move the liftgate, then tighten the nuts securely. To adjust the liftgate to the left or right, or up or down, in relation to the body, loosen the hinge bolts on the liftgate, carefully move the liftgate the direction you want to go, then tighten the bolts securely. To adjust the liftgate striker, loosen the striker screws and carefully tap the striker with a small plastic hammer until the liftgate lock and the striker are engaging properly, then tighten the striker screws securely. The liftgate-to-body gap is adjusted by loosening the mounting bolts and moving the stoppers in increments until the proper gap is achieved.

Liftgate spoiler

7   Remove the liftgate trim panels around the window by prying them loose using a trim tool.
8   Locate the two hole covers and pry them out.
9   Remove the rear spoiler mounting nuts through the holes.
10   Working on the outside of the liftgate, carefully pry the spoiler outwards, disengaging the clips and pins on the spoiler until it can be removed.
11   Installation is the reverse of removal.

17   Liftgate glass and regulator (Sequoia models) - removal and installation

1   If possible, lower the liftgate glass half way down, then disconnect the negative cable from the battery.
Note: *The liftgate glass must be lowered to access the glass mounting bolts.*
2   Although it's not absolutely necessary, you may want to remove the liftgate (see Section 16) before removing the liftgate glass. Servicing the liftgate components is easier with the liftgate on the floor.

3   Lift open the pull strap fastener cover then remove the fastener and pull strap.
4   To remove the trim panels, pry loose the trim panel clips. For access to the inner door, carefully peel back the plastic watershield.
5   Pull the outer weather-stripping from the liftgate and disconnect the electrical connectors on the right side of the liftgate.
Note: *It may be necessary to remove the ECU from the liftgate before the window or the regulator can be removed.*
6   Hold the glass to prevent it from falling into the liftgate, then remove the four mounting bolts and the liftgate glass.
7   Disconnect the electrical connection from the window regulator.
8   Remove the regulator mounting bolts and regulator from the liftgate.
9   Installation is the reverse of removal.

18   Liftgate power door unit (Sequoia models) - removal and installation

1   Disconnect the negative cable from the battery.
2   To remove the luggage compartment box and covers from inside the rear of the vehicle, pry the fastener cover caps off and remove the fasteners.
3   Remove the rear seat mounting bolts and rear seats (see Section 28).
4   Pry the rear scuff plates from each door opening.
5   Remove the floor hook fasteners and hooks from the floor and the floor mat support plate.
6   Remove the rear quarter trim panel fasteners, then remove the trim panels by carefully prying around the edge to detach the retaining clips. Pull it away from the quarter panel and disconnect the electrical connections.
7   Carefully pry around the edge to detach the retaining clips and remove the D-pillar trim panels. Allow the seat belt to slide through the panel.
8   While supporting the liftgate, detach the liftgate power door rods from the liftgate and power door unit **(see illustration 7.4)**.
9   Disconnect the electrical connector, then remove the power door unit mounting bolts and power unit from the vehicle.
10   Installation is the reverse of removal.

19   Door window glass - removal and installation

Warning: *Safety glasses and gloves should be worn when performing this procedure.*
1   Lower the glass fully in the door. On models equipped with power windows and/or door locks, disconnect the cable from the negative battery terminal.
2   Remove the door trim panel and watershield (see Section 11).
3   Remove the outside mirror (see Section 12).
4   Pry out the window weather-strip in the top of the door opening,

Front door

Refer to illustration 19.5

5   Remove the door glass bolts **(see illustration)**.
6   Rotate the back of the glass up slightly and remove the door glass.
7   Installation is the reverse of removal. Tighten all fasteners securely.

Rear door (Sequoia)

8   Remove the door speaker (see Chapter 12).
9   Pry out the door frame window trim panels.
10   Remove the front and rear window channel bolts and channels.
11   Remove the door glass bolts, rotate the back of the glass up slightly and remove the door glass.
12   Installation is the reverse of removal.

Tundra access door

13   Remove the rear window channel bolts and channel.
14   Remove the door glass bolts, rotate the back of the glass up slightly and remove the door glass.
Note: *On double cab models, slide the window glass back, then rotate the rear of the glass up and out of the door.*
15   Installation is the reverse of removal.

20   Door window regulator - removal and installation

Refer to illustrations 20.4 and 20.5
Note: *This procedure applies to front and rear doors.*
1   On models equipped with power windows and door locks, disconnect the cable from the negative battery terminal.
2   Remove the door trim panel and watershield (see Section 11).
3   Remove the door glass (see Section 19).
4   Unplug the electrical connector from the motor (power windows), remove the retaining bolts **(see illustration)** and maneuver the regulator out through the access hole.
5   On power window models, remove the power window motor fasteners and separate

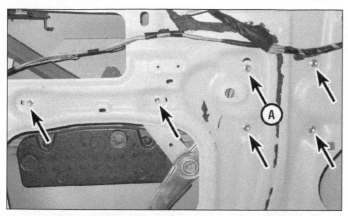

20.4 To remove the window regulator, loosen bolt (A) and remove the remaining bolts, raise the regulator up to allow bolt (A) to come through the door and remove it through the access hole (Tundra model shown, other models similar)

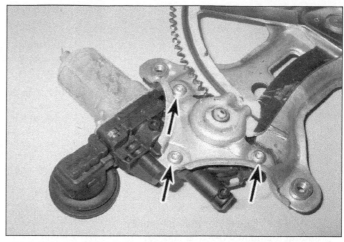

20.5 Remove the power window motor mounting fasteners (but only after securing the sector gear to the regulator frame)

the motor from the regulator **(see illustration).** **Warning:** *Before loosening the window motor mounting bolts, clamp the moveable portion of the regulator to the regulator frame to prevent the spring force from snapping the two components together.*

6   Installation is the reverse of removal. Lubricate all regulator rollers with multi-purpose grease. Tighten all fasteners securely. On rear doors, tighten the regulator bolts in a criss-cross pattern.

21 Door latch, lock cylinder and handles - removal and installation

1   On models equipped with power windows and door locks, disconnect the cable from the negative battery terminal.
2   Remove the door trim panel and the plastic watershield (see Section 11).

Door latch

Refer to illustrations 21.3 and 21.5

3   Working through the large access hole,

disengage the rod to the outside handle and the lock cylinder **(see illustration).**
4   The door lock rod is attached by plastic clips. The plastic clips can be removed by unsnapping the portion engaging the connecting rod, then pulling the rod out of its locating hole.
5   Remove the screws securing the latch to the door **(see illustration),** then pull the latch back enough to disconnect the electrical connector at the latch. Remove the latch assembly through the door opening with the two cables from the inside door handle still attached to the latch.
6   Installation is the reverse of removal.

Door lock cylinder

Refer to illustrations 21.9a and 21.9b

7   To remove the outside handle and lock cylinder assembly, raise the window before removing the door trim panel and watershield (see Section 11).
Caution: *Take care not to scratch the paint on the outside of the door. Wide masking tape applied around the handle opening before beginning the procedure can help avoid scratches.*

21.3 Rotate the retaining clip up and disconnect the rod from the outside handle

8   Remove an access plug at the rear of the door (jamb side).
9   Working through the access hole, loosen the TORX screw **(see illustration)** and remove the lock cylinder cover and the door lock key cylinder as a unit **(see illustration).**
Note: *The TORX screw can't be fully removed, it is a part of the outside handle.*

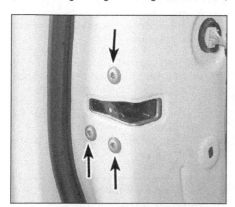

21.5 To detach the latch assembly from the door, remove these three screws

21.9a Remove the rubber plug at the end of the door and use a Torx tool through the hole to loosen the lock cylinder bolt (the bolt does not come out)

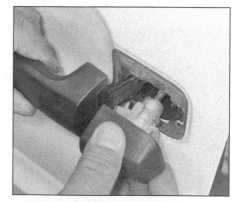

21.9b Slide the lock cylinder out

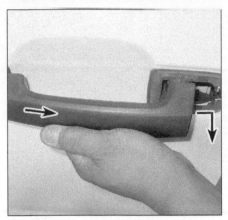

21.12 Remove the exterior door handle by sliding the handle rewards and out

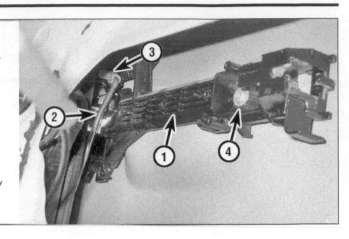

21.14 Outside door handle assembly details

1   Outside door handle frame assembly
2   Actuating rod
3   Actuating rod mounting clip
4   Mounting screw

10   Once the lock cylinder is out, disengage the lock cover clips using a screwdriver, and separate the cover from the lock cylinder.

11   Installation is the reverse of removal.

Outside door handle

Refer to illustrations 21.12 and 21.14

Outside handle

12   Once the lock cylinder has been removed, slide the exterior handle reward and out to remove **(see illustration)**.

Outside handle assembly

13   Working from the outside of the door, remove the handle assembly exterior mounting screw.

14   From inside of the door, disconnect the electrical connector, remove the mounting screw, disconnect the lock rod clip and move the lock rod away from the handle **(see illustration)**. **Note:** *If the door latch was removed, leave the lock rod attached to the door handle assembly.*

15   Installation is the reverse of removal.

Inside door handle

Refer to illustrations 21.17 and 21.18

16   Remove the door trim panel (see Section 11).

21.17 Detach the cables from the back of the inside handle

17   Disengage the handle-to-latch cables and remove the cables from the handle **(see illustration)**.

18   Remove the handle retaining screws and disengage the handle from the door trim panel **(see illustration)**.

19   Installation is the reverse of removal.

---

22   Rear cab window, glass and regulator (Tundra models) - removal and installation

---

Window glass assembly

Removal

1   Lower the window halfway down, then disconnect the cable from the negative battery terminal.

2   Place tape over the paint of the cab, then carefully pry the rear roof drip molding off both sides.

3   Remove the rear seats (see Section 28).

4   Using a trim tool, remove the rear window trim.

5   Remove the package holder net fasteners and remove the holders (if equipped).

6   Remove the rear seat belt lower mounting bolts.

7   Using a trim tool, remove the back trim panels, starting with the right side first.

8   Using a trim tool, pry the rear side trim panel out and remove the panels.

21.18 Remove the inside handle retaining screws

9   Remove the rear assist handle retainer caps, then remove the fasteners and handles.

10   Starting from the rear, carefully pull the rear headliner down, disconnecting the rear clips.

11   Remove the rear access mounting fasteners and remove the access panel.

12   Using a screwdriver, pry the outer back weather strip off, then pull the inner weather strip or channel out from the frame opening.

13   Remove the window glass-to-regulator mounting bolts.

14   Raise the window by hand and tape the window into place. Disconnect the rear window defogger connections.

15   Starting from the top, lift the assembly out at an angle.

Installation

16   Clean the window opening and apply new butyl seal to the window assembly.

17   Lift the window assembly into place, onto the regulator, and tighten the fasteners securely.

18   The remainder of installation is the reverse of removal.

Window regulator

Removal

18   Remove the rear window assembly (see Steps 1 through 15).

19   Disconnect the power window motor regulator electrical connections.

20   Remove the window motor regulator mounting fasteners and remove the regulator and motor assembly.

Installation

21   Install the regulator and motor assembly into the opening and lower the assembly.

22   Connect the electrical connections and install the mounting fasteners. Tighten the fasteners securely.

23   The remainder of installation is the reverse of removal.

---

23   Steering column covers - removal and installation

Refer to illustrations 23.5, 23.6a and 23.6b

Warning: *The models covered by this*

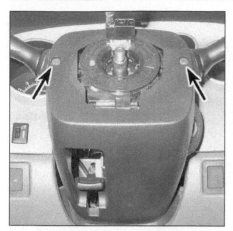

23.5 Remove the three screws and detach the steering column cover halves

23.6a Detach the four clips and separate the instrument cluster seal from the upper cover half

23.6b Pry the upper cover out, making sure to disengage the center retaining clip

manual are equipped with Supplemental Restraint systems (SRS), more commonly known as airbags. Always disable the airbag system before working in the vicinity of any airbag system component to avoid the possibility of accidental deployment of the airbag(s), which could cause personal injury (see Chapter 12).

1   On models equipped with power tilt/telescopic columns, fully extend and lower the steering wheel.

2   Wait at least 1 minute after turning the key to the OFF position, then disconnect the cable from the negative battery terminal (see Chapter 5).

3   Disable the airbag system (see Chapter 12).

4   Remove the steering wheel (see Chapter 10).

5   Remove the steering column cover screws **(see illustration)**.

6   Separate the cover halves and remove them from the column **(see illustrations)**.

7   Installation is the reverse of removal.

24   Console - removal and installation

Front console (floor shift models)

Refer to illustrations 24.3, 24.4a, 24.4b, 24.5, 24.6 and 24.7

Warning: *The models covered by this manual are equipped with Supplemental Restraint systems (SRS), more commonly known as airbags. Always disable the airbag system before working in the vicinity of any airbag system component to avoid the possibility of accidental deployment of the airbag(s), which could cause personal injury (see Chapter 12).*

1   Disconnect the negative battery cable.

2   Insert the key into the shift lock slot, place the lever in the neutral position and unscrew the manual shift lever knob (see Chapter 7A).

3   Carefully pry the cup holder console panel, disengaging the clips along the perimeter of the panel **(see illustration)**.

24.3 Detach the cup holder from the center console - 2013 and earlier models shown, later models similar

4   Remove the plastic Phillips head screw in the center, pry the clip out of the shifter trim panel **(see illustration)** then remove the shifter trim panel **(see illustration)**.

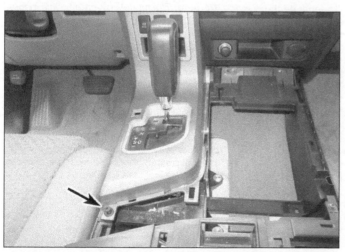

24.4a Remove the plastic Phillips head screw in the center of the retaining clip - 2013 and earlier models shown, later models similar

24.4b Pry the shifter trim panel up, disengaging the clips along the perimeter - 2013 and earlier models shown, later models similar

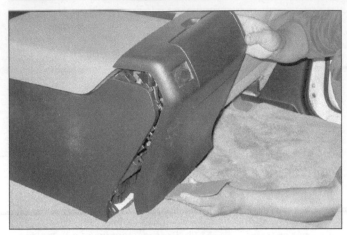

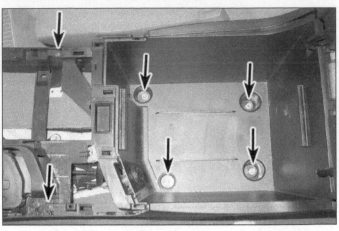

24.5 Pry the end panel back to disengage the clips from the rear section of the console - 2013 and earlier models shown, later models similar

24.6 Remove the console mounting fasteners and lift out the rear section of the console - 2013 and earlier models shown, later models similar

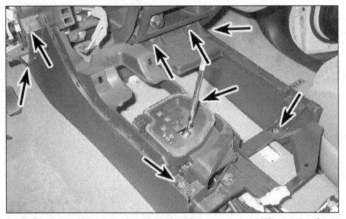

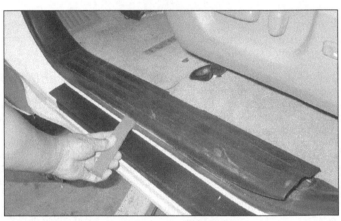

24.7 Remove the console mounting fasteners and lift out the front section of the console - 2013 and earlier models shown, later models similar

25.2 Pry up the scuff plate

5   Pry the end panel off of the console **(see illustration)** and disconnect the electrical connectors.

6   Open the arm rest and remove the carpet then remove the console mounting fasteners and rear section of the console **(see illustration)**.

7   Remove the console fasteners and remove the front section of the console **(see illustration)**.

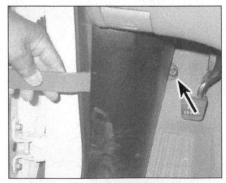

25.3 Remove the kick panel fastener, then pry the panel out

8   Installation is the reverse of removal.

Rear console (Sequoia models)

9   Use a trim tool to disengage the clips and retainer, then carefully separate the cup holder panel from the rear console. Lift the cup holder panel up, then disconnect the electrical connectors to the panel and remove the panel.

10   Open the rear console and remove the storage tray and the mat from the bottom of the console.

11   Remove the mounting bolts, then lift the rear of console up to disengage the clamp and remove the rear console.

12   Installation is the reverse of removal.

Roof console (Tundra models)

13   On long console types, use a plastic trim tool to carefully pry out the rear light lens.

14   On short console types, open the storage lid and sunglass lid.

15   Remove the front and rear fasteners, then lower the console and disconnect the wiring.

16   Installation is the reverse of removal.

25   Dashboard trim panels - removal and installation

Warning: *The models covered by this manual are equipped with Supplemental Restraint systems (SRS), more commonly known as airbags. Always disable the airbag system before working in the vicinity of any airbag system component to avoid the possibility of accidental deployment of the airbag(s), which could cause personal injury (see Chapter 12).*
Note: *Several types and designs of mounting clips and hooks are used to secure the trim panels to the instrument panel. Care should be taken when trying to remove the panel.*

1   Disconnect the cable from the negative battery terminal (see Chapter 5).

Lower left finish panel

Refer to illustration 25.2, 25.3, 25.4a and 25.4b

2   Pry the scuff plate up and off from the floorboard **(see illustration)**.

3   Remove the left side kick panel plastic retainer and pry the panel out **(see illustration)** from the side.

25.4a Remove the lower left finish panel fasteners

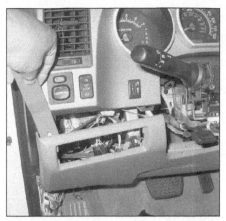

25.4b Pry the lower left finish panel out and lower it from the instrument panel

25.5 Pry the end cap off of the instrument panel

25.7a Disconnect the left instrument cluster bezel harness connector . . .

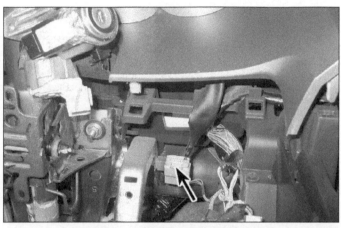

25.7b . . . then the right connector

4   Remove the lower left finish panel retaining screws **(see illustrations)**, pull off the lower left finish panel and detach the hood release cable (see Section 8). Installation is the reverse of removal.

Instrument cluster bezel

Refer to illustration 25.5, 25.7a, 25.7b and 25.8

5   Pry the end cap off of the instrument panel **(see illustration)**.
6   Remove the steering column covers (see Section 23).
7   Disconnect the two harness connectors **(see illustrations)** and leave all the switches connected to the panel.
8   Carefully pry the instrument cluster bezel out along the edges to disengage the retaining clips and pull it out of the dashboard.
Note: *On 2014 and later Tundra models, the driver's side air register is separate from the instrument cluster and must be pried out using a trim tool to remove the instrument cluster bezel.*
9   Installation is the reverse of removal.

Center lower switch panel

Refer to illustration 25.11

10   Remove the center console (see Section 24).
11   Carefully pry the switch panel from the

instrument panel, disengaging the clips **(see illustration)**.
12   Lower the panel and disconnect the electrical connectors to the power outlets or switches.
13   Installation is the reverse of removal.

25.8 Starting from the left side and working to the right, pry the bezel out from the instrument panel

25.11 Pry the center lower switch panel from the instrument panel

25.14a Disengage the lower sub panel retaining clips . . .

25.14b . . . and lower the sub panel

of the glove box **(see illustration)**.

19   Open the compartment door, remove the mounting screws **(see illustration)** then detach the glove compartment. Lower the glove box and disconnect the electrical connector. Installation is the reverse of removal.

Storage box compartment

Refer to illustrations 25.21 and 25.22

20   Remove the glove box (see Steps 17 to 19).

21   Open the compartment door, remove the mounting screws **(see illustration)**.

22   Detach the storage box compartment and lower it from the dashboard **(see illustration)**.

23   Installation is the reverse of removal.

Lower right finish panel

Refer to illustrations 25.14a, 25.14b and 25.15

14   Disengage the lower sub panel retaining clips and lower the panel **(see illustrations)**.

15   Remove the screws, then detach the panel **(see illustration)** and pull the lower right finish panel out

16   Installation is the reverse of removal.

Glove compartment

Refer to illustrations 25.17, 25.18 and 25.19

17   Pry the end cap off of the instrument panel **(see illustration)**.

18   Remove the fasteners from the outside

26   Cowl vent grille - removal and installation

Refer to illustrations 26.2, 26.4 and 26.5

1   Remove the wiper arm mounting nuts, mark the position of the wiper arm to wiper link, and detach the wipers (see Chapter 12).

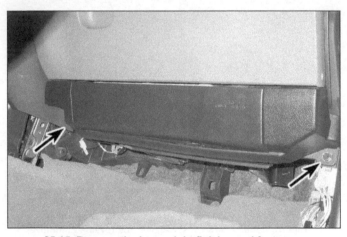

25.15 Remove the lower right finish panel fasteners

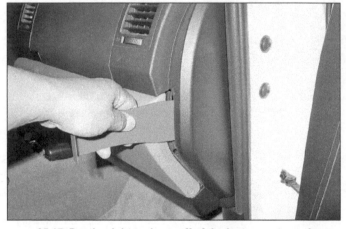

25.17 Pry the right end cap off of the instrument panel

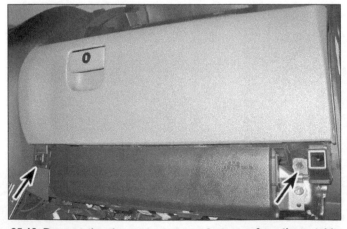

25.18 Remove the glove compartment fasteners from the outside

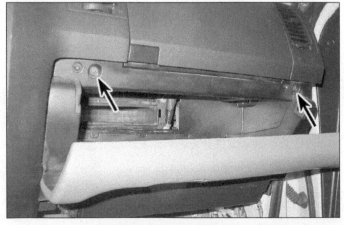

25.19 Remove the glove compartment fasteners from inside the compartment

25.21 Open the glove compartment door and remove the mounting screws

25.22 Detach the storage box from the instrument panel

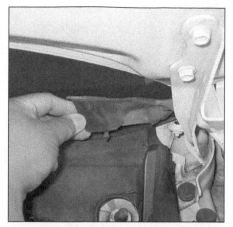

26.2 Remove the side seals from the cowl – left side shown right side similar

2   Remove the cowl side seals from both ends of the cowl (see illustration).
3   Pull the weather strip off along the front of the cowl.
4   Remove the plastic Phillips head screw in the center, then pry the clip out from each side (see illustration).
5   Disengage the sixteen clips, slide the cowl back and remove the cowl (see illustration).
6   Installation is the reverse of removal. Align the wiper blades with the marks made during removal.

27 Instrument panel - removal and installation

Refer to illustrations 27.8, 27.10a, 27.10b, 27.12a, 27.12b, 27.12c and 27.12d

Warning: Models covered by this manual are equipped with a Supplemental Restraint System (SRS), more commonly known as airbags. Always disable the airbag system before working in the vicinity of any airbag system component to avoid the

possibility of accidental deployment of the airbag, which could cause personal injury (see Chapter 12).
Note: This is a difficult procedure for the home mechanic. There are many hidden fasteners, difficult angles to work in and many electrical connectors to tag and disconnect/connect. We recommend that this procedure be done only by an experienced do-it-yourselfer.
Note: During removal of the instrument panel, make careful notes of how each piece comes off, where it fits in relation to other pieces and what holds it in place. If you note how each part is installed before removing it, getting the instrument panel back together again will be much easier.
Note: It is not necessary, but it is suggested to remove both front seats to allow additional working space and lessen the chance of damage to the seats during this procedure.
1   Disconnect the cable from the negative battery terminal (see Chapter 5).
2   Remove the dashboard trim panels (see Section 25) and the center floor console (see Section 24).

Note: Disconnect the driver's side knee airbag (if equipped), remove the mounting bolts and remove the airbag (see Chapter 12).
3   Remove the instrument cluster (see Chapter 12), the glove compartment and the storage box (see Section 24).
4   Carefully pry out the instrument panel lower side, console trim and remove the trim panels.
5   Remove the A-pillar trim panels, disconnect the passenger's side airbag, remove the mounting fasteners, and remove the airbag (see Chapter 12).
6   Remove the audio unit from the center of the dashboard (see Chapter 12).
7   Remove the air conditioning control panel (see Chapter 3).
8   Pry off the end caps (see illustration) for access to the reinforcement tube mounting bolts.
9   Detach the bolts securing the steering column and lower it away from the instrument panel (see Chapter 10).
10   A number of electrical connectors must be disconnected in order to remove the instrument panel. Most are designed so that they

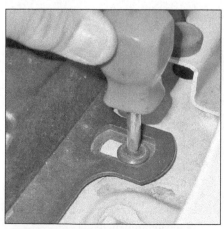

26.4 Remove cowl vent grille plastic mounting screws from each side

26.5 Pry the cowl grille up, disengaging the clips until all the clips are free – three of sixteen clips shown

27.8 Pry off the end caps and remove the bolts

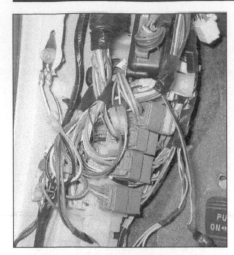

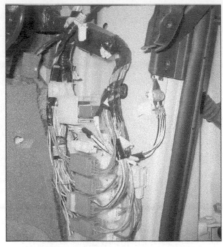

27.10a Label and disconnect the main instrument panel electrical connectors at the left . . .

27.10b . . . and the right sides, below the instrument panel

will only fit on the matching connector (male or female), but if there is any doubt, mark the connectors with masking tape and a marking pen before disconnecting them **(see illustrations)**.

11   Pry up the speaker trim covers and the defroster trim panel (at the top of the instrument panel) (see Chapter 12), then disconnect the optical sensor electrical connector.

12   Remove all of the fasteners (bolts, screws and nuts) holding the instrument panel to the body **(see illustrations)**. Once all are removed, lift the panel, then pull it away from the windshield, and take it out through the driver's door opening **(see illustration)**.

13   If you're also removing the instrument panel reinforcement tube, disconnect any electrical connectors that might interfere with the removal of the reinforcement tube, then detach the wire harness and remove the fasteners securing the tube, then take it out through the driver's door opening.

14   Installation is the reverse of removal.

27.12a Remove the instrument panel fasteners to the support tube, on the left . . .

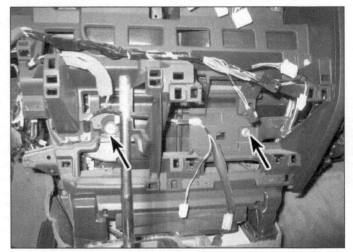

27.12b . . . the center . . .

27.12c . . . and the right

27.12d Remove the instrument panel

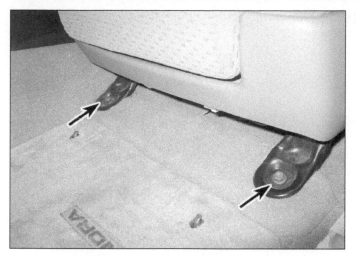

28.2a Remove the bolts from the front side

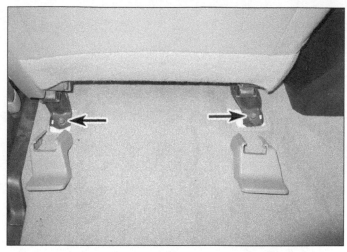

28.2b Remove the covers and the bolts from the back side, then remove the seat

28   Seats - removal and installation

Warning: *Some later models covered by this manual are equipped with Supplemental Restraint system (SRS), more commonly known as airbags, within the seats. Always disable the airbag system before working in the vicinity of any airbag system component to avoid the possibility of accidental deployment of the airbag(s), which could cause personal injury (see Chapter 12).*

Warning: *The front seat belts on these models may be equipped with pre-tensioners, which are pyrotechnic (explosive) devices designed to retract the seatbelts in the event of a collision. On models equipped with pre-tensioners, do not remove the front seatbelt retractor assemblies, and do not disconnect the electrical connectors leading to the assemblies. Problems with the pre-tensioners will turn on the SRS (airbag) warning light on the dash. If any pre-tensioner problems are suspected, take the vehicle to a dealer service department or other qualified repair shop.*

Warning: *Do not use a memory saving device to preserve the ECM's memory when working on or near restraint system components.*

Front seat

Refer to illustrations 28.2a, 28.2b and 28.3

1   Position the seat all the way forward, then all the way to the rear to access the front seat retaining bolts.

2   Detach any bolt trim covers and remove the retaining bolts **(see illustrations)**.

3   Tilt the seat upward to access the underside, disconnect any electrical connectors **(see illustration)**, and lift the seat from the vehicle.

Note: *To disconnect the electrical connector on models equipped with seat airbags, depress the connector lock then slide the lock to the rear of the connector and separate the connectors.*

4   Installation is the reverse of removal.

Rear seat(s)

Tundra models

5   Unlatch the seat cushions and lift them up.

6   On 2014 and later Crew Max models, remove the seat belt anchor bolts and pull the seat belt through the opening in the seats.

7   Unscrew the mounting bolts, pull the seat up and remove it from the vehicle.

8   Installation is the reverse of removal.

Sequoia models

Middle seat(s)

9   Slide the seat all the way to the rear and remove the trim caps and retaining bolts.

10   Slide the seat all the way to the front and carefully pry the track trim cover out,

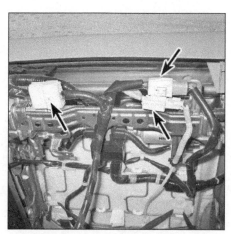

28.3 Tilt the seat upwards and disconnect any electrical connectors

then remove the retaining bolts.

11   Installation is the reverse of removal.

Back seat

12   Remove the luggage compartment box and cover. On power seat models, disconnect the electrical connectors from under the seat.

13   Remove the covers and unscrew the bolts from the seat frame. Remove the seat from the vehicle.

14   Installation is the reverse of removal.

Notes

Chapter 12
Chassis electrical system

Contents

1 General information

The electrical system is a 12-volt, negative ground type. Power for the lights and all electrical accessories is supplied by a lead/acid-type battery that is charged by the alternator.

This Chapter covers repair and service procedures for the various electrical components not associated with the engine. Information on the battery, alternator, ignition system and starter motor can be found in Chapter 5.

It should be noted that when portions of the electrical system are serviced, the negative cable should be disconnected from the battery to prevent electrical shorts and/or fires.

2 Electrical troubleshooting - general information

Refer to illustrations 2.5a, 2.5b, 2.6 and 2.9

A typical electrical circuit consists of an electrical component, any switches, relays, motors, fuses, fusible links or circuit breakers related to that component and the wiring and connectors that link the component to both the battery and the chassis. To help you pinpoint an electrical circuit problem, wiring diagrams are included at the end of this Chapter.

Before tackling any troublesome electrical circuit, first study the appropriate wiring diagrams to get a complete understanding of what makes up that individual circuit. Trouble spots, for instance, can often be narrowed down by noting if other components related to the circuit are operating properly. If several components or circuits fail at one time, chances are the problem is in a fuse or ground connection, because several circuits are often routed through the same fuse and ground connections.

Electrical problems usually stem from simple causes, such as loose or corroded connections, a blown fuse, a melted fusible link or a failed relay. Visually inspect the condition of all fuses, wires and connections in a problem circuit before troubleshooting the circuit.

If test equipment and instruments are going to be utilized, use the diagrams to plan ahead of time where you will make the necessary connections in order to accurately pinpoint the trouble spot.

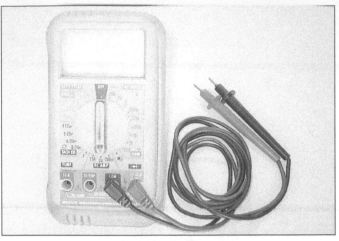

2.5a The most useful tool for electrical troubleshooting is a digital multimeter that can check volts, amps, and test continuity

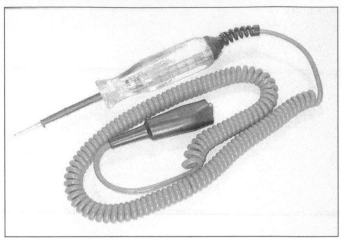

2.5b A simple test light is a very handy tool for testing voltage

The basic tools needed for electrical troubleshooting include a circuit tester or voltmeter (a 12-volt bulb with a set of test leads can also be used), a continuity tester, which includes a bulb, battery and set of test leads, and a jumper wire, preferably with a circuit breaker incorporated, which can be used to bypass electrical components **(see illustrations)**. Before attempting to locate a problem with test instruments, use the wiring diagram(s) to decide where to make the connections.

Voltage checks

Voltage checks should be performed if a circuit is not functioning properly. Connect one lead of a circuit tester to either the negative battery terminal or a known good ground. Connect the other lead to a connector in the circuit being tested, preferably nearest to the battery or fuse **(see illustration)**. If the bulb of the tester lights, voltage is present, which means that the part of the circuit between the connector and the bat-

tery is problem free. Continue checking the rest of the circuit in the same fashion. When you reach a point at which no voltage is present, the problem lies between that point and the last test point with voltage. Most of the time the problem can be traced to a loose connection. **Note:** *Keep in mind that some circuits receive voltage only when the ignition key is in the Accessory or Run position.*

Finding a short

One method of finding shorts in a circuit is to remove the fuse and connect a test light or voltmeter in place of the fuse terminals. There should be no voltage present in the circuit. Move the wiring harness from side-to-side while watching the test light. If the bulb goes on, there is a short to ground somewhere in that area, probably where the insulation has rubbed through. The same test can be performed on each component in the circuit, even a switch.

Ground check

Perform a ground test to check whether a component is properly grounded. Disconnect the battery and connect one lead of a continuity tester or multimeter (set to the ohms scale), to a known good ground. Connect the other lead to the wire or ground connection being tested. If the resistance is low (less than 5 ohms), the ground is good. If the bulb on a self-powered test light does not go on, the ground is not good.

Continuity check

A continuity check is done to determine if there are any breaks in a circuit - if it is passing electricity properly. With the circuit off (no power in the circuit), a self-powered continuity tester or multimeter can be used to check the circuit. Connect the test leads to both ends of the circuit (or to the power end and a good ground), and if the test light comes on the circuit is passing current properly **(see illustration)**. If the resistance is low (less than 5 ohms), there is continuity; if the reading is 10,000 ohms or higher, there is a break somewhere in the circuit. The same procedure can be used to test a switch, by connecting the continuity tester to the switch terminals. With the switch turned On, the test light should come on (or low resistance should be indicated on a meter).

Finding an open circuit

When diagnosing for possible open circuits, it is often difficult to locate them by sight because the connectors hide oxidation or terminal misalignment. Merely wiggling a connector on a sensor or in the wiring harness may correct the open circuit condition. Remember this when an open circuit is indicated when troubleshooting a circuit. Intermittent problems may also be caused by oxidized or loose connections.

Electrical troubleshooting is simple if you keep in mind that all electrical circuits are basically electricity running from the battery, through the wires, switches, relays, fuses

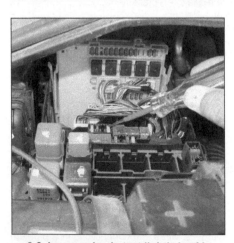

2.6 In use, a basic test light's lead is clipped to a known good ground, then the pointed probe can test connectors, wires or electrical sockets - if the bulb lights, the circuit being tested has battery voltage

2.9 With a multimeter set to the ohms scale, resistance can be checked across two terminals - when checking for continuity, a low reading indicates continuity, a high reading or infinity indicates lack of continuity

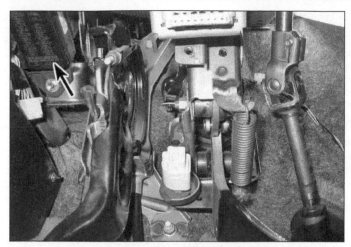

3.1a The passenger compartment fuse/relay block is located in the under left end of the instrument panel to the left of the brake pedal

3.1b The engine compartment fuse/relay block is located on the driver's side of the engine compartment - the various circuits are identified on the fuse panel label on the inside of the cover

and fusible links to each electrical component (light bulb, motor, etc.) and to ground, from which it is passed back to the battery. Any electrical problem is an interruption in the flow of electricity to and from the battery.

3 Fuses and fusible links - general information

Fuses

Refer to illustrations 3.1a, 3.1b and 3.3

The electrical circuits of the vehicle are protected by a combination of fuses, circuit breakers and fusible links. The main fuse/relay panel is in the engine compartment, while the interior fuse/relay panel is located inside on the driver's side of the instrument panel **(see illustrations)**. Each of the fuses is designed to protect a specific circuit, and the various circuits are identified on the fuse panel itself.

Several sizes of fuses are employed in the fuse blocks. There are small, medium and large sizes of the same design, all with the same blade terminal design. The medium and

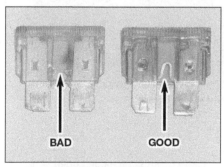

3.3 When a fuse blows, the element between the terminals burns - the fuse on the left is blown, the fuse on the right is good

large fuses can be removed with your fingers, but the small fuses require the use of pliers or the small plastic fuse-puller tool found in most fuse boxes.

If an electrical component fails, always check the fuse first. The best way to check the fuses is with a test light. Check for power at the exposed terminal tips of each fuse. If power is present at one side of the fuse but not the other, the fuse is blown. A blown fuse can also be identified by visually inspecting it **(see illustration)**.

Be sure to replace blown fuses with the correct type. Fuses (of the same physical size) of different ratings may be physically interchangeable, but only fuses of the proper rating should be used. Replacing a fuse with one of a higher or lower value than specified is not recommended. Each electrical circuit needs a specific amount of protection. The amperage value of each fuse is molded into the top of the fuse body.

If the replacement fuse immediately fails, don't replace it again until the cause of the problem is isolated and corrected. In most cases, this will be a short circuit in the wiring caused by a broken or deteriorated wire.

Fusible links

Some circuits are protected by fusible links. The links are used in circuits which are not ordinarily fused, or which carry high current, such as the circuit between the alternator and the battery. Fusible links, which are usually several wire gauges smaller in size than the circuit that they protect, are designed to melt if the circuit is subjected to more current than it was designed to carry. If you have to replace a blown fusible link, make sure that you replace it with one of the same specification. If the replacement fusible link blows in the same circuit, make sure that you troubleshoot the circuit in which the fusible link melted BEFORE installing another fusible link.

4 Circuit breakers - general information

Circuit breakers protect certain circuits, such as the power windows or heated seats. Depending on the vehicle's accessories, there may be one or two circuit breakers, located in the fuse/relay box in the engine compartment.

Because the circuit breakers reset automatically, an electrical overload in a circuit breaker-protected system will cause the circuit to fail momentarily, then come back on. If the circuit does not come back on, check it immediately.

For a basic check, pull the circuit breaker up out of its socket on the fuse panel, but just far enough to probe with a voltmeter. The breaker should still contact the sockets. With the voltmeter negative lead on a good chassis ground, touch each end prong of the circuit breaker with the positive meter probe. There should be battery voltage at each end. If there is battery voltage only at one end, the circuit breaker must be replaced.

Some circuit breakers must be reset manually.

5 Relays - general information

Several electrical accessories in the vehicle, such as the fuel injection system, horns, starter, and fog lamps use relays to transmit the electrical signal to the component. Relays use a low-current circuit (the control circuit) to open and close a high-current circuit (the power circuit). If the relay is defective, that component will not operate properly. Most relays are mounted in the engine compartment and interior fuse/relay boxes **(see illustrations 3.1a and 3.1b).**

6   Electrical connectors - general information

Most electrical connections on these vehicles are made with multiwire plastic connectors. The mating halves of many connectors are secured with locking clips molded into the plastic connector shells. The mating halves of some large connectors, such as some of those under the instrument panel, are held together by a bolt through the center of the connector.

To separate a connector with locking clips, use a small screwdriver to pry the clips apart carefully, then separate the connector halves. Pull only on the shell, never pull on the wiring harness, as you may damage the individual wires and terminals inside the connectors. Look at the connector closely before trying to separate the halves. Often the locking clips are engaged in a way that is not immediately clear. Additionally, many connectors have more than one set of clips.

Each pair of connector terminals has a male half and a female half. When you look at the end view of a connector in a diagram, be sure to understand whether the view shows the harness side or the component side of the connector. Connector halves are mirror images of each other, and a terminal shown on the right side end-view of one half will be on the left side end-view of the other half.

It is often necessary to take circuit voltage measurements with a connector connected. Whenever possible, carefully insert a small straight pin (not your meter probe) into the rear of the connector shell to contact the terminal inside, then clip your meter lead to the pin. This kind of connection is called "backprobing." When inserting a test probe into a terminal, be careful not to distort the terminal opening. Doing so can lead to a poor connection and corrosion at that terminal later. Using the small straight pin instead of a meter probe results in less chance of deforming the terminal connector.

Electrical connectors

Most electrical connectors have a single release tab that you depress to release the connector

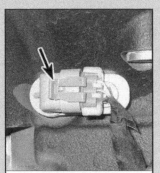

Some electrical connectors have a retaining tab which must be pried up to free the connector

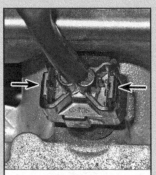

Some connectors have two release tabs that you must squeeze to release the connector

Some connectors use wire retainers that you squeeze to release the connector

Critical connectors often employ a sliding lock (1) that you must pull out before you can depress the release tab (2)

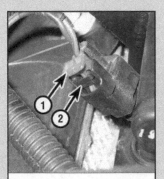

Here's another sliding-lock style connector, with the lock (1) and the release tab (2) on the side of the connector

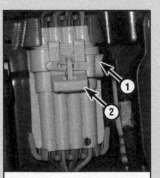

On some connectors the lock (1) must be pulled out to the side and removed before you can lift the release tab (2)

Some critical connectors, like the multi-pin connectors at the Powertrain Control Module employ pivoting locks that must be flipped open

7 Turn signal/hazard flasher switch - replacement

Refer to illustrations 7.1 and 7.2

Warning: *The models covered by this manual are equipped with Supplemental Restraint Systems (SRS), more commonly known as airbags. Always disable the airbag system before working in the vicinity of any airbag system components to avoid the possibility of accidental deployment of the airbag(s), which could cause personal injury (see Section 28).*

2013 and earlier models

1   Carefully pry the switch out of the instrument cluster trim panel **(see illustration)**.
Note: *If the switch can't be pried out, the trim panel will need to be removed (see Chapter 11) and the switch pushed out from the back side of the panel.*
2   Disconnect the electrical connector to the switch **(see illustration)**.
3   Installation is the reverse of removal.

2014 and later models

4   On these models the hazard flasher switch is an integral component of the air conditioning control assembly. If there is a problem with the switch, the entire air conditioning control assembly must be replaced (see Chapter 3).

8 Headlight/turn signal switch and tilt/telescopic manual switch - replacement

Refer to illustrations 8.5 and 8.8

Warning: *The models covered by this manual are equipped with Supplemental Restraint Systems (SRS), more commonly known as airbags. Always disable the airbag system before working in the vicinity of any airbag system components to avoid the possibility of accidental deployment of the airbag(s), which could cause personal injury (see Section 28).*

7.1 Carefully pry the switch from the trim panel

1   Disconnect the cable from the negative terminal of the battery (see Chapter 5).
2   Disable the airbag system (see Section 28).
3   Remove the steering wheel (see Chapter 10).
4   Remove the steering column covers (see Chapter 11).

Headlight/turn signal switch

5   Unplug the electrical connector from the turn signal switch/headlight switch **(see illustration)**.
6   Remove the windshield wiper switch (see Section 11).
7   Remove the airbag clockspring (see Chapter 10, Section 17).
8   Using a pair of pliers squeeze the clamp ends together **(see illustration)** and slide the switch off of the steering column.
9   Installation is the reverse of removal. Verify that the clockspring is correctly centered before installing the steering wheel (see Chapter 10).

Tilt/telescopic manual switch

Note: *It is not necessary to remove the headlight/turn signal switch to remove the tilt/telescopic manual switch.*

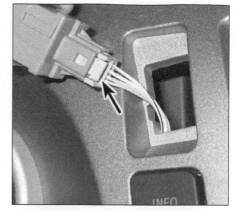

7.2 Depress the locking tab and disconnect the electrical connector

10   Disconnect the electrical connector to the tilt/telescopic manual switch.
11   Using a small screwdriver, carefully push the telescopic switch retaining claw upwards and pull the switch out from the steering column.
12   Insert the switch back into the opening, making sure the retaining claw is locked into place. The remainder of installation is the reverse of removal.

9 Ignition switch and key lock cylinder - replacement

Warning: *The models covered by this manual are equipped with Supplemental Restraint Systems (SRS), more commonly known as airbags. Always disable the airbag system before working in the vicinity of any airbag system components to avoid the possibility of accidental deployment of the airbag(s), which could cause personal injury (see Section 28).*

1   Disconnect the cable from the negative terminal of the battery (see Chapter 5).
2   Remove the left-side under-dash panel (see Chapter 11).

8.5 Unplug the electrical connector from the turn signal switch

8.8 Use a pair of pliers to squeeze the retaining clamp ends together and remove the switch

10.4 Push the power mirror switch out from the backside of the panel

10.7 Push the illumination rheostat switch from the backside of the panel

10.10 Carefully pry the switch panel from the instrument panel

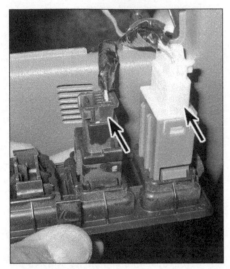

10.11 Disconnect the electrical connector to the switch that is going to be replaced

Ignition switch

3   Remove the headlight/turn signal switch (see Section 8).
4   Unplug the electrical connectors from the rear of the ignition switch.
5   Disengage the two mounting clips and detach the switch from the lock cylinder housing.
6   Installation is the reverse of removal.

Lock cylinder

7   Use a small screwdriver to push the locking tabs outwards on the transponder key amplifier or the key light assembly, then slide the unit off the end of the key lock cylinder housing.
8   Insert the ignition key and turn it to the ACC position.
9   Insert a small screwdriver into the hole in the bottom of the ignition switch casting and press the release button while pulling the lock cylinder straight out.
10   Installation is the reverse of removal.

10   Instrument panel switches - replacement

Warning: *The models covered by this manual are equipped with Supplemental Restraint Systems (SRS), more commonly known as airbags. Always disable the airbag system before working in the vicinity of any airbag system components to avoid the possibility of accidental deployment of the airbag(s), which could cause personal injury (see Section 28).*
1   Disconnect the cable from the negative terminal of the battery (see Chapter 5).
2   Disable the airbag system (see Section 28).
3   Remove the instrument cluster trim panel (see Chapter 11).

Power mirror switch

Refer to illustration 10.4
4   Push the switch out from the backside of the panel **(see illustration)**.
5   Unplug the electrical connector from the switch.
6   Installation is the reverse of removal.

Instrument panel illumination rheostat

Refer to illustration 10.7
7   Push the switch out from the backside of the panel **(see illustration)**.
8   Unplug the electrical connector from the switch.
9   Installation is the reverse of removal.

Vehicle Skid Control (VSC) and power rear window switch

Refer to illustrations 10.10 and 10.11
10   Pry the switch panel from the instrument panel **(see illustration)**.
11   Disconnect the electrical connector(s) from the switch(es) **(see illustration)**.
12   Carefully pry the switch(es) from the switch panel.
13   Installation is the reverse of removal.

11   Windshield wiper switch - replacement

Refer to illustration 11.4
Warning: *The models covered by this manual are equipped with Supplemental Restraint Systems (SRS), more commonly known as airbags. Always disable the airbag system before working in the vicinity of any airbag system components to avoid the possibility of accidental deployment of the airbag(s), which could cause personal injury (see Section 28).*
1   Disconnect the cable from the negative terminal of the battery (see Chapter 5).
2   Disable the airbag system (see Section 28).
3   Remove the steering column covers (see Chapter 11).
4   Unplug the electrical connectors **(see illustration)**.
5   Disengage the locking tab and detach the windshield wiper switch **(see illustration 11.4)**.
6   Installation is the reverse of removal.

11.4 Unplug the electrical connectors from the windshield wiper switch (A), then disengage the locking tab (B) by pressing it back and remove the switch

**12.1 Unplug the electrical connector from the headlight bulb;
low beam (A) and high beam (B)**

**12.3 Turn the bulb body counterclockwise and remove
it from the housing**

12 Headlight bulb - replacement

Refer to illustrations 12.1 and 12.3

Caution: *Don't touch the bulb with your fingers. If you do, clean it with rubbing alcohol (the oil from your skin can cause the bulb to overheat and fail).*

Note: *2018 and later Sequoia models and some 2018 and later Tundra models use LED headlights and are not serviceable. If there is a problem with a LED headlight the entire headlight assembly must be replaced.*

1   Unplug the electrical connector from the headlight bulb holder **(see illustration)**.
2   Pull the rubber watershield off the bulb assembly, if equipped.
3   Turn the bulb body counterclockwise and remove it from the housing **(see illustration)**.
4   Installation is the reverse of removal.

13 Headlights - adjustment

Refer to illustrations 13.1 and 13.3

Note: *The headlights must be aimed correctly. If adjusted incorrectly they could blind*

the driver of an oncoming vehicle and cause a serious accident or seriously reduce your ability to see the road. The headlights should be checked for proper aim every 12 months and any time a new headlight is installed or front end body work is performed. It should be emphasized that the following procedure is only an interim step that will provide temporary adjustment until a properly equipped shop can adjust the headlights.

1   The headlights have an adjusting screw on the headlight housing to control up-and-down (vertical) movement **(see illustration)**.
2   There are several methods of adjusting the headlights. The simplest method requires a blank wall 25-feet in front of the vehicle and a level floor.
3   Position masking tape vertically on the wall in reference to the vehicle centerline and the centerlines of both headlights **(see illustration)**.

13.1 Insert a Phillips head screwdriver into the hole to make vertical headlight adjustments

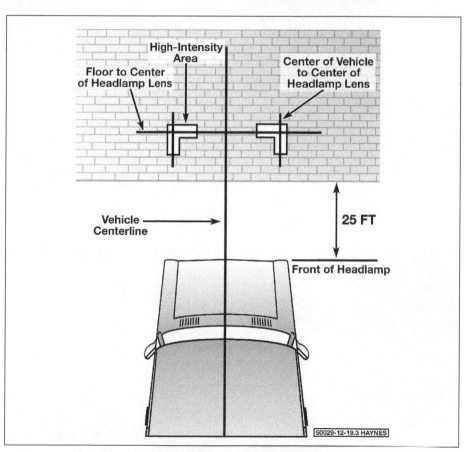

13.3 Headlight adjustment details

Bulb removal

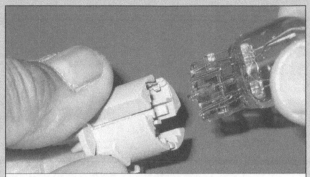

To remove many modern exterior bulbs from their holders, simply pull them out

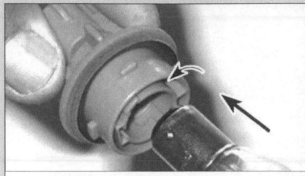

On bulbs with a cylindrical base ("bayonet" bulbs), the socket is spring-loaded; a pair of small posts on the side of the base hold the bulb in place against spring pressure. To remove this type of bulb, push it into the holder, rotate it 1/4-turn counterclockwise, then pull it out

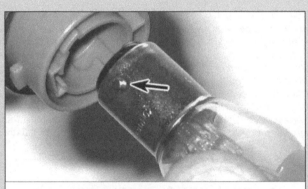

If a bayonet bulb has dual filaments, the posts are staggered, so the bulb can only be installed one way

To remove most overhead interior light bulbs, simply unclip them

4   Position a horizontal tape line in reference to the centerline of all the headlights. **Note:** *It may be easier to position the tape on the wall with the vehicle parked only a few inches away.*

5   Adjustment should be made with the vehicle sitting level, the gas tank half-full and no unusually heavy load in the vehicle.

6   Starting with the low beam adjustment, position the high intensity zone so it's two inches below the horizontal line and two inches to the side of the headlight vertical line away from oncoming traffic. Adjustment is made by turning the top adjusting screw clockwise to raise the beam and counterclockwise to lower the beam.

7   With the high beams on, the high intensity zone should be vertically centered with the exact center just below the horizontal line. **Note:** *It may not be possible to position the headlight aim exactly for both high and low beams. If a compromise must be made, keep in mind that the low beams are the most used and have the greatest effect on driver safety.*

8   Have the headlights adjusted by a dealer service department or service station at the earliest opportunity.

14   Headlight housing - removal and installation

Refer to illustrations 14.2a, 14.2b and 14.3

1   On 2014 and later models, remove the radiator grille (see Chapter 11).

2   Remove the two headlight housing trim clips **(see illustration)** then carefully pry the trim panel out, disengaging the two remaining clips on the back of the panel **(see illustration).**

3   Remove the headlight housing retaining fasteners **(see illustration)** and pull out the

14.2a Remove the two headlight housing trim clips

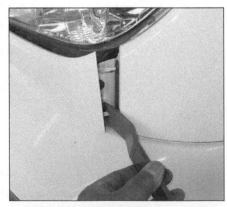

14.2b Carefully pry the trim panel out to disengage the two remaining clips

14.3 Remove the fasteners and detach the headlight housing

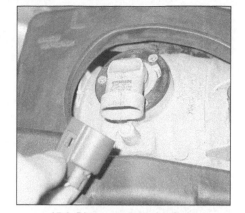

15.1 Disconnect the fog light electrical connector

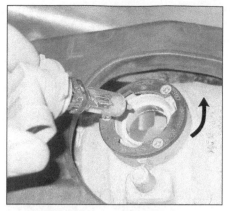

15.2 Turn the bulb holder counter-clockwise and remove it from the fog light housing

assembly. **Note:** *On 2015 and later models, the headlight housing uses clips and retaining screws to hold it in place.*

4   Unplug the electrical connector from the headlight bulb and the adjuster motor, if equipped.

5   Installation is the reverse of removal. Check and adjust the headlights (see Section 13).

15   Bulb replacement

Fog lights

Note: *2018 and later Sequoia models and some 2018 and later Tundra models use LED fog lights and are not serviceable. If there is a problem with a LED fog light the entire fog light assembly must be replaced.*

Halogen bulb types

Refer to illustrations 15.1 and 15.2

1   Disconnect the electrical connector from the fog light bulb holder **(see illustration)**.

2   Turn the bulb holder counterclockwise and remove it from the housing **(see illustration)**.

3   To remove the bulb from the bulb holder, pull it straight out.

4   Installation is the reverse of removal.

LED types

5   Remove the bumper cover (see Chapter 11).

Sequoia models

6   Remove and replace the retaining clip from the back of the bumper cover.

7   Place tape on the bumper cover around the edge of the fog light cover.

8   Using a trim tool, carefully pry up the fog light cover and disengage the retaining clips until the light cover can be removed from the bumper cover.

9   Remove the fog light mounting bolts and remove the light from the bumper cover.

10   Installation is the reverse of removal.

Tundra models

11   Remove the mounting screw from the upper corner of the fog light assembly, then detach the fog light from the two guides from the opposite side and remove the fog light.

12   Installation is the reverse of removal

Front side marker/turn signal bulb

Refer to illustrations 15.14a, 15.14b, 15.15a and 15.15b

13   The side marker/turn signal bulbs are located in the headlight housing.

14   Open the hood and disconnect the electrical connector to the bulb **(see illustrations)**. **Note:** *There is not much room to reach the side marker bulb; if you cannot reach the bulb, remove the headlight housing (see Section 14) to access the bulbs.*

15   Turn the bulb holder counterclockwise and remove it from the headlight housing **(see illustrations)**.

15.14a Disconnect the electrical connector to the side marker light

15.14b Unplug the electrical connector from the bulb holder for the turn signal bulb

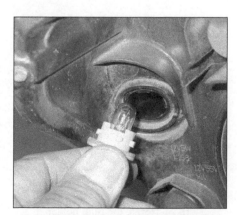

15.15a Turn the side marker bulb holder counter-clockwise and pull it out

15.15b Turn the turn signal bulb holder counter-clockwise and pull it out

15.18a To detach the Tundra tail light housing, remove the screws . . .

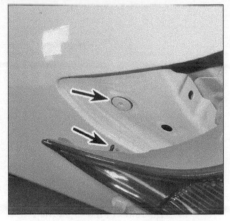

15.18b . . . then pry the retaining pin out from the corner, and rotate the housing out

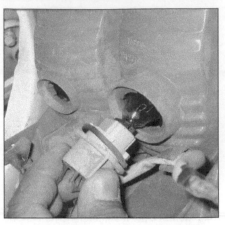

15.21 To replace a bulb, turn the holder counterclockwise, pull it out of the tail light housing, then pull the bulb straight out of the holder

16   To remove a bulb from the bulb holder, pull it straight out.

17   Installation is the reverse of removal. When installing the retaining screw, place it in position and press straight down until it clicks in place.

Tail light/brake light/turn signal

Refer to illustrations 15.18a, 15.18b and 15.21

18   On Tundra models, open the tailgate, remove the two bolts and rotate the tail light housing out for access to the bulbs **(see illustrations)**.

19   On Sequoia models, open the liftgate and remove the screws along the edge of the stop/tail and backup light housing, then detach the light housing from the vehicle.

20   On the liftgate-mounted tail light housings, remove the two retaining nuts, then use a screwdriver with a taped tip to carefully pry the housing off.

21   Remove the bulb holder by rotating it counterclockwise, then remove the bulb by pulling it straight out **(see illustration)**.

22   Installation is the reverse of removal.

License plate light

Tundra

Refer to illustrations 15.23 and 15.24

23   On models with side-mounted lights, remove the screw and detach the housing from the bumper **(see illustration)**. On center mounted lights, squeeze the tabs together and lower the bulb housing.

24   To remove the bulb holder, turn it counterclockwise and pull it out **(see illustration)**.

25   To remove the bulb from the holder, pull it straight out.

Sequoia

26   Remove the retaining screws and detach the lens. Remove the bulb from the holder by pulling it straight out of the housing.

All models

27   Installation is the reverse of removal.

High-mounted brake and cargo light

Refer to illustrations 15.28 and 15.29

28   Remove the screws and detach the lens for bulb access **(see illustration)**.

29   Pull out the defective bulb **(see illustration)**.

30   Replace the bulb and install the lens and screws.

Dome light bulb

Refer to illustration 15.31 and 15.32

31   Detach the dome light lens **(see illustration)**.

32   Grasp the bulb securely and detach it from the spring clips at each end **(see illustration)**.

33   Installation is the reverse of removal.

Overhead personal light bulbs (Sequoia models)

34   Use a small screwdriver and pry the light lens out of the overhead console.

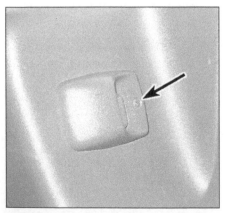

15.23 To remove the license plate bulb holder on the side-mounted license plate lights, remove the screw and detach the housing from the bumper (Tundra)

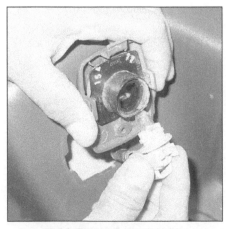

15.24 Rotate the bulb holder counterclockwise and withdraw it from the housing

15.28 To access the cargo light (or high-mounted brake light) bulbs, remove the screws and detach the lens

15.29 Grasp the bulb securely and pull it out

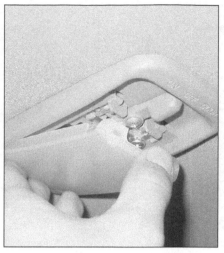

15.31 Detach the dome light lens

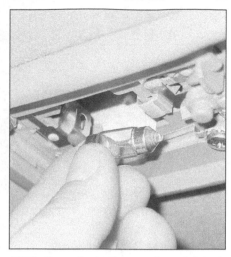

15.32 Grasp the bulb securely and detach it from the spring clip at each end

35   Grasp the bulb securely and pull it out.

36   Installation is the reverse of removal.

---

16 Radio and speakers - removal and installation

Warning: *The models covered by this manual are equipped with Supplemental Restraint Systems (SRS), more commonly known as airbags. Always disable the airbag system before working in the vicinity of any airbag system components to avoid the possibility of accidental deployment of the airbag(s), which could cause personal injury (see Section 28).*

Note: *On 2015 and later models equipped with a navigation system or rear seat entertainment system, once the ignition switch is turned to the OFF position wait at least 1 minute before disconnecting the cable from the negative battery terminal to prevent damage to the navigation system.*

1   Detach the cable from the negative terminal of the battery.

Radio

Refer to illustration 16.4

2   Remove the driver's side lower trim panel, instrument cluster trim panel, lower instrument trim panel and the cup holder panel. On floor shift models, remove the shift knob by turning the handle counterclockwise then remove the console trim shift panels (see Chapter 11).

3   On 2014 and later models, remove the air conditioning control assembly (see Chapter 3).

4   Remove the radio fasteners **(see illustration)** and pull out the radio, then disconnect the antenna lead and the electrical connector.

5   Remove the radio from the instrument panel.

6   Installation is the reverse of removal.

Navigation receiver

7   Remove the driver's side lower trim panel, instrument cluster trim panel, lower instrument trim panel and the cup holder panel (see Chapter 11).

8   On floor shift models, remove the shift knob by turning the handle counterclockwise, then remove the console trim shift panels (see Chapter 11).

9   On 2014 and later models, remove the air conditioning control assembly (see Chapter 3).

10   Remove the navigation receiver fasteners and pull out the receiver and bracket assembly, then disconnect the antenna lead and the electrical connector.

11   Remove the receiver and bracket assembly from the instrument panel, then remove the bracket fasteners and separate the bracket from the receiver.

12   Installation is the reverse of removal.

Speakers

Door speakers

Refer to illustration 16.14

13   All models have at least one speaker in each door. Tundra (Crew Max) or Sequoia models are also equipped with a pair of rear speakers. To remove any door-mounted speaker, remove the door trim panel (see Chapter 11).

14   To remove the lower speaker, remove retaining screws. To remove the upper speaker (Crew Max and Sequoia models), disengage the three mounting clips. On all speakers, disconnect the electrical connector and remove the speaker(s) from the door **(see illustration)**.

15   Installation is the reverse of removal.

Dash mounted speakers

Refer to illustrations 16.17 and 16.18

16   To remove the speakers from each side, remove the A-pillar trim panel (see Chapter 11).

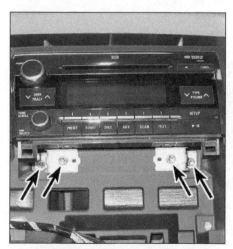

16.4 Remove the mounting bolts, detach the radio from the instrument panel and disconnect the electrical connectors

16.14 To remove a door-mounted lower speaker, remove the retaining screws, pull out the speaker and unplug the electrical connector

16.17 **Carefully pry up the speaker grille**

16.18 **Remove the mounting bolts, detach the speaker from the dash and disconnect the electrical connectors**

17 On all dash speakers, pry up the speaker grille **(see illustration)**.

18 Remove the speaker retaining screws, unplug the electrical connector and remove the speaker from the dash **(see illustration)**.

19 Installation is the reverse of removal.

Floor mounted speakers (CrewMax only)

20 Remove the rear seats (see Chapter 11) then remove the back panel fasteners and pry the panel out.

Note: *The panels overlap, so remove the right side first.*

21 Remove the speaker box retaining bolts, unplug the electrical connector and remove the speakers.

22 Installation is the reverse of removal.

Rear seat entertainment (Sequoia models only)

23 Using a plastic trim stick, disengage the four clips by carefully prying each corner of the overhead light assembly down. Once all four clips have been released, disconnect the electrical connector to the light assembly and remove the assembly.

24 Disengage the wiring harness retainer, then disconnect the electrical connector to the video display assembly and move the harness out of the way.

25 While supporting the rear video display, remove the four mounting bolts, then disengage the retaining clips from each side and remove the display assembly.

26 If necessary to remove the video display assembly mounting bracket, remove the four mounting bolts then disengage the retaining clips and remove the bracket.

27 Installation is the reverse of removal, making sure to tighten the mounting bolts securely.

17 Antenna - removal and installation

Refer to illustration 17.2

Note: *The following procedure applies to conventional external type antennas. It does not apply to the glass-printed antennas used on some Sequoia models. Glass-printed antennas should be serviced by a dealer service department or other qualified repair shop.*

1 Remove the radio and unplug the antenna cable (see Section 16). Trace the routing of the cable and detach all cable clamps and/or clips. Attach about four or five feet of string or wire to the antenna lead to help aid the installation of the new antenna lead.

2 On non-power antenna models, use a small open-end wrench to unscrew the antenna mast **(see illustration)**.

3 On power antenna models, push the ornament trim ring upwards from the bottom until it releases from the fender.

4 Remove the inner fender well splash shield (see Chapter 11).

5 Remove the antenna base mounting bolt inside the fender well, detach the antenna base, and pull out the antenna lead until the string or wire is exposed.

6 Installation is the reverse of removal, making sure the antenna lead is routed back into the vehicle.

17.2 **Use a small open-end wrench to unscrew the antenna mast**

18.1 **Before removing the wiper arm, mark the relationship of the arm to the linkage**

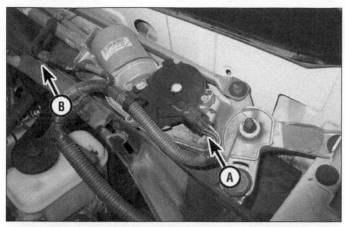

18.3 Disconnect the electrical connector to the wiper motor (A), then disconnect the harness (B) to allow the linkage to be easily removed

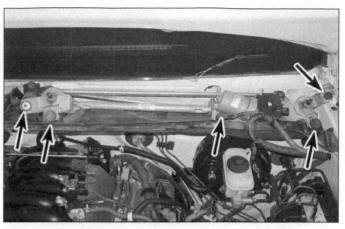

18.4 Remove the mounting bolts and lift the linkage and motor out as a unit

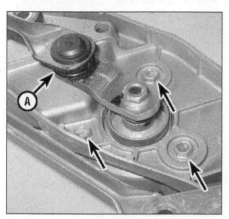

18.5 Pop the linkage free from the wiper motor crank arm (A) then remove the wiper motor-to-linkage mounting bolts

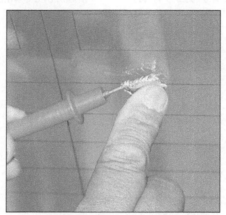

20.4 When measuring the voltage at the rear window defogger grid, wrap a piece of aluminum foil around the positive probe of the voltmeter and press the foil against the wire with your finger

20.5 To determine if a heating element has broken, check the voltage at the center of each element - if the voltage is 6-volts, the element is unbroken - if the voltage is 12-volts, the element is broken between the center and the ground side - if there's no voltage, the wire is broken between the center of the wire and the power side

18   Front wiper motor - replacement

Refer to illustrations 18.1, 18.3, 18.4 and 18.5

1   Remove the wiper arm mounting nut, then make an alignment mark from the arm to the linkage **(see illustration)**.
2   Remove the cowl, then the front fender-to-cowl seal (see Chapter 11).
3   Disconnect the electrical connector from the wiper motor and remove the wiper motor mounting bolts **(see illustration)**.
4   Remove the linkage and motor bolts and remove the assembly from the cowl **(see illustration)**.
5   Using a screwdriver, pop the linkage free from the wiper motor crank arm, remove the wiper motor-to-linkage mounting bolts, and separate the motor **(see illustration)**.
6   Installation is the reverse of removal.

19   Rear wiper motor (Sequoia models) - replacement

1   To detach the wiper arm, flip up the cover and remove the nut and wiper arm.

2   Open the liftgate and support the liftgate in the open position. Remove the liftgate support struts, power door rod (if equipped) and the trim panels (see Chapter 11).
3   Remove the three retaining bolts, unplug the electrical connector, then lift the wiper motor from the vehicle.
4   Installation is the reverse of removal.

20   Rear window defogger (Sequoia models) - check and repair

1   The rear window defogger consists of a number of horizontal elements baked onto a glass surface.
2   Small breaks in the element can be repaired without removing the rear window.

Check

Refer to illustrations 20.4, 20.5 and 20.7

3   Turn the ignition switch and defogger system switches to the ON position. Using a voltmeter, place the positive probe against the defogger grid positive terminal and the nega-

tive lead against the ground terminal. If battery voltage is not indicated, check the fuse, defogger switch and related wiring.
4   When measuring voltage during the next two tests, wrap a piece of aluminum foil around the tip of the voltmeter positive probe and press the foil against the heating element with your finger **(see illustration)**.
5   Check the voltage at the center of each heating element **(see illustration)**. If the voltage is 6-volts, the element is okay (there is no break). If the voltage is 12-volts, the element is broken between the center of the element and the ground side. If the voltage is 0-volts the element is broken between the center of the element and positive side.
6   If none of the elements are broken, connect the negative lead to a good body ground. the voltage reading should stay the same, if it doesn't the ground connection is bad.

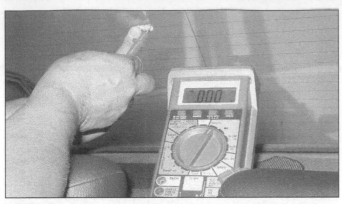

20.7 To find the break, place the voltmeter negative lead against the defogger ground terminal, place the voltmeter positive lead with the foil strip against the heating element at the positive terminal end and slide it toward the negative terminal end - the point at which the voltmeter deflects from 12-volts to zero volts is the point at which the element is broken

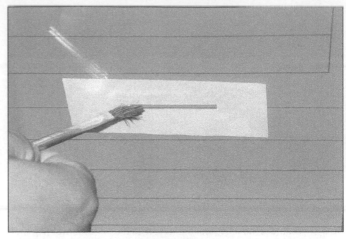

20.13 To use a defogger repair kit, apply masking tape to the inside of the window at the damaged area, then brush on the special conductive coating

7   To find the break, place the voltmeter negative lead against the defogger ground terminal. Place the voltmeter positive lead with the foil strip against the heating element at the positive terminal end and slide it toward the negative terminal end. The point at which the voltmeter deflects from several volts to zero is the point at which the heating element is broken **(see illustration)**.

Repair

Refer to illustration 20.13

8   Repair the break in the element using a repair kit specifically recommended for this purpose, such as Dupont paste No. 4817 (or equivalent). Included in this kit is plastic conductive epoxy.
9   Prior to repairing a break, turn off the system and allow it to cool off for a few minutes.
10   Lightly buff the element area with fine steel wool, then clean it thoroughly with rubbing alcohol.
11   Use masking tape to mask off the area being repaired.
12   Thoroughly mix the epoxy, following the instructions provided with the repair kit.
13   Apply the epoxy material to the slit in the masking tape, overlapping the undamaged area about 3/4-inch on either end **(see illustration)**.
14   Allow the repair to cure for 24 hours before removing the tape and using the system.

21   Instrument cluster - removal and installation

Refer to illustrations 21.4a and 21.4b
Warning: *The models covered by this manual are equipped with Supplemental Restraint Systems (SRS), more commonly known as airbags. Always disable the airbag system before working in the vicinity of any airbag system components to avoid the possibility of accidental deployment of the airbag(s), which could cause personal injury (see Section 28).*
1   On models equipped with power tilt and power telescopic steering wheels, fully extend and lower the steering column before disconnecting the battery.
2   Disconnect the cable from the negative terminal of the battery (see Chapter 5).
Note: *On 2015 and later models equipped with a navigation system or rear seat entertainment system, once the ignition switch is turned to the OFF position, wait at least 1 minute before disconnecting the cable from the negative battery terminal to prevent damage to the navigation system.*
3   Remove the instrument cluster trim bezel (see Chapter 11).
4   Remove the instrument cluster retaining screws **(see illustration)**, pull out the cluster and unplug the electrical connectors from the cluster **(see illustration)**.
5   Installation is the reverse of removal.

22   Horn - replacement

Refer to illustration 22.2
1   The horns are located on the radiator support.

21.4a Instrument cluster retaining screws locations

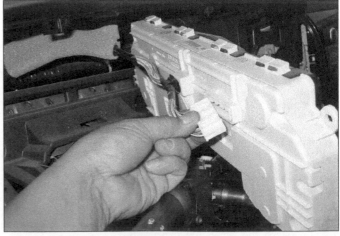

21.4b Pull out the cluster and unplug the electrical connectors

2   On 2014 and later models, remove the radiator grille (see Chapter 11).

3   To replace a horn, unplug the electrical connector and remove the bracket bolt **(see illustration)**.

4   Unbolt the bracket from the old horn and bolt it onto the new unit.

5   Installation is the reverse of removal.

23   Power mirror control system - description and check

1   Electric rear view mirrors use two motors to move the glass; one for up and down adjustments and one for left-right adjustments.

2   The control switch has a selector portion which sends voltage to the left or right side mirror. With the ignition ON but the engine OFF, roll down the windows and operate the mirror control switch through all functions (left-right and up-down) for both the left and right side mirrors.

3   Listen carefully for the sound of the electric motors running in the mirrors.

4   If the motors can be heard but the mirror glass doesn't move, there's probably a problem with the drive mechanism inside the mirror.

5   If the mirrors do not operate and no sound comes from the mirrors, check the fuse (see Section 3).

6   If the fuse is OK, remove the mirror control switch from its mounting without disconnecting the wires attached to it. Turn the ignition ON and check for voltage at the switch. There should be voltage at one terminal. If there's no voltage at the switch, check for an open or short in the circuit between the fuse panel and the switch.

7   If the mirror motor fails to operate as described, replace the mirror assembly (see Chapter 11).

24   Cruise control system - description and check

1   All models have an electronically-controlled throttle body - there is no accelerator cable (or cruise control cable). When you select the speed that you want to maintain, the PCM controls vehicle speed by opening and closing the throttle plate by means of a computer-controlled solenoid (motor) inside the throttle body.

2   The diagnostic procedures for troubleshooting the cruise control system are beyond the scope of this manual, but if the system can't be set, or the set speed doesn't cancel when the brake pedal is depressed, check the fuses. Start with the fuses in the engine compartment fuse and relay box, then check the fuses in the under-dash fuse and relay box. If the set speed doesn't cancel when the CANCEL button is depressed, check the fuse for that circuit.

3   Other than checking the fuses, the diagnostic procedures for troubleshooting the cruise

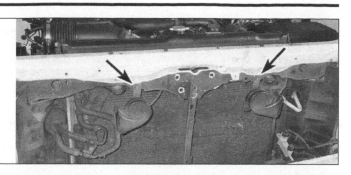

22.2 To detach a horn from the body, remove the bracket bolt

control system on these models are beyond the scope of this manual. A dealer service department should handle any further testing.

25   Power window system - description and check

1   The power window system operates electric motors, mounted in the doors, which lower and raise the windows. The system consists of the control switches, relays, the motors, regulators, glass mechanisms and associated wiring.

2   The power windows can be lowered and raised from the master control switch by the driver or by remote switches located at the individual windows. Each window has a separate motor which is reversible. The position of the control switch determines the polarity and therefore the direction of operation.

3   The circuit is protected by a fuse and a circuit breaker. Each motor is also equipped with an internal circuit breaker; this prevents one stuck window from disabling the whole system.

4   The power window system will only operate when the ignition switch is ON. In addition, many models have a window lockout switch at the master control switch which, when activated, disables the switches at the rear windows and, sometimes, the switch at the passenger's window also. Always check these items before troubleshooting a window problem.

5   These procedures are general in nature, so if you can't find the problem using them, take the vehicle to a dealer service department or other properly equipped repair facility.

6   If the power windows won't operate, always check the fuse and circuit breaker first.

7   If only the rear windows are inoperative, or if the windows only operate from the master control switch, check the rear window lockout switch for continuity in the unlocked position. Replace it if it doesn't have continuity.

8   Check the wiring between the switches and fuse panel for continuity. Repair the wiring, if necessary.

9   If only one window is inoperative from the master control switch, try the other control switch at the window.

Note: *This doesn't apply to the drivers door window.*

10   If voltage is reaching the motor, discon-

nect the glass from the regulator (see Chapter 11). Move the window up and down by hand while checking for binding and damage. Also check for binding and damage to the regulator. If the regulator is not damaged and the window moves up and down smoothly, replace the motor. If there's binding or damage, lubricate, repair or replace parts, as necessary.

11   If voltage isn't reaching the motor, check the wiring in the circuit for continuity between the switches and motors. You'll need to consult the wiring diagram for the vehicle. If the circuit is equipped with a relay, check that the relay is grounded properly and receiving voltage.

12   Test the windows after you are done to confirm proper repairs.

26   Power door lock system - description and check

Description

1   A power door lock system operates the door lock actuators mounted in each door. The system consists of the switches, actuators, the Body Control Module and associated wiring. Diagnosis can usually be limited to simple checks of the wiring connections and actuators for minor faults that can be easily repaired.

2   Power door lock systems are operated by bi-directional solenoids located in the doors. The lock switches have two operating positions: Lock and Unlock. When activated, the switch sends a ground signal to the door lock control unit to lock or unlock the doors. Depending on which way the switch is activated, the control unit reverses polarity to the solenoids, allowing the two sides of the circuit to be used alternately as the feed (positive) and ground side.

3   Some vehicles may have an anti-theft system incorporated into the power locks. If you are unable to locate the trouble using the following general Steps, consult a dealer service department or other qualified repair shop.

4   Always check the circuit protection first. Some vehicles use a combination of circuit breakers and fuses.

5   Operate the door lock switches in both directions (Lock and Unlock) with the engine off. Listen for the click of the solenoids operating.

6   Test the switches for continuity. Remove the switches and have them checked by a dealer service department or other qualified automobile repair facility.

7   Check the wiring between the switches, control unit and solenoids for continuity. Repair the wiring if there's no continuity.

8   Check for a bad ground at the switches or the control unit.

9   If all but one lock solenoids operate, remove the trim panel from the affected door (see Chapter 11) and check for voltage at the solenoid while the lock switch is operated. One of the wires should have voltage in the Lock position; the other should have voltage in the Unlock position.

10   If the inoperative solenoid is receiving voltage, replace the solenoid.

11   If the inoperative solenoid isn't receiving voltage, check for an open or short in the wire between the switch and the Body Control Module (see the wiring diagrams at the end of this Chapter).

Remote keyhead transmitters

General information

12   Here's how the transmitter inside the keyhead should work:

13   When you press the UNLOCK button, the driver's door unlocks. If you press the UNLOCK button a second time within four seconds, all the doors unlock.

14   Pressing the lock button sets the alarm and locks all of the doors.

Battery replacement

15   When the transmitter becomes weak, operation will become intermittent and require you to be closer to the vehicle for it to work. Eventually it won't work at all.

16   To replace the transmitter battery, carefully pry down to open the transmitter body by inserting a small screwdriver into the slot in the body of the transmitter and separate the halves of the transmitter body.

17   Using a small screwdriver, carefully pry out the old battery. Installation is the reverse of removal, making sure the positive (+) side of the battery is facing up. Make sure the two halves of the cover snap together tightly to keep out dirt, dust, humidity and rain.

---

27   Daytime Running Lights (DRL) - general information

The Daytime Running Lights (DRL) system used on some models illuminates the headlights whenever the engine is running. The only exception is with the engine running and the parking brake engaged. Once the parking brake is released, the lights will remain on as long as the ignition switch is on, even if the parking brake is later applied.

The DRL system supplies reduced power to the headlights so they won't be too bright for daytime use, while prolonging headlight life.

---

28   Airbags - general information

These models are equipped with a Supplemental Restraint System (SRS), more commonly known as airbags. This system is designed to protect the driver and the front seat passenger from serious injury in the event of a head-on or frontal collision. It consists of an airbag module in the center of the steering wheel and another airbag module on the right side of the instrument panel plus, on some and later models, side airbags and curtain shield airbags designed to protect the occupants in a side impact and a sensing/diagnostic module which is mounted in the center of the vehicle below the instrument panel. These models are also equipped with a pair of impact sensors that are located at the front of the vehicle.

Some later models are equipped with seatbelt pre-tensioners, also part of the airbag system. The pre-tensioners are pyrotechnic (explosive) devices designed to retract the seat belts in the event of a collision.

On models equipped with pre-tensioners, do not remove the front seat belt retractor assemblies. Problems with the pre-tensioners will turn on the SRS (airbag) warning light on the dash. If any pre-tensioner problems are suspected, take the vehicle to a dealer service department.

Airbag module

Steering wheel-mounted

The airbag inflator module contains a housing incorporating the cushion (airbag) and inflator unit, mounted in the center of the steering wheel. The inflator assembly is mounted on the back of the housing over a hole through which gas is expelled, inflating the bag almost instantaneously when an electrical signal is sent from the system. A spiral cable assembly on the steering column under the module carries this signal to the module. This spiral cable assembly can transmit an electrical signal regardless of steering wheel position.

Instrument panel-mounted

The passenger side airbag is mounted above the glove compartment and designated by the letters SRS (Supplemental Restraint System). It consists of an inflator containing an igniter, a bag assembly, a reaction housing and a trim cover.

The passenger airbag is considerably larger than the steering wheel-mounted unit and is supported by the steel reaction housing. The trim cover has a molded seam which splits when the bag inflates.

Sensing and diagnostic module

The sensing and diagnostic module supplies the current to the airbag system in the event of the collision, even if battery power is cut off. It checks this system every time the vehicle is started, causing the "AIR BAG" light to go on then off, if the system is operating properly. If there is a fault in the system, the light will go on and stay on, flash, or the dash will make a beeping sound. If this happens,

the vehicle should be taken to your dealer immediately for service.

Side and curtain airbags

The passenger side airbag and inflator modules are mounted on the sides of the front seats contain an inflator containing an igniter and bag assembly. The curtain shield airbag assemblies run along the interior of the roof from the front A-pillar to the rear of the passenger compartment. In the event of a side impact both airbag assemblies are activated by the sensors mounted at the base of the center pillar behind the seats.

Precautions

Disabling the SRS system

Warning: *Failure to follow these precautions could result in accidental deployment of the airbag and personal injury.*

Warning: *Never install a memory-saver device, used to preserve PCM memory and radio station presets, when working on or around any of the airbag system components.*

Whenever working in the vicinity of the steering wheel, instrument panel or any of the other SRS system components, the system must be disarmed. To disarm the system:

a) *Point the wheels straight ahead and turn the ignition key to the LOCK position.*

b) *Disconnect the cable from the negative terminal of the battery.*

c) *Wait at least two minutes for the back-up power supply capacitor to be depleted.*

Whenever handling an airbag module, always keep the airbag opening (trim side) pointed away from your body. Never place the airbag module on a bench or other surface with the airbag opening facing the surface. Always place the airbag module in a safe location with the airbag opening (trim side) facing up.

Never measure the resistance of any SRS component. An ohmmeter has a built-in battery supply that could accidentally deploy the airbag.

Never use electrical welding equipment on a vehicle equipped with an airbag without first disconnecting the negative battery cable.

Never dispose of a live airbag module. Return it to your dealer for safe deployment, using special equipment, and disposal.

---

29   Wiring diagrams - general information

Since it isn't possible to include all wiring diagrams for every year covered by this manual, the following diagrams are those that are typical and most commonly needed.

Prior to troubleshooting any circuits, check the fuse and circuit breakers (if equipped) to make sure they're in good condition. Make sure the battery is properly charged and check the cable connections (see Chapter 1).

When checking a circuit, make sure that all connectors are clean, with no broken or loose terminals. When unplugging a connector, do not pull on the wires. Pull only on the connector housings themselves.

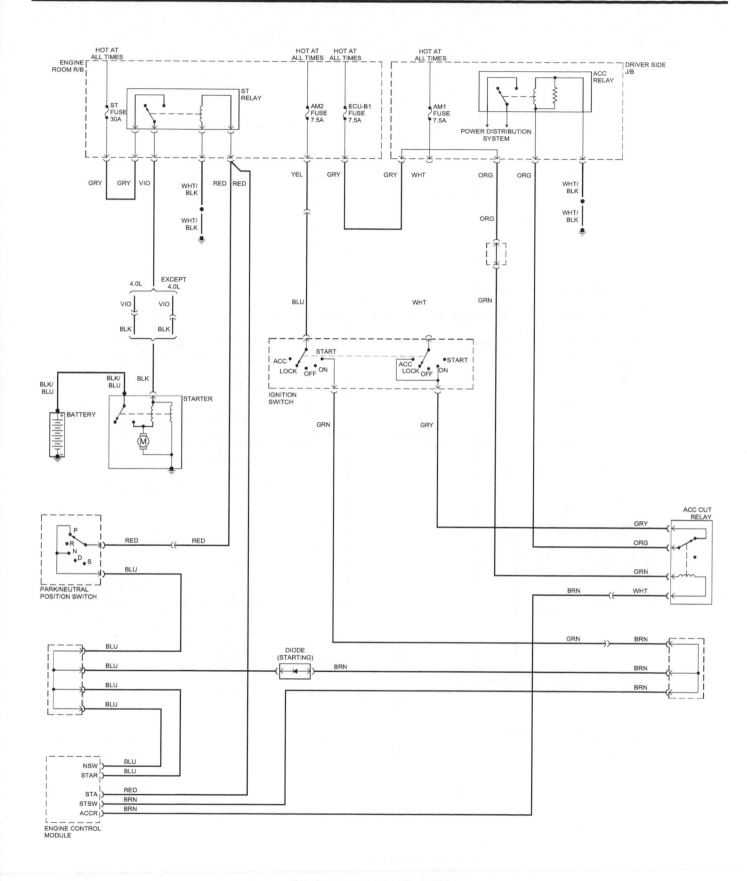

Starting system - Tundra models

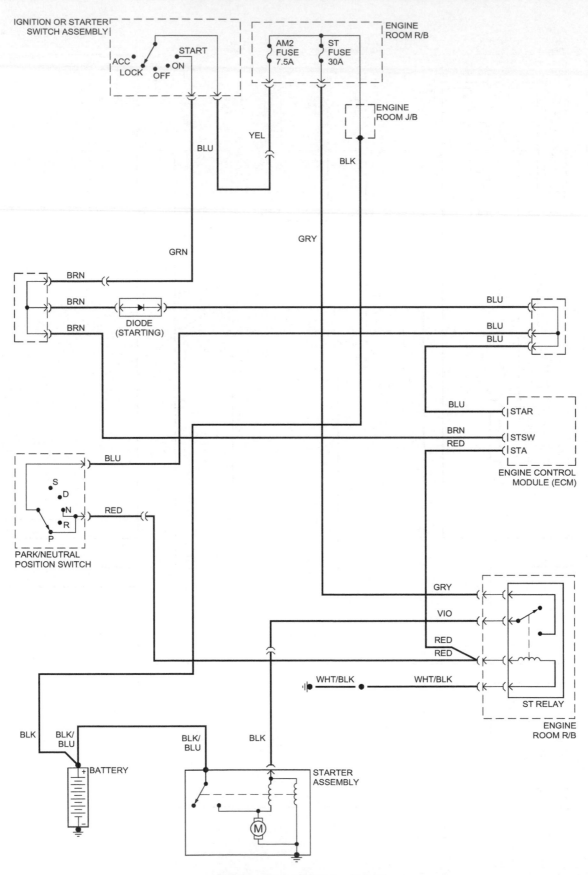

Starting system - Sequoia models

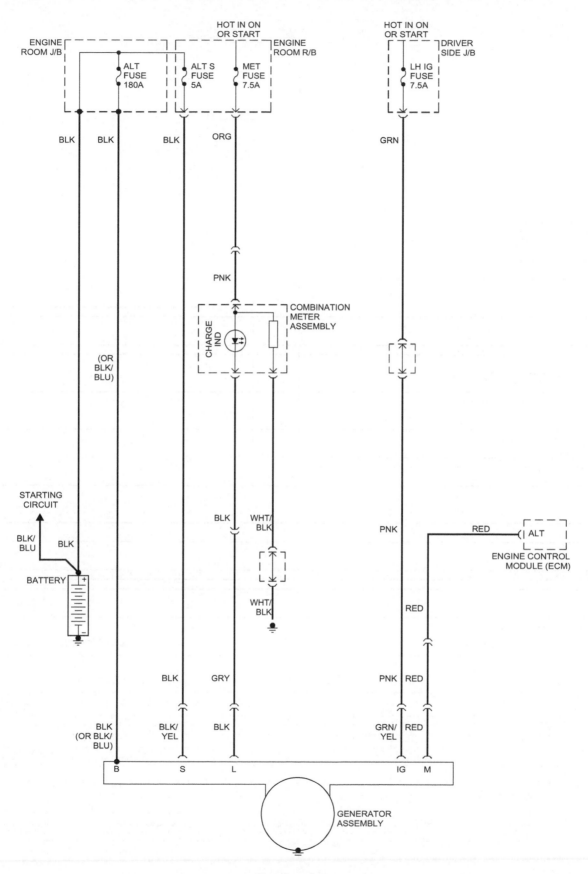

Charging system

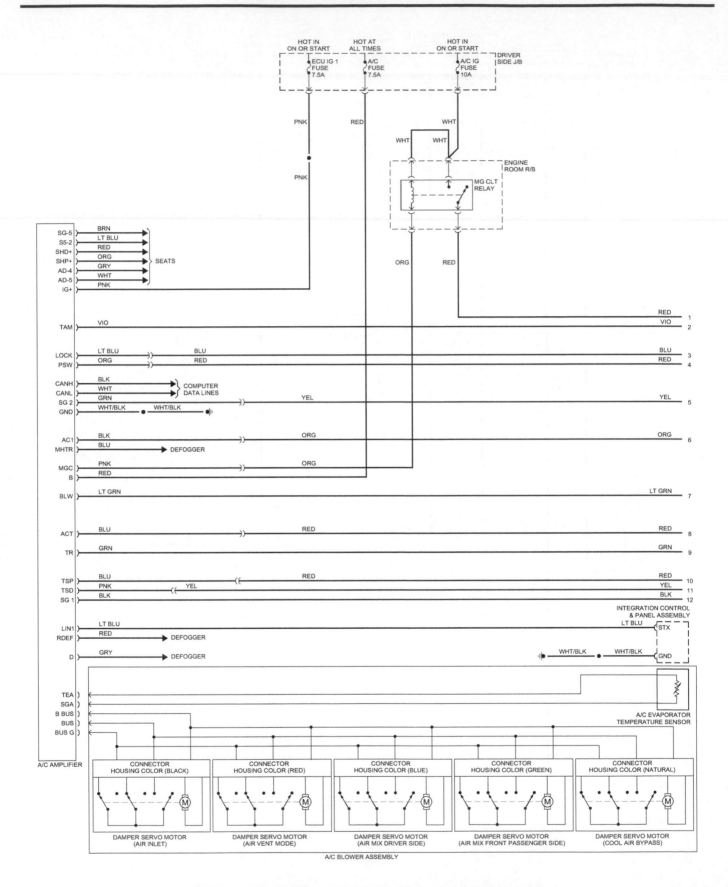

Heating and air condition system (automatic) - Tundra models (1 of 2)

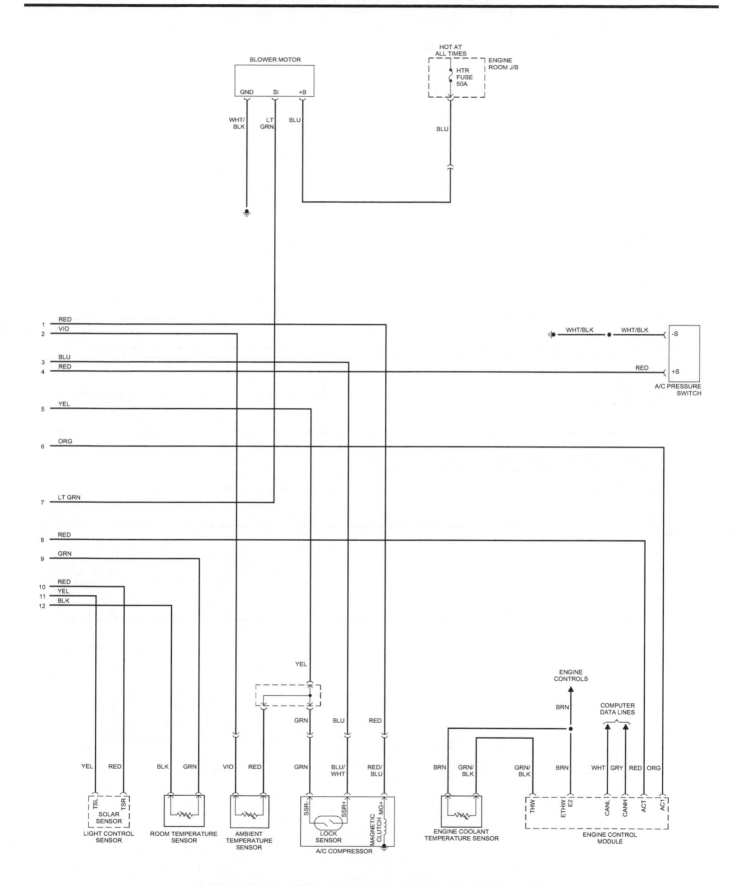

Heating and air condition system (automatic) - Tundra models (2 of 2)

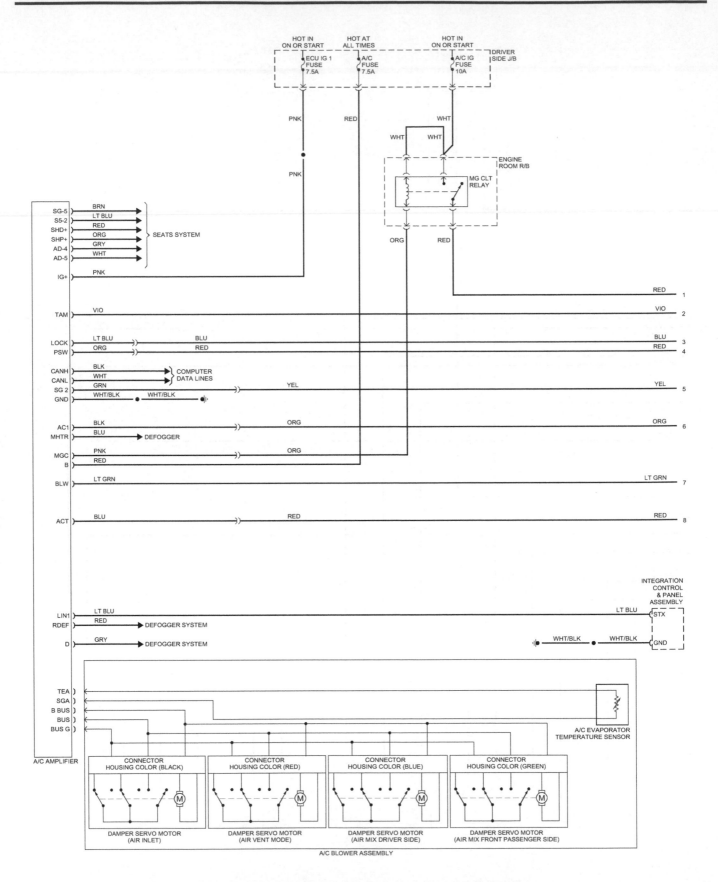

Heating and air condition system (manual) - Tundra models (1 of 2)

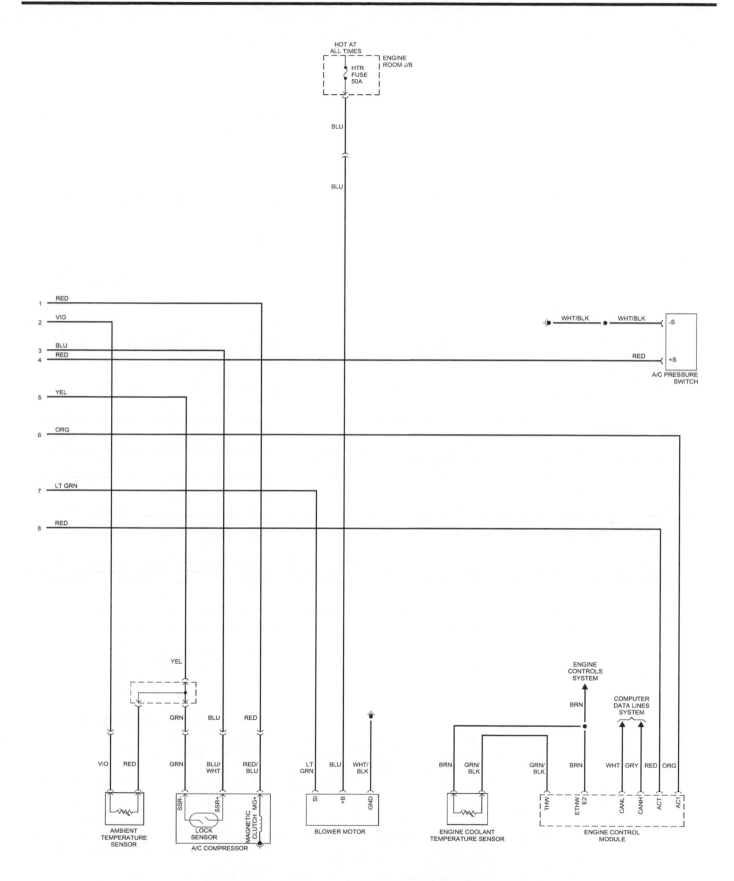

Heating and air condition system (manual) - Tundra models (2 of 2)

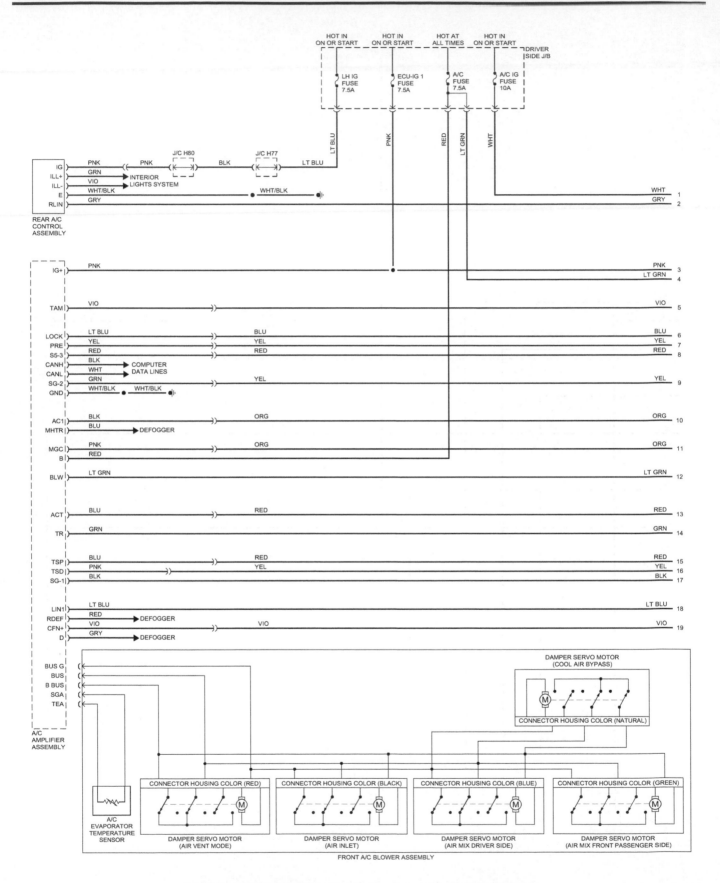

Heating and air condition system (automatic) - Sequoia models (1 of 3)

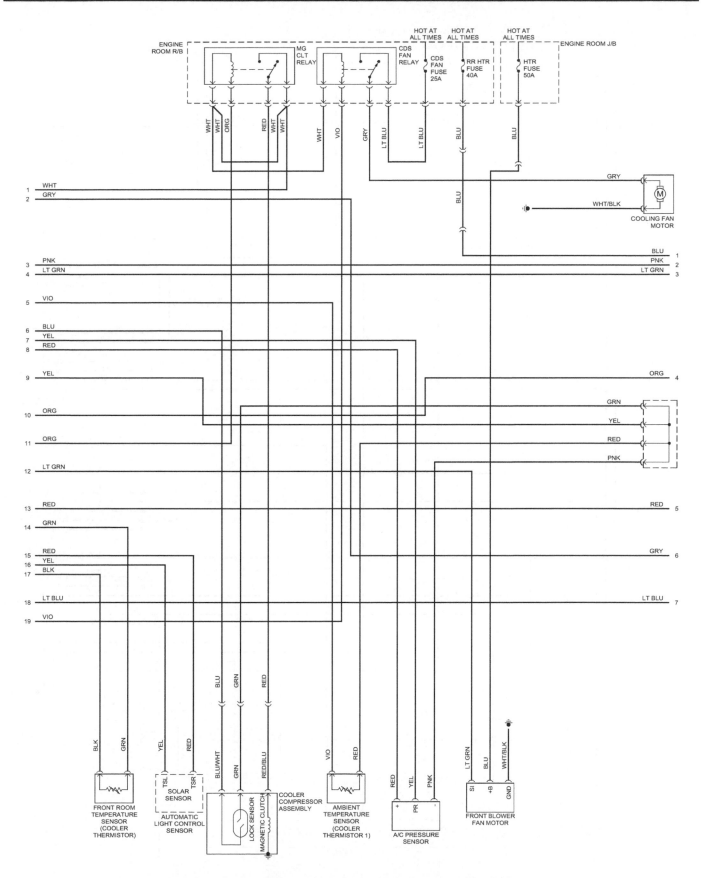

Heating and air condition system (automatic) - Sequoia models (2 of 3)

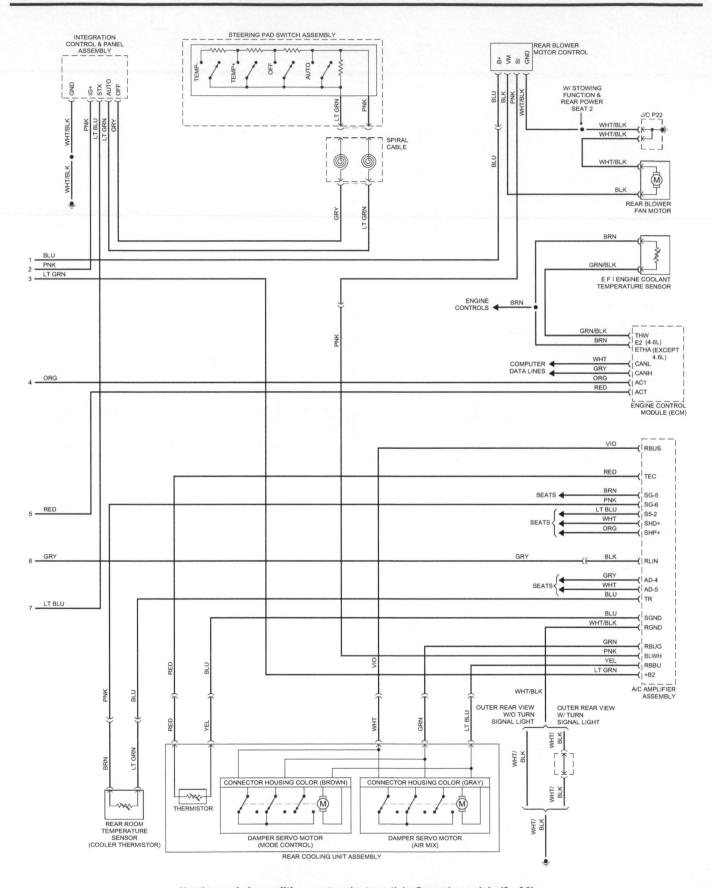

Heating and air condition system (automatic) - Sequoia models (3 of 3)

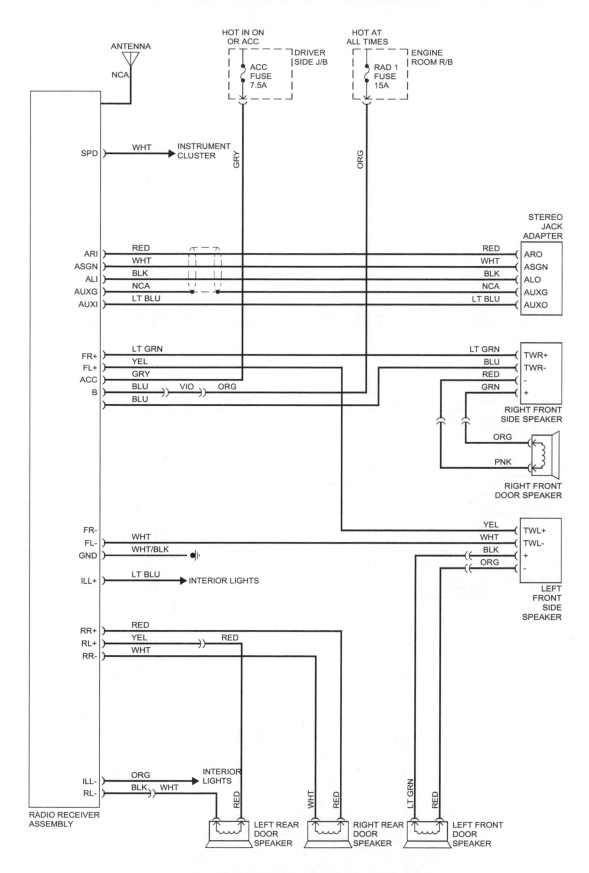

Audio system - 2007 through 2009 Tundra models

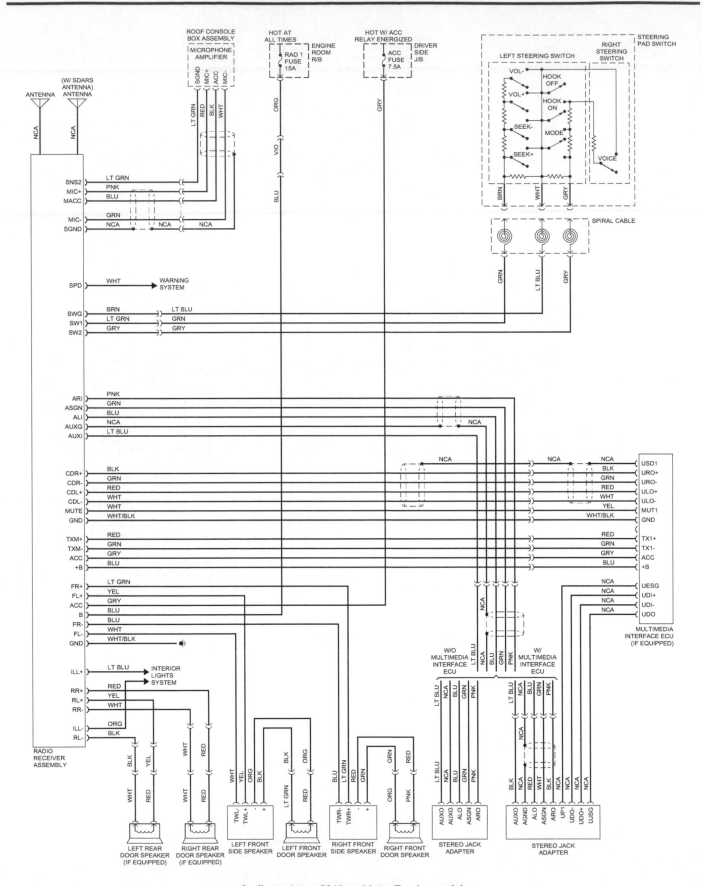

Audio system - 2010 and later Tundra models

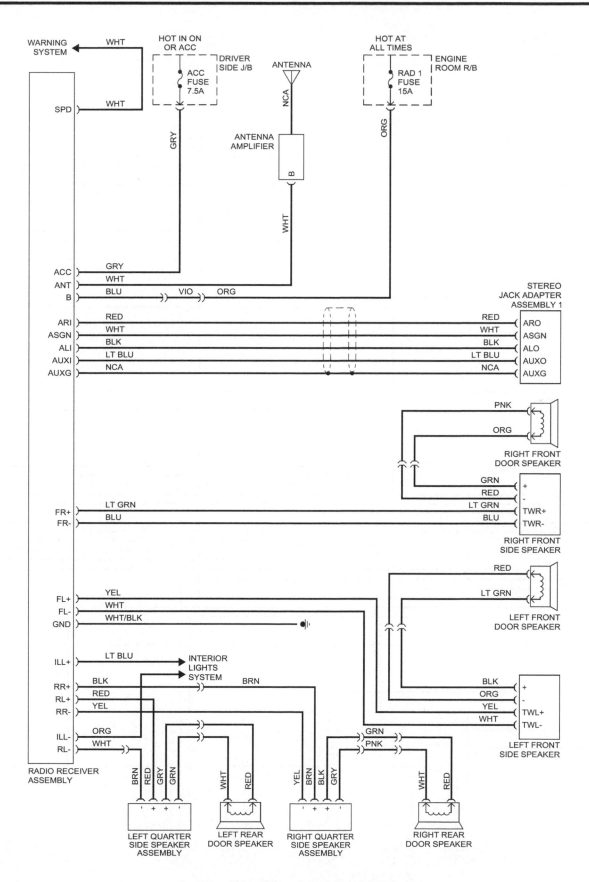

Audio system - 2008 and 2009 Sequoia models

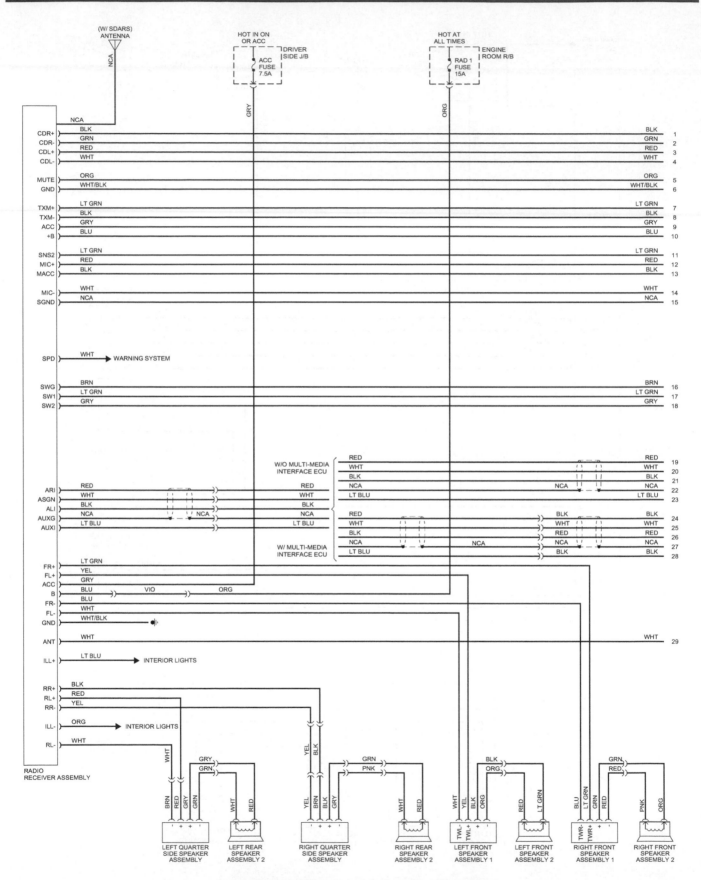

Audio system - 2010 and later Sequoia models (1 of 2)

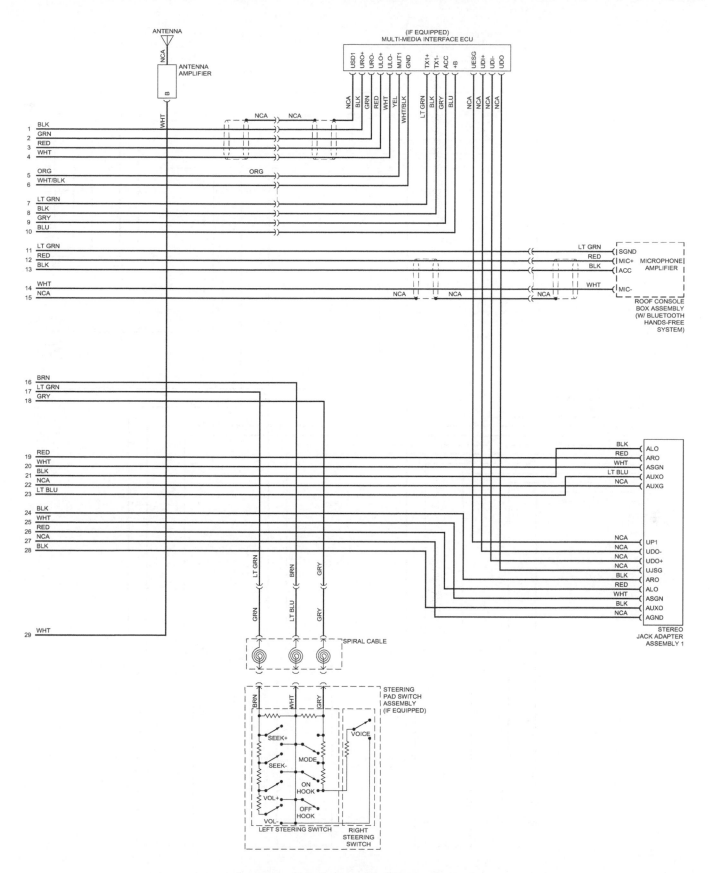

Audio system - 2010 and later Sequoia models (2 of 2)

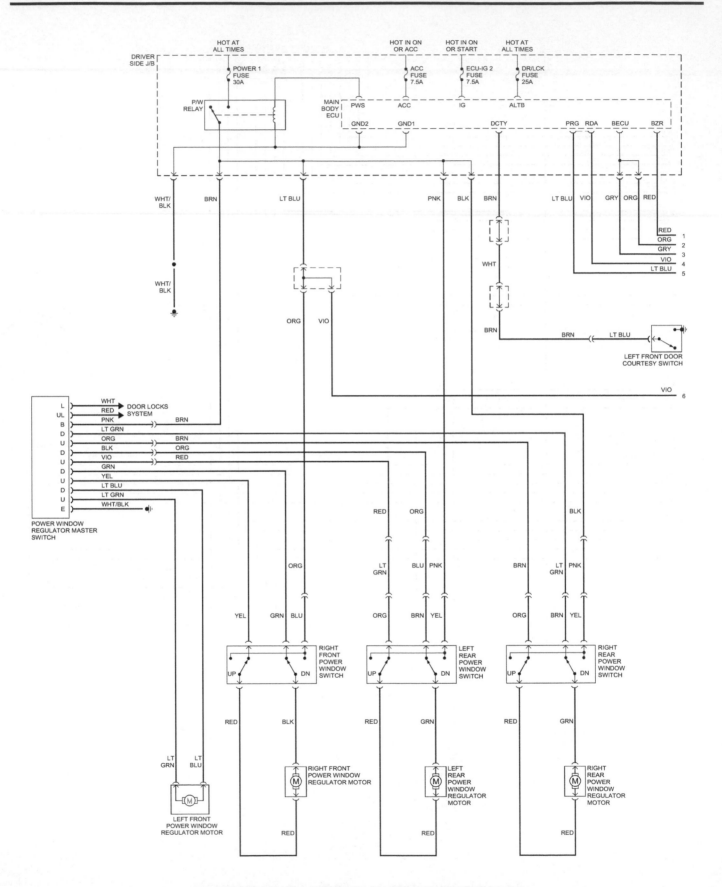

Power window system - 2007 through 2009 Tundra models (1 of 2)

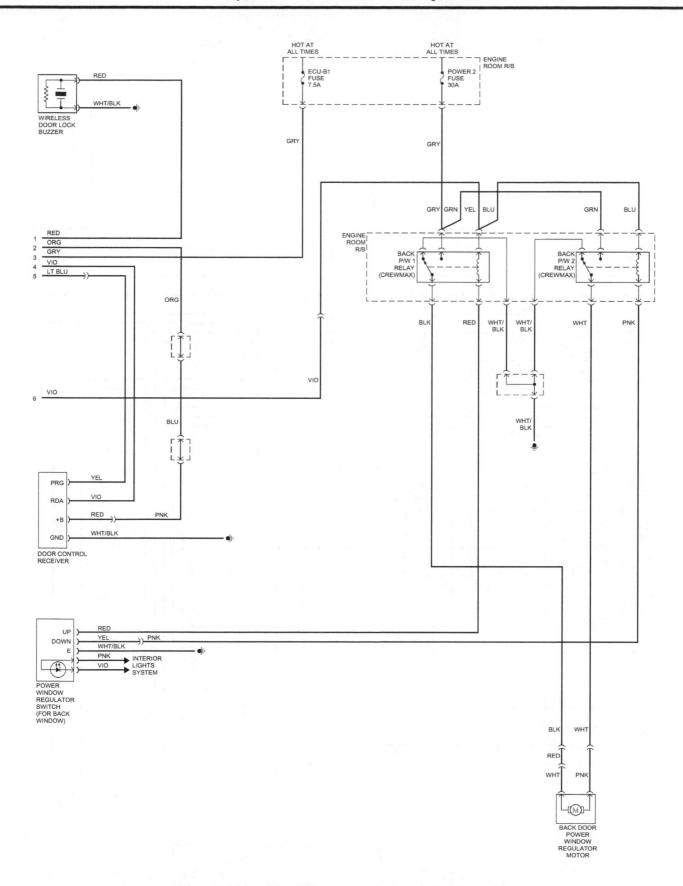

Power window system - 2007 through 2009 Tundra models (2 of 2)

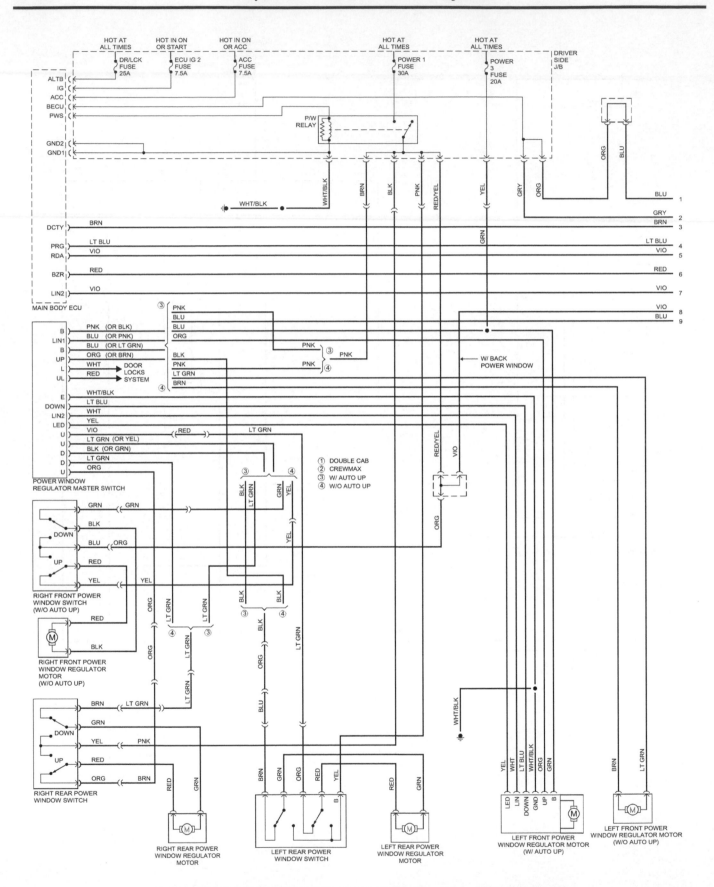

Power window system - 2010 and later Tundra models, except regular cab (1 of 2)

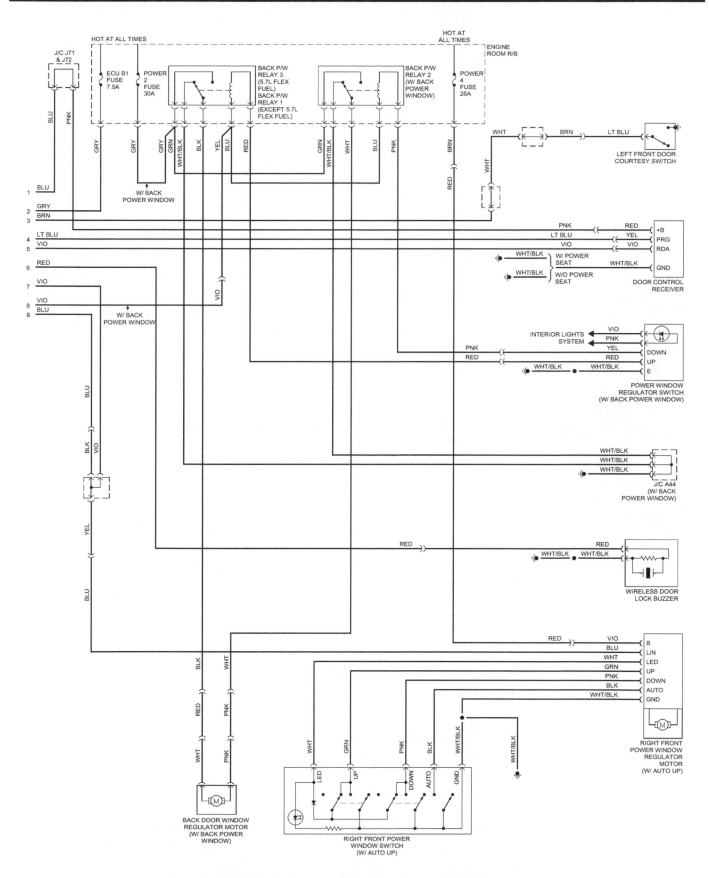

Power window system - 2010 and later Tundra models, except regular cab (2 of 2)

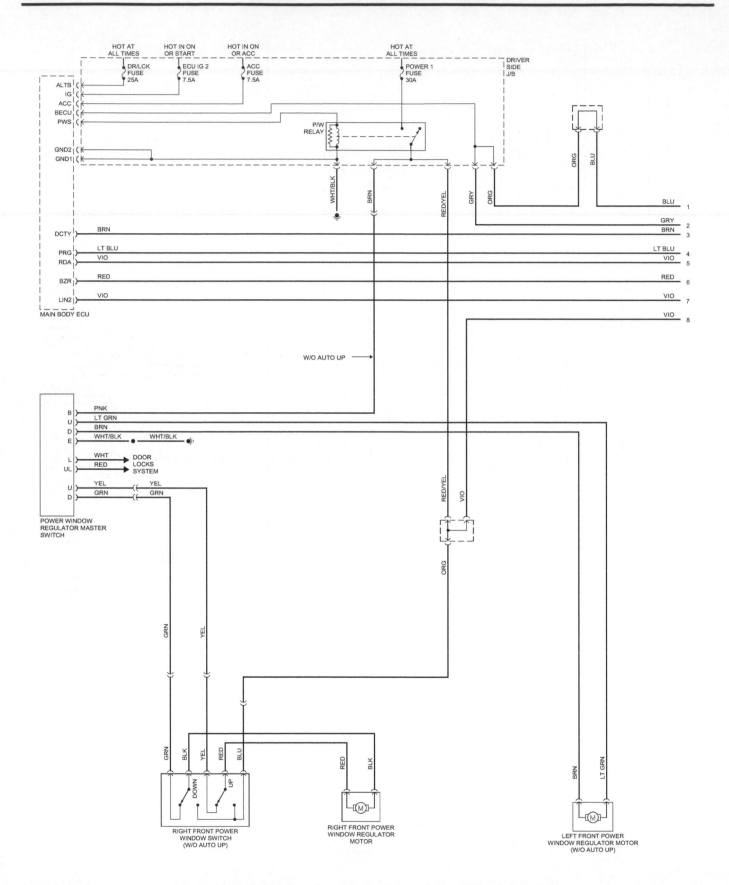

Power window system - 2010 and later Tundra models, regular cab (1 of 2)

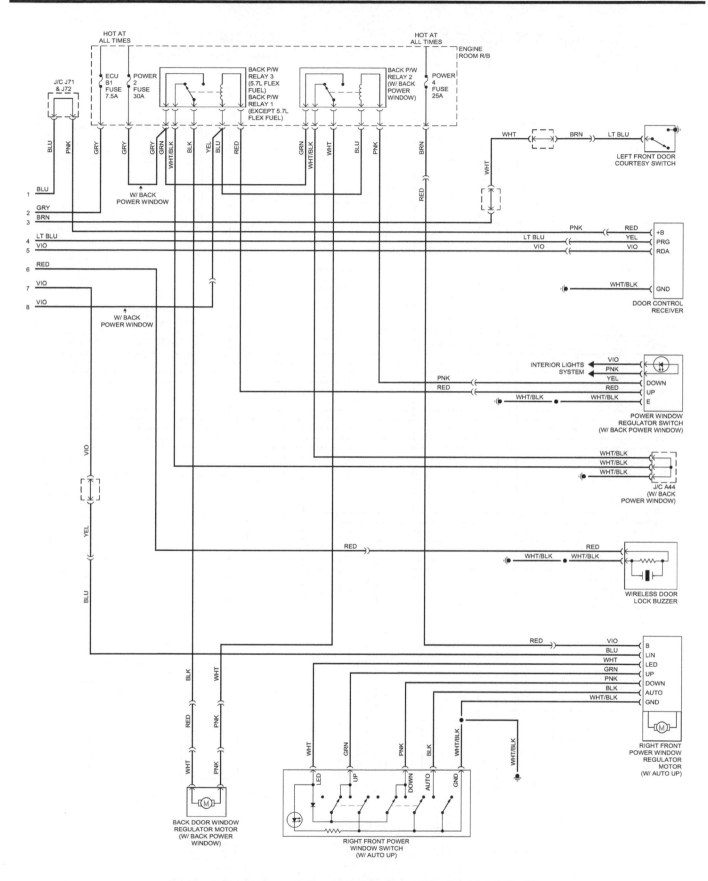

Power window system - 2010 and later Tundra models, regular cab (2 of 2)

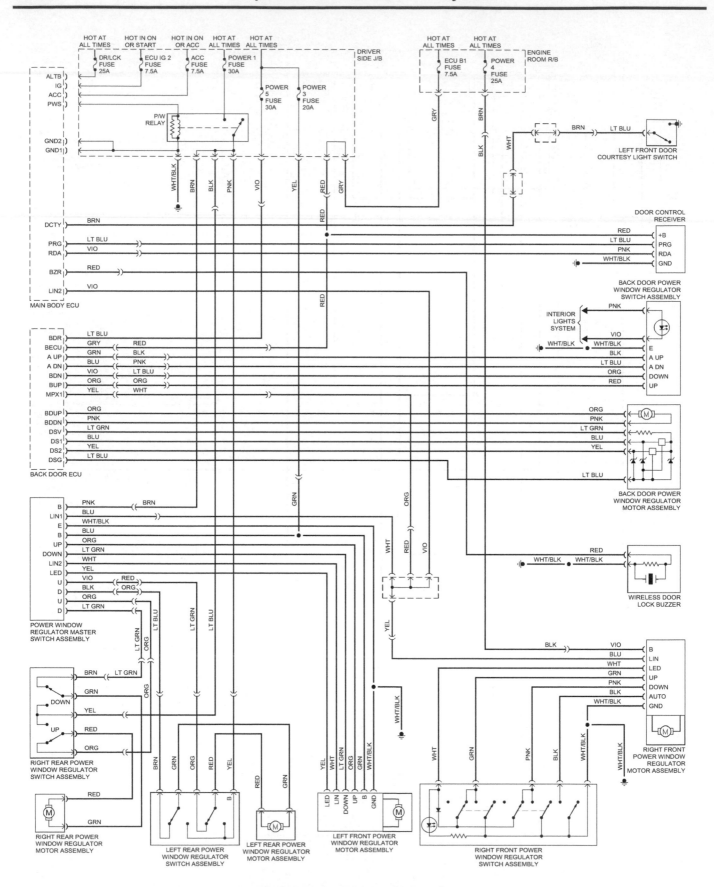

Power window system - Sequoia models

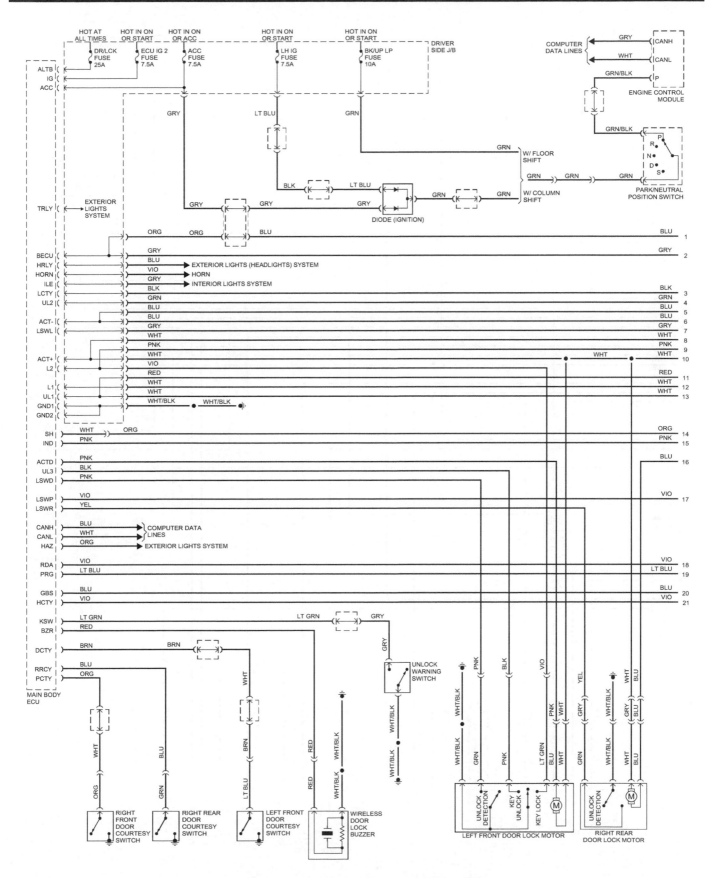

Power door lock system - Tundra models (1 of 2)

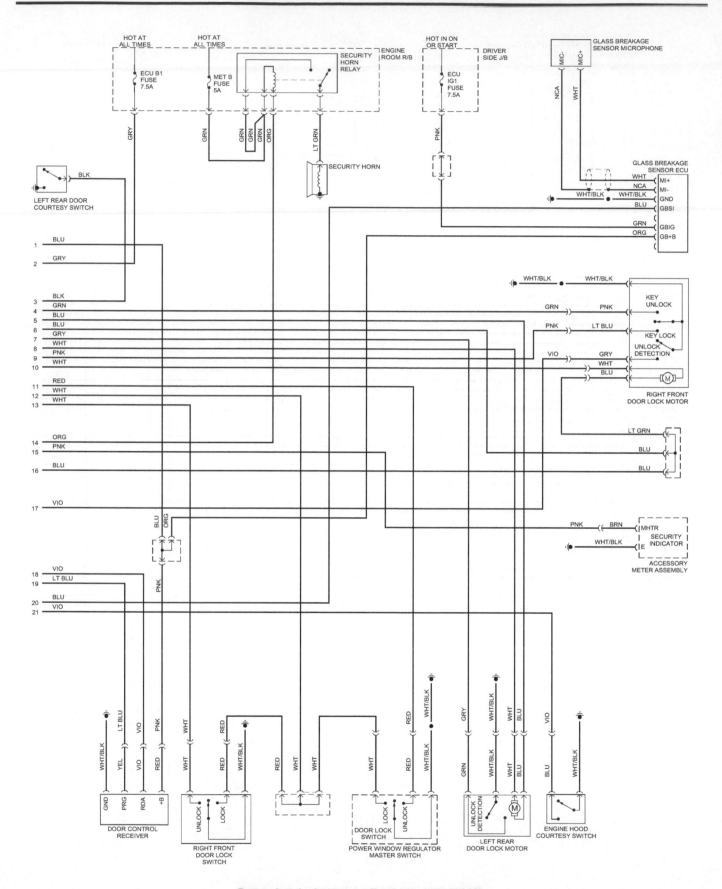

Power door lock system - Tundra models (2 of 2)

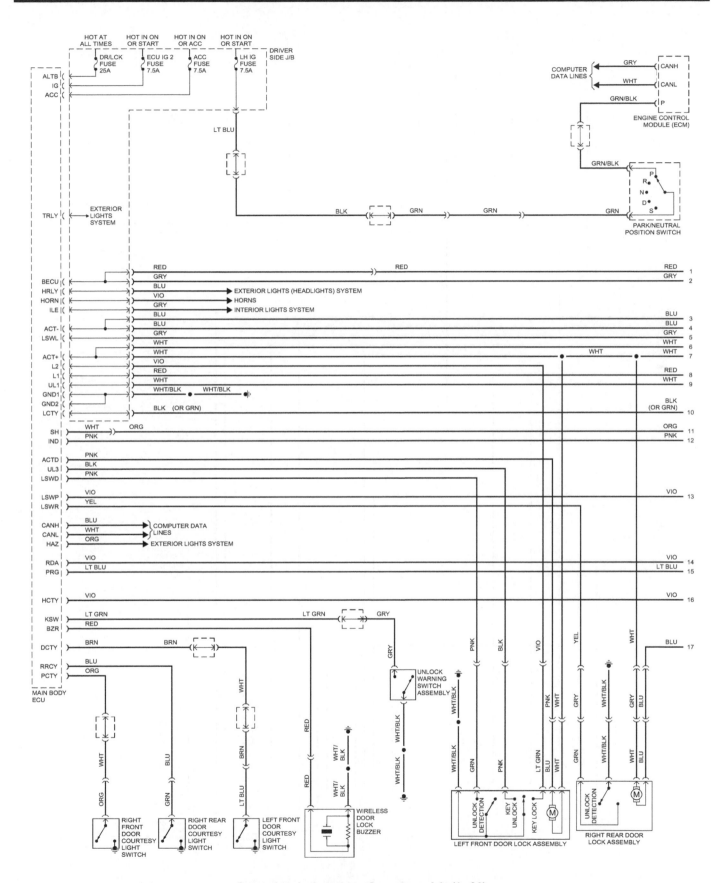

Power door lock system - Sequoia models (1 of 2)

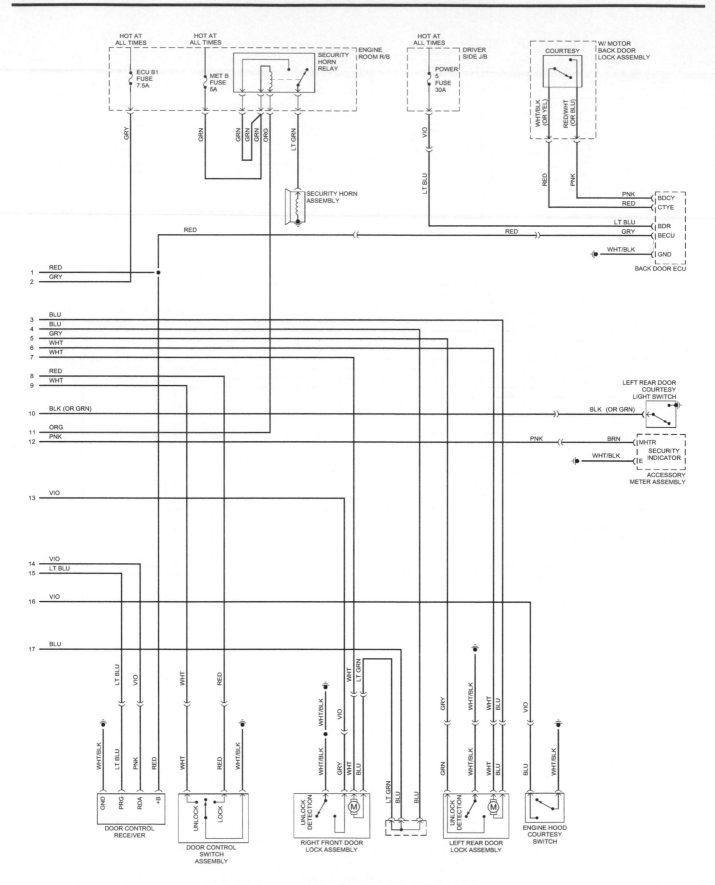

Power door lock system - Sequoia models (2 of 2)

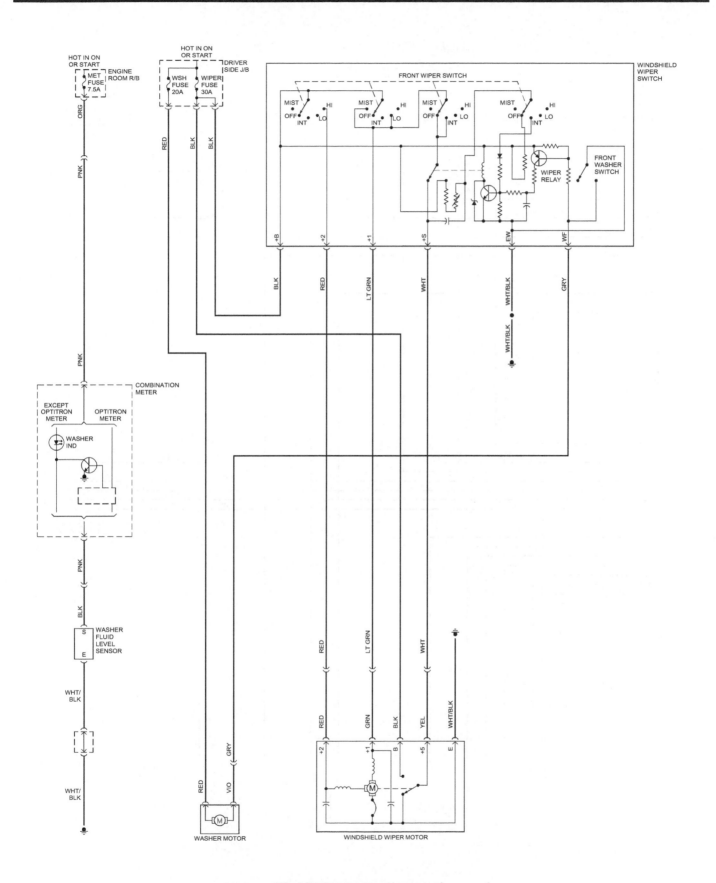

Windshield wiper/washer system

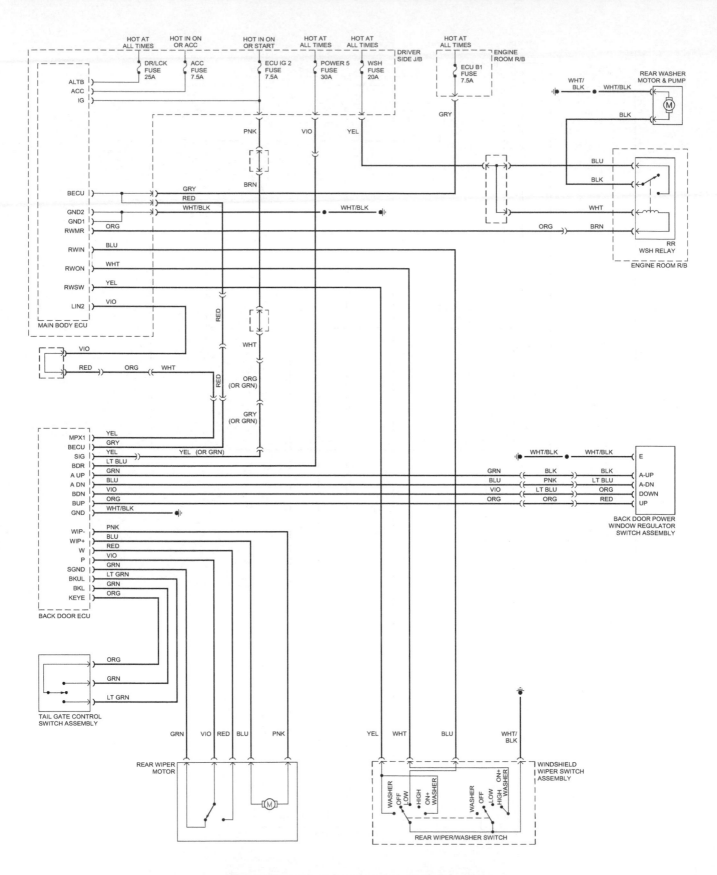

Rear window wiper/washer system (Sequoia models)

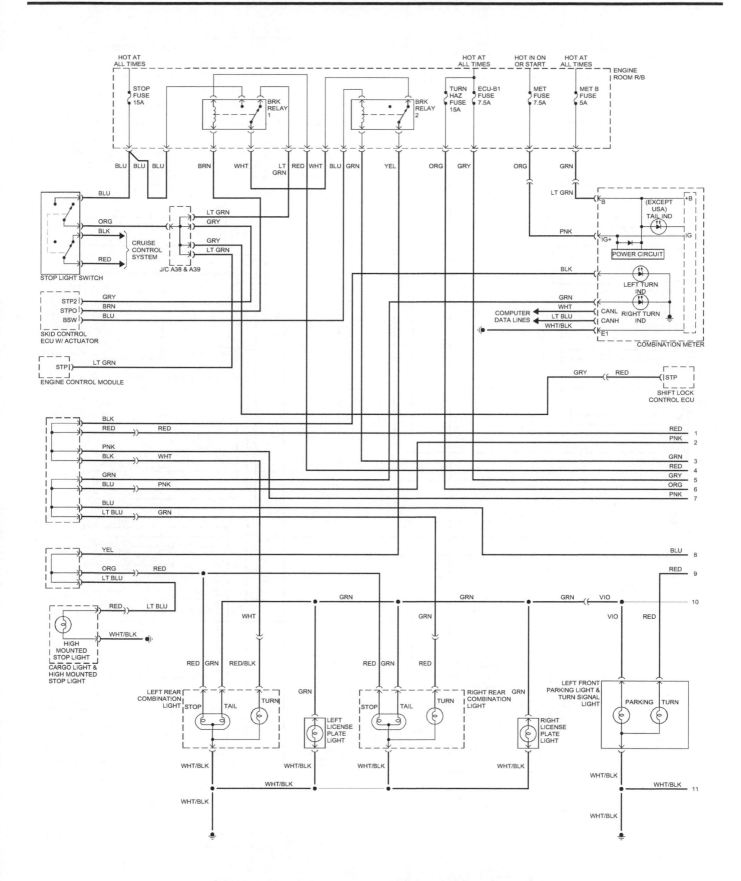

Exterior lighting system (except headlights) - Tundra models (1 of 2)

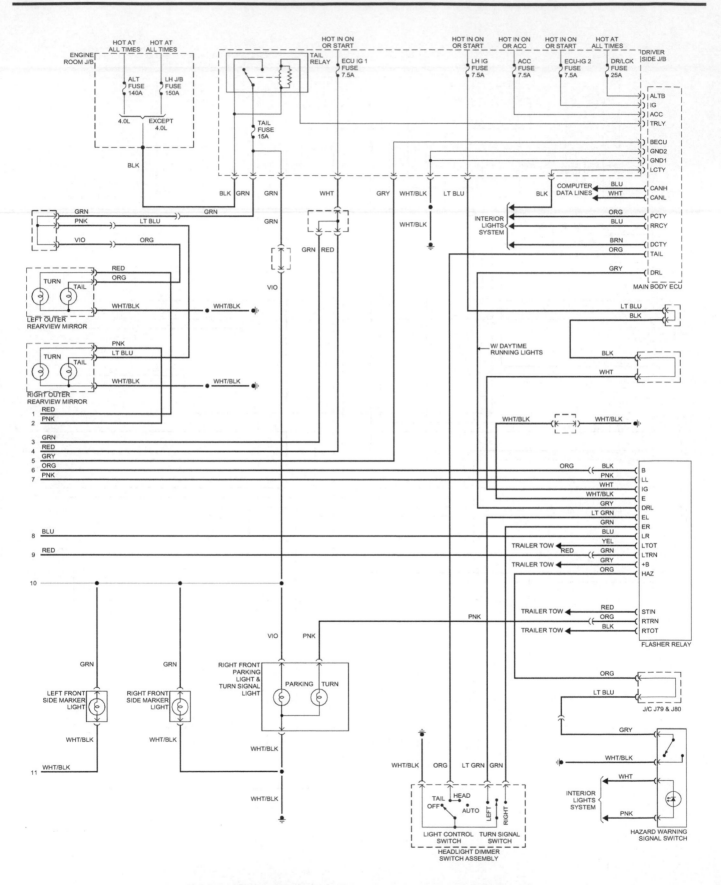

Exterior lighting system (except headlights) - Tundra models (2 of 2)

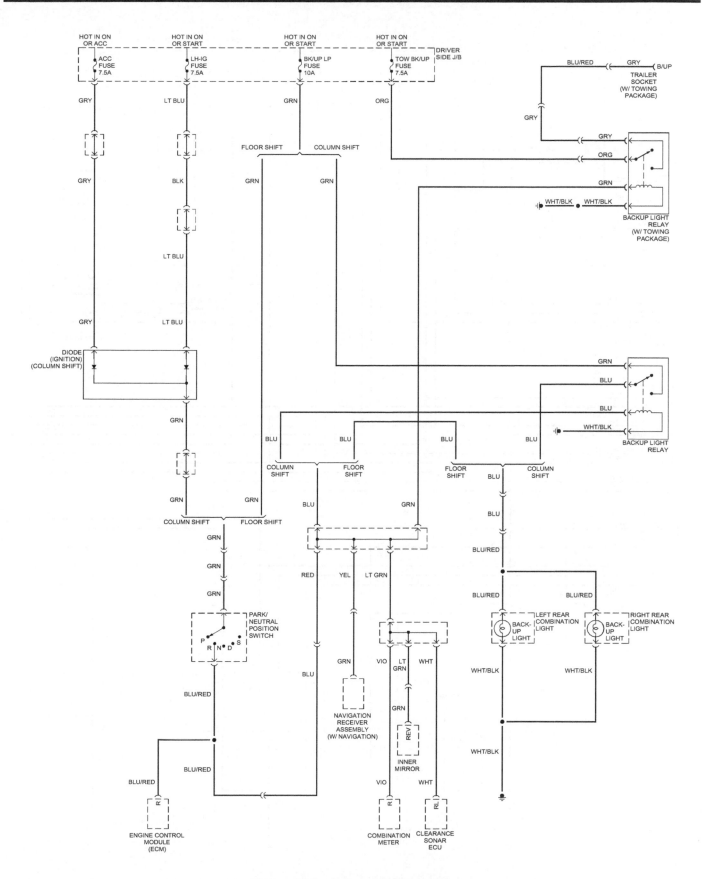

Back-up lights system - Tundra models

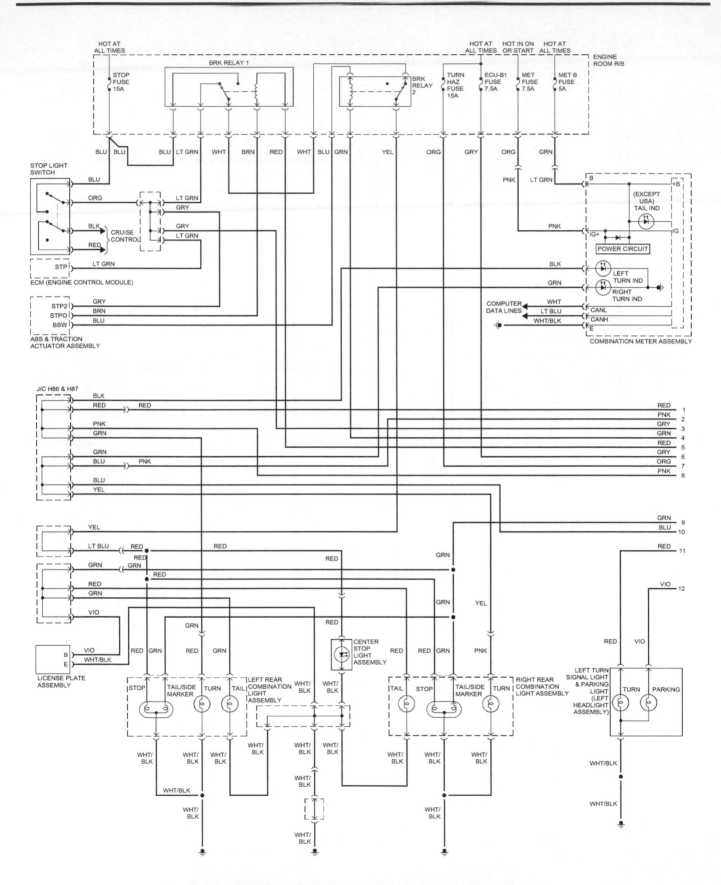

Exterior lighting system (except headlights) - Sequoia models (1 of 2)

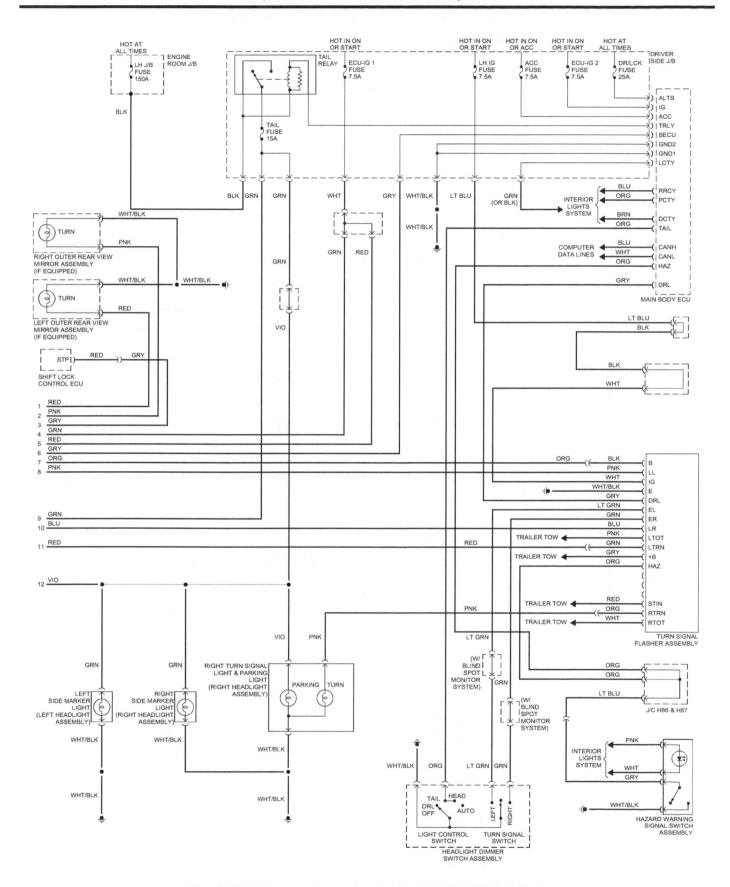

Exterior lighting system (except headlights) - Sequoia models (2 of 2)

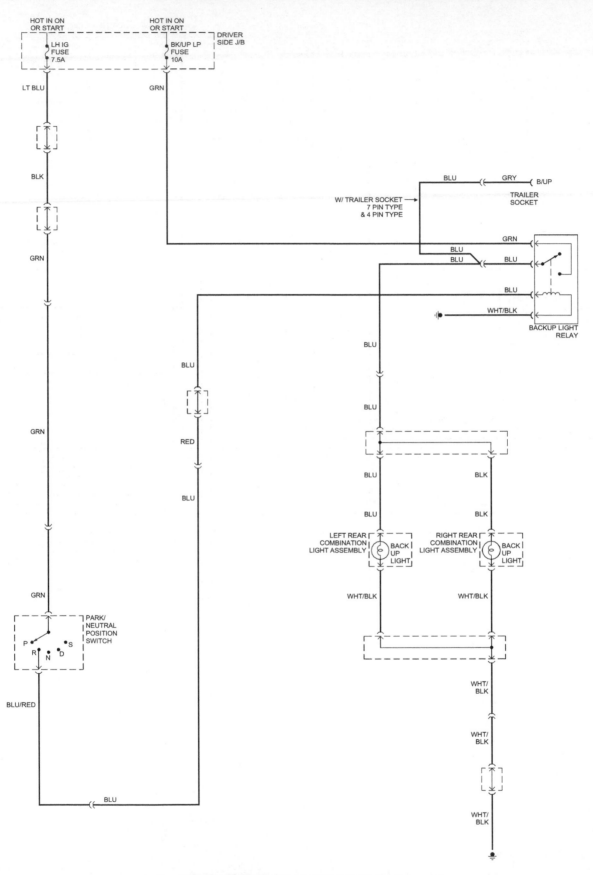

Back-up lights system - Sequoia models

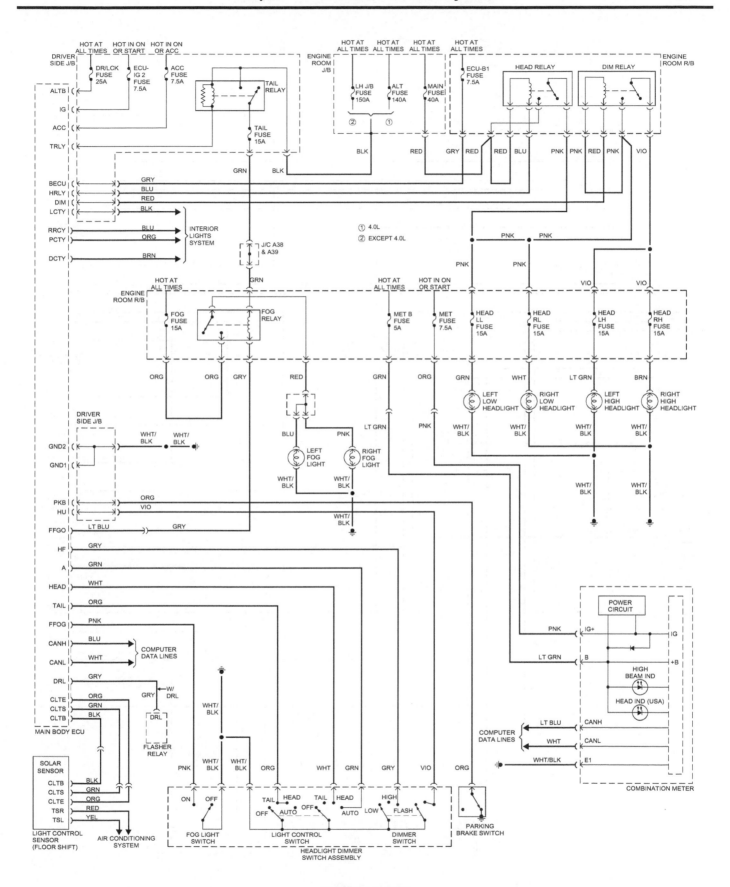

Headlight system

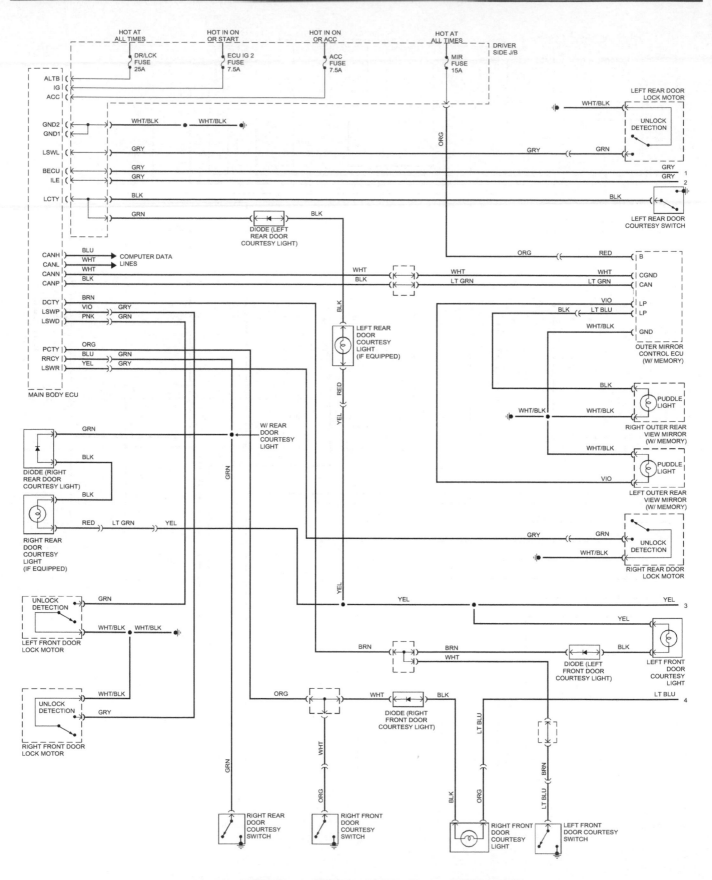

Courtesy light system - 2007 through 2010 Tundra models (1 of 2)

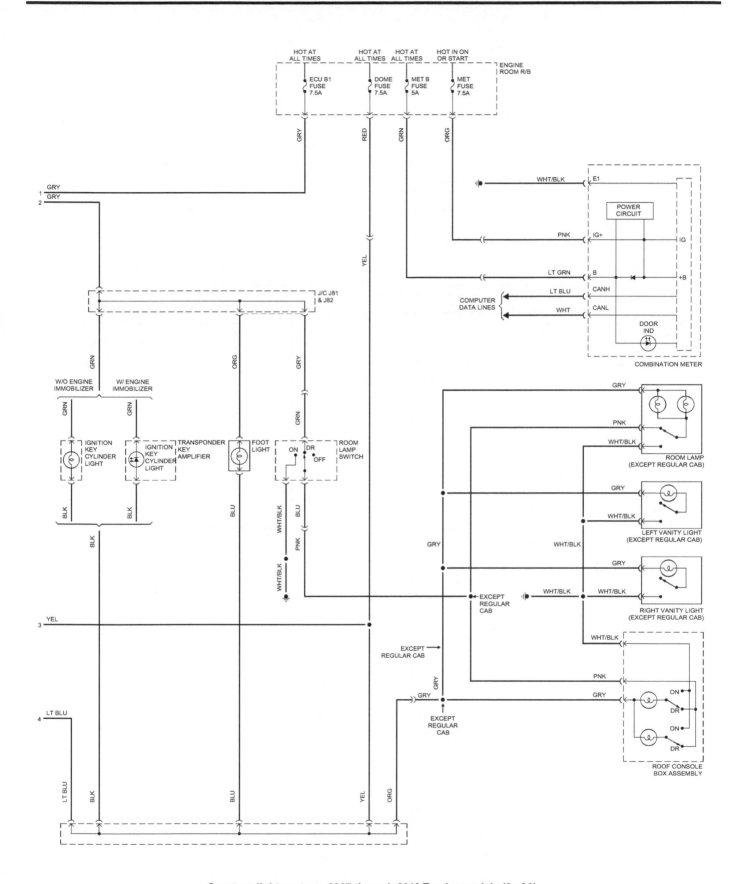

Courtesy light system - 2007 through 2010 Tundra models (2 of 2)

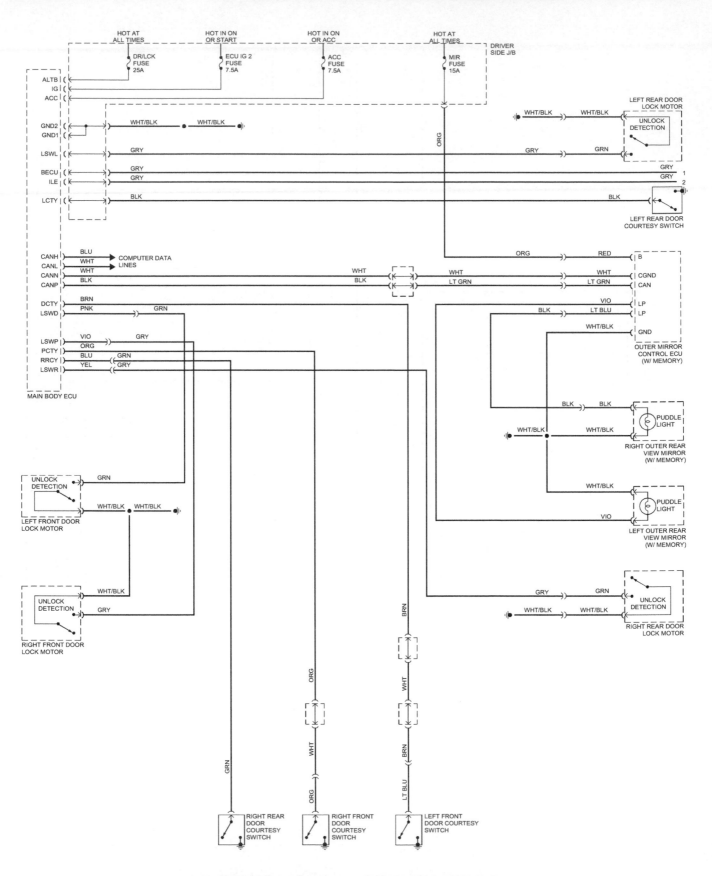

Courtesy light system - 2011 and later Tundra models (1 of 2)

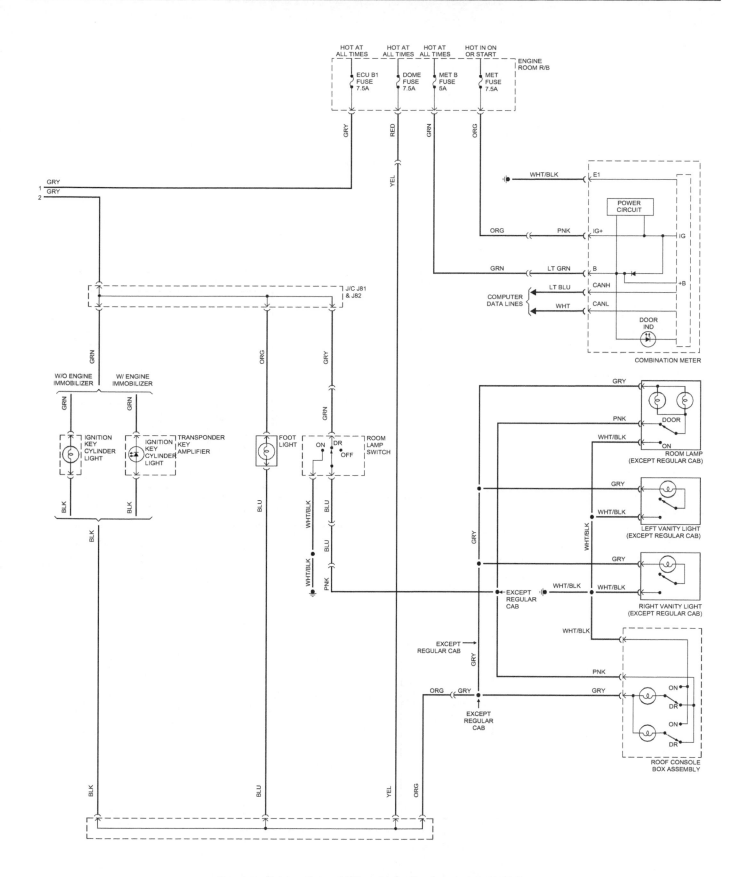

Courtesy light system - 2011 and later Tundra models (2 of 2)

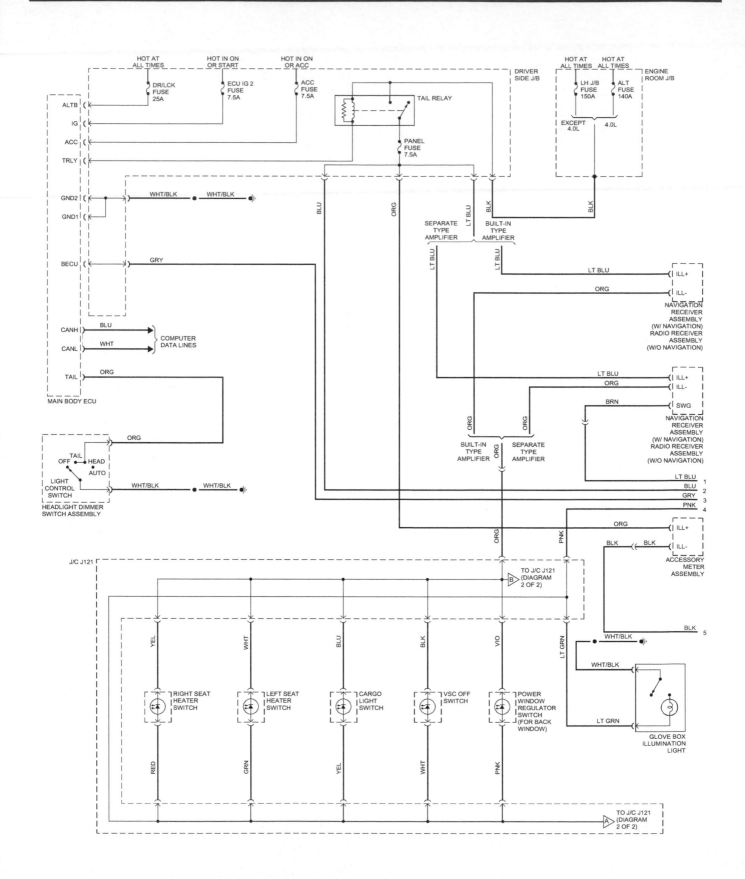

Instrument panel and switch illumination - Tundra models (1 of 2)

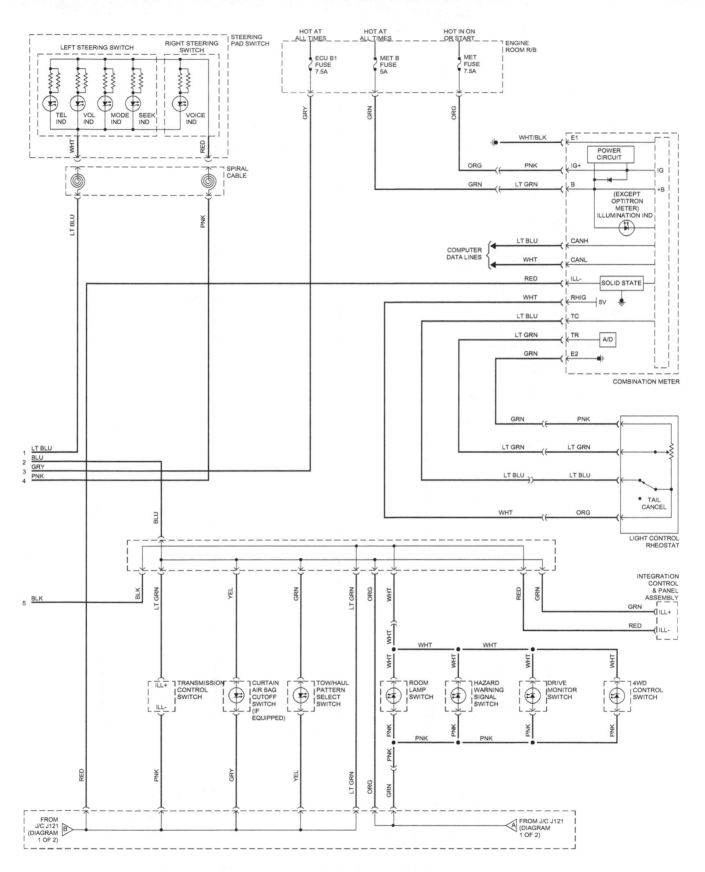

Instrument panel and switch illumination - Tundra models (2 of 2)

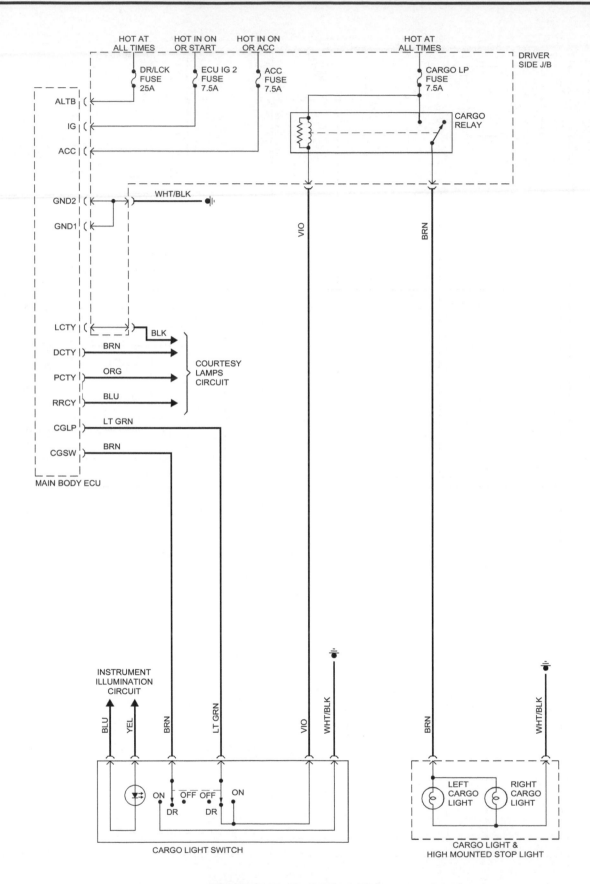

Cargo light circuit - Tundra models

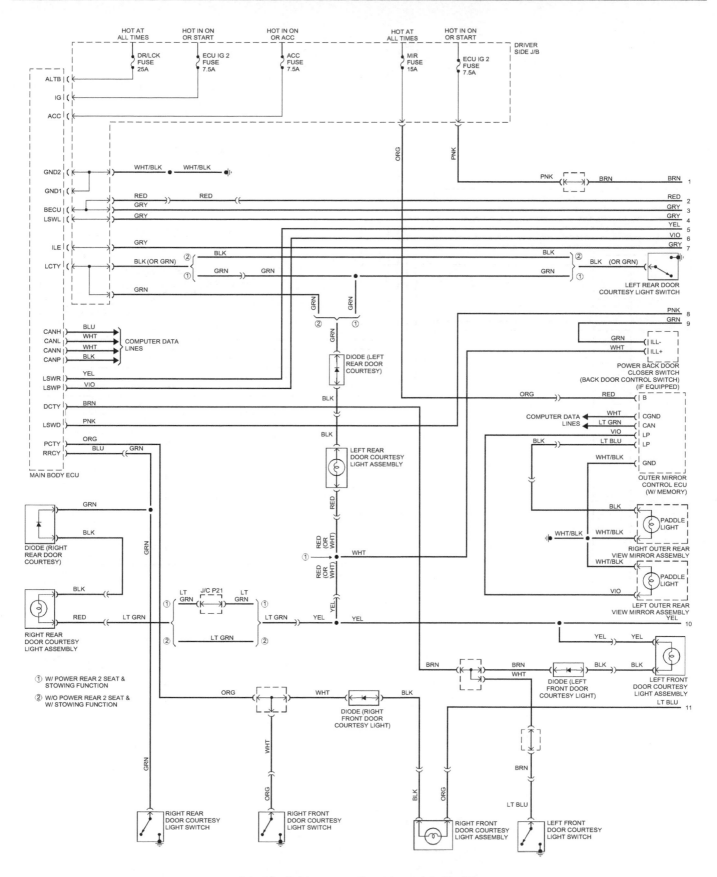

Courtesy light system - Sequoia models (1 of 3)

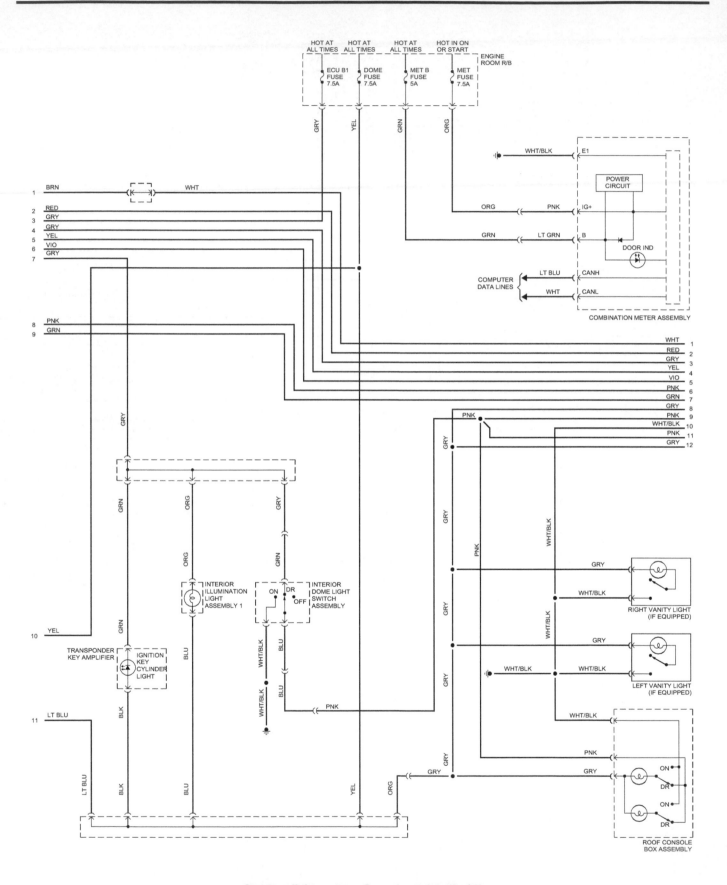

Courtesy light system - Sequoia models (2 of 3)

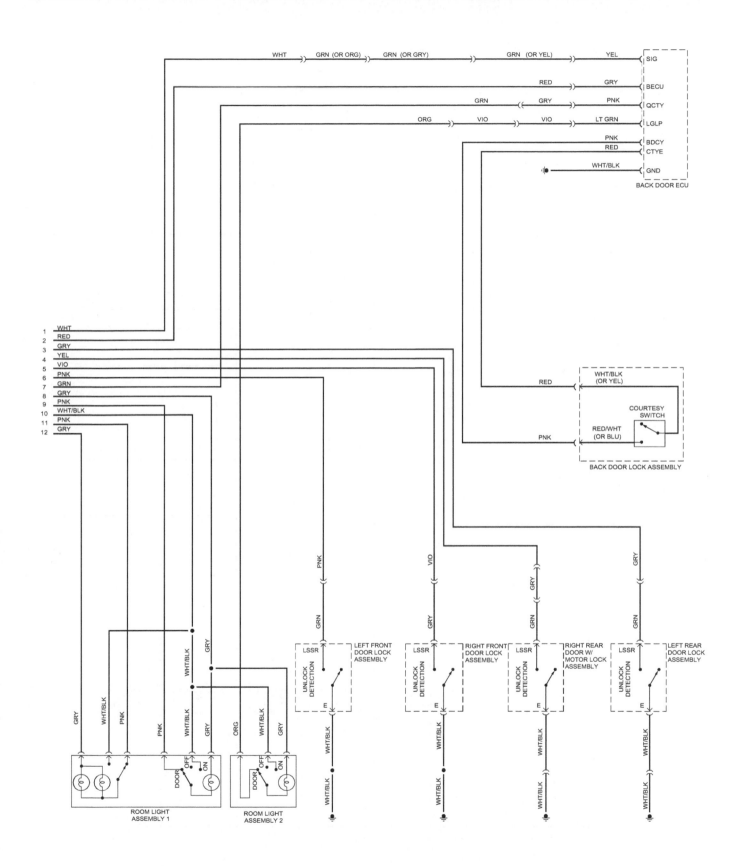

Courtesy light system - Sequoia models (3 of 3)

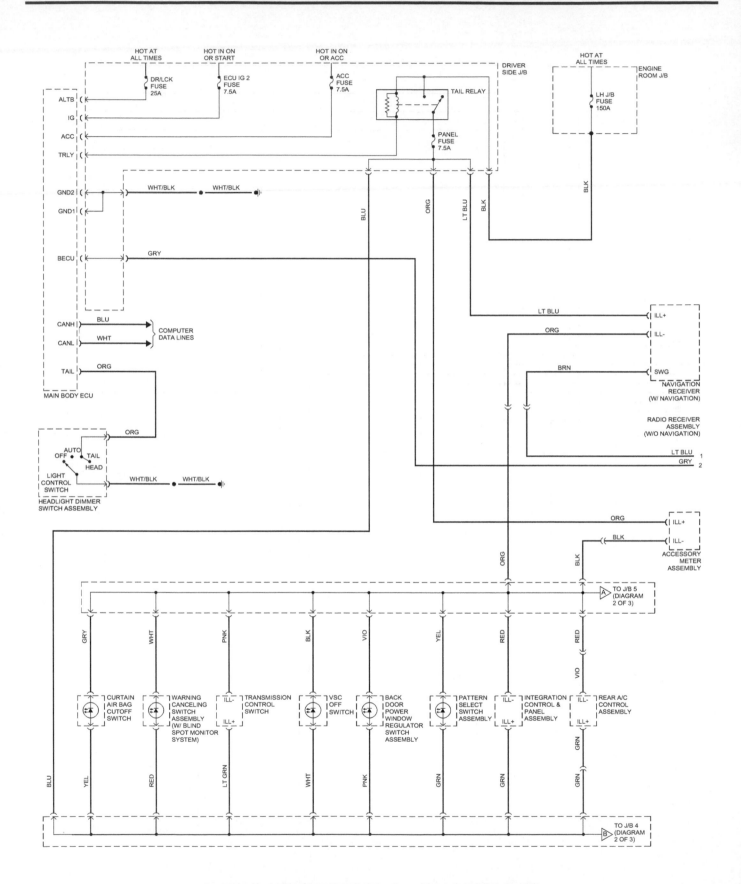

Instrument panel and switch illumination - Sequoia models (1 of 3)

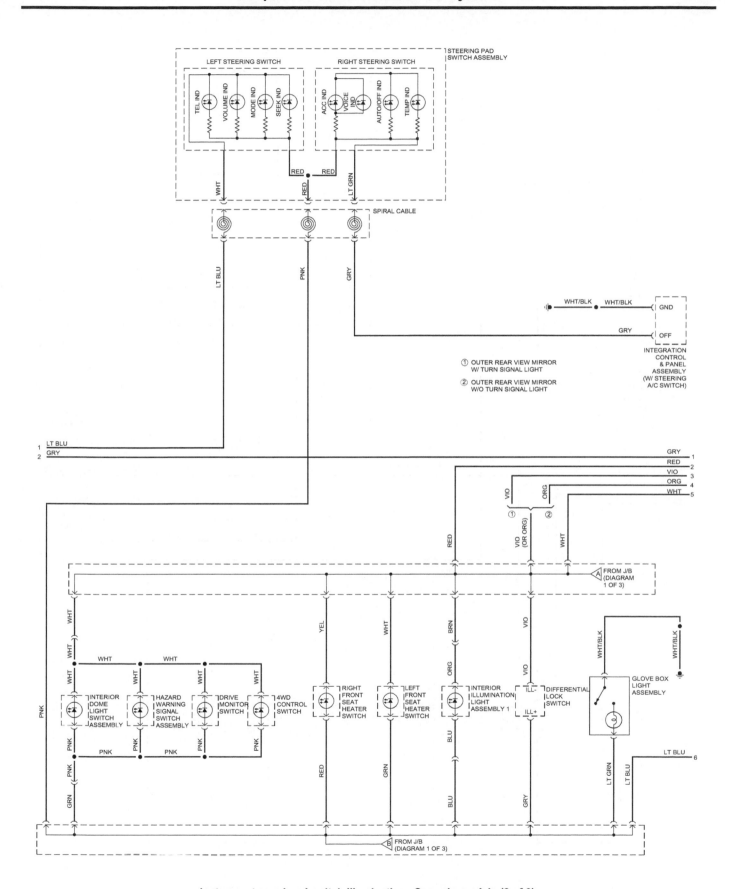

Instrument panel and switch illumination - Sequoia models (2 of 3)

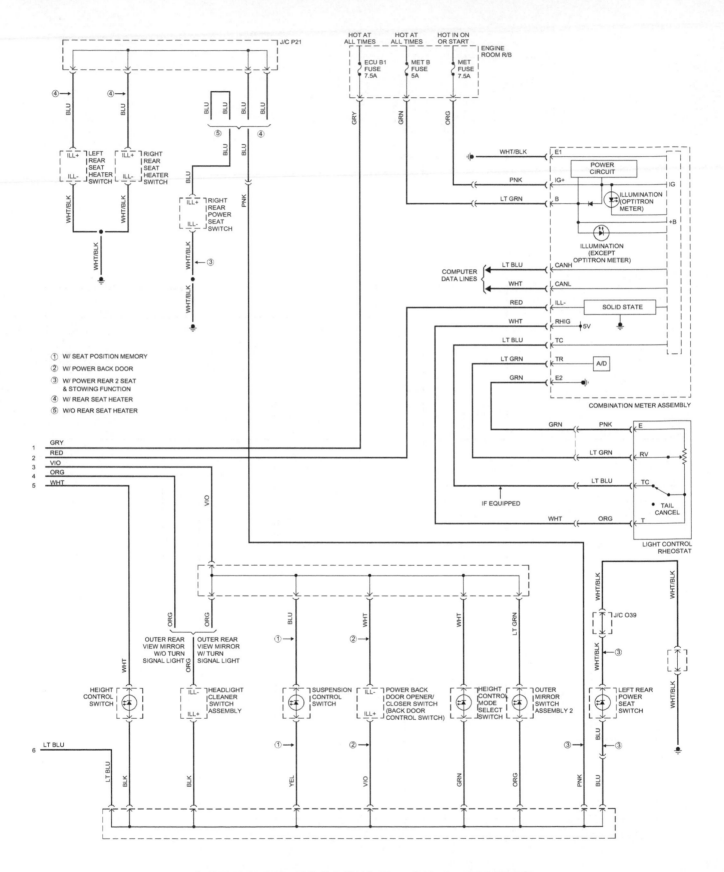

Instrument panel and switch illumination - Sequoia models (3 of 3)

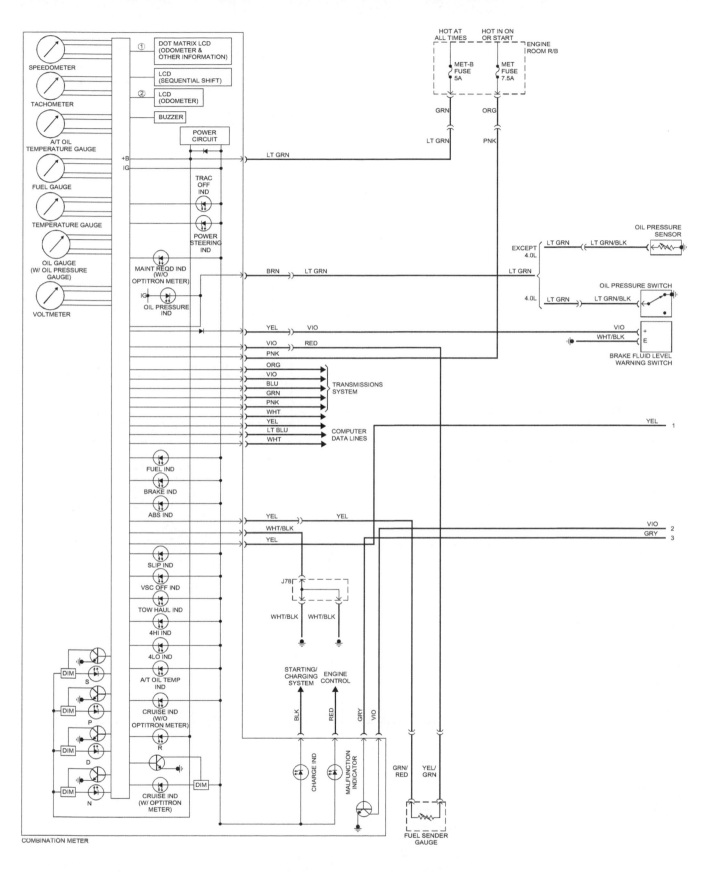

Warning lights and gauges system (1 of 2)

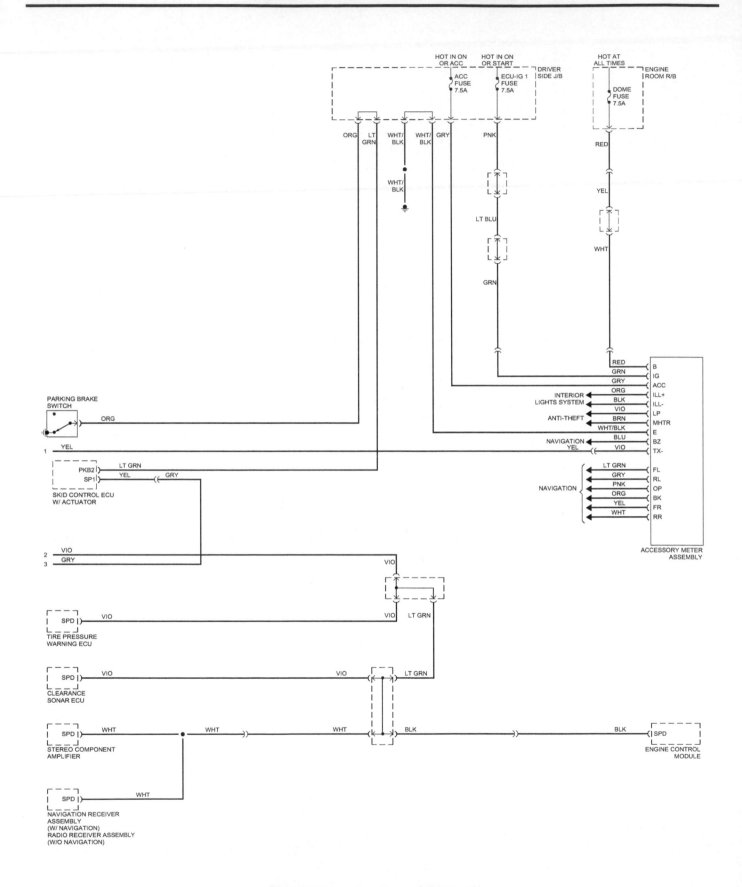

Warning lights and gauges system (2 of 2)

Index

Haynes Automotive Manuals

ACURA
12020 Integra '86 thru '89 & Legend '86 thru '90
12021 Integra '90 thru '93 & Legend '91 thru '95
Integra '94 thru '00 - see HONDA Civic (42025)
MDX '01 thru '07 - see HONDA Pilot (42037)
12050 Acura TL all models '99 thru '08

AMC
14020 Mid-size models '70 thru '83
14025 (Renault) Alliance & Encore '83 thru '87

AUDI
15020 4000 all models '80 thru '87
15025 5000 all models '77 thru '83
15026 5000 all models '84 thru '88
Audi A4 '96 thru '01 - see VW Passat (96023)
15030 Audi A4 '02 thru '08

AUSTIN-HEALEY
Sprite - see MG Midget (66015)

BMW
18020 3/5 Series '82 thru '92
18021 3-Series incl. Z3 models '92 thru '98
18022 3-Series incl. Z4 models '99 thru '05
18023 3-Series '06 thru '14
18025 320i all 4-cylinder models '75 thru '83
18050 1500 thru 2002 except Turbo '59 thru '77

BUICK
19010 Buick Century '97 thru '05
Century (front-wheel drive) - see GM (38005)
19020 Buick, Oldsmobile & Pontiac Full-size
(Front-wheel drive) '85 thru '05
Buick Electra, LeSabre and Park Avenue;
Oldsmobile Delta 88 Royale, Ninety Eight
and Regency; Pontiac Bonneville
19025 Buick, Oldsmobile & Pontiac Full-size
(Rear wheel drive) '70 thru '90
Buick Estate, Electra, LeSabre, Limited,
Oldsmobile Custom Cruiser, Delta 88,
Ninety-eight, Pontiac Bonneville,
Catalina, Grandville, Parisienne
19027 Buick LaCrosse '05 thru '13
Enclave - see GENERAL MOTORS (38001)
Rainier - see CHEVROLET (24072)
Regal - see GENERAL MOTORS (38010)
Riviera - see GENERAL MOTORS (38030, 38031)
Roadmaster - see CHEVROLET (24046)
Skyhawk - see GENERAL MOTORS (38015)
Skylark - see GENERAL MOTORS (38020, 38025)
Somerset - see GENERAL MOTORS (38025)

CADILLAC
21015 CTS & CTS-V '03 thru '14
21030 Cadillac Rear Wheel Drive '70 thru '93
Cimarron - see GENERAL MOTORS (38015)
DeVille - see GENERAL MOTORS (38031 & 38032)
Eldorado - see GENERAL MOTORS (38030)
Fleetwood - see GENERAL MOTORS (38031)
Seville - see GM (38030, 38031 & 38032)

CHEVROLET
10305 Chevrolet Engine Overhaul Manual
24010 Astro & GMC Safari Mini-vans '85 thru '05
24013 Aveo '04 thru '11
24015 Camaro V8 all models '70 thru '81
24016 Camaro all models '82 thru '92
24017 Camaro & Firebird '93 thru '02
Cavalier - see GENERAL MOTORS (38016)
Celebrity - see GENERAL MOTORS (38005)
24018 Camaro '10 thru '15
24020 Chevelle, Malibu & El Camino '69 thru '87
Cobalt - see GENERAL MOTORS (38017)
24024 Chevette & Pontiac T1000 '76 thru '87
Citation - see GENERAL MOTORS (38020)
24027 Colorado & GMC Canyon '04 thru '12
24032 Corsica & Beretta all models '87 thru '96
24040 Corvette all V8 models '68 thru '82
24041 Corvette all models '84 thru '96
24042 Corvette all models '97 thru '13
24044 Cruze '11 thru '19
24045 Full-size Sedans Caprice, Impala, Biscayne,
Bel Air & Wagons '69 thru '90
24046 Impala SS & Caprice and Buick Roadmaster
'91 thru '96
Impala '00 thru '05 - see LUMINA (24048)
24047 Impala & Monte Carlo all models '06 thru '11
Lumina '90 thru '94 - see GM (38010)
24048 Lumina & Monte Carlo '95 thru '05
Lumina APV - see GM (38035)
24050 Luv Pick-up all 2WD & 4WD '72 thru '82
24051 Malibu '13 thru '19
24055 Monte Carlo all models '70 thru '88
Monte Carlo '95 thru '01 - see LUMINA (24048)
24059 Nova all V8 models '69 thru '79

24060 Nova and Geo Prizm '85 thru '92
24064 Pick-ups '67 thru '87 - Chevrolet & GMC
24065 Pick-ups '88 thru '98 - Chevrolet & GMC
24066 Pick-ups '99 thru '06 - Chevrolet & GMC
24067 Chevrolet Silverado & GMC Sierra '07 thru '14
24068 Chevrolet Silverado & GMC Sierra '14 thru '19
24070 S-10 & S-15 Pick-ups '82 thru '93,
Blazer & Jimmy '83 thru '94,
24071 S-10 & Sonoma Pick-ups '94 thru '04,
including Blazer, Jimmy & Hombre
24072 Chevrolet TrailBlazer, GMC Envoy &
Oldsmobile Bravada '02 thru '09
24075 Sprint '85 thru '88 & Geo Metro '89 thru '01
24080 Vans - Chevrolet & GMC '68 thru '96
24081 Chevrolet Express & GMC Savana
Full-size Vans '96 thru '19

CHRYSLER
10310 Chrysler Engine Overhaul Manual
25015 Chrysler Cirrus, Dodge Stratus,
Plymouth Breeze '95 thru '00
25020 Full-size Front-Wheel Drive '88 thru '93
K-Cars - see DODGE Aries (30008)
Laser - see DODGE Daytona (30030)
25025 Chrysler LHS, Concorde, New Yorker,
Dodge Intrepid, Eagle Vision, '93 thru '97
25026 Chrysler LHS, Concorde, 300M,
Dodge Intrepid, '98 thru '04
25027 Chrysler 300 '05 thru '18, Dodge Charger
'06 thru '18, Magnum '05 thru '08 &
Challenger '08 thru '18
25030 Chrysler & Plymouth Mid-size
front wheel drive '82 thru '95
Rear-wheel Drive - see Dodge (30050)
25035 PT Cruiser all models '01 thru '10
25040 Chrysler Sebring '95 thru '06, Dodge Stratus
'01 thru '06 & Dodge Avenger '95 thru '00
25041 Chrysler Sebring '07 thru '10, 200 '11 thru '17
Dodge Avenger '08 thru '14

DATSUN
28005 200SX all models '80 thru '83
28012 240Z, 260Z & 280Z Coupe '70 thru '78
28014 280ZX Coupe & 2+2 '79 thru '83
300ZX - see NISSAN (72010)
28018 510 & PL521 Pick-up '68 thru '73
28020 510 all models '78 thru '81
28022 620 Series Pick-up all models '73 thru '79
720 Series Pick-up - see NISSAN (72030)

DODGE
400 & 600 - see CHRYSLER (25030)
30008 Aries & Plymouth Reliant '81 thru '89
30010 Caravan & Plymouth Voyager '84 thru '95
30011 Caravan & Plymouth Voyager '96 thru '02
30012 Challenger & Plymouth Sapporro '78 thru '83
30013 Caravan, Chrysler Voyager &
Town & Country '03 thru '07
30014 Grand Caravan &
Chrysler Town & Country '08 thru '18
30016 Colt & Plymouth Champ '78 thru '87
30020 Dakota Pick-ups all models '87 thru '96
30021 Durango '98 & '99 & Dakota '97 thru '99
30022 Durango '00 thru '03 & Dakota '00 thru '04
30023 Durango '04 thru '09 & Dakota '05 thru '11
30025 Dart, Demon, Plymouth Barracuda,
Duster & Valiant 6-cylinder models '67 thru '76
30030 Daytona & Chrysler Laser '84 thru '89
Intrepid - see CHRYSLER (25025, 25026)
30034 Neon all models '95 thru '99
30035 Omni & Plymouth Horizon '78 thru '90
30036 Dodge & Plymouth Neon '00 thru '05
30040 Pick-ups full-size models '74 thru '93
30042 Pick-ups full-size models '94 thru '08
30043 Pick-ups full-size models '09 thru '18
30045 Ram 50/D50 Pick-ups & Raider and
Plymouth Arrow Pick-ups '79 thru '93
30050 Dodge/Plymouth/Chrysler RWD '71 thru '89
30055 Shadow & Plymouth Sundance '87 thru '94
30060 Spirit & Plymouth Acclaim '89 thru '95
30065 Vans - Dodge & Plymouth '71 thru '03

EAGLE
Talon - see MITSUBISHI (68030, 68031)
Vision - see CHRYSLER (25025)

FIAT
34010 124 Sport Coupe & Spider '68 thru '78
34025 X1/9 all models '74 thru '80

FORD
10320 Ford Engine Overhaul Manual
10355 Ford Automatic Transmission Overhaul
11500 Mustang '64-1/2 thru '70 Restoration Guide
36004 Aerostar Mini-vans all models '86 thru '97
36006 Contour & Mercury Mystique '95 thru '00
36008 Courier Pick-up all models '72 thru '82

36012 Crown Victoria &
Mercury Grand Marquis '88 thru '11
36014 Edge '07 thru '19 & Lincoln MKX '07 thru '18
36016 Escort & Mercury Lynx all models '81 thru '90
36020 Escort & Mercury Tracer '91 thru '02
36022 Escape '01 thru '17, Mazda Tribute '01 thru '11,
& Mercury Mariner '05 thru '11
36024 Explorer & Mazda Navajo '91 thru '01
36025 Explorer & Mercury Mountaineer '02 thru '10
36026 Explorer '11 thru '17
36028 Fairmont & Mercury Zephyr '78 thru '83
36030 Festiva & Aspire '88 thru '97
36032 Fiesta all models '77 thru '80
36034 Focus '00 thru '11
36035 Focus '12 thru '14
36045 Fusion '06 thru '14 & Mercury Milan '06 thru '11
36048 Mustang V8 all models '64-1/2 thru '73
36049 Mustang II 4-cylinder, V6 & V8 models '74 thru '78
36050 Mustang & Mercury Capri '79 thru '93
36051 Mustang all models '94 thru '04
36052 Mustang '05 thru '14
36054 Pick-ups & Bronco '73 thru '79
36058 Pick-ups & Bronco '80 thru '96
36059 F-150 '97 thru '03, Expedition '97 thru '17,
F-250 '97 thru '99, F-150 Heritage '04
& Lincoln Navigator '98 thru '17
36060 Super Duty Pick-ups & Excursion '99 thru '10
36061 F-150 full-size '04 thru '14
36062 Pinto & Mercury Bobcat '75 thru '80
36063 F-150 full-size '15 thru '17
36064 Super Duty Pick-ups '11 thru '16
36066 Probe all models '89 thru '92
Probe '93 thru '97 - see MAZDA 626 (61042)
36070 Ranger & Bronco II gas models '83 thru '92
36071 Ranger '93 thru '11 & Mazda Pick-ups '94 thru '09
36074 Taurus & Mercury Sable '86 thru '95
36075 Taurus & Mercury Sable '96 thru '07
36076 Taurus '08 thru '14, Five Hundred '05 thru '07,
Mercury Montego '05 thru '07 & Sable '08 thru '09
36078 Tempo & Mercury Topaz '84 thru '94
36082 Thunderbird & Mercury Cougar '83 thru '88
36086 Thunderbird & Mercury Cougar '89 thru '97
36090 Vans all V8 Econoline models '69 thru '91
36094 Vans full size '92 thru '14
36097 Windstar '95 thru '03, Freestar & Mercury
Monterey Mini-van '04 thru '07

GENERAL MOTORS
10360 GM Automatic Transmission Overhaul
38001 GMC Acadia '07 thru '16, Buick Enclave
'08 thru '17, Saturn Outlook '07 thru '10
& Chevrolet Traverse '09 thru '17
38005 Buick Century, Chevrolet Celebrity,
Oldsmobile Cutlass Ciera & Pontiac 6000
all models '82 thru '96
38010 Buick Regal '88 thru '04, Chevrolet Lumina
'88 thru '04, Oldsmobile Cutlass Supreme
'88 thru '97 & Pontiac Grand Prix '88 thru '07
38015 Buick Skyhawk, Cadillac Cimarron,
Chevrolet Cavalier, Oldsmobile Firenza,
Pontiac J-2000 & Sunbird '82 thru '94
38016 Chevrolet Cavalier & Pontiac Sunfire '95 thru '05
38017 Chevrolet Cobalt '05 thru '10, HHR '06 thru '11,
Pontiac G5 '07 thru '09, Pursuit '05 thru '06
& Saturn ION '03 thru '07
38020 Buick Skylark, Chevrolet Citation,
Oldsmobile Omega, Pontiac Phoenix '80 thru '85
38025 Buick Skylark '86 thru '98, Somerset '85 thru '87,
Oldsmobile Achieva '92 thru '98, Calais '85 thru '91,
& Pontiac Grand Am all models '85 thru '98
38026 Chevrolet Malibu '97 thru '03, Classic '04 thru '05,
Oldsmobile Alero '99 thru '03, Cutlass '97 thru '00,
& Pontiac Grand Am '99 thru '03
38027 Chevrolet Malibu '04 thru '12, Pontiac G6
'05 thru '10 & Saturn Aura '07 thru '10
38030 Cadillac Eldorado, Seville, Oldsmobile
Toronado & Buick Riviera '71 thru '85
38031 Cadillac Eldorado, Seville, DeVille, Fleetwood,
Oldsmobile Toronado & Buick Riviera '86 thru '93
38032 Cadillac DeVille '94 thru '05, Seville '92 thru '04
& Cadillac DTS '06 thru '10
38035 Chevrolet Lumina APV, Oldsmobile Silhouette
& Pontiac Trans Sport all models '90 thru '96
38036 Chevrolet Venture '97 thru '05, Oldsmobile
Silhouette '97 thru '04, Pontiac Trans Sport
'97 thru '98 & Montana '99 thru '05
38040 Chevrolet Equinox '05 thru '17, GMC Terrain
'10 thru '17 & Pontiac Torrent '06 thru '09

GEO
Metro - see CHEVROLET Sprint (24075)
Prizm - '85 thru '92 see CHEVY (24060),
'93 thru '02 see TOYOTA Corolla (92036)
40030 Storm all models '90 thru '93
Tracker - see SUZUKI Samurai (90010)

(Continued on other side)

Haynes Automotive Manuals (continued)

NOTE: If you do not see a listing for your vehicle, please visit **haynes.com** for the latest product information and check out our **Online Manuals!**

GMC
    Acadia - *see GENERAL MOTORS (38001)*
    Pick-ups - *see CHEVROLET (24027, 24068)*
    Vans - *see CHEVROLET (24081)*

HONDA
42010 **Accord CVCC** all models '76 thru '83
42011 **Accord** all models '84 thru '89
42012 **Accord** all models '90 thru '93
42013 **Accord** all models '94 thru '97
42014 **Accord** all models '98 thru '02
42015 **Accord** '03 thru '12 & **Crosstour** '10 thru '14
42016 **Accord** '13 thru '17
42020 **Civic 1200** all models '73 thru '79
42021 **Civic 1300 & 1500 CVCC** '80 thru '83
42022 **Civic 1500 CVCC** all models '75 thru '79
42023 **Civic** all models '84 thru '91
42024 **Civic & del Sol** '92 thru '95
42025 **Civic** '96 thru '00, **CR-V** '97 thru '01
       & Acura Integra '94 thru '00
42026 **Civic** '01 thru '11 & **CR-V** '02 thru '11
42027 **Civic** '12 thru '15 & **CR-V** '12 thru '16
42030 **Fit** '07 thru '13
42035 **Odyssey** all models '99 thru '10
       Passport - *see ISUZU Rodeo (47017)*
42037 **Honda Pilot** '03 thru '08, **Ridgeline** '06 thru '14
       & Acura MDX '01 thru '07
42040 **Prelude CVCC** all models '79 thru '89

HYUNDAI
43010 **Elantra** all models '96 thru '19
43015 **Excel & Accent** all models '86 thru '13
43050 **Santa Fe** all models '01 thru '12
43055 **Sonata** all models '99 thru '14

INFINITI
    G35 '03 thru '08 - *see NISSAN 350Z (72011)*

ISUZU
    Hombre - *see CHEVROLET S-10 (24071)*
47017 **Rodeo** '91 thru '02, **Amigo** '89 thru '94 & '98 thru '02
       & Honda Passport '95 thru '02
47020 **Trooper** '84 thru '91 & **Pick-up** '81 thru '93

JAGUAR
49010 **XJ6** all 6-cylinder models '68 thru '86
49011 **XJ6** all models '88 thru '94
49015 **XJ12 & XJS** all 12-cylinder models '72 thru '85

JEEP
50010 **Cherokee, Comanche & Wagoneer Limited**
       all models '84 thru '01
50011 **Cherokee** '14 thru '19
50020 **CJ** all models '49 thru '86
50025 **Grand Cherokee** all models '93 thru '04
50026 **Grand Cherokee** '05 thru '19
       & Dodge Durango '11 thru '19
50029 **Grand Wagoneer & Pick-up** '72 thru '91
       Grand Wagoneer '84 thru '91, Cherokee &
       Wagoneer '72 thru '83, Pick-up '72 thru '88
50030 **Wrangler** all models '87 thru '17
50035 **Liberty** '02 thru '12 & **Dodge Nitro** '07 thru '11
50050 **Patriot & Compass** '07 thru '17

KIA
54050 **Optima** '01 thru '10
54060 **Sedona** '02 thru '14
54070 **Sephia** '94 thru '01, **Spectra** '00 thru '09,
       Sportage '05 thru '20
54077 **Sorento** '03 thru '13

LEXUS
    ES 300/330 - *see TOYOTA Camry (92007, 92008)*
    ES 350 - *see TOYOTA Camry (92009)*
    RX 300/330/350 - *see TOYOTA Highlander (92095)*

LINCOLN
    MKX - *see FORD (36014)*
    Navigator - *see FORD Pick-up (36059)*
59010 **Rear-Wheel Drive Continental** '70 thru '87,
       Mark Series '70 thru '92 & **Town Car** '81 thru '10

MAZDA
61010 **GLC (rear-wheel drive)** '77 thru '83
61011 **GLC (front-wheel drive)** '81 thru '85
61012 **Mazda3** '04 thru '11
61015 **323 & Protegé** '90 thru '03
61016 **MX-5 Miata** '90 thru '14
61020 **MPV** all models '89 thru '98
       Navajo - *see Ford Explorer (36024)*
61030 **Pick-ups** '72 thru '93
       Pick-ups '94 thru '09 - *see Ford Ranger (36071)*
61035 **RX-7** all models '79 thru '85
61036 **RX-7** all models '86 thru '91
61040 **626 (rear-wheel drive)** all models '79 thru '82
61041 **626 & MX-6 (front-wheel drive)** '83 thru '92
61042 **626** '93 thru '01 & **MX-6/Ford Probe** '93 thru '02
61043 **Mazda6** '03 thru '13

MERCEDES-BENZ
63012 **123 Series Diesel** '76 thru '85
63015 **190 Series** 4-cylinder gas models '84 thru '88
63020 **230/250/280** 6-cylinder SOHC models '68 thru '72
63025 **280 123 Series** gas models '77 thru '81
63030 **350 & 450** all models '71 thru '80
63040 **C-Class:** C230/C240/C280/C320/C350 '01 thru '07

MERCURY
64200 **Villager & Nissan Quest** '93 thru '01
       All other titles, see FORD Listing.

MG
66010 **MGB** Roadster & GT Coupe '62 thru '80
66015 **MG Midget, Austin Healey Sprite** '58 thru '80

MINI
67020 **Mini** '02 thru '13

MITSUBISHI
68020 **Cordia, Tredia, Galant, Precis & Mirage** '83 thru '93
68030 **Eclipse, Eagle Talon & Plymouth Laser** '90 thru '94
68031 **Eclipse** '95 thru '05 & **Eagle Talon** '95 thru '98
68035 **Galant** '94 thru '12
68040 **Pick-up** '83 thru '96 & **Montero** '83 thru '93

NISSAN
72010 **300ZX** all models including Turbo '84 thru '89
72011 **350Z & Infiniti G35** all models '03 thru '08
72015 **Altima** all models '93 thru '06
72016 **Altima** '07 thru '12
72020 **Maxima** all models '85 thru '92
72021 **Maxima** all models '93 thru '08
72025 **Murano** '03 thru '14
72030 **Pick-ups** '80 thru '97 & **Pathfinder** '87 thru '95
72031 **Frontier** '98 thru '04, **Xterra** '00 thru '04,
       & Pathfinder '96 thru '04
72032 **Frontier & Xterra** '05 thru '14
72037 **Pathfinder** '05 thru '14
72040 **Pulsar** all models '83 thru '86
72042 **Roque** all models '08 thru '20
72050 **Sentra** all models '82 thru '94
72051 **Sentra & 200SX** all models '95 thru '06
72060 **Stanza** all models '82 thru '90
72070 **Titan pick-ups** '04 thru '10, **Armada** '05 thru '10
       & Pathfinder Armada '04
72080 **Versa** all models '07 thru '19

OLDSMOBILE
73015 **Cutlass V6 & V8** gas models '74 thru '88
       For other OLDSMOBILE titles, see BUICK,
       CHEVROLET or GENERAL MOTORS listings.

PLYMOUTH
    For PLYMOUTH titles, see DODGE listing.

PONTIAC
79008 **Fiero** all models '84 thru '88
79018 **Firebird V8** models except Turbo '70 thru '81
79019 **Firebird** all models '82 thru '92
79025 **G6** all models '05 thru '09
79040 **Mid-size Rear-wheel Drive** '70 thru '87
       Vibe '03 thru '10 - *see TOYOTA Corolla (92037)*
       For other PONTIAC titles, see BUICK,
       CHEVROLET or GENERAL MOTORS listings.

PORSCHE
80020 **911 Coupe & Targa** models '65 thru '89
80025 **914** all 4-cylinder models '69 thru '76
80030 **924** all models including Turbo '76 thru '82
80035 **944** all models including Turbo '83 thru '89

RENAULT
    Alliance & Encore - *see AMC (14025)*

SAAB
84010 **900** all models including Turbo '79 thru '88

SATURN
87010 **Saturn** all S-series models '91 thru '02
       Saturn Ion '03 thru '07 - *see GM (38017)*
       Saturn Outlook - *see GM (38001)*
87020 **Saturn L-series** all models '00 thru '04
87040 **Saturn VUE** '02 thru '09

SUBARU
89002 **1100, 1300, 1400 & 1600** '71 thru '79
89003 **1600 & 1800** 2WD & 4WD '80 thru '94
89080 **Impreza** '02 thru '11, **WRX** '02 thru '14,
       & WRX STI '04 thru '14
89100 **Legacy** all models '90 thru '99
89101 **Legacy & Forester** '00 thru '09
89102 **Legacy** '10 thru '16 & **Forester** '12 thru '16

SUZUKI
90010 **Samurai/Sidekick & Geo Tracker** '86 thru '01

TOYOTA
92005 **Camry** all models '83 thru '91
92006 **Camry** '92 thru '96 & **Avalon** '95 thru '96
92007 **Camry, Avalon, Solara, Lexus ES 300** '97 thru '01
92008 **Camry, Avalon, Lexus ES 300/330** '02 thru '06
       & Solara '02 thru '08
92009 **Camry, Avalon & Lexus ES 350** '07 thru '17
92015 **Celica Rear-wheel Drive** '71 thru '85
92020 **Celica Front-wheel Drive** '86 thru '99
92025 **Celica Supra** all models '79 thru '92
92030 **Corolla** all models '75 thru '79
92032 **Corolla** all rear-wheel drive models '80 thru '87
92035 **Corolla** all front-wheel drive models '84 thru '92
92036 **Corolla & Geo/Chevrolet Prizm** '93 thru '02
92037 **Corolla** '03 thru '19, **Matrix** '03 thru '14,
       & Pontiac Vibe '03 thru '10
92040 **Corolla Tercel** all models '80 thru '82
92045 **Corona** all models '74 thru '82
92050 **Cressida** all models '78 thru '82
92055 **Land Cruiser FJ40, 43, 45, 55** '68 thru '82
92056 **Land Cruiser FJ60, 62, 80, FZJ80** '80 thru '96
92060 **Matrix** '03 thru '11 & **Pontiac Vibe** '03 thru '10
92065 **MR2** all models '85 thru '87
92070 **Pick-up** all models '69 thru '78
92075 **Pick-up** all models '79 thru '95
92076 **Tacoma** '95 thru '04, **4Runner** '96 thru '02
       & T100 '93 thru '98
92077 **Tacoma** all models '05 thru '18
92078 **Tundra** '00 thru '06 & **Sequoia** '01 thru '07
92079 **4Runner** all models '03 thru '09
92080 **Previa** all models '91 thru '95
92081 **Prius** all models '01 thru '12
92082 **RAV4** all models '96 thru '12
92085 **Tercel** all models '87 thru '94
92090 **Sienna** all models '98 thru '10
92095 **Highlander** '01 thru '19
       & Lexus RX330/330/350 '99 thru '19
92179 **Tundra** '07 thru '19 & **Sequoia** '08 thru '19

TRIUMPH
94007 **Spitfire** all models '62 thru '81
94010 **TR7** all models '75 thru '81

VW
96008 **Beetle & Karmann Ghia** '54 thru '79
96009 **New Beetle** '98 thru '10
96016 **Rabbit, Jetta, Scirocco & Pick-up**
       gas models '75 thru '92 & Convertible '80 thru '92
96017 **Golf, GTI & Jetta** '93 thru '98, **Cabrio** '95 thru '02
96018 **Golf, GTI, Jetta** '99 thru '05
96019 **Jetta, Rabbit, GLI, GTI & Golf** '05 thru '11
96020 **Rabbit, Jetta & Pick-up** diesel '77 thru '84
96021 **Jetta** '11 thru '18 & **Golf** '15 thru '19
96023 **Passat** '98 thru '05 & **Audi A4** '96 thru '01
96030 **Transporter 1600** all models '68 thru '79
96035 **Transporter 1700, 1800 & 2000** '72 thru '79
96040 **Type 3 1500 & 1600** all models '63 thru '73
96045 **Vanagon Air-Cooled** all models '80 thru '83

VOLVO
97010 **120, 130 Series & 1800 Sports** '61 thru '73
97015 **140 Series** all models '66 thru '74
97020 **240 Series** all models '76 thru '93
97040 **740 & 760 Series** all models '82 thru '88
97050 **850 Series** all models '93 thru '97

TECHBOOK MANUALS
10205 **Automotive Computer Codes**
10206 **OBD-II & Electronic Engine Management**
10210 **Automotive Emissions Control Manual**
10215 **Fuel Injection Manual** '78 thru '85
10225 **Holley Carburetor Manual**
10230 **Rochester Carburetor Manual**
10305 **Chevrolet Engine Overhaul Manual**
10320 **Ford Engine Overhaul Manual**
10330 **GM and Ford Diesel Engine Repair Manual**
10331 **Duramax Diesel Engines** '01 thru '19
10332 **Cummins Diesel Engine Performance Manual**
10333 **GM, Ford & Chrysler Engine Performance Manual**
10334 **GM Engine Performance Manual**
10340 **Small Engine Repair Manual, 5 HP & Less**
10341 **Small Engine Repair Manual, 5.5 thru 20 HP**
10345 **Suspension, Steering & Driveline Manual**
10355 **Ford Automatic Transmission Overhaul**
10360 **GM Automatic Transmission Overhaul**
10405 **Automotive Body Repair & Painting**
10410 **Automotive Brake Manual**
10411 **Automotive Anti-lock Brake (ABS) Systems**
10420 **Automotive Electrical Manual**
10425 **Automotive Heating & Air Conditioning**
10435 **Automotive Tools Manual**
10445 **Welding Manual**
10450 **ATV Basics**

Over a 100 Haynes
motorcycle manuals
also available

10/22